UNDERSTANDING AND AVOIDING THE OIL CURSE IN RESOURCE-RICH ARAB ECONOMIES

For over eighty years the Arab region has derived massive wealth from its natural resources, yet the region's economies remain little diversified, while the oil market is experiencing major structural shifts with the advent of shale gas. Moreover, the resource itself is eventually exhaustible. Under these conditions economic prosperity cannot be sustainable. The critical question is how can the countries of this region escape the "oil curse"?

In this volume, leading economists argue that the curse is not a predestined outcome but a result of weak institutions and bad governance. A variety of analytical perspectives and examination of various international case studies leads to the conclusion that natural resources can only spur economic development when combined with sound political institutions and effective economic governance.

This volume, with its unique focus on the Arab region, will be an important reference for researchers and policy makers alike.

Ibrahim Elbadawi has been Director, Macroeconomics Research Department, Dubai Economic Council, since March 2009. Until February 2009 he was Lead Economist at the Development Research Group of the World Bank, which he joined in 1989; and before that he was Associate Professor of economics at the University of Gezira in Sudan. He holds a PhD in economics and statistics from North Carolina State and Northwestern universities in the USA.

Hoda Selim is an economist at the Economic Research Forum in Cairo, a position she has held since 2011. Previously, she worked for five years at the World Bank's Cairo Office in the Poverty Reduction and Economic Monitoring department. Her recent research focuses on the macroeconomics of oil management and the political economy of development, with particular focus on Arab economies. She holds an MA and PhD from Sciences Po in Paris.

Understanding and Avoiding the Oil Curse in Resource-rich Arab Economies

Edited by

IBRAHIM ELBADAWI

Dubai Economic Council, Dubai, and Economic Research Forum, Egypt

HODA SELIM

Economic Research Forum, Egypt

ECONOMIC RESEARCH FORUM

منتدى البحوث الاقتصادية

CAMBRIDGE UNIVERSITY PRESS

CAMBRIDGE
UNIVERSITY PRESS

One Liberty Plaza, New York NY 10006

Cambridge University Press is part of the University of Cambridge.

It furthers the University's mission by disseminating knowledge in the pursuit of education, learning, and research at the highest international levels of excellence.

www.cambridge.org
Information on this title: www.cambridge.org/9781107141728

First published 2016

A catalog record for this publication is available from the British Library.

Library of Congress Cataloging in Publication Data
Names: Elbadawi, Ibrahim, editor. | Selim, Hoda, editor.
Title: Understanding and avoiding the oil curse in resource-rich Arab economies / Ibrahim Elbadawi, Hoda Selim.
Description: New York, NY : Cambridge University Press, [2016] | Includes bibliographical references and index.
Identifiers: LCCN 2016000953 | ISBN 9781107141728 (Hardback)
Subjects: LCSH: Natural resources–Arab countries. | Economic development–Arab countries. | Arab countries–Economic policy.
Classification: LCC HC498.Z65 U53 2016 | DDC 333.8/230953–dc23 LC record available at http://lccn.loc.gov/2016000953

ISBN 978-1-107-14172-8 Hardback

Contents

List of Figures *page* vii
List of Tables xiii
List of Contributors xvii
Foreword by Ahmed Galal xix
Acknowledgments xxi

1 Overview of Context, Issues and Summary 1
Ibrahim Elbadawi and Hoda Selim

2 The Global Impact of the Systemic Economies and MENA Business Cycles 16
Paul Cashin, Kamiar Mohaddes and Mehdi Raissi

3 Real Exchange Rates and Export Performance in Oil-dependent Arab Economies 44
Ibrahim Elbadawi and Linda Kaltani

4 Dutch Disease in the Services Sector: Evidence from Oil Exporters in the Arab Region 82
Ndiamé Diop and Jaime de Melo

5 The Political Economy of Public Sector Employment in Resource-dependent Countries 103
Omer Ali and Ibrahim Elbadawi

6 The Oil Curse and Labor Markets: The Case of Saudi Arabia 149
Hend Al-Sheikh and S. Nuri Erbas

7 Resource Rents, Political Institutions and Economic Growth 187
Ibrahim Elbadawi and Raimundo Soto

8 Fiscal Institutions in Resource-rich Economies: Lessons from Chile and Norway 225
Klaus Schmidt-Hebbel

9 Savings and Investment Decisions from Natural Resource Revenues: Implications for Arab Development 284
Paul Collier

10 Labor Market Heterogeneity and Optimal Exchange Rate Regimes in Resource-rich Arab Countries 301
Almukhtar Saif Al-Abri

11 The Institutional Curse of Natural Resources in the Arab World 322
Hoda Selim and Chahir Zaki

12 Has the UAE Escaped the Oil Curse? 373
Raimundo Soto and Ilham Haouas

13 Understanding and Avoiding the Oil Curse in Sudan 421
Kabbashi M. Suliman

Index 461

Figures

2.1 Four Quarters Cumulated Impulse Responses of Output to a Negative GDP Shock in the United States (Relative to the United States). *page* 29
2.2 Four Quarters Cumulated Impulse Responses of Oil Prices and Supply. 30
2.3 Four Quarters Cumulated Impulse Responses of Output to a Negative GDP Shock in the Euro Area (Relative to the Euro Area) 33
2.4 Impulse Responses of a Negative Unit Shock to Euro Area Output 35
2.5 Four Quarters Cumulated Impulse Responses of Output to a Negative GDP Shock in China (Relative to China). 36
2.6 Four Quarters Cumulated Impulse Responses of Output to a Positive GDP Shock in the GCC Region (Relative to the GCC). 38
2.7 Four Quarters Cumulated Impulse Responses of Output to a Positive GDP Shock in the MENA Oil Exporters (Relative to the MENAEX). 40
3.1 Export Concentration Index (HHI): (2000–2013) 49
3.2 Degree of Export Sophistication (EXPY: 2007) 53
3.3a Export Concentration (HHI) and the Share of Hydrocarbon Fuel Exports 54
3.3b Export Sophistication (EXPY) and the Share of Hydrocarbon Fuel Exports 54
3.4 RER Undervaluation – Oil Exporting vs. Non-oil Middle Income 60

3.5 RER Undervaluation – Arab Oil Economies 60
3.6 RER Undervaluation – GCC vs. Populous Arab Oil Economies 61
3.7 Government Effectiveness in Arab Oil and Comparator Countries 68
3.8 Regulatory Quality in the Arab Oil and Comparator Economies (2012) 68
4.1 Services Value-added Growth and GDP Growth in the Arab Region 85
4.2 Composition of Exports in the Arab Region and South Asia, 2008. 86
4.3 Changes in the Composition of GDP: 1980–1983 to 2007–2010 87
4.4 Services Share in GDP by Level of Income 87
4.5 Services Share in GDP, Arab World vs. Rest of the World 88
4.6 Share of Services in Non-mining GDP 89
4.7 Share of Consumption of Services in GDP (%) 90
4.8 Share of Imported Services in GDP (%) 91
4.9 Restrictiveness of Services Trade Policies and Share of Services in GDP, Arab Region, GCC and other Regions 96
5.1 The Extensive Form of the Game between the Elite and Citizens 111
5.2 The Extensive Form of the Game between the Elite and Citizens When the Elite Have an Option to Form a State Security Apparatus that Removes the Option of Citizens to Revolt 115
5.3 The Optimal Policy of the Elite as R/N Increases 117
5.4 Resource Rent Per Capita Across the Globe (In Current US Dollars: Average 2000–2007) 119
5.5 The Income and Resource Rents Scatter 124
5.6 The Extensive Form of the Game between the Elite and Citizens with Imperfect Repression 132
7.1 Scatter of Political Variables Average 1975–2009 212
8.1 Selected Countries: Government Budget Transparency and Accountability, 2010 237
8.2 Transparency of Sovereign Wealth Funds, 2011 238
8.3 Chile: Trend GDP Growth and Output Gap Projection by MoF in 2010, 1990–2015 (%) 248
8.4 Chile: Committee's Projections, Average Actual Future GDP Growth and Actual GDP Growth, 2002–2010 (%) 249

8.5 Chile: Committee's Projected Copper Price, Average Actual Copper Price and Actual Copper Price, 2001–2011 250
8.6 Chile: Actual and Cyclically Adjusted Central Government Balance, 1990–2010 (Ratios to GDP, %) 252
8.7 Chile: Central Government Expenditure and Revenue, 1990–2010 (Ratios to GDP, %) 253
8.8 Chile: Central Government Revenue and Expenditure Growth Rates, 1991–2010 (%) 253
8.9 Chile: Government Revenue, 2005–2010 (Ratio to GDP, %) 254
8.10 Chile: Gross Assets, Gross Liabilities and Net Assets of the Government, 1990–2010 (Ratios to GDP, %) 256
8.11 Chile: Sovereign Risk Measures: International Country Risk Guide Index and EMBI Spread, 1990–2010 256
8.12 Chile: Government Foreign Assets Held in CRCF, ESSF and PRF, 2001–2010 (Ratios to GDP, %) 260
8.13 Chile: Current International Portfolio Composition of Sovereign Wealth Funds (PRF and ESSF) and Recommended Composition for PRF (%) 261
8.14 Norway: Past (1970–2010) and Projected (2011–2030) Oil Production 267
8.15 Norway: Government Budget and Pension Fund Performance 271
8.16 Norway: Government Pension Fund Global (GPFG) Investment Portfolio, 1997–2011 (Ratio to GDP, %) 273
8.17 Norway: Government Pension Fund Norway (GPFN) Investment Portfolio, 1997–2011 (Ratio to GDP, %) 274
11.1 Natural Resource Rents, Percent of GDP 1970–2011 328
11.2 Natural Resource Rent Per Capita (Current US$) 330
11.3 Structure of Value-added and Exports, 2011 or 2012 331
11.4 Percent of Total Fiscal Revenues, Average (1992–2012) 331
11.5 Hydrocarbon Revenues during Booms and Busts, 1998 and 2008, Percent of GDP 332
11.6 Fiscal Balances, 1992–2012 (Average), Percent of GDP 333
11.7 Fiscal Balances during Booms and Busts, 1998 and 2008, Percent of GDP 333
11.8 Natural Resources and Freedom House Scores, 2013 335
11.9 Natural Resource Rents and Long-run GDP Per Capita Growth, Average 1981–2011 341

11.10 Natural Resource Rents and Volatility of GDP Per Capita Growth, 1981–2011 341
11.11 Employment of Non-nationals in the Private Sector, Share in Total Private Sector Employment 345
11.12 Manufacturing and Rents, 1980–2011 346
11.13 Manufactures Exports and Rents, 1980–2011 347
11.14 Value-added Manufacturing (Percent of GDP) 347
11.15 Manufactures Exports (% of Merchandise Exports), 1975 and 2011 348
11.16 Hydrocarbons Rents and Public Consumption, Percent of GDP, 1980–2011 350
11.17 Hydrocarbons Rents and Domestic Investments, 1980–2011 351
11.18 Hydrocarbons Rents and Financial Sector Development, 1980–2011 352
11.19 Loan-to-Deposit Ratios in Resource-rich Arab Countries, 1980–2011 353
11.20 Marginal Effect of Natural Resources Rents with Political Institutions on Growth 359
11.21 Country Correlations between the Cyclical Components of Real Government Spending and Real GDP, 1970–2012 361
11.22 Composition of Public Spending in 2008 and 2009, Percent of GDP 362
12.1 UAE: Real GDP (at PPP Prices) 383
12.2 UAE: Share of Oil in GDP, Exports and Government Revenues 385
12.3 UAE: Real GDP, Capital Stock and Employment (1975 = 100) 387
12.4 Average Labor Productivity (Real GDP per Worker) 388
12.5 UAE: GDP per Worker and Total Factor Productivity (Normalized so that 1975 = 100) 391
12.6 UAE: Total Factor Productivity by Economic Sector 392
12.7 UAE: Overall Fiscal Balance (% of GDP) and Non-oil Real GDP Growth (%) 397
12.8 UAE: Consolidated Government Revenue and Oil Price 398
12.9 UAE: Real Effective Exchange Rate (Normalized 2000 = 100) 401
12.10 UAE: Value-added, Employment and Average Labor Productivity in the Public Sector (Normalized 1988 = 100) 404

12.11 Total Factor Productivity in the Non-oil Sectors of Abu Dhabi and Dubai (Normalized 1988 =100) 411
13.1 Real GDP Growth, Inflation 1960–2010 425
13.2 Exports, Imports and FDI Net Flows 1970–2010 426
13.3 Contribution of Oil and Cotton to Exports and Oil GDP 1996–2010 427
13.4 Per Capita Income 1996–2010 427
13.5 Non-oil Exports, Crops Exports and RER Overvaluation 1996–2010 428
13.6 Quantity of Cotton Exported and Free Market Price 1980–2010 428
13.7 Agricultural Exports 1970–2010 429
13.8 Non-oil GDP: Contribution to Growth 1999–2010 430
13.9 Oil and Non-oil Revenues (Percent of GDP) 1999–2010 431
13.10 Tax, Non-tax and Oil Revenue (Percent GDP) 1970–2010 431
13.11 Government Expenditure (Percent of GDP) 1999–2010 432
13.12 The Budget Position (Percent of GDP) 1999–2020 433
13.13 Oil Price Index 1999–2010 433
13.14 GDP, Non-oil GDP Growth and Ratio of Expending to Revenue 1998–2010 434
13.15 Cyclical Revenue, Expenditure and Output for Sudan 1970–2010 448
13.16 Cochrane's Variance Ratio (Several Samples) 452

Tables

2.1 Countries and Regions in the GVAR Model Including MENA *page* 19
2.2 MENA Trade Weights 24
2.3 Variables Specification of the Country-specific VARX* Models 27
3.1 Industrial Performance in Developing Regions (2005) 51
3.2 The Long and Short-run Determinants of the Real Exchange Rate 58
3.3 Resource Rents, RER Undervaluation and Manufacturing Exports (1980–2011) 64
3.4 Resource Rents, RER Undervaluation and Exports Sophistication of Goods (1980–2011) 66
3.5 Resource Rents, RER Undervaluation and Exports Concentration (1980–2011) 67
3.6 HP-gap RER Undervaluation and Export Performance (1980–2011) 69
4.1 Share of Rents from Natural Resources in GDP% 92
4.2 Correlates of the Share of Services in GDP 93
4.3 Determinants of the Services Share in GDP 94
5.1 Public Wage and Correlates in Resource-rich and other Countries 105
5.2 Public Sector Employment vs. Political Repression (Annual Average: 2000–2007) 106
5.3 Probability of Regime Change (Full Sample) 123
5.4 Probability of Regime Change (Restricted Sample: Initial Polity ≤6) 125

5.5 Public Sector Wages Per Capita (Log_pw_pc) (Full Sample) 128
5.6 Public Sector Wages Per Capita (Log_pw_pc) (Restricted Sample: Initial Polity ≤6) 130
A.1 Definition of Variables 140
A.2 Data Summary (1970–2007) 141
A.3 Public Wage and Resource Rents by Country (1970–2007): In Real PPP Dollars 142
A.4 Probability of Regime Change (Full Sample) 146
6.1 Saudi Arabia: Summary Economic Indicators 155
6.2 Saudi Arabia: Fiscal Indicators (2003–2010 Averages) 156
6.3 Saudi Arabia: Demographic Characteristics 163
6.4 Saudi Arabia: Labor Force Characteristics 165
6.5 Saudi Arabia: Saudi and Non-Saudi Employment and Unemployment, 2007–2009 167
6.6 Saudi Arabia: Unemployment Rates and Shares of Saudi and Non-Saudi Workers in Saudi, Non-Saudi and Total Labor Force by Age Group, 2009 168
6.7 Saudi Arabia: Private Sector and Government Sector Employment Growth and Shares 170
6.8 Saudi Arabia: Composition of Private Sector Employment by Occupation, 2008–2009 Average 171
6.9 Saudi Arabia: Weighted Average Monthly Salaries in Private Sector, 1996–2010 173
6.10 Saudi Arabia: Weighted Average Government Sector Compensation, 2005–2010 175
6.11 Saudi Arabia: Higher Education Degrees Obtained by Saudi Males and Females, 2007–2009, and, Saudi Educational Preferences, 2010 177
7.1 Econometric Results: Long-run Growth Determinants Dependent Variable: Real GDP Per Capita 201
7.2 Econometric Results: Testing for Cross-country Correlations 208
7.3 Econometric Results: Long-run Growth Determinants Allowing for Common Correlated Effects 210
7.4 Estimated Long-run Effect of Resource Rents on Economic Growth Allowing for Common Correlated Effects 213
7.5 Democracy, Checks and Balances as Indicators of Credibility and Intertemporal Commitments 215
7.6 Econometric Results: Long-run Growth Determinants Allowing for Transmission Channels Low Democracy and Low Checks and Balances 217

8.1 Central Government Accounts, Growth and Copper Price, 1990–2010 244
8.2 Correlations between Committee Forecasts and Actual Variables, 1990–2011 251
8.3 Correlations between Government Budget Variables, GDP Growth and Copper Price, 1990–2010 and Sub-periods 255
8.4 Correlations between Macroeconomic Variables, 1990–2010 (Quarterly Data) 258
8.5 Gross Foreign Assets of the Central Government by Funds, 2000–2011 (US$ m.) 259
8.6 Asset Composition (%) and US$ Returns (%) of Selected Sovereign Wealth Funds, 2007–2010 262
8.7 Norway: Average Rates of Return on GPGF and GPFN Investment Portfolios, 1998–2010 (%) 273
10.1 Natural Resources Abundance and Export Diversification in the Arab Region 304
10.2 Selected Indicators of the Arab Region 305
A.1 The Central Bank's Loss Function under Alternative Exchange Rate Regimes and Different Nominal Wage Indexation 319
11.1 Proven Oil Reserves and Production, 2013 326
11.2 Proven Natural Gas Reserves and Production, 2013 327
11.3 Basic Statistics 329
11.4 Government Revenues and Expenditures in Selected Resource-rich Arab Countries, Averages 1997–2012, Percent 338
11.5 Long-term Per Capita GDP Growth in Resource-rich Arab Economies, 1981–2011 342
11.6 Selected Macroeconomic Indicators during Booms and Busts 342
11.7 Deviations from Equilibrium Real Exchange Rate, 1975–2005, Percent 344
11.8 Empirical Results of the Growth Equation 356
11.9 Empirical Results of the Growth Equation 357
12.1 Average Natural Resource Rents (As Share of GDP, Percent) 378
12.2 Real GDP Annual Growth Rate (%) 384
12.3 UAE: Annual Earnings and Hours of Work in Different Sectors in 2008 402
12.4 UAE: Economic Indicators by Emirate, 2009 408

12.5 Average Annual GDP Growth Rate by Economic Sectors in Each Emirate, 1997–2009 409
13.1 Growth Accounting Decomposition for Sudan 1961–2010 444
13.2 Aggregate Revenue and Expenditure Out-turns Compared to Original Approved Budget 446
13.3 Causality Test Based on ECM 449
13.4 Forecast Error Variance from Revenue and Expenditure Equations PP-period Ahead 450
13.5 Budget Allocation by Selected Spending Categories 453

Contributors

Omer Ali, University of California, Los Angeles

Hend Al-Sheikh, Institute of Public Administration, Saudi Arabia

Paul Cashin, International Monetary Fund

Paul Collier, Oxford University

Ndiamé Diop, The World Bank

Ibrahim Elbadawi, Dubai Economic Council and Economic Research Forum

S. Nuri Erbas, International Monetary Fund

Ilham Haouas, Abu Dhabi University

Linda Kaltani, International Monetary Fund

Jaime de Melo, University of Geneva

Kamiar Mohaddes, University of Cambridge

Mehdi Raissi, International Monetary Fund

Almukhtar Saif Al-Abri, Sultan Qaboos University

Klaus Schmidt-Hebbel, Catholic University of Chile

Hoda Selim, Economic Research Forum

Raimundo Soto, Catholic University of Chile

Kabbashi M. Suliman, University of Khartoum

Chahir Zaki, Cairo University

Foreword

One of the most distinguishing features of the Arab region is its endowment with tremendous hydrocarbon wealth. The magnitude of resource rents, that is, the difference between revenues and extraction costs, was estimated in 2012 at around US$892 billion annually, or around a third of the Arab region's GDP. With such resources, the region has the opportunity to invest in the future and join the advanced world. But, as many would argue, this outcome is not inevitable and there is a real fear of the so-called natural resource curse. This volume contributes to this debate with a view to "avoiding the natural resource curse" in natural-resource-rich Arab countries.

The literature dealing with the "resource curse" has evolved over time. The initial interpretation of the curse was derived from economics, arguing that large resource windfalls lead to, among other ills, Dutch Disease, overall macroeconomic volatility and debt overhang. This explanation fell short of explaining the successful economic performance of developed countries like Norway, the world's seventh largest oil exporter and fourteenth largest oil producer and that of Chile, an emerging country, the world's biggest producer and exporter of copper. Capitalizing on its riches, Chile has succeeded in diversifying its economy and reducing poverty levels to a historic minimum. In comparison to the experiences of Chile and Norway, the economic performance of most resource-rich Arab countries has been much less impressive, to say the least. The question is, why? Or rather, what is so different about the Arab region that has prevented it from benefiting from its abundant natural resources?

To address this question, the work in this volume was initiated by the editors as part of a broader research agenda by the Economic Research Forum on different countries in the region. The volume comprises the work of a group of distinguished scholars. Topics include typical economic concerns, such as the impact of natural resources on the region's business cycle, Dutch Disease and the impact of real exchange rate adjustment on growth and economic diversification. It also includes discussions of the labor markets in Arab countries, which act as a mechanism for state patronage through the provision of well remunerated public jobs to nationals, particularly in the GCC countries.

Equally important, this volume helps dispel the myth of the resource curse as an economic phenomenon. This volume's most powerful message is that the resource curse in the Arab region, to the extent that it does exist, is rooted in the quality of institutions, both political and economic. In a context where governments do not face significant restraints on their power and enjoy ample access to rents, it is not surprising that some rulers may use the rents to further private ends at the expense of the social good, ultimately leading to a lack of sustainable development.

To be sure, the abundance of natural resources in the region, over the past eighty years or so, has brought about economic growth, improvements in the standards of living, large savings, significant reserves and external surpluses. Thanks to oil money, many governments have also embarked on massive investment programs in basic infrastructure. However, "avoiding the oil curse" and achieving sustainable development remain far from being achieved. I hope this book will allow us to make significant strides toward the achievement of these objectives.

Ahmed Galal

Acknowledgments

The achievement of this work would not have been possible without the contribution and support of many people whom we have had the honor and pleasure to collaborate with during the past few years.

The research presented in this volume was initially conceived as part of a broader research agenda on natural resources and economic diversification. The editors are deeply grateful for the guidance of eminent colleagues who helped refine this research agenda and our thinking about the oil curse in the Arab World. They include Hasan Ersel, Mustapha Nabli, Lant Pritchett, Khalid Sekkat, James Robinson, Nisreen Salti, Nadeem Ul-Haque and Anthony Venables. Our profound appreciation goes to Ahmed Galal, ERF's Managing Director, who through his intellectual leadership, wise counsel, encouragement and guidance helped shape this work.

We would like to warmly acknowledge the hard work of the contributors who give this volume its real value. Although each is credited with their own chapters, we are indebted to all of them for their timely response to our feedback and deadlines and patient work to bring this volume into existence.

Moreover, we would like to thank three anonymous referees who reviewed our proposal submitted to Cambridge University Press (CUP). We truly believe that their excellent comments contributed to improving the overall structure of the volume and final output.

Drafts of the chapters in this volume were presented in a workshop held in Cairo and disseminated in a large policy conference held in Kuwait. The chapters benefited greatly from the comments and suggestions provided by discussants and participants, who generously gave their time and expertise. In particular, we recognize several distinguished individuals whose views were of critical value. They include Rania Al-Mashat, Rabah Arezki, Sambit Bhattacharyya, David Cobham, Mahmoud El-Gamal, Alan Gelb,

Thorvaldur Gylfason, Clement Henry and Ragnar Torvik. The editors are extremely appreciative of the generous organizational efforts extended by the Arab Fund for Economic and Social Development (AFESD) in Kuwait, which proved critical in making the policy conference a success.

We are also deeply grateful to CUP, particularly Karen Maloney and Kate Gavino, who provided excellent advice throughout. Our warm thanks are also addressed to Mahmoud El-Gamal at Rice University for having introduced us to CUP.

Last but not least, we are deeply grateful to the ERF family, especially Hoda Azmi and Namees Nabil for well-organized events, as well as to Yasmine Fahim for her tireless support and advice. In particular, we thank Jala Youssef for help with revising the proof edits and Namees for coordinating the overall publication process with CUP. Mohamed Youssri and Sherif Ossama were invaluable on financial management behind the scenes.

The research presented in this publication is the outcome of an ERF project that benefited from the financial support of the AFESD and the World Bank – within the context of the Research Initiative for Arab Development (RIAD) – support without which this work would never have materialized.

1

Overview of Context, Issues and Summary

Ibrahim Elbadawi and Hoda Selim

ABSTRACT

Despite being blessed with immense oil resources, Arab countries have neither achieved economic prosperity nor became developed countries. This chapter provides a brief overview of issues related to understanding the origins and symptoms of the resource curse and the challenges triggered by oil dependency in the Arab World including the traditional issues related to economic diversification, Dutch Disease etc. Among the novel issues explored are the interactions between oil windfalls and weak institutions (including fiscal institutions and optimal exchange rate regimes) as well as the role of political economy factors of the impact of resource rents on the strategies of the ruling elites to fend-off potential revolts that might remove them from power.

Keywords: Oil Curse, rentier–state, institutions, Arab World, GCC

The Arab region is blessed with large hydrocarbon endowments, with resource discoveries taking place since the first half of the twentieth century. Currently, 11 of the 22 members of the League of Arab States are natural resource exporters and 7 are members of the Organization of the Petroleum Exporting Countries (OPEC). The region holds close to half of global oil reserves and a quarter of natural gas reserves. It controls close to a third of oil production and 14% of natural gas production. The hydrocarbon sector dominates most of these economies, accounting for 50% of GDP on average. Fuel exports represent around three-quarters of merchandise exports. Moreover, these countries derive at least two-thirds of their fiscal revenues from hydrocarbons. Furthermore, resource dependency casts a much longer shadow through its indirect impact on the non-resource, labor-importing Arab economies. For example, large financial surpluses in the Gulf Cooperation Council (GCC) countries, accumulated from oil booms, were channeled into non-resource economies in the form of remittances and capital inflows. Most resource-poor Arab countries are relatively more dependent on remittances than on FDI inflows as a source of external financing.

Despite this blessing, resource-rich Arab countries have neither achieved economic prosperity nor become developed countries. And even though the per capita income of some such countries is high, their growth performance has been extremely volatile and raises the question of whether their current income levels could be sustained in the future.

One of the explanations put forward to explain why economic development in resource-rich countries of the region proved elusive is that of the "resource curse".[1] The early literature considered natural resource abundance as a source of economic development because of its ability to generate income, savings and investment, and, therefore, growth, thus enabling governments to provide public goods (Nurkse 1953). Unfortunately, the received literature suggests that avoiding the curse is, in fact, the exception rather than the rule in the development experiences of oil-rich countries. This is because for the case of non-renewable depletable resources, such as oil, the underlying income is temporary and is also unreliable because of its high price volatility. Therefore, consequences of failure to properly manage the volatility of oil income, or to effectively use it to accumulate large and sufficiently diverse stocks of tangible (e.g., infrastructure) and intangible (e.g., human and knowledge capital and good institutions) types of capital, are likely to be extremely dire. Country experiences have, by and large, lent support to this pessimistic view, in that resource abundance is associated with poor development outcomes and slower economic growth when compared to resource-poor countries (Gelb 1988).

The early strand of the cross-country empirical literature, pioneered by the seminal work of Sachs and Warner (1995), emphasized the failures of economic policy in explaining why resource-abundant countries have slower growth than resource-poor countries. They find that lack of openness and real exchange rate appreciation are the main culprits. Others focused on the role of volatility. For example, Hausmann and Rigobon (2003), who estimate that a one standard deviation shock to the price of oil (of 30%–35%) can generate an income shock as high as 6% of GDP in an economy where oil accounts for 20% of GDP, argue that such extreme

[1] The name resource curse is attributed to Auty (1993).

volatility, coupled with financial-sector underdevelopment and real exchange rate appreciation, could lead to inefficient specialization and post-boom growth collapse. Also, Van der Ploeg and Poelhekke (2009) find that volatility of natural resource prices is a major determinant of growth performance and that the direct positive effect of resource dependence on growth is swamped by its adverse effect via increased volatility. However, exploiting more sophisticated panel data econometrics, more recent empirical literature provides more articulated evidence. It finds that the curse is real, but is conditional on bad governance. Moreover, conditional on bad governance, this literature suggests that the curse is transmitted through three channels: excessive domestic consumption, debt overhang and real exchange rate overvaluation (e.g., Collier and Goderis 2007 and Elbadawi and Soto in this volume). The empirical significance of these channels, and their being confined to badly managed economies, corroborates the recent political economy literature, which predicts that resource booms, when government accountability is lacking, allow politicians to expand public-sector employment or directly boost private consumption to enhance their popularity (e.g., Robinson and Torvik 2005 and Robinson et al. 2006). In the context of the Arab World, Galal and Selim (2013) argue that weak political institutions in the region seem to have predated the discovery of oil. In other words, the institutional set-up was initially weak and its effects have lingered over time.

Unfortunately, as stressed earlier, the received literature suggests that avoiding the oil curse is, in fact, the exception rather than the rule in the development experiences of oil-rich countries. For the Arab World, avoiding the oil curse is, therefore, not only critical for its development, but also extremely challenging for its public policy and institutions. However, it is important, at this juncture, to draw a sharp distinction between the GCC group [comprising Bahrain, Kuwait, Oman, Saudi Arabia, Qatar and the United Arab Emirates (UAE)], which is endowed with large natural resource rents resulting in high rent per capita, and the populous oil-rich Arab economies (like Algeria, Sudan, Syria and Yemen), which have lesser rents and larger populations, resulting in significantly lower rent per capita.

On the one hand, GCC growth seems more exposed to the ebbs and flows of the oil price cycle than the growth of the populous group. They tend to experience very high growth spells during booms, but their GDP growth is almost halved and non-hydrocarbon GDP growth is lowered by a third during busts. Nevertheless, oil booms have enabled them to achieve large savings and investment ratios as well as accumulate significantly higher reserves and external surpluses. In recent decades, the GCC states have been able to channel some of their oil wealth into massive investment programs, with priority given to basic infrastructure. On the other hand, the impact of oil price fluctuations on the domestic economies of the populous group seems to be somewhat mitigated, except for one aspect: these countries tend to engage in excessive borrowing during oil busts, with external debt stocks exceeding 100% of GNI. Moreover, unlike the GCC countries, they have so far failed to use their natural capital to develop the required physical capital to promote the much needed economic diversification of their economies. Nevertheless, all of the oil-dependent Arab economies are subject to the curse, despite the differences between the GCC and others in terms of the extent and manifestations of the phenomenon. As we will show in this volume, the GCC group, especially the relatively populous Saudi Arabia, has been subject to more subtle forms of the resource curse, such as the one associated with the labor market.

In this context, the Economic Research Forum (ERF) initiated, in 2011, a major research project on understanding and avoiding the oil curse in the Arab World. This research aimed at understanding the origins and symptoms of the curse, the challenges triggered by oil dependency, as well as suggesting the macroeconomic policies that might, hopefully, turn this curse into a blessing. To this effect, thematic papers not only examined the implications of large oil windfalls for long-term optimal investment-savings decisions, but also on growth, Dutch Disease and economic diversification. This research also made an attempt to account for the interactions between oil windfalls and institutions. In particular, it assessed the role of fiscal institutions in improving the management of oil revenue and investigated optimal exchange rate regimes for resource-rich economies. It also comprised a political economy analysis of the impact of resource rents on the strategies of the ruling elites to fend-off potential revolts that might remove them from power. Finally, the discoveries from within and outside this research were selectively deployed in writing some case studies of resource-dependent Arab countries, with the twin objectives of explaining the economic and political developments in these countries as well as proposing strategies for turning the resources into a boon for development in these countries.

This volume provides a critical mass of scholarly work in the field of natural resource economics, focusing on the unique context of the Arab World, which would complement recent academic books and several articles on the subject. For example, the recent book by Ross (2012) has mainly focused on issues of democracy and conflicts, though it also addresses other topics related to the oil curse and its consequences for growth. Also, Humphreys et al. (2007) have emphasized institutional and contractual issues related to oil extraction and management. Moreover, these books do not address Arab-specific issues, though their analysis is inherently relevant to the Arab region, which makes this volume among the very few in the field of natural resource economics in the context of the Arab region. One Arab-specific book (Elbadawi and Makdisi 2011) almost exclusively focuses on the role of oil dependency in explaining the Arab democracy deficit. Another recent report by Diop et al. (2012) has solely focused on diversification and structural transformation in countries in the Middle East and North Africa, choosing to focus on the role of macroeconomic factors and policies, including the real exchange rate and fiscal policy, in promoting (or hindering) diversification away from oil.

This volume, we would argue, provides a new and complementary contribution to the existing literature by applying frontier research to issues that are directly relevant to the highly resource-dependent, but substantially under-researched, Arab region. Therefore, this volume should be an important reference for researchers and policy makers alike.

The three main questions addressed in this volume are: (i) Is there a resource curse in the Arab World and what are the symptoms? (ii) How do large oil windfalls interact with institutions and how can the latter help avoid the curse? (iii) What type of policies can help improve macroeconomic management in Arab resource-rich economies?

An attempt to answer these questions will be presented throughout the remaining twelve chapters of this volume. To the extent possible, the analysis tries to distinguish between two groups of countries: the six richer GCC countries versus the populous countries.

The book is divided into four main parts. Parts I and II of the book describe the symptoms of the curse in Arab resource-rich economies. Part I addresses questions related to business cycle linkages of Arab economies with the rest of the world and economic diversification. In Chapter 2, Cashin, Mohaddes and Raissi argue that while the international business cycle is very important for the Arab region's economic performance; macroeconomic and political developments in this region also have large consequences for the rest of the world, due to the abundance of natural resources. They are able to disentangle the speed and transmission of inward and outward macroeconomic shocks from and to the Arab region and to and from the rest of the world (China, the Euro Area and the United

States). Their results show that whereas the impact of a negative GDP shock from the Euro Area and the United States is modest, the impact of a shock to output in China is more substantial, reflecting the direction of evolving trade patterns, and China's growing role in the global economy and the world oil market.

Chapter 3 describes an important symptom of the curse. Indeed, Elbadawi and Kaltani provide new evidence about the impact of oil rents and real exchange rate undervaluation on various measures of exports. The results suggest that real exchange rate undervaluation can ameliorate the negative impact of oil rents on exports and can be particularly effective in countries with underdeveloped financial markets or low institutional development. In light of these findings, the chapter argues that a strategy of depreciating the real currency can be a viable, albeit second best, industrial policy choice in order to promote export diversification, technical upgrading and export sophistication in institutionally deficient oil and mineral-dependent economies. Moreover, this type of public policy can minimize the reliance on traditional vertical industrial policy, which usually requires high initial institutional capacity in order to succeed.

Finally, in Chapter 4, Diop and de Melo provide empirical evidence that natural resource rents in Arab countries are correlated with a declining share of services to GDP, in other words, that services sectors are a victim of a Dutch Disease phenomenon. On the basis that services can no longer be considered non-tradable because of recent revolutions in technology, transportability and tradability, they explain that rents from natural resources tend to inflate wages and non-tradable prices in resource-rich countries, thereby appreciating the real exchange rate and discouraging domestic production of tradable goods and services. This explains why Arab resource-rich countries have become large importers of tradable services and why only domestic production of non-tradable services (such as real estate, retail trade, hotels and restaurants) has really developed.

Part II addresses another symptom of the curse, related to the labor markets, where, more in the GCC than in the populous group, natural resources act as a mechanism for patronage through the provision of well remunerated public jobs to nationals. Indeed, in Chapter 5, Ali and Elbadawi argue, from a political economy perspective, that it is optimal for resource-rich authoritarian regimes that rule over small populations to offer public-sector jobs in an inflated bureaucracy to effectively remove incentives to revolt. Alternatively, when natural resource endowments are limited, it is optimal for governments to set up a repressive security apparatus and employ a smaller proportion of the population. Illustrating the case of Saudi Arabia in Chapter 6, Al-Sheikh and Erbas show how oil income enables the government to hire nationals in the civil service at very high wages and how this skews Saudi worker preferences for government-sector employment and increases their reservation wages for private-sector employment. This labor market segmentation crowds out private business, contributing to high unemployment.

Part III of the volume generally argues that institutions and sound macroeconomic management could contribute to avoiding the curse. In Chapter 7, Elbadawi and Soto argue that the oil curse is not predestined and that poor development outcomes depend on weak institutions and bad governance. They provide empirical evidence that resource-rich economies with a high degree of inclusiveness (a measure of democracy) and strong political checks and balances turn the resource curse into a blessing. In Chapter 8, Schmidt-Hebbel reviews recent international evidence on fiscal policies and outcomes in resource-rich economies at large, and Arab oil-exporting countries in particular. The chapter highlights the importance of developing counter-cyclical fiscal policies to stem excessive volatility and stabilize the non-resource economy. In comparing the decade-long experiences of Chile and Norway in setting up new fiscal institutions and rules to manage their resource rents, Schmidt-Hebbel recommends two institutional reforms: the adoption of fiscal rules and the creation of Sovereign Wealth Funds (SWF).

Chapter 9, by Paul Collier, discusses the principles and rules for optimal investment decisions in the Arab region. Collier shows that long-term sustainability is a major concern for the region. Despite some oil depletion since the 1970s, the GCC group still enjoys more resources than the populous group. Consequently, natural resources will deplete more rapidly in the latter group and so, despite being much poorer, they need higher saving rates of their resource revenues in order to convert their unsustainable resource incomes into sustainable consumption. To manage this process of high domestic investment successfully, their priority should be a phase of "investing-in-investing."

Finally, in Chapter 10, Al-Abri discusses alternative exchange rate regime arrangements for natural-resource-rich economies in the region. In his analysis, he evaluates the loss of flexibility or the gain in creditability with alternative assumptions about labor market flexibility. The analysis suggests that in countries with relatively more flexible labor markets, fixed exchange regimes are more desirable in responding to shocks, as their creditability gains might outweigh the loss of flexibility in monetary policy. However, with rigid labor markets, flexible exchange regimes are more desirable in responding to shocks.

The final part of the book includes a comparative chapter that contrasts the experience of oil countries in the region: the GCC versus populous economies, in addition to two case studies. In Chapter 11, Selim and Zaki argue that the resource curse in the Arab World is primarily an "institutional curse," even though it has several macroeconomic manifestations. In other words, while macroeconomic mismanagement and oil abundance are important determinants of performance, these factors are shaped primarily by the prevailing political institutions, which predate resource discovery. Over time, the interaction between these factors became intertwined, preventing these countries from embarking on a sustainable development path. While weak governance is responsible for the region's lack of development, the curse has operated in different ways in both groups. On the one hand, in the GCC, large rents per capita have been utilized to increase government legitimacy and foster regime stability. Indeed, the curse is expressed subtly through a clear segmentation of the labor markets, which act as an efficient mechanism of rent distribution in the form of well-remunerated public-sector jobs and other generous social welfare schemes to national citizens. In contrast, populous countries are much poorer *rentier* states and rents are not sufficient to provide a stipend to citizens. These countries have experienced conflict, violence and social unrest. Moreover, limited resources may have somewhat shielded the populous economies from extreme volatility, but seem to have led to more dire economic consequences. Resource busts tend to drive them to engage in excessive borrowing, while booms seem to have almost eliminated their manufacturing sectors. Most importantly, in a context of low rent per capita, excessive consumption resulted in massive deficiencies in infrastructure investments and an underdeveloped financial sector.

The last two chapters in this part provide two illustrations of the oil curse in the Arab World: the UAE (Chapter 12) and Sudan (Chapter 13). Soto and Haouas argue that while the UAE has not been immune to the oil curse, it has managed to make the benefits outweigh negative outcomes. Nevertheless, some symptoms of the curse remain in three areas: very low growth in labor productivity, absence of counter-cyclical policies and massive over-employment and declining productivity in the public sector. Furthermore, they show that Dubai was able to overcome the dependency on oil exports and diversify the economy. In Sudan, Suliman shows that the Dutch Disease and the fiscal linkages are the main mechanisms that transmitted the negative effects of oil booms. Moreover, the current distributive politics have tended to magnify the impact of shocks and it worsened the pro-cyclicality of fiscal policy and contributed to the excessive currency appreciation.

1.1 Lessons and Policy Implications

Natural resources in the Arab World have contributed to better standards of living and have most certainly widened the set of policy choices: countries did do better than they would have done without resource rents. Notwithstanding this improvement, the economic performance of Arab resource-rich countries has been disappointing, even in the absence of a counterfactual. For the Arab World, avoiding the Oil Curse is critical for its development. The main insight from this volume is that the interdependence between politics and resource management is the essence of the resource curse, where politics affects exploitation of resource wealth and in turn rents influence politics (Collier 2010). The main policy implication from this research is that for *Arab countries* to achieve sustained growth and prosperity, they *must adopt meaningful political reforms,* encompassing a strong system of political checks and balances. Though this remains a challenge, it is hoped that with adherence to the Extractive Industries Transparency Initiative (EITI) and the Natural Resource Charter (NRC) there is the potential of triggering some reforms on political accountability and of instituting more transparent and effective management of natural resources.

Assuming progress is made on this front, countries should first start by addressing *sustainability issues,* which means transforming their exhaustible natural resources into assets – human, domestic, private capital and foreign financial assets – that will generate future income and support sustained development. Addressing sustainability goes *hand in hand with reforming fiscal institutions.* Without sound fiscal frameworks, Arab countries will constantly be exposed to the risk of easily squandering their natural riches. It is hoped that the following reforms would contribute to higher savings and larger and more efficient investments, and put public finances on a sustainable track while increasing the ability of countries to adopt counter-cyclical fiscal policies.

Countries of the populous group are immediately confronted with shorter resource horizons and greater uncertainty about production volumes, with drastic implications on public spending sustainability once resources are depleted. Moreover, they have so far not been successful in effectively deploying their rents to build a broad-based domestic capital base. In addition, they have so far saved very little. Moreover, political instability in some countries and the violent conflicts that afflict some countries like Sudan and Yemen would tend to limit their access to international financial markets for borrowing purposes. To address this fundamental and pressing challenge, *countries of the populous group need both Sovereign Wealth Funds (SWF)* that would save or invest a small proportion of resource revenues in external financial assets abroad, in addition to a *Sovereign Development Fund (SDF)* that would invest the majority of their wealth in domestic infrastructure, either in the form of physical assets or in human capital through improving the health care and education of citizens (Collier, this volume). Moreover, these countries would need to closely monitor their public finances. They could benefit from *setting a target for the non-resource fiscal balance* (which is an indicator of the capacity of the economy to absorb resource revenues without causing inflation and a large current account deficit) at a level that can be maintained after resource revenues run out. This would avoid the need for abrupt breaks in government expenditures or tax increases after the natural resources have been depleted, given that the non-resource fiscal balance gradually converges to the overall balance as resource revenues decline (Daniel et al. 2013).

It is often argued that the use of SDFs to finance domestic infrastructure poses significant challenges. Administrative systems may lack the technical ability to scale-up expenditure rapidly (to identify, implement, budget as well as monitor and evaluate projects) and coordinate between government entities. In low-capacity and low-governance environments, like those present in the populous group, this may lead to, among other things, wasteful expenditure and budget fragmentation. In addition, even though the SDF also has the potential of acting as a means to improve the quality of public spending and even to crowd in private investors to strengthen investment discipline, it could become a vehicle for politically driven "investments" in a context of elite capture (Gelb et al. 2014). In order to bypass this problem and ensure that public spending is allocated towards high-quality public investment projects, *a more gradual increase in spending may be advisable for the populous group,* with an initial focus on investing resources to remove existing bottlenecks – a process sometimes called "investing in investing," as defined by Collier et al. (2010).

For countries with long resource horizons like the GCC, the main challenge should be to manage revenue price volatility. Collier (this volume) judges that *the creation of Sovereign Wealth Funds (SWF) is sufficient to ride out revenue volatility for the GCC,* which also have few opportunities for domestic investments. By accounting for year-to-year fluctuations in resource prices, the adoption of *fiscal rules* may smooth revenues and allow the authorities to determine how much of their resource revenues they can safely spend through the annual budget. In addition, they could potentially mitigate discretionary interventions by governments (Daniel et al. 2013). Unlike the populous economies, which need to monitor their non-resource balance, the GCC would benefit from setting a target for the structural budget balance on the basis of an estimate of the long-term oil price. This indicator would net out the cyclical impact of the level of economic activity and hydrocarbon prices that affect central government income. Thus, the structural balance equals structural revenues plus interest on net government assets minus actual expenditures on goods and services. It thus reflects the financial balances that the central government would have shown if GDP had been at its trend level and hydrocarbon prices had been running at their long-term level.

In addition, *more transparency with respect to fiscal policy and public procurement is also warranted.* In particular, fiscal transparency in the collection and utilization of natural resource revenues through appropriate accounting, reporting and auditing is important. Meanwhile, in terms of public procurement, it is better if governments solicit interest from oil multinational companies to extract resources through open and transparent processes, including competitive bidding in exploration and extraction. In order to increase transparency, it is important that Arab resource-rich economies follow good practices, including a clear assignment of roles and responsibilities of different government entities, establishment of an open budget process, making information publicly available and providing assurances of data integrity. Fiscal transparency could be a good way to combat corruption. The EITI provides a global standard for transparency in the oil, gas and mining industries, while the NRC, which builds on the transparency initiative, offers more comprehensive principles for governments and societies on how to best harness the opportunities for development generated by extractive commodity windfalls. Those initiatives can serve as anchors for enhancing transparency and accountability in commodity-rich countries. More specifically, open publication of public procurement contracts can help improve investment quality and reduce contract costs and cost overruns (Kenny and Karver 2012).

Better savings and more efficient public spending are expected to release resources for diversification of the non-resource sector. While GCC governments have become increasingly aware of the need for economic diversification in recent years, the main challenge they face is to undertake comprehensive reforms to re-orient their economies toward greater private-sector employment of nationals. Controlling public-sector employment should be accompanied by raising educational quality and skills development to raise productivity in order to make nationals more competitive for private-sector hiring. As for the populous countries, it is hoped that the creation of an SDF would lead to a more efficient utilization of their resources and most importantly it would promote the non-resource economy. In parallel, a strategy of depreciating the real currency can be a viable, albeit second best, industrial policy choice in order to promote export diversification, technical upgrading and export sophistication. Moreover, this type of public policy can minimize the reliance on traditional vertical industrial policy, which usually requires high initial institutional capacity in order to succeed.

1.2 Further Research

Despite the comprehensive nature of this volume, we had to leave out several important research and policy issues, partly due to limitation of space. However, more importantly, most of these issues are currently being considered by the ERF in the context of ongoing research projects; hence, are not yet sufficiently developed for the case of the resource-rich Arab economies, while a few others might be considered new frontier areas of research, such as the potential implications of the US-originated hydrocarbon "fracking" revolution or the welfare consequences and political economy of unconditional cash transfers in resource-rich economies.

i. ***Industrial Policy, Product Space and Economic Diversification.*** The main issues to be addressed by such a topic might include construction and analysis of the product spaces (PS) of oil-dependent economies in order to provide the lay of the land in terms of establishing the need for, as well as the requirements for success of, industrial policy (IP). In particular, following Hausmann and Klinger (2007), for example, this research would aim at assessing the potential of PS-based IP, aimed at informing when and how the government should mount IP for the promotion of new areas of comparative advantage that are feasible, given the prevailing structure of production in the these economies.

ii. ***Managing Real Estate Booms and Busts and the Oil Cycle.*** The strength of the feedback channels among oil cycles, credit growth and crunch and the real estate cycle has been a major focus of recent research, especially after the recent great economic recession. However, partly due to lack of institutional information and relative unavailability of data, there is a dearth of research on the Arab World, especially the GCC economies. Subscribing to the received literature (e.g., Crowe et al. 2011), some of the questions that could be addressed by this research would include: What kind of indicators should trigger policy intervention to control a real estate boom? If policymakers were fairly certain that intervention were warranted, what would be the policy tools at their disposal? What are the likely impacts of these interventions: negative side effects and limitations? What practical issues would limit the use of these policy tools?

iii. ***Fiscal and Monetary Policy Institutions.*** Most Arab commodity-exporting economies rely on a tightly managed peg as a nominal anchor for inflation expectations. Consequently, monetary policy is not autonomous, but credibility is imported from a strong foreign currency. However, if and when these economies decide to exit the peg, monetary policy would need to rely on a domestic anchor, which would require strong domestic institutions, including rule-based fiscal institutions. Therefore, among other issues, this research would analyze the institutional and political economy requirements for better coordination between fiscal and monetary policy as well as enhanced credibility and independence of monetary policy.

iv. ***Sovereign Wealth Funds: Stabilization, Investment Strategies and Lessons for Oil-rich Arab Countries.*** In the aftermath of the global crisis new opportunities will be opened to SWFs in and outside the Arab World, but they will also be bound to face major risks. Analysis of these issues would address some key positivist and normative dimensions arising from the new challenges and opportunities facing sovereign investors, especially SWFs, including: SWFs and their links with fiscal policies and country features; macroeconomics of SWFs and their relation to fiscal policies; and SWF investment policies and practice. The issues considered relate to domestic (fund owners) and foreign (investment recipient) economies alike, and could help glean lessons for the Arab World, including for countries that have already established SWFs as well as those that have not.

v. ***The Political Economy of Rent Allocation in Oil-rich Societies.*** Resource rents have always been the prerogative of the public sector, especially with regard to the favored option of investing the rents. The claim of governments to the monopoly of collecting and spending public funds, including natural resource rents, is premised on the view that only they can coordinate a public investment path that will deliver a first-best growth strategy. In most growth models that seek to maximize the sum of citizens' future consumption per head, the shadow value of sound investment in infrastructure and other vital public goods would probably be higher than that of private consumption. But the investment path also represents a riskier strategy in terms of the indirect governance effects; in a government without strong internal checks and accountability, there are few guarantees that the government will use the funds appropriately (Gelb and Majerowicz 2011). Therefore, recent policy research on natural resource rents and governance has presented the rival option focusing on the distribution and consumption of the rents through direct cash transfers to citizens in resource endowed societies. Therefore, arguably, under such volatile economic and political institutions, distribution of the oil rents could be an effective way to mitigate the resource curse that afflicts these countries. Also, very importantly, cash transfers out of resource rents open up possibilities for building up a tax system that can produce a more accountable delivery of public services (including the much needed infrastructure). Unfortunately, in most resource-rich countries such institutional requirements for successful investment-driven approaches do not exist, especially in the case of the populous Arab oil-rich countries, which tend to be ruled by unstable and understandably myopic regimes. Subscribing to this overview, future research in this area would address some of the following key issues: How well has the oil resource rent been managed? What are the political economy considerations for rent allocation, including those pertaining to cash-transfer programs? What should be the optimum design for such programs, given the prevailing political economy? Should cash transfers complement presumably "reformed" existing food and fuel subsidy programs or, instead, be substitutes for these programs? And, for the GCC, how effective can cash transfers be in reducing labor market segmentation between nationals and foreigners and, hence, enhancing employment of nationals in the private sector?

References

Auty, R. M. (1993). *Sustaining Development in Mineral Economies: The Resource Curse Thesis*. London: Routledge.

Collier, P. (2010). The Political Economy of Natural Resources. *Social Research* 77(4): 1105–1132.

Collier, P., F. van der Ploeg, M. Spence and A. Venables. (2010). Managing Resource Revenues in Developing Economies. *IMF Staff Papers* 57(1): 84–118.

Collier, P. and B. Goderis. (2007). Commodity Prices, Growth, and the Natural Resource Curse: Reconciling a Conundrum. Economics Series Working Papers WPS, University of Oxford, Department of Economics.

Crowe, C., G. Dell'Ariccia, D. Igan and P. Rabanal. (2011). Policies for Macrofinancial Stability: Options to Deal with Real Estate Booms. IMF staff discussions note (SDN/11/02), International Monetary Fund, Washington DC.

Daniel, P. S. G., T. Mattina and A. Segura-Ubiergo. (2013). Extracting Resource Revenue. Finance and Development. Washington, D.C.: IMF.

Diop, N., D. Marotta and J. de Melo. (2012). *Natural Resource Abundance, Growth, and Diversification in the Middle East and North Africa: The Effects of Natural Resources and the Role of Policies*. Washington, D.C.: World Bank.

Elbadawi, I. and S. Makdisi. Eds. (2011). *Democracy in the Arab World: Explaining the Deficit*. New York and Canada: Routledge and International Development Centre.

Galal, A. and H. Selim. (2013). The Elusive Quest for Economic Development in the Arab Countries. *Middle East Development Journal* 05(01):1–29.

Gelb, A., S. Tordo and H. Hallan. (2014). *Sovereign Wealth Funds and Domestic Investment in Resource-Rich Countries: Love Me, or Love Me Not? World Bank Economic Premise Note Series 133*. Washington, D.C.: World Bank.

Gelb, A. and S. Majerowicz. (2011). Oil for Uganda-or Ugandans? Can Cash Transfers Prevent the Resource Curse? Working Paper 261, Center for Global Development, Washington DC.

Gelb, A. (1988). *Oil Windfalls: Blessing or Curse*. New York: Oxford University Press (for the World Bank).

Hausmann, R. and B. Klinger (2007). The Structure of the Product Space and the Evolution of Comparative Advantage. CID Working Paper No. 146, Harvard University.

Hausmann, R. and R. Rigobon. (2003). An Alternative Interpretation of the 'Resource Curse': Theory and Policy Implications. NBER Working Paper 9424, Cambridge, Mass.

Humphreys, M., J.D. Sachs and J.E. Stiglitz. Eds. (2007). *Escaping the Resource Curse*. New York: Columbia University Press.

Kenny, C. and J. Karver. (2012). Publish What You Buy: The Case for Routine Publication of Government Contracts. CGD Policy Paper 011, Center for Global Development, Washington DC.

Nurkse, R. 1953. *Problems of Capital Formation in Underdeveloped Countries*. New York: Oxford University Press.

Van der Ploeg, F. and S. Poelhekke. (2009). Volatility and the Natural Resource Curse. *Oxford Economic Papers* 61(4): 727–760.

Robinson, J., R. Torvik and T. Verdier. (2006). Political Foundations of the Resource Curse. *Journal of Development Economics* 79(2): 447–468.

Robinson, J. and R. Torvik. (2005).White Elephants. *Journal of Public Economics* 89 (2–3): 197–210.

Rodrik, D. (2008). The Real Exchange Rate and Economic Growth. *Brookings Papers on Economic Activity* 39(2): 365–439.

Ross, M. (2012). *The Oil Curse: How Petroleum Wealth Shapes the Development of Nations*. Princeton and Oxford: Princeton University Press.

Sachs, J. and A. Warner. (1995). Natural Resource Abundance and Economic Growth. NBER Working Paper 5398. Cambridge, Mass.

2

The Global Impact of the Systemic Economies and MENA Business Cycles[1]

Paul Cashin[a], Kamiar Mohaddes[b,2] and Mehdi Raissi[a]

ABSTRACT

This chapter analyzes spillovers from macroeconomic shocks in systemic economies (China, the Euro Area and the United States) to the Middle East and North Africa (MENA) region as well as outward spillovers from a GDP shock in the Gulf Cooperation Council (GCC) countries and MENA oil exporters to the rest of the world. This analysis is based on a global vector autoregression (GVAR) model, estimated for thirty-eight countries/regions over the period 1979Q2 to 2011Q2. Spillovers are transmitted across economies via trade, financial and commodity price linkages. The results show that the MENA countries are becoming more sensitive to developments in China than to shocks in the Euro Area or the United States, in line with the direction of evolving trade patterns and the emergence of China as a key driver of the global economy. Outward spillovers from the GCC region and MENA oil exporters are likely to be stronger in their immediate geographical proximity, but also have global implications.

JEL Classifications: C32, E17, E32, F44, O53, Q41

[a] International Monetary Fund, Washington DC, USA
[b] Faculty of Economics and Girton College, University of Cambridge, UK
[1] We are grateful to Alberto Behar, Rishi Goyal, Malhar Nabar and Alasdair Scott, as well as seminar participants at the IMF, the European Bank for Reconstruction and Development and the University of St Andrews for constructive comments and suggestions. We would also like to thank the Editors and two anonymous referees for most helpful suggestions. An earlier version of this chapter was circulated under the title "A Global VAR Model of MENA Business Cycles." The views expressed in this chapter are those of the authors and do not necessarily represent those of the International Monetary Fund or IMF policy.
[2] Corresponding author. Email address: km418@cam.ac.uk.

Keywords: Global VAR (GVAR), interconnectedness, global macroeconomic modeling, impulse responses, macroeconomic shocks, international business cycle, oil exporters, GCC, Arab region.

2.1 Introduction

The GVAR literature almost exclusively focuses on business cycle linkages among advanced and major emerging market economies, with limited attention to growth spillovers to/from the Middle East and North Africa (MENA) countries, in particular the Gulf Cooperation Council (GCC) region. While the international business cycle is very important for the MENA region's economic performance, macroeconomic and political developments in this region also have large consequences for the rest of the world, due to the abundance of natural resources in MENA. We use a GVAR model to disentangle the size and speed of the transmission of different macroeconomic shocks originating from three systemic countries to the Maghreb (Algeria, Libya, Mauritania, Morocco and Tunisia), Mashreq (Egypt, Jordan and Syria) and GCC (Bahrain, Kuwait, Oman, Qatar, Saudi Arabia and the UAE) regions, as well as outward spillovers from the MENA region to the rest of the world. We also focus on the emergence of China as a global force in the world economy and study how changes in trade patterns between China and the rest of the world may have affected the transmission of the international business cycle to MENA countries and other systemic economies. We also discuss the differential effects of demand-driven oil price shocks and supply-driven oil price disturbances on the MENA region.

Our approach uses a dynamic multi-country framework for the analysis of the international transmission of shocks and is based on the model developed in Cashin et al. (2014). The framework comprises thirty-eight country/region-specific models, among which is a single Euro Area block (comprising eight of the eleven countries that joined the Euro on January 1, 1999) as well as the GCC region. These individual models are solved in a global setting where the core macroeconomic variables of each economy are related to corresponding foreign variables (constructed exclusively to match the international trade pattern of the country under consideration). The model has both real and financial variables: real GDP, inflation, real equity price, real effective exchange rate, short and long-term interest rates, a measure of global oil production and the price of oil. This framework is able to account for various transmission channels, including not only trade relationships but also financial and commodity price linkages. Compared to Dees et al. (2007), this chapter advances the work on GVAR modeling in the following directions: (i) we extend the geographical coverage of the GVAR model to the Middle East and North Africa region as well as to other major oil exporters; (ii) we add a measure of global oil production to the GCC model to account for supply side factors in the world oil market as many supply shortfalls originate in the MENA region (for instance, the more recent Arab Spring and associated supply shortfalls from Libya, or the effects of sanctions on Iran and the resulting drop in its oil exports); and (iii) we investigate the growing impact of China's macroeconomic shocks on other systemic economies, the MENA region in general and major oil exporters in particular.

We estimate the GVAR model based on two sets of fixed trade weights at different points in time, being twenty years apart. Specifically, we make use of a set of weights averaged over 1986 and 1988 and another between 2006 and 2008. This allows us to study how the transmission of shocks has changed following the emergence of China as a major driver of the world economy. Our results, using quarterly data between 1979Q2 to 2011Q2, indicate that the impact of a Chinese GDP shock on a typical MENA economy, as well as on other systemic countries and oil exporters, has increased significantly since the mid-1980s. A negative GDP shock in China (using the 2006–2008 weights) would have major global repercussions, especially for less-diversified commodity exporters. The effects on other systemic countries are smaller, but not trivial. At the same time, the impact of a US GDP shock on a typical MENA economy is large, and has not changed significantly since the mid-1980s. We also find that outward spillovers from the GCC and MENA oil exporters are likely to be stronger in their immediate geographical proximity, but also have implications for systemic economies and other major oil exporters through the commodity channel.

The rest of the chapter is organized as follows. Section 2.2 outlines our modeling approach. Section 2.3 focuses on the potential macroeconomic consequences of a GDP shock in systemic countries. Section 2.4 investigates the extent to which the macroeconomic conditions in the GCC region and MENA oil exporters affect, and are affected by, the global economy. Finally, Section 2.5 concludes.

2.2 A Global VAR Model Including the MENA Region

We extend the country coverage of the GVAR dataset used in Dees et al. (2007) by adding fourteen countries located in the Middle East and North

Table 2.1: *Countries and Regions in the GVAR Model Including MENA*

MENA Oil Exporters	**Systemic Countries**	**Other Oil Exporters**
Algeria*	China	Canada
GCC Countries	Euro Area	Ecuador*
*Bahrain**	*Austria*	Indonesia
*Kuwait**	*Belgium*	Mexico
*Oman**	*Finland*	Nigeria*
*Qatar**	*France*	Norway
Saudi Arabia	*Germany*	Venezuela*
*UAE**	*Italy*	
Iran*	*Netherlands*	**Latin America**
Libya*	*Spain*	Argentina
	Japan	Brazil
MENA Oil Importers	United Kingdom	Chile
Egypt*	United States	Peru
Jordan*		
Mauritania*	**Emerging Asia**	**Rest of the World**
Morocco*	Korea	Australia
Syria*	Malaysia	India
Tunisia*	Philippines	New Zealand
Turkey	Singapore	South Africa
	Thailand	Sweden
		Switzerland

Notes: * indicates that the country has been added to the Smith and Galesi (2010) database. Countries in italics are included in a region for estimation purposes.

Africa region as well as three other Organization of the Petroleum Exporting Countries (OPEC) members (see Table 2.1). Thus, our version of the GVAR model includes fifty countries, covering over 90% of world GDP, as opposed to the "standard" thirty-three country set-up used in the literature, see Smith and Galesi (2010). Of the fifty countries included in our sample, seventeen are oil exporters, of which ten are current OPEC members and one is a former member (Indonesia left OPEC in January 2009). We were not able to include Angola and Iraq, the remaining two OPEC members, due to the lack of sufficiently long time series data. We therefore extend the country coverage both in terms of major oil exporters and also by including an important region of the world when it comes to oil supply, the MENA region.

For empirical application, we create two regions, one of which comprises the six Gulf Cooperation Council (GCC) countries: Bahrain, Kuwait,

Oman, Qatar, Saudi Arabia and the UAE; and the other is the Euro Area block comprising eight of the eleven countries that initially joined the Euro in 1999: Austria, Belgium, Finland, France, Germany, Italy, Netherlands and Spain. The time series data for the GCC block and the Euro Area block are constructed as cross-sectionally weighted averages of the domestic variables (described in detail below), using purchasing power parity GDP weights, averaged over the 2006–2008 period. Thus, as displayed in Table 2.1, the GVAR model that we specify includes thirty-eight country/region-specific VARX* models.[3]

2.2.1 Variables

The macroeconomic variables included in the individual VARX* models depend on both the modeling strategy employed as well as whether data on a particular variable is available. Each country-specific model has a maximum of six domestic (endogenous) variables and five foreign (exogenous) variables. We also include two global variables, each of which is treated endogenously in only one country, while being weakly exogenous in the remaining thirty-seven country models. Below, we describe the different variables included in our model and provide justification for our modeling specification. For various data sources used to build the quarterly GVAR dataset, covering 1979Q2 to 2011Q2, see the Data Appendix of Cashin et al. (2012).

2.2.1.1 Domestic Variables

Real GDP, y_{it}, the rate of inflation, π_{it}, short-term interest rate, r_{it}^S, long-term interest rate, r_{it}^L, real equity prices, eq_{it} and real effective exchange rate, $reer_{it}$, are the six domestic variables that are included in our model. These variables are constructed as:

$$\begin{array}{l} y_{it} = \ln(\mathrm{GDP}_{it}), \quad \pi_{it} = p_{it} - p_{it-1}, \\ p_{it} = \ln(\mathrm{CPI}_{it}), \quad eq_{it} = \ln(\mathrm{EQ}_{it}/\mathrm{CPI}_{it}), \\ e_{it} = \ln(\mathrm{E}_{it}), \quad reer_{it} = \left(e_{it} - p_{it}\right) - \left(e_{it}^* - p_{it}^*\right), \\ r_{it}^S = 0.25\ln\left(1 + R_{it}^S/100\right), \quad r_{it}^L = 0.25\ln\left(1 + R_{it}^L/100\right), \end{array} \tag{1}$$

where GDP_{it} is the real gross domestic product at time t for country i, CPI_{it} is the consumer price index, EQ_{it} is a nominal equity price index, E_{it} is the

[3] Section II of Cashin et al. (2012) describes the GVAR methodology in detail.

nominal exchange rate and $R_{it}^S\left(R_{it}^L\right)$ is the short-term (long-term) interest rate. The log of foreign price level $\left(p_{it}^*\right)$ and exchange rate $\left(e_{it}^*\right)$ are calculated as the weighted sum of log price level indices $\left(p_{jt}\right)$ and log of nominal exchange rate $\left(e_{jt}\right)$ of country i's trading partners respectively. More specifically, $p_{it}^* = \sum_{j=0}^N w_{ij}p_{jt}$ and $e_{it}^* = \sum_{j=0}^N w_{ij}e_{jt}$, where w_{ij} is the trade share of country j for country i (see below for more details), where $j = 0, \ldots, N$. In the current application, we have a total of thirty-six countries and two regions in our model, $N = 37$.

2.2.1.2 Foreign Variables

We include five foreign variables in our model. In particular, all domestic variables, except for that of the real effective exchange rate, have corresponding foreign variables. The exclusion of $reer_{it}^*$ is simply because $reer_{it}$ already includes both domestic, $e_{it} - p_{it}$, and foreign, $e_{it}^* - p_{it}^*$, nominal exchanges rates deflated by the appropriate price levels, see Equation (2.1). Therefore, $reer_{it}^*$ does not by itself have any economic meaning. The foreign variables are all computed as:

$$\begin{aligned}
y_{it}^* &= \sum_{j=0}^N w_{ij}y_{jt}, \quad eq_{it}^* = \sum_{j=0}^N w_{ij}eq_{jt}, \\
p_{it}^* &= \sum_{j=0}^N w_{ij}p_{jt}, \quad \pi_{it}^* = p_{it}^* - p_{it-1}^*, \\
r_{it}^{S*} &= \sum_{j=0}^N w_{ij}r_{jt}^S, \quad r_{it}^{L*} = \sum_{j=0}^N w_{ij}r_{jt}^L.
\end{aligned} \tag{2}$$

The trade weights, w_{ij}, are computed as a three-year average to reduce the impact of individual yearly movements on the weights[4]:

$$w_{ij} = \frac{T_{ij,t-2} + T_{ij,t-1} + T_{ij,t}}{T_{i,t-2} + T_{i,t-1} + T_{i,t}}, \tag{3}$$

where T_{ijt} is the bilateral trade of country i with country j during a given year t and is calculated as the average of exports and imports of country i with j and $T_{it} = \sum_{j=0}^N T_{ijt}$ (the total trade of country i). The trade shares used to construct the foreign variables are given in the 38×38 matrices provided in Tables 7 and 8 of Cashin et al. (2012).

[4] A similar approach has also typically been followed in Global VAR models estimated in the literature. See, for example, Dees et al. (2007).

2.2.1.3 Global Variables

Given the importance of oil price and production for the MENA region, we also include nominal oil prices (in US dollars), P_t^{oil}, as well as the quantity of oil produced in the world, Q_t^{oil}, in our model. As is now standard in the literature, we include log oil prices, p_t^{oil}, as a "global variable" determined in the US VARX* model; that is, the price of oil is included in the US model as an endogenous variable while it is treated as weakly exogenous in the model for all other countries. The main reason for this is that the US is the world's largest oil consumer. On average, about 27% of the world oil in 1979–2010 was consumed by the US, which is far larger compared to the other three major oil importers in the world (China, Euro area and Japan), even when combined.

On the other hand, the GCC countries produce more than 22% of world oil and export around 30% of the world total. They also possess 36% of the world's proven oil reserves, and Saudi Arabia, by itself, has the largest spare capacity in the world. Thus, we include log of oil production, q_t^{oil}, as an endogenous variable in the GCC block, and as a weakly exogenous variable in all other countries. The variable q_t^{oil} is therefore the second "global variable" in our model.[5]

Making one region out of Bahrain, Kuwait, Oman, Qatar, Saudi Arabia and the UAE, is not without economic reasoning. The rationale is that these countries have in recent decades implemented a number of policies and initiatives to foster economic and financial integration in the region with a view of establishing a monetary union (loosely based on that of the Euro Area). Abstracting from their level of success with the above objectives, the states of the GCC are relatively similar in structure, though in the short term they may face some difficulties in meeting the convergence criteria they have set for economic integration based on those of the European Union (EU). Inflation rates vary significantly across these countries and fiscal deficits, which have improved since the start of the oil boom in 2003, are about to re-emerge in some countries. However, these economies already peg their currencies to the US dollar, except for Kuwait, which uses a dollar-dominated basket of currencies, and are accustomed to outsourcing their interest-rate policy. They also have relatively open capital accounts, and hence, it is reasonable to group these countries as one region.[6]

[5] For a more detailed discussion of oil supply and price modeling in a GVAR model see Cashin et al. (2014).

[6] See Mohaddes and Williams (2013) for more details.

2.2.2 MENA Trade Weights

The Middle East and North African countries are globally less competitive relative to their peers. The Middle East accounts for less than 1% of world non-fuel exports, compared with 4% from Latin America. Of its limited global export share, inter-regional trade accounts for less than a tenth, barely more than in 1960. The usual explanation for the poor trade performance in the region is its reliance on crude oil exports, and hence, its little success in developing significant merchandise exports. Furthermore, since most countries in the region export the same products – oil and gas – they naturally do not tend to actively trade with each other (see Table 2.2). More trade would enable firms to reap greater economies of scale, increase returns to investment, adopt superior technology and promote growth[7].

Table 2.2b shows that the Euro Area is the most important trading partner for the Maghreb countries (Algeria, Libya, Mauritania, Morocco and Tunisia). More than 48% of their trade originates in or is destined for the Euro Area. The US and China are also large trading partners for the Maghreb, with the weights ranging between 3% and 22% and 2% and 25% with the US and China, respectively.[8] However, the Maghreb's trade with the GCC is generally limited to less than 1% of total trade in all countries except for Morocco (for which it is 6%).

On the other hand, the Mashreq countries (Egypt, Jordan and Syria) trade much more with the GCC, where the shares are between 13% and 28%.[9] The Euro Area is nevertheless very important for the region, as between 16% and 32% of Mashreq trade is destined for or originates in the eight Euro Area countries in our sample. Europe is also an important trading partner for Turkey (45%), as compared to China, the GCC, the UK and the US, where the individual trade weights are just above 6%. Iran's largest trading partner is the Euro Area (24%), but it also trades substantially with China, the GCC countries and Japan (all exceeding 12%).

Comparing with other countries in the MENA region, the GCC's trade is less concentrated on one country/region, trading more than 10% with China, the Euro Area, Japan and the US individually. However, as

7 See Behar and Freund (2011) for an extensive discussion of the trade performance of the MENA region over the past fifteen years.

8 As a general caveat, we should mention that bilateral trade data are subject to underreporting and measurement errors.

9 See Mohaddes and Raissi (2013) for more details regarding the link between the Mashreq countries and the GCC.

Table 2.2: *MENA Trade Weights*
(a) Averages over 1986–1988

	Algeria	Egypt	GCC	Iran	Jordan	Libya	Morocco	Mauritania	Syria	Tunisia	Turkey
Algeria	0	0.00	0.00	0.00	0.00	0.00	0.00	0.04	0.01	0.03	0.01
China	0.00	0.01	0.01	0.01	0.03	0.00	0.02	0.03	0.01	0.01	0.01
Egypt	0.00	0	0.00	0.00	0.02	0.00	0.00	0.00	0.00	0.00	0.01
Euro Area	0.68	0.48	0.25	0.39	0.29	0.75	0.60	0.63	0.53	0.75	0.47
GCC	0.00	0.03	0	0.04	0.17	0.00	0.05	0.00	0.05	0.02	0.06
Iran	0.00	0.00	0.01	0	0.00	0.00	0.00	0.00	0.08	0.00	0.06
Japan	0.03	0.07	0.27	0.15	0.08	0.04	0.03	0.21	0.06	0.01	0.05
Jordan	0.00	0.01	0.01	0.00	0	0.00	0.00	0.00	0.02	0.00	0.01
Libya	0.00	0.00	0.00	0.00	0.00	0	0.02	0.00	0.03	0.00	0.02
Morocco	0.00	0.00	0.00	0.00	0.00	0.01	0	0.00	0.00	0.01	0.00
Mauritania	0.00	0.00	0.00	0.00	0.00	0.00	0.00	0	0.00	0.00	0.00
Syria	0.00	0.00	0.00	0.01	0.02	0.01	0.00	0.00	0	0.00	0.00
Tunisia	0.01	0.00	0.00	0.00	0.00	0.00	0.01	0.00	0.00	0	0.00
Turkey	0.02	0.02	0.01	0.07	0.04	0.04	0.01	0.00	0.04	0.02	0
UK	0.02	0.05	0.06	0.06	0.08	0.06	0.03	0.02	0.04	0.02	0.06
USA	0.15	0.17	0.16	0.05	0.12	0.00	0.07	0.02	0.06	0.05	0.11
Other Oil Exporters	0.02	0.01	0.01	0.02	0.02	0.01	0.06	0.01	0.01	0.03	0.02
Latin America	0.02	0.02	0.02	0.06	0.03	0.00	0.02	0.00	0.02	0.01	0.01
Emerging Asia	0.00	0.03	0.11	0.07	0.03	0.03	0.01	0.02	0.00	0.00	0.01
Rest of the World	0.03	0.10	0.07	0.09	0.08	0.04	0.07	0.01	0.04	0.03	0.07

(b) *Averages over 2006–2008*

	Algeria	Egypt	GCC	Iran	Jordan	Libya	Morocco	Mauritania	Syria	Tunisia	Turkey
Algeria	0	0.01	0.00	0.00	0.01	0.00	0.02	0.01	0.00	0.02	0.02
China	0.04	0.07	0.10	0.15	0.09	0.05	0.05	0.25	0.08	0.02	0.07
Egypt	0.01	0	0.01	0.00	0.04	0.01	0.01	0.01	0.02	0.01	0.01
Euro Area	0.48	0.30	0.15	0.24	0.16	0.66	0.56	0.48	0.32	0.70	0.45
GCC	0.00	0.13	0	0.12	0.28	0.01	0.06	0.01	0.25	0.01	0.06
Iran	0.00	0.00	0.02	0	0.00	0.00	0.02	0.00	0.04	0.00	0.04
Japan	0.02	0.04	0.18	0.12	0.03	0.01	0.02	0.04	0.02	0.01	0.02
Jordan	0.00	0.01	0.01	0.00	0	0.00	0.00	0.00	0.02	0.00	0.00
Libya	0.00	0.01	0.00	0.00	0.00	0	0.00	0.00	0.03	0.05	0.01
Morocco	0.01	0.01	0.00	0.01	0.00	0.00	0	0.01	0.00	0.01	0.01
Mauritania	0.00	0.00	0.00	0.00	0.00	0.00	0.00	0	0.00	0.00	0.00
Syria	0.00	0.01	0.01	0.01	0.04	0.01	0.00	0.00	0	0.00	0.01
Tunisia	0.01	0.00	0.00	0.00	0.00	0.03	0.01	0.01	0.00	0	0.00
Turkey	0.04	0.03	0.02	0.06	0.02	0.03	0.03	0.01	0.05	0.02	0
UK	0.03	0.03	0.03	0.02	0.02	0.03	0.04	0.02	0.02	0.03	0.07
USA	0.22	0.11	0.11	0.00	0.11	0.07	0.05	0.05	0.02	0.03	0.07
Other Oil Exporters	0.07	0.04	0.02	0.01	0.02	0.01	0.02	0.03	0.01	0.01	0.03
Latin America	0.04	0.04	0.01	0.01	0.02	0.02	0.04	0.03	0.02	0.02	0.01
Emerging Asia	0.02	0.05	0.20	0.11	0.05	0.02	0.02	0.03	0.04	0.01	0.04
Rest of the World	0.03	0.10	0.13	0.14	0.09	0.05	0.06	0.03	0.04	0.04	0.08

Notes: Trade weights are computed as shares of exports and imports, displayed in columns by country/region (such that a column, but not a row, sum to 1).
Source: International Monetary Fund, *Direction of Trade Statistics*.

mentioned earlier, this is mainly due to oil exports to different regions rather than having a more diversified export basket/market.

Comparing the more recent trade weights, averaged over 2006–2008, with those from twenty years ago in Table 2.2a, we see that for the MENA region as a whole, trade with the Euro Area has fallen, but trade with China has increased many fold for all countries. On the other hand, trade with the US has increased for some, but decreased for others, while trade between the region and the GCC has remained more or less stable, except for Egypt, Iran, Jordan and Syria for which it has increased between 8 (Iran) and 20 (Syria) percentage points, respectively.

Overall, Table 2.2 illustrates the continuing importance of the Euro Area countries in our sample for the MENA region, but also shows that both China and the US are important for the region. We will therefore focus on spillovers from these three systemic economies to MENA countries in Section 2.3. Moreover, given the emergence of China in the world economy and its increasing importance for the MENA region, and for the largest oil exporters in the world in particular, we shall also illustrate how China's impact on the region and other systemic economies has changed over the past two decades.

Table 2.2 also illustrates that trade between the GCC and other MENA countries are large for a few economies but small for others. This is also the case for overall trade between the GCC and the rest of the world: the GCC trade weights for China, the Euro Area and the US are between 3% and 4%, although trade with India is more than 20% and the trade shares with Japan and Korea are more than 12%, see Table 7 of Cashin et al. (2012). However, given the importance of the Persian Gulf in determining oil supply (and oil prices), we expect the GCC's performance to have a global impact through the commodity channel rather than purely via the trade one. Thus, we look at spillovers from the GCC to the rest of the world in Section 2.4.

2.2.3 Model Specification

Given the discussion in Section 2.2.1, we specify three different sets of individual country-specific models. The first specification is common across all countries apart from the United States and the GCC block. These thirty-six VARX* models include six endogenous/domestic variables, when available, five country-specific foreign variables, and two global variables (see Table 2.3).

The second specification relates to the GCC block only, for which the log of oil production, q_t^{oil}, is included in the model endogenously in addition to

Table 2.3: *Variables Specification of the Country-specific VARX* Models*

The US Model		The GCC Model		All Other Models	
Domestic	Foreign	Domestic	Foreign	Domestic	Foreign
$y_{US,t}$	$y^*_{US,t}$	$y_{GCC,t}$	$y^*_{GCC,t}$	y_{it}	y^*_{it}
$\pi_{US,t}$	$\pi^*_{US,t}$	$\pi_{GCC,t}$	$\pi^*_{GCC,t}$	π_{it}	π^*_{it}
$eq_{US,t}$	–	–	$eq^*_{GCC,t}$	eq_{it}	eq^*_{it}
$r^S_{US,t}$	$r^{*S}_{US,t}$	–	$r^{*S}_{GCC,t}$	r^S_{it}	r^{*S}_{it}
$r^L_{US,t}$	–	–	$r^{*L}_{GCC,t}$	r^L_{it}	r^{*L}_{it}
–	$e^*_{US,t} - p^*_{US,t}$	$reer_{GCC,t}$	–	$reer_{it}$	–
p^{oil}_t	–	–	p^{oil}_t	–	p^{oil}_t
–	q^{oil}_t	q^{oil}_t	–	–	q^{oil}_t

Notes: See Equation (2.1) for the definition of the variables.

the three domestic variables, while the log of nominal oil prices, p^{oil}_t, is included as weakly exogenous variables.

Finally, the US model is specified differently from the others, mainly because of the dominance of the US in the world economy. First, based on the above discussion regarding oil consumption, the price of oil is included in the model endogenously. Second, given the importance of US financial variables in the global economy, the US-specific foreign financial variables, $eq^*_{US,t}$ and $r^{*L}_{US,t}$, are not included in this model. The exclusion of these two variables was also confirmed by our preliminary analysis, in which the weak exogeneity assumption was rejected for $eq^*_{US,t}$ and $r^{*L}_{US,t}$ in the US model. Finally, since e_{it} is expressed as domestic currency price of a US dollar, $e_{US,t} - p_{US,t}$, it is by construction determined outside this model. Thus, instead of the real effective exchange rate, we included $e^*_{US,t} - p^*_{US,t}$ as a weakly exogenous foreign variable in the US model.

For brevity, we have not presented the country-specific estimates and tests, but these are available in the working paper version of this chapter. In particular, see Section III of Cashin et al. (2012) for discussions about lag order selection, cointegrating relations and persistence profiles. We also provide evidence for the weak exogeneity assumption of the foreign variables and discuss the issue of structural breaks in the context of our GVAR model in Section III.

2.3 Inward Spillovers

This section studies whether the increasing economic integration at the world level and the resulting emergence of large economic players, such as China, have weakened the role of the US economy or the Euro Area as drivers of global growth. To do so, we look at the effects of negative US, Euro Area and Chinese real output shocks on the MENA region, other oil exporters and systemic economies.[10]

2.3.1 Shock to US GDP

As a result of the dominance of the United States in the world economy, any slowdown in this country can bring about negative spillovers to other economies. As the recent global economic crisis has shown, the history of past US recessions usually coincides with significant reductions in global growth. Furthermore, the continuing dominance of US debt and equity markets, backed by the still-strong global role of the US dollar, also plays an important role. The results of our GVAR model, presented in Figure 2.1b, show first that countries with a substantial trade exposure to the US economy have a relatively large sensitivity to US developments. Specifically, in response to a 1% decline in US GDP, Canadian (70%), Mexican (69%) and Nigerian (36%) real outputs fall by 0.37%, 0.56% and 0.66% respectively, with this effect being statistically significant [the numbers in brackets are corresponding trade weights which are reported in Table 7 of Cashin et al. (2012)].

However, even countries that do not trade as much with the US are largely influenced by its dominance through other partners' trade. For instance, following a negative US GDP shock, the Euro Area (16%), Norway (6%) and UK (13%) real outputs fall by between 0.16% and 0.29% after a year, with these median effects being statistically significant. Overall, the influence of the US on other economies remains larger than direct trade ties would suggest, owing to third-market effects together with increased financial integration that tends to foster the international transmission of business cycles.

In general, lower demand for commodities is another channel through which a negative US shock affects countries. In particular, around 27% of world oil demand comes from the US, so it is not surprising that in response to the US shock, both oil prices and production levels decline, with the latter effect being statistically significant (see Figure 2.2). The oil

[10] Due to model and parameter uncertainties, and the possibility of measurement errors in the data (for the MENA countries in particular), the confidence intervals produced for different MENA countries are generally wide. In this case, the median responses are mainly used for inference as they contain useful information about the direction of the responses and their magnitudes.

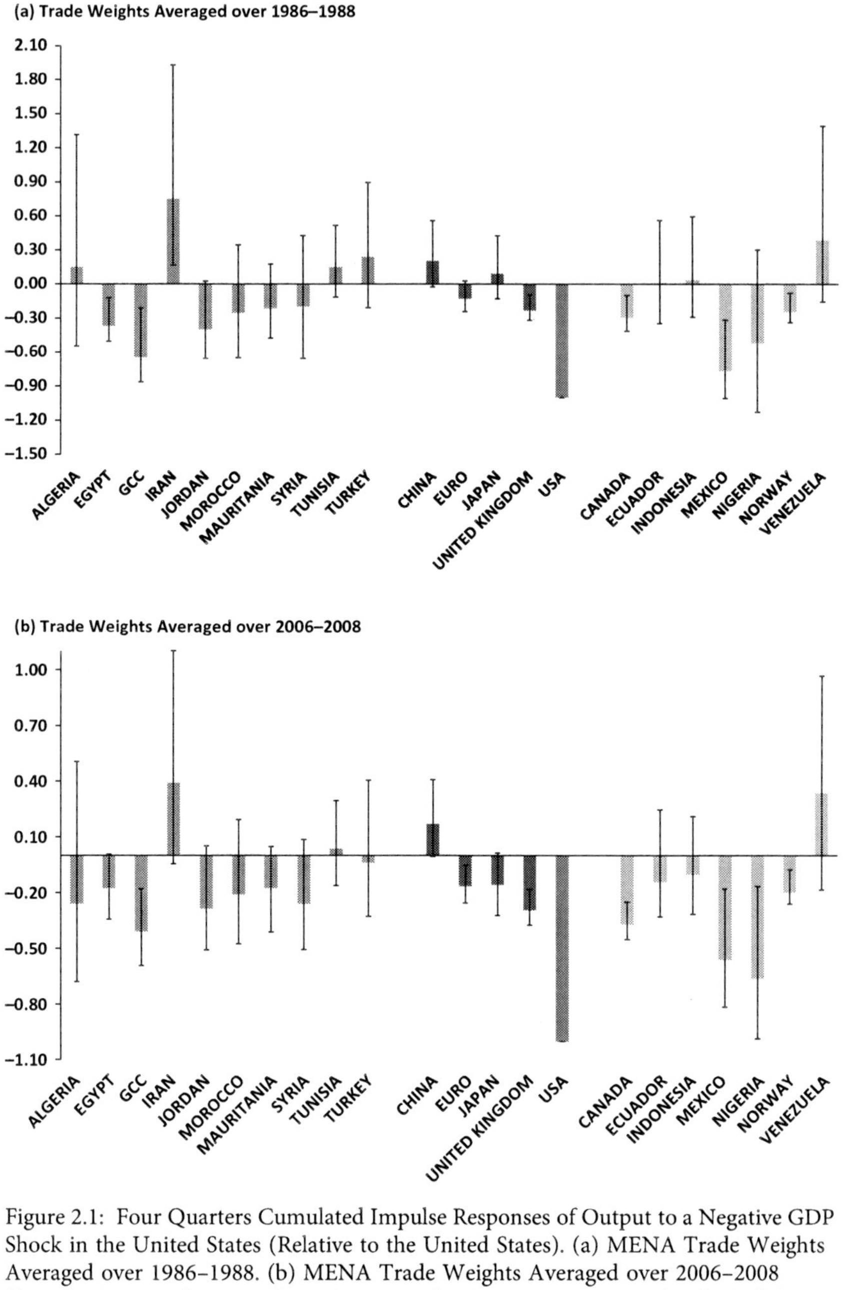

Figure 2.1: Four Quarters Cumulated Impulse Responses of Output to a Negative GDP Shock in the United States (Relative to the United States). (a) MENA Trade Weights Averaged over 1986–1988. (b) MENA Trade Weights Averaged over 2006–2008

Notes: Depicts annual percent change in output of a given country associated with 1% decline in US GDP, together with the 16th and 84th percentile error bands.

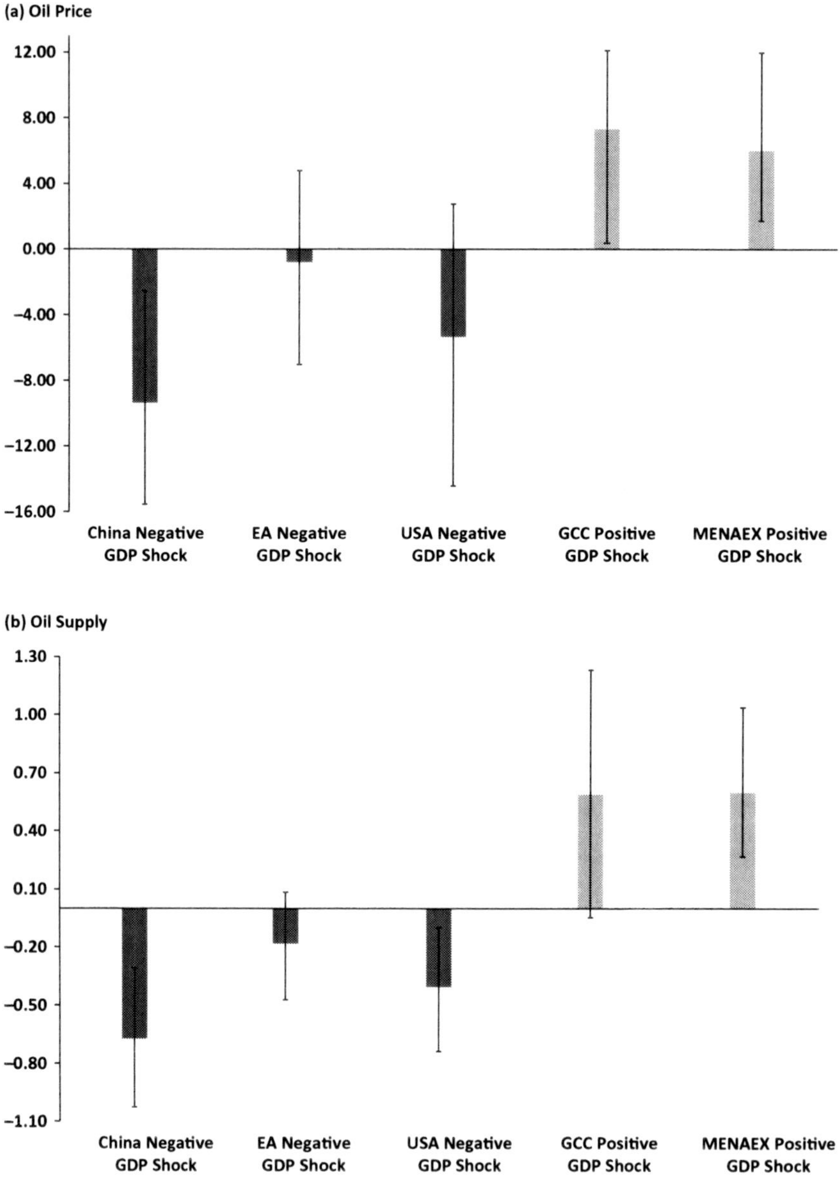

Figure 2.2: Four Quarters Cumulated Impulse Responses of Oil Prices and Supply. (a) Oil Price. (b) Oil Supply

Notes: Depicts annual percent change in oil price and production associated with a negative/positive unit shock (equal to one standard error) to the GDP of corresponding economy or region, together with the 16th and 84th percentile error bands. These responses are based on trade weights averaged over 2006–2008.

channel has a negative impact on the MENA countries, where on average, their GDPs fall between 0.17% and 0.29% after one year. For the GCC, exporting around 30% of world oil, this effect is larger and statistically significant: real output declines by as much as 0.41%. Note that the negative income effect of the US GDP shock on the GCC countries is somewhat dampened by a monetary channel that arises from the special role of the US dollar in international markets (as the oil invoicing currency). The US dollar's depreciation (in response to a negative US GDP shock) entails monetary easing in the GCC countries as well as in China, as their currencies are pegged/anchored to the US dollar. The resulting lower real interest rates, higher liquidity levels and increased demand for oil from China can stimulate aggregate demand in the GCC countries and alleviate the initial negative income effect of lower oil prices and production levels. The median effects of a negative US output shock for other systemic countries and major oil exporters are generally negative, with those few that have positive median impacts being statistically insignificant.[11]

To investigate whether the global impact of a US negative output shock has changed over the past two decades, we re-estimated our model using trade weights averaged over 1986 to 1988 [see Table 8 of Cashin et al. (2012)]. Comparing these results, as illustrated in Figure 2.1a, with those from our original specification using trade weights between 2006 and 2008 in Figure 2.1b, we note that the impact of this shock has remained very similar over the past twenty years. This finding suggests that the influence of the US on the global economy remains very prominent.

These results are robust to different ways of constructing the weights, w_{ij}. In particular, we experimented with using exports weights and found the impulse responses to be very similar to those with trade weights used above. We also estimated our model with the foreign variables computed using trade weights averaged over 1981–1983 and 2001–2003 and obtained very similar results to the benchmark weights (1986–1988 and 2006–2008). Therefore, as is now standard in the literature, we only report the results with the weights calculated as the average of exports and imports of country i with j [Tables 7 and 8 of Cashin et al. (2012)].

[11] The output response of China to a negative US shock is not statistically different from zero. This response could arise from the influence of third country (indirect) effects emanating from non-US trading partners of China. Identification of the direct effect of the US shock on China is beyond the scope of this chapter, but is an issue for further investigation.

2.3.2 Shock to Euro Area GDP

We initially shocked the Euro Area GDP, but obtained extremely small (and not statistically different from zero) median responses on both oil prices and production levels. Figure 2.2 shows that the responses of these two variables to negative GDP shocks in the US and China are much larger (and statistically significant) compared with that of an Euro Area shock. However, given that the Euro Area consumes around 15% of world oil, we would expect a negative shock to its output to be associated with lower oil prices/production levels in line with that of a Chinese shock, in particular given that over the same period, China's consumption of oil was around 5% of the world total (one-third that of the Euro Area). To obtain reasonable estimates, in line with economic theory, we imposed sign restrictions on the generalized impulse responses of the oil price and supply variables, such that a negative Euro Area shock is contemporaneously associated with a decline in both of these variables. For more details on sign restrictions within a GVAR model, see Cashin et al. (2014).

The four quarters impulse responses of output to one percent negative GDP shock in the Euro Area, reported in Figure 2.3, are most significant for Maghreb countries, reflecting their geographical proximity to the Euro Area, and the strength of their trade linkages with Europe in general. Maghreb countries rely heavily on Europe as a market for exports (nearly 60% of Maghreb's exports are destined for Europe), as well as tourism, workers' remittances and foreign direct investment.

Growth spillovers vary greatly from country to country. The highest dependencies are observed for Algeria and Tunisia, with annual output elasticity of more than a half (Figure 2.3). Algeria is adversely affected via both trade and commodity price channels. Although the country's share of proven oil reserves in the world's total is only about 1% at the end of 2011, it is highly dependent on oil exports (98% of its exports still come from the hydrocarbon sector), rendering it extremely vulnerable to a Euro Area shock. Specifically, Algeria exports around 42% of its oil to Europe, and given our assumption about the slowdown in the Euro Area, demand for Algerian oil from Europe declines. However, Figure 2.3 also shows that the rest of the world (including North America to which 35% of the Algerian oil is destined) experiences a drag on its output as well. Considering this worldwide fall in oil demand, Algeria is not able to readily shift its commodity exports (which are predominantly destined for Europe and North America) to other countries (such as China). The country therefore experiences

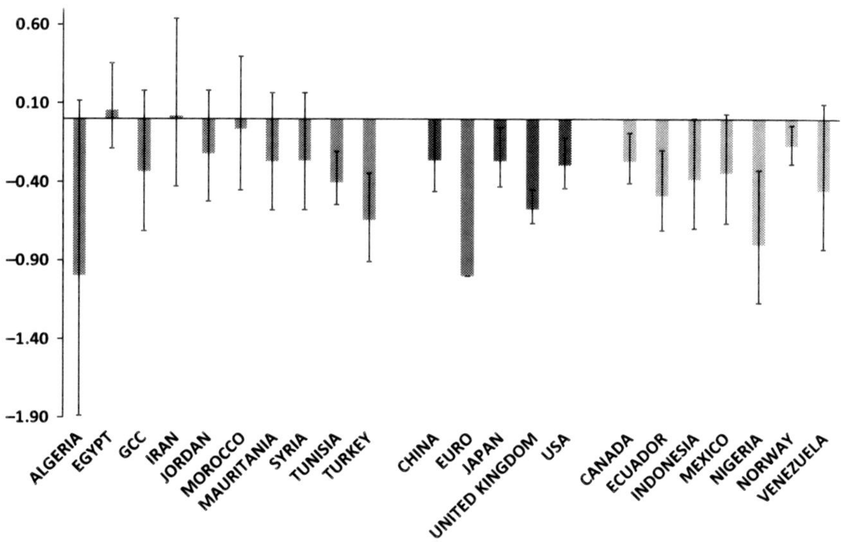

Figure 2.3: Four Quarters Cumulated Impulse Responses of Output to a Negative GDP Shock in the Euro Area (Relative to the Euro Area)

Notes: Depicts annual percent change in output of a given country associated with 1% decline in Euro Area GDP, together with the 16th and 84th percentile error bands.

a fall in oil revenues, which has a direct and large negative impact on its economy. This commodity channel applies equally to other oil exporters in our sample, which is why we note that a negative Euro Area shock has an adverse impact on economic activity in these countries. Moreover, given that 68% of Algerian trade is with the Euro Area, a fall in aggregate demand in this region has a negative impact on Algerian growth through the direct trade channel.

In the case of Mashreq countries, Syria is the most affected by a downturn in the Euro Area, while the impacts on Egypt and Jordan are moderate due to their larger regional ties with the GCC. Moreover, following a 1% decline in the Euro Area GDP, Turkish output falls by 0.64%, with this effect being highly significant, illustrating the close trade linkages between Turkey and the eight Euro Area countries in our sample. As for the region's oil exporters, a negative GDP shock in the Euro Area affects their economies mainly through its impact on oil prices and production, lowering their overall growth.

Estimated spillovers from the Euro Area to the other systemic countries, which abstract from financial contagion and may therefore understate the magnitude of true spillovers, are nevertheless of meaningful size with output elasticities ranging between 0.26 and 0.57 (see Figure 2.3). The response is especially interesting in the case of the UK, which has close trade linkages with continental Europe as illustrated by a trade weight of 0.52, for which output falls by 0.57% following a 1% drop in Euro Area GDP.

Moving beyond the systemic countries, oil exporters outside of the MENA are also adversely affected by the Euro Area's economic downturn through its impact on commodity prices. More specifically, the elasticities for this group of countries range between 0.17 (Norway) and 0.80 (Nigeria). Figure 2.4 also reports the generalized impulse responses of a negative Euro Area shock for the set of focus countries over a 40-quarter horizon and illustrates how the responses reported in Figure 2.3 evolve over the long run.

2.3.3 Shock to Chinese GDP

A negative GDP shock in China affects the economies of oil exporters in our sample mainly through its impact on global demand for oil and associated prices. This is clearly shown in Figure 2.2, as both oil prices and production levels fall significantly in response to a negative output shock. For oil exporters, the slowdown in China translates into lower overall economic growth (see Figure 2.5b). In particular, those countries with large commodity export exposures to China are most vulnerable to a slowdown in this country; for example, Ecuador and Venezuela both experience a statistically significant fall in their outputs, corresponding to 0.34% and 0.33%, respectively. In contrast, larger commodity exporters with more diversified economies do not seem to suffer as much; an example of the latter is Norway for which the response is −0.04% and is not statistically different from zero.[12]

Turning to the systemic countries, Figure 2.5b shows that following a negative GDP shock in China, the output of the other four systemic countries falls, with the average effect being −0.15% and statistically significant. Moreover, China has a large (significant) and growing impact on MENA countries and other oil exporters.[13] This finding is expected given the emergence of China as a key driver of the global economy over

[12] See also Roache (2012) and International Monetary Fund (2012) for a detailed discussion on the outward spillovers from China through commodity price channels.

[13] The results for the other countries in our sample, listed in Table 2.1, are not reported here, but are available on request.

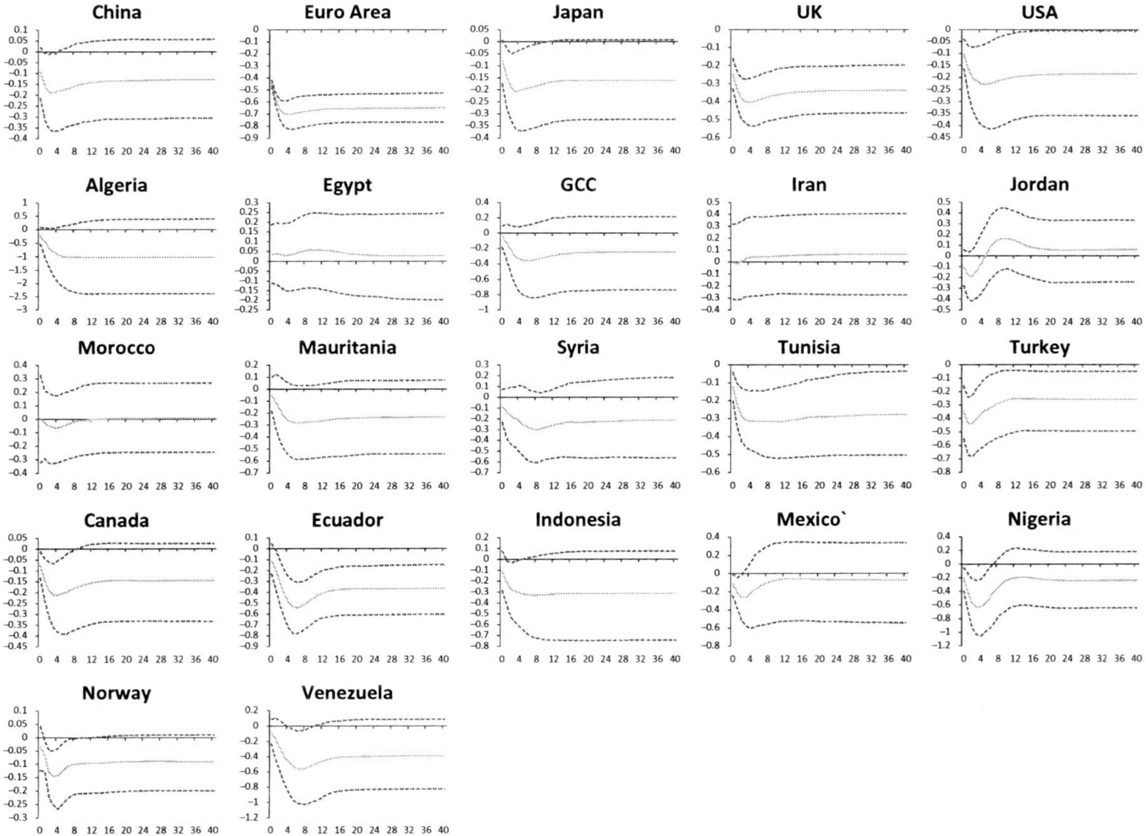

Figure 2.4: Impulse Responses of a Negative Unit Shock to Euro Area Output

Notes: Figures are median impulse responses to a one standard deviation negative shock to Euro Area GDP with sign restrictions on both oil prices and supply, together with the 16th and 84th percentile error bands. The impact is in percentage points and the horizon is quarterly.

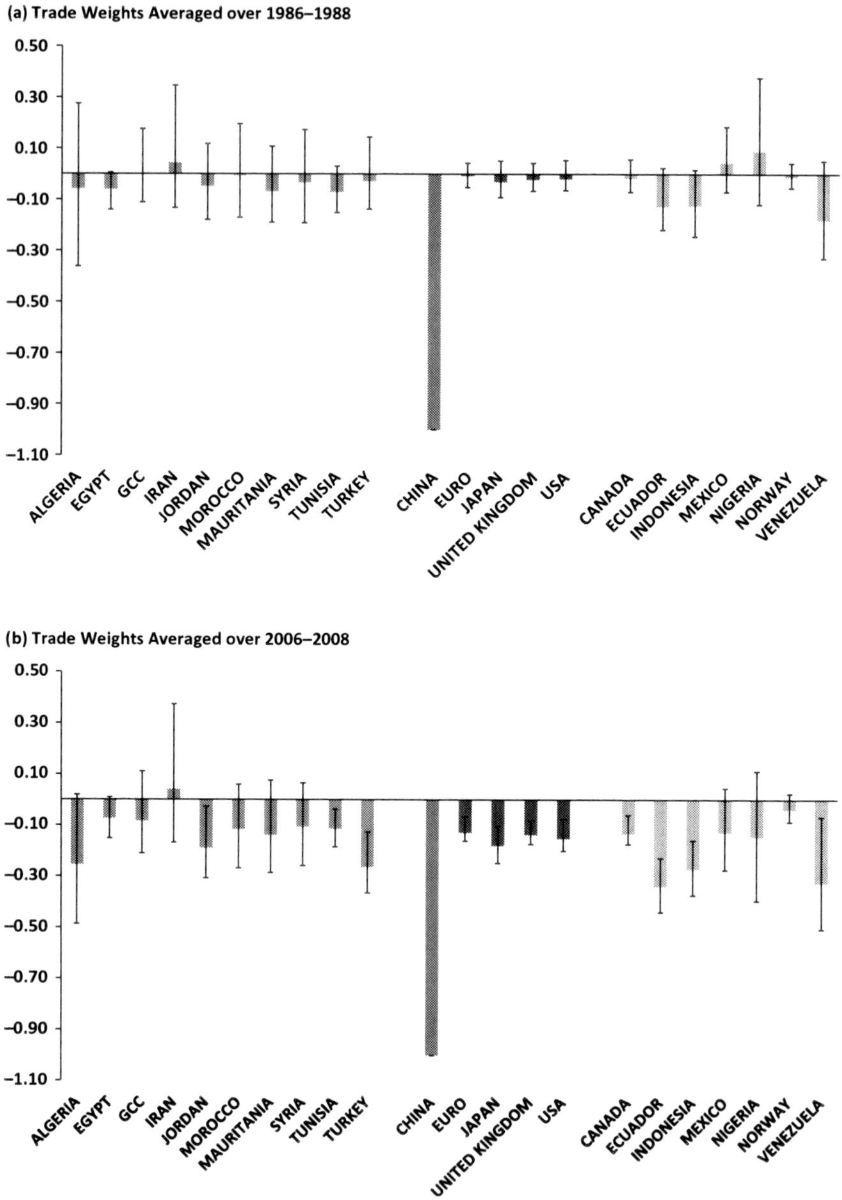

Figure 2.5: Four Quarters Cumulated Impulse Responses of Output to a Negative GDP Shock in China (Relative to China). (a) Trade Weights Averaged over 1986–1988. (b) Trade Weights Averaged over 2006–2008

Notes: Depicts annual percent change in output of a given country associated with 1% decline in Chinese GDP, together with the 16th and 84th percentile error bands.

recent decades. In fact, re-estimating our model with trade weights averaged over 1986–1988, we note that twenty years ago, a negative Chinese output shock would not have had a statistically significant effect on either the systemic economies, major oil exporters or the MENA region (see Figure 2.5a). Comparing the results in Figure 2.5, we note that not only does a Chinese GDP shock affect the global economy in a much more prominent way, but the median effects are generally much larger than two decades ago.

These results are consistent with the findings of Cesa-Bianchi et al. (2012), who argue that the reason why Latin American economies recovered from the recent global crisis much faster than initially anticipated was due to their increasing trade linkages with China. We also argue that the emergence of China as a driver of growth in the world economy might help to explain the "lower-than-expected" effect of the global crisis on MENA countries and other emerging economies. Indeed, we show that the MENA countries' growth is not only dependent on the United States and Europe, but also on the fast-growing Chinese economy.

2.4 Outward Spillovers

A positive GDP shock in the GCC region generates significant output gains in Jordan and Syria, 0.43% and 0.34% after a year, respectively, together with mild to moderate output spillovers to the rest of the MENA countries, the effect ranging from a low of 3% (Tunisia) to a high of 43% (Algeria) – see Figure 2.6b. Spillovers from the GCC to the wider MENA region are transmitted via trade, remittances, foreign direct investment (FDI) and commodity price channels. The macroeconomic situation in Jordan, for example, is closely tied to those of other countries in the Middle East. Remittances from Jordanians working in the region are an important source of national income (equivalent to 15%–20% of GDP); the Persian Gulf region is the primary destination for Jordanian exports, and in turn, supplies most of its energy requirements; furthermore, the country receives substantial grants and FDI from other states in the region [see Mohaddes and Raissi (2013)].

Subject to data availability, it is of course relatively straightforward to augment the country-specific VARX* models with other aggregate variables such as consumption and investment, or particular variables of interest, for instance remittances, FDI and grants. However, the inclusion of these variables is unlikely to alter the long-run relationship that we have estimated between the variables in our model if each of consumption,

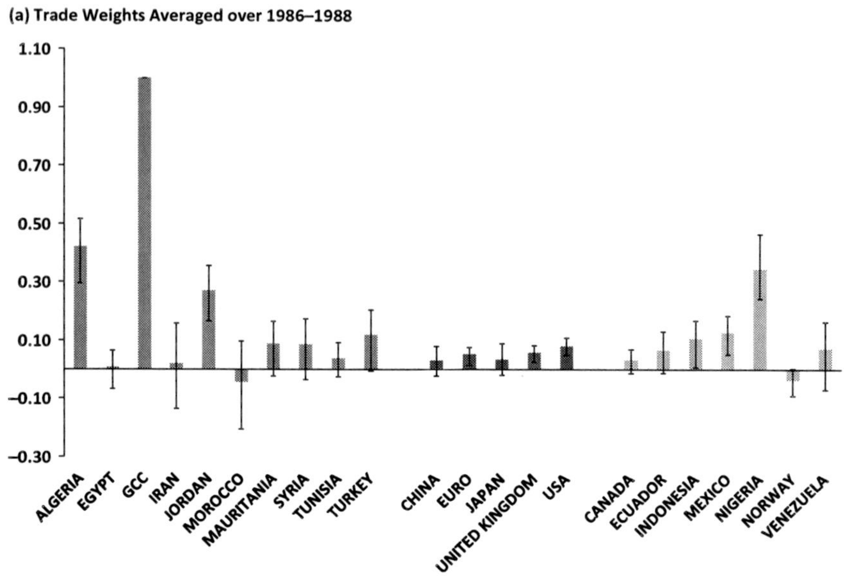

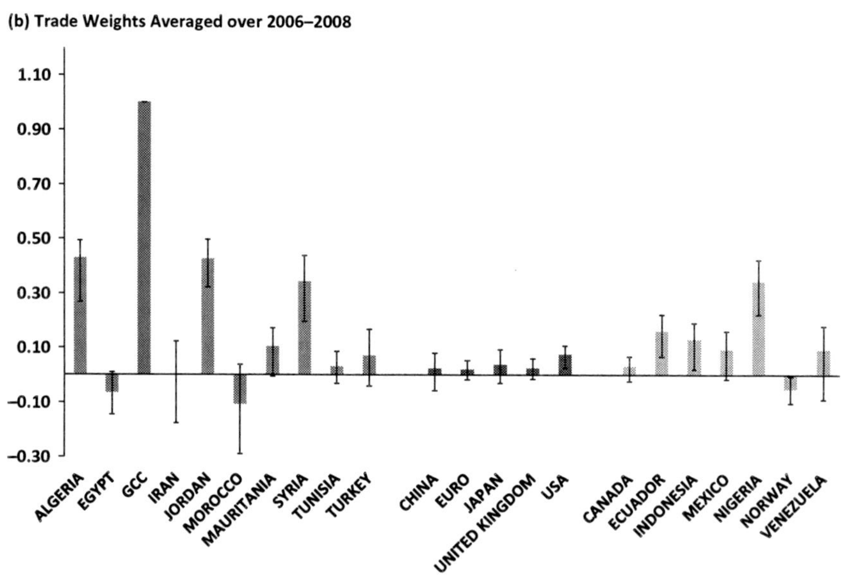

Figure 2.6: Four Quarters Cumulated Impulse Responses of Output to a Positive GDP Shock in the GCC Region (Relative to the GCC). (a) Trade Weights Averaged over 1986–1988. (b) Trade Weights Averaged over 2006–2008

Notes: Depicts annual percent change in output of a given country associated with 1% increase in GCC countries' GDP, together with the 16th and 84th percentile error bands.

investment, remittances, FDI and grants are cointegrated with, for instance, output or oil prices. This is because any linear combination of cointegrating relations will also be cointegrated. To illustrate this point, Esfahani et al. (2013) estimate a cointegrating VAR(2) model for investment and oil export revenues for Iran and find that the hypothesis that the long-run elasticity of investment to real oil income is unity cannot be rejected. They also show that the exactly identified cointegrating relation between log real output and consumption is given by $c_t = y_t + \xi_t$, where ξ_t is a mean zero stationary process. Similarly, Mohaddes and Raissi (2013) estimate a cointegrating VAR(2) model for external income (the sum of remittances, grants and foreign direct investment) and oil prices, and find that the latter represent a good proxy for external income in the Jordanian economy. The above results show that, from a long-run perspective, only one of the variables in the corresponding cointegrating relation needs to be included in the country-specific models.

Figure 2.6b also shows the extent to which the output of the GCC affects, and is affected by the global economy, in particular systemic countries but also other oil exporters.[14] Specifically, the oil market provides an important channel of impact: for example, Saudi Arabia, being part of the GCC, is currently the largest oil exporter in the world and is at present the only producer with significant spare capacity that can be used to stabilize global energy markets. While the level of oil supply from the GCC has significant macroeconomic effects on developed and emerging economies, raising global growth prospects has an important impact on the demand for oil and hence the economic performance of the GCC. Given a near-vertical global oil supply curve, the increase in output in the GCC region is mainly induced by rising oil prices (Figure 2.2). This increase in p_t^{oil} coincides with higher outputs in systemic countries, reflecting a demand-driven oil price spike, and higher GDP levels in other commodity producers, most of which are statistically significant (see Figure 2.6b).

We performed the same exercise as above, but now adding Algeria, Iran and Libya to the GCC block, calling the nine combined countries MENA oil exporters (MENAEX). These countries include seven out of the current twelve OPEC member states, supply over 41% of world oil exports and possess a majority of the total proven oil reserves in the world. As Figure 2.7b shows, the results from a positive MENAEX output shock

[14] For panel applications studying the growth effects of higher commodity prices, see Cavalcanti et al. (2011a, 2011b, 2012).

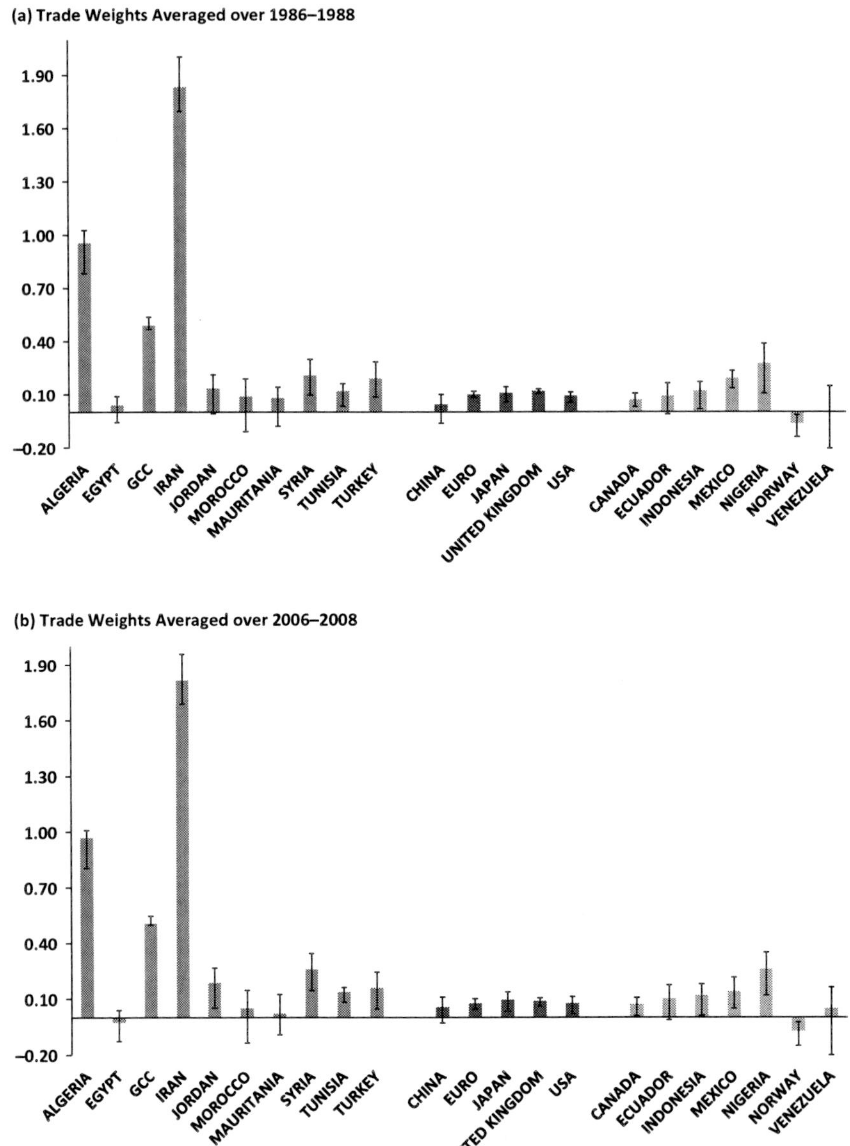

Figure 2.7: Four Quarters Cumulated Impulse Responses of Output to a Positive GDP Shock in the MENA Oil Exporters (Relative to the MENAEX). (a) Trade Weights Averaged over 1986–1988. (b) Trade Weights Averaged over 2006–2008

Notes: Depicts annual percent change in output of a given country associated with 1% increase in MENA oil exporters' GDP, together with the 16th and 84th percentile error bands.

are more pronounced when compared with a GCC shock, but the channels of impact are unchanged.

To see whether the impact of a positive GDP shock in the GCC and the MENA oil exporters have changed over the past few decades, we re-estimated our model using 1986–1988 trade weights. Figures 2.6a and 2.7a show that the effect of these shocks for the region, as well as for the global economy, has not differed much over the past twenty years. This is perhaps not surprising, given that crude oil was and remains the largest export and driver of growth in the region, see for instance Esfahani et al. (2014).

Given the close association of commodity price swings and the economic performance of oil-producing countries, one could also investigate the impact of an oil-price shock on the global economy using the GVAR model developed in this chapter. However, Cashin et al. (2014) show that the underlying source of the oil-price shock is crucial in determining its macroeconomic consequence for oil-importing countries as well as major commodity exporters many of which are located in the MENA region. Their results – using the same set-up as in this chapter but imposing sign restrictions on the generalized impulse responses – show that while oil importers typically experience a long-lived fall in economic activity in response to a supply-driven surge in oil prices, the impact is positive for energy-exporting countries that possess large proven oil/gas reserves. Cross-country differences are absent though when it comes to the demand side of the global crude oil market. In response to an oil-demand disturbance, almost all countries experience a short-run increase in real output and face additional inflationary pressures (which is similar to the case of a positive GCC/MENAEX GDP shock as in Figures 2.6 and 2.7).

2.5 Concluding Remarks

In this study, we estimated a GVAR model for thirty-eight countries/regions (including the GCC countries in particular and the MENA region in general, as well as other major oil exporters) over the period 1979Q2–2011Q2 to analyze the inwards/outwards output spillovers to/from the MENA countries. We also investigated the growing impact of China's macroeconomic shocks (compared to the US and the Euro Area) on other systemic economies, the MENA region and other major oil exporters; and examined how the transmission of shocks has changed following the emergence of China as a major driver of the world economy.

The results show that output spillovers from China, the Euro Area and the US to other systemic economies, as well as the MENA region and other oil exporters, are meaningful. Specifically, the impact of a negative Euro Area GDP shock on MENA economies are modest, and on par with a shock to the GDP of the United States. However, the impact of a shock to output in China is more substantial (being statistically significant in more cases), reflecting the direction of evolving trade patterns, and China's growing role in the global economy and the world oil market. We also find that outward spillovers from the GCC and MENA oil exporters are likely to be stronger in their immediate geographical proximity, although they also have implications for the systemic economies and the rest of the oil exporters.

The sensitivity of MENA countries to economic developments in China, Euro Area and the United States raises the question of which policies and institutions are needed in response to such shocks. Since oil prices and production levels are significantly affected by a drag on global economic activity, countercyclical fiscal policies are needed to insulate small commodity exporters from an adverse economic volatility in systemic countries as well as commodity price fluctuations. In addition to building fiscal buffers, the other priority for commodity exporters should be to enhance their macroeconomic policy frameworks and institutions (such as more autonomy in conducting the monetary and exchange rate policies).

References

Behar, A. and C. Freund (2011). The Trade Performance of the Middle East and North Africa. World Banks Middle East and North Africa Working Paper Series No. 53.

Cashin, P., K. Mohaddes and M. Raissi (2012). The Global Impact of the Systemic Economies and MENA Business Cycles. IMF Working Paper WP/12/255.

Cashin, P., K. Mohaddes, M. Raissi and M. Raissi (2014). The Differential Effects of Oil Demand and Supply Shocks on the Global Economy. *Energy Economics* 44, 113–134.

Cavalcanti, T. V. d. V., K. Mohaddes and M. Raissi (2011a). Growth, Development and Natural Resources: New Evidence Using a Heterogeneous Panel Analysis. *The Quarterly Review of Economics and Finance* 51(4), 305–318.

(2011b). Does Oil Abundance Harm Growth? *Applied Economics Letters* 18(12), 1181–1184.

(2012). Commodity Price Volatility and the Sources of Growth. *Journal of Applied Econometrics* 30(6), 857–873.

Cesa-Bianchi, A., M. H. Pesaran, A. Rebucci and T. Xu (2012). Chinas Emergence in the World Economy and Business Cycles in Latin America. Economia, *The Journal of LACEA* 12, 1–75.

Dees, S., F. di Mauro, M. H. Pesaran and L. V. Smith (2007). Exploring the International Linkages of the Euro Area: A Global VAR Analysis. *Journal of Applied Econometrics* 22, 1–38.
Esfahani, H. S., K. Mohaddes and M. H. Pesaran (2013). Oil Exports and the Iranian Economy. *The Quarterly Review of Economics and Finance* 53(3), 221–237.
(2014). An Empirical Growth Model for Major Oil Exporters. *Journal of Applied Econometrics* 29(1), 1–21.
International Monetary Fund. (2012). 2012 Spillover Report. International Monetary Fund, Washington D.C., USA.
Mohaddes, K. and M. Raissi (2013). Oil Prices, External Income, and Growth: Lessons from Jordan. *Review of Middle East Economics and Finance* 9(2), 99–131.
Mohaddes, K. and O. H. Williams (2013). Inflation Differentials in the GCC: Does the Oil Cycle Matter? *Middle East Development Journal* 5(2), 1–23.
Roache, S. K. (2012). Chinas Impact on World Commodity Markets. IMF Working Paper WP/12/115.
Smith, L. and A. Galesi (2010). GVAR Toolbox 1.1. University of Cambridge: Judge Business School.

3

Real Exchange Rates and Export Performance in Oil-dependent Arab Economies*

Ibrahim Elbadawi and Linda Kaltani

ABSTRACT

Diversifying oil economies toward manufacturing and more sophisticated products and services is one of the most pressing public policy challenges facing these countries. This paper provides new evidence about the impact of oil rents and real exchange rate (RER) undervaluation on various measures of exports, using a global sample spanning 1980–2011. Our results suggest that RER undervaluation can ameliorate the negative impact of oil rents on exports, and that it can be particularly effective in countries with underdeveloped financial markets or low institutional development. In the light of these findings, the paper argues that a strategy of depreciating the real currency can be a viable, albeit second best, industrial policy choice in order to promote export diversification, technical upgrading and export sophistication in institutionally-deficient oil and mineral-dependent economies. Moreover, this type of public policy can minimize the reliance on traditional vertical industrial policy, which usually requires high initial institutional capacity in order to succeed.

JEL Classification: O5, E60, F31, O43, E44, F14, O14

Keywords: Arab world, oil and resource rents, real exchange rate undervaluation, institutions, financial development, export diversification, export sophistication, manufacturing

* The views expressed in this chapter are not necessarily those of the Dubai Economic Council or the International Monetary Fund, their Board of Directors or affiliated institutions. We would like to thank Nikola Spatafora and Saurabh Mishra of the World Bank for sharing their data of export sophistication with us. The authors would also like to acknowledge the able research support by Eman Refaat and Mireille Ntsama.

3.1 Introduction

Hydrocarbon-dependent Arab countries are characterized by disappointing growth and employment performance, a low degree of diversification and sophistication of exports and limited manufacturing sectors. These countries make up eleven of the twenty-two members of the League of the Arab States that constitute the Arab World. They account for approximately 55% of global oil reserves and 29% of natural gas reserves. Naturally, the hydrocarbon sector dominates these economies, where it contributes around 50% to GDP and 80% to government revenues. Moreover, the rest of the Arab economies, namely the non-oil ones, have also been substantially influenced by oil due to their strong economic linkages with the highly endowed oil economies of the Gulf Cooperation Council (GCC). The GCC has been a major source of foreign investment and tourism as well as a host to large numbers of workers from these countries (Selim and Zaki in this volume).

It is not surprising, therefore, that resource dependency has been singled out as the main culprit behind the development problems of the region. While oil resources have provided a huge opportunity to the Arab World to finance accelerated development, they have, nevertheless, complicated the development process in the region. In particular, there appears to be a wide consensus in the received literature that oil rents impede economic diversification, penalize manufacturing growth by generating Dutch Disease and extreme volatility, promote bad governance and complicate transition to transparent and accountable democratic rule. In addition, the oil sector tends to be located at the periphery of the product space, which makes it difficult for the economy to move into new and more sophisticated lines of products and services (e.g., Elbadawi and Gelb 2010).

However, the received literature also suggests that the role of oil wealth in an economy's performance is not predetermined and that its ultimate impact on development hinges on the underlying institutional and policy environment (e.g., Collier and Goderis 2009; Elbadawi and Soto in this volume). Subscribing to this view, this chapter contributes to this literature by assessing the effectiveness of competitive real exchange rates (RER) in stemming the potential negative impact of oil rents on the capacity of oil and other point source-dependent economies to diversify and modernize their exports. The received literature suggests that countries that have achieved high and sustained growth, by and large, are characterized by diversified economies and are endowed with good economic governance, and most, though not all, have large and dynamic manufacturing sectors.[1]

The recent growth literature already provides ample evidence of the central role of competitive real exchange rates and robust institutions in averting the resource curse, as measured by the post-boom collapse of economic growth. In particular, this literature finds that maintaining the RER close to its equilibrium level is a necessary condition for sustained growth and that countries that avoided overvaluation have been associated with sustained export-led growth and substantial export diversification (e.g., Elbadawi and Helleiner 2004).Moreover, not only is avoiding overvaluation necessary for growth, but a mild undervaluation may be good for growth (e.g., Aguirre and Calderón 2005).

It has been recently argued (most notably by Rodrik 2008) that empirical findings like the ones discussed earlier are, in fact, a reflection of a deeper causal effect that promises to open a new set of ideas for thinking about a growth agenda in which the RER takes center stage.[2] According to Rodrik, countries that have managed to engineer an RER undervaluation appear to have resolved deep institutional constraints.[3] First, "weak institutions" create a wedge between private and social returns, which is different from simply having a low endowment of an input. Second, to the extent that tradables may be more "complex" and entail more transaction-intensive activities, the wedge between private and social returns may be more severe in tradable than non-tradable economic activities, and can lead to static misallocation of resources in favor of the latter and greater dynamic distortions in the former. When the tradable sector is more dynamic, as would be expected in many low-income, small economies, an increase in the relative prices of tradables to non-tradables can improve static efficiency and enhance growth in a second-best fashion.

Therefore, RER undervaluation can be the most feasible and effective approach for alleviating such institutional weaknesses. Another theoretical justification for engineering an RER undervaluation strategy is based on the view that tradables (particularly new and non-traditional tradables) are subject to a variety of market imperfections, such as information externalities (learning and cost-discovery externalities) and coordination

[1] See for example, Imbs and Wacziarg (2003); Hausmann, Hwang and Rodrik (2006) and UNIDO (2009).

[2] For other works in the literature on the role of RER undervaluation in promoting growth and export diversification, see Williamson (1997) and Elbadawi and Helleiner (2004).

[3] Country examples from Rodrik (2008) are China, Republic of Korea, Taiwan, Uganda and Tanzania.

externalities. These imperfections keep output and investment in tradable sectors at sub-optimal levels. Again, by raising the profitability of tradable sectors, an RER undervaluation can be an effective strategy in a second-best world. In particular, it can be an effective substitute to a traditional "industrial policy" and all the well-known limitations associated with it.

An important qualification, however, is provided by Eichengreen (2007), who argues that targeting certain sophisticated export activities through certain domestic policies, including those that promote RER undervaluation, may merely play a role of "facilitating" channels to permit the realization of certain favorable conditions. For example, he argues that to the extent that Chinese firms rely on their links to overseas Chinese investors and importers or to their proximity to Japan and Korea, RER undervaluation or other domestic policies may not be enough for other countries that do not possess such an advantage.

Section 3.2 analyzes export performance of Arab oil economies, in terms of their overall export concentration (the reverse of export diversification), the extent of the sophistication of their exports and per capita manufacturing exports. Section 3.3 reports estimation results for the determinants of the RER, based on a world sample of fifty countries, spanning 1980–2011. This allows us to subsequently derive measures of RER equilibrium and RER undervaluation, a key determinant of export performance as discussed earlier. Section 3.4 estimates a plethora of empirical exports performance models to assess the impact of resource rents and RER undervaluation, among other standard controls, on the aforementioned three indicators and to also investigate the channels through which undervaluation can spur export performance. The model is estimated for an extensive panel dataset (ranging from 66 to 118 countries depending on the dependent variable) spanning the period 1980–2011. Section 3.5 concludes.

3.2 Exports of Oil-dependent Arab Economies

Diversification to new products is correlated with economic development, especially sustained export-oriented growth. Imbs and Wacziarg (2003) study the patterns of sectoral concentration across countries and time and find that economies become more diversified as their income increases, but eventually specialize again at high income levels.[4] Moreover, successful

[4] They find that the diversification process goes on until countries attain the income level of Ireland. Therefore, growth is associated with a high degree of specialization only at relatively high levels of income.

export-oriented strategies that led to major economic transformations have been associated with significant economic diversification. For example, the empirical literature finds that non-traditional exports are characterized by higher income elasticities, less volatile terms of trade and higher prospects of dynamic productivity gains (Elbadawi 2002; Sekkat and Varoudakis 1998). Also, following on the work of Haussman et al. (2005), Johnson et al. (2007) find that growth accelerations are associated with structural changes in manufacturing.

On the other hand, empirical research has found that countries specialized in primary product exports tend to grow more slowly than economies with diversified export bases. Sachs and Warner (1997) have shown that the 1970 share of primary exports of GDP is negatively correlated with GDP growth in a large sample of eighty-three countries over the period 1965–1990. Sala-i-Martin (1997) has found a similar result for the 1970 share of primary products in total exports. In the special case of "point-source" natural resources (those extracted from a narrow geographic or economic base such as oil and other minerals), the received literature suggests that prudent and development-oriented management of oil and mineral booms has been the exception rather than the rule, with many countries "inefficiently" specializing in the resource sector and other non-tradable activities that are likely to be dependent on it, which might lead to the collapse of output after the end of the boom – the so-called "resource curse" (e.g., Hausmann and Rigobon 2003).

Very importantly from the perspective of industrial development, Imbs and Wacziarg also find that the diversification process does not only hold when economies transform from agriculture to manufacturing, but also within manufacturing.[5] This evidence is consistent with a long-held view in the development literature, which argues that enhancing the productive capabilities over a large range of manufactured goods – including the production of new ones – is an "integral" part of development (e.g., Rodrik, 2006). More recently, further support to this evidence is provided by the new research pioneered by Hausmann et al. (2006), which find a strong association between the degree of export sophistication and subsequent economic growth. In this case, however, the positive association between sophistication of exports and development is not just confined to manufacturing but also includes services.

[5] See also Klinger and Lederman (2004) and Carrere et al. (2007), who confirm Imbs and Wacziarg's pioneer work using more recent datasets.

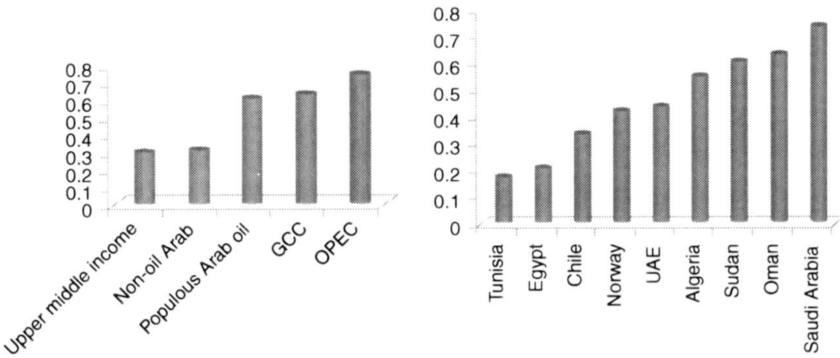

Figure 3.1: Export Concentration Index (HHI): (2000–2013)
Note: Each bar represents average values.

3.2.1 Manufacturing and Export Concentration

As discussed, the hydrocarbon sector looms very large in the oil-dependent Arab economies, especially in the export sector. It is not surprising, therefore, that exports from these countries tend to be highly concentrated. For example, in 2000–2013, the degree of export concentration for the median GCC or populous oil Arab economies was close to 70% in the (0–100%) scale of the Herfindahl–Hirschman index (HHI).[6] Though this is still less than the OPEC median of about 80%, it is nevertheless, twice the median HHI for the non-oil Arab or upper middle income country (Figure 3.1). In terms of individual country experiences, Saudi Arabia's exports are the most concentrated, with a staggering HHI index of more than 75%, followed by Oman, Sudan and Algeria, where all three, including the emerging oil economy of Sudan, have relatively high concentration indexes above 55%. On the other hand, the UAE has achieved a considerable degree of economic diversification, which puts it on par with Norway and close to the export diversification of Chile. This is quite impressive for a major oil-exporting economy like the UAE. Also, the two non-oil middle income Arab countries of Egypt and Tunisia have

[6] The empirical literature uses a variety of measures to capture export diversification. Elbadawi (2002) uses a measure that is the residual of exports after the ten largest three-digit commodity groups have been accounted for. Imbs and Wacziarg (2003) capture concentration (the inverse of diversification) through the use of a Herfindahl–Hirschman index (HHI), coefficients of variation of sector shares and maximum–minimum spreads. Maloney and Lederman (2007) also make use of HHI as well as of the share of natural resources in total exports.

achieved substantial degrees of export diversification, with the HHI being well below 30%.

The export concentration story of the Arab World is mirrored by the limited role of manufacturing in the Middle Eastern economies, which are dominated by the Arab countries, especially when Turkey is excluded. The UNIDO Industrial Development Report (2009) contains extensive data on six components of industrial performance: manufacturing value added (MVA) per capita; manufactured exports per capita; share of MVA in GDP; share of medium/high technology production in MVA; manufactured exports in total exports. The evidence presented in the report indicates that for the first five indicators, the Middle East and North Africa region (MENA) uniformly underperformed relative to Latin America and the Caribbean and the East Asia and the Pacific, the two regions with comparable levels of development, and only slightly overperformed relative to the vastly poorer regions of South Asia and Sub-Saharan Africa. However, for the last indicator (the share of manufactured exports in total exports), which reflects the direct impact of dependency on the hydrocarbon resource exports in the region, MENA underperformed relative to all four developing regions (Table 3.1).[7]

In addition to their failure to penetrate the global markets for low technology labor-intensive manufacturing exports, the aforementioned evidence also reveals an equally disappointing performance for the Arab countries (MENA outside Turkey) with regard to the share of medium and high technology components of manufacturing. This, we will argue, should be particularly worrisome for these countries. In fact, the MENA countries display limited comparative advantage in basic low technology manufacturing compared to very low wage SSA South Asia and, especially China – the latter having essentially defined the frontier for labor-intensive manufacturing.

3.2.2 Export Sophistication

Recent new research pioneered by Hausmann et al. (2006) finds a strong association between the degree of export sophistication and subsequent economic growth. This evidence, argues Rodrik (2006), suggests that "industrial upgrading is a leading indicator of economic performance" (p. 10) and that productivity levels associated with a country's exports

[7] However, the share manufactured to total exports for SSA reported in this table is much larger than other estimates, including those of the IMF (2009).

Table 3.1: *Industrial Performance in Developing Regions (2005)*

Region/ Components	Manufacturing Value Added (MVA) per Capita*	Manufactured Exports per Capita*	Share of MVA in GDP (Percentage)	Share of Manufactured Exports in Total Exports	Share of Medium/ High-Technology Production in MVA (Percentage)	Share of Medium/High-Technology Exports in Manufactured Exports (Percentage)
Industrialized economies	4,771.0	5,428.2	16.8	85.7	75.2	66.1
Sub-Saharan Africa	63.6	81.2	10.9	62	41.3	32.4
excluding South Africa	29.2	38.5	7.6	54.9	17.5	13.3
South Asia	82.1	74.3	14.5	86.3	18.3	20.2
excluding India	79.6	51.9	15.9	84.6	5.2	8.0
Middle East and North Africa	398.1	474.7	12.5	31.7	33.3	27.9
excluding Turkey	381.4	367.1	12.1	22.7	19.2	20.0
Latin America and the Caribbean	761.2	642.2	18.2	63.4	47.2	55.9
excluding Mexico	703.2	400.2	18.8	51.9	20.9	36.8
East Asia and the Pacific	582.3	885.6	29.5	91.9	97.5	64.1
excluding China	750.0	1,524.9	25.2	89.9	32.8	68.6

Source: Table 2.1 of Elbadawi and Gelb (2010).
*MVA is in constant 2000 US dollars.

are not fully captured by factor endowments, such as human capital or institutional quality. The new index developed by these authors ranks traded goods in terms of their implied productivity. Thus, for each country j, the index is given by:

$$EXP\,Y_j = \sum_l \frac{x_{jl}}{X_j} . PROD\ Y_l \tag{1}$$

where x_{jl} is the exports of product l by country j; $X_j = \sum_l x_{jl}$ is the total exports of country j; and $PROD\ Y_l$ is the weighted sum of the per capita GDP of countries exporting a given product, where the weights reflect the revealed comparative advantage of each country in that product:

$$PROD\ Y_l = \sum_j \frac{x_{jl}/X_j}{\sum_j \left(x_{jl}/X_j\right)} . Y_j \tag{2}$$

The higher this index is, the higher the content of "rich-country products" in exports. This index is motivated by the view that "not all goods are alike in terms of their consequences for economic performance," and that specializing in some products will bring higher growth than specializing in others. In this setting, government policy has a potentially important positive role to play in shaping the production structure. Everything else being the same, countries that specialize in the types of goods that rich countries export are likely to grow faster than countries that specialize in other goods. Rich countries are those that have latched on to "rich-country products," while countries that continue to produce "poor-country" goods remain poor.

Figure 3.2 presents estimates of EXPY in 2007 (the last year of the more recent EXPY series calculated by Anand et al. 2012) for a few Arab countries and other comparators. Again, given its relatively high income, the aforementioned evidence makes clear that exports from the Arab World are characterized by relatively low sophistication. For example, with only slightly more than half of Saudi Arabia's income per capita, Malaysia exports much more sophisticated products (of about 17,500 units in the EXPY scale compared to about 1,400 for Saudi Arabia). Moreover, China provides even more spectacular contrasts where with an income per capita of less than one fourth of Saudi Arabia's, the sophistication level of its exports was comparable to that of Malaysia; and with an income approximately equal to the median for the non-oil middle income Arab group, China's EXPY index was 170% of the median score for the group. Finally,

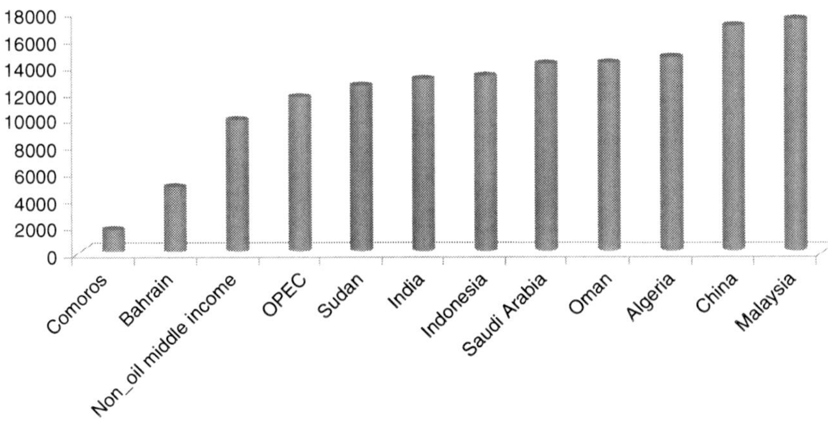

Figure 3.2: Degree of Export Sophistication (EXPY: 2007)

and despite having a 2.8 times larger income per capita, Algeria's export sophistication was only 12% higher than that of India.

At this juncture, and not notwithstanding perhaps a few notable success stories,[8] we pose to ask the question as to why the Arab region's export performance has been so disappointing as captured by the various indicators discussed so far.

As a first approximation, it is natural to think that the dominance of the hydrocarbon sector must be an important explanatory factor. Controlling for income per capita and country fixed effects, the share of hydrocarbon fuels to total exports is positively associated with export concentration (HHI), while it has a robust negative impact on export sophistication (EXPY). Moreover, the residuals for large oil exporters fall above (below) the line for the case of HHI (EXPY), indicating that their performance could not be explained by their level of development and country fixed effects (Figures 3.3a and 3.3b).

In Section 3.4, we assess the impact of oil rents on export performance in a more formal model that also accounts for other controls, most notably the real exchange rate undervaluation. Since the latter is a constructed variable that, in turn, requires the estimation of the equilibrium RER and the construction of the RER misalignment relative to the estimated equilibrium, we turn next to this task in the following section.

[8] Tunisia, for example, has been frequently referred to as an example of successful industrial transformation in the Arab region.

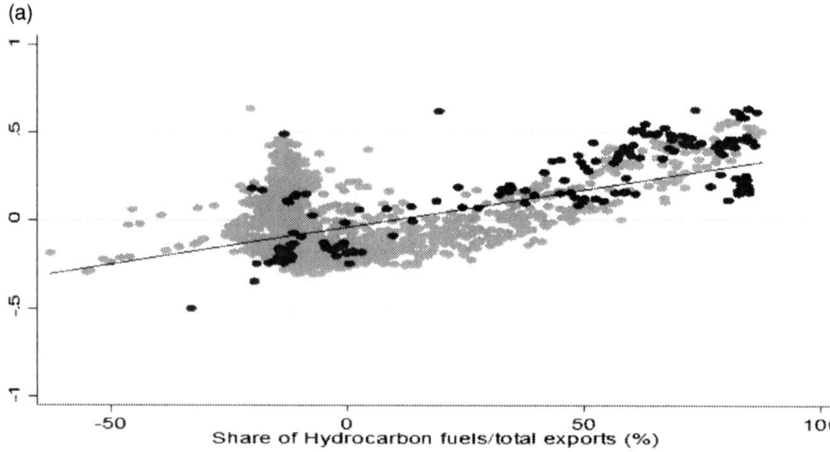

Figure 3.3a: Export Concentration (HHI) and the Share of Hydrocarbon Fuel Exports
Note:
1. Authors' elaboration using EXPY data due to Anand et al. (2012).
2. Data points in red refer to oil and mineral-dependent countries.

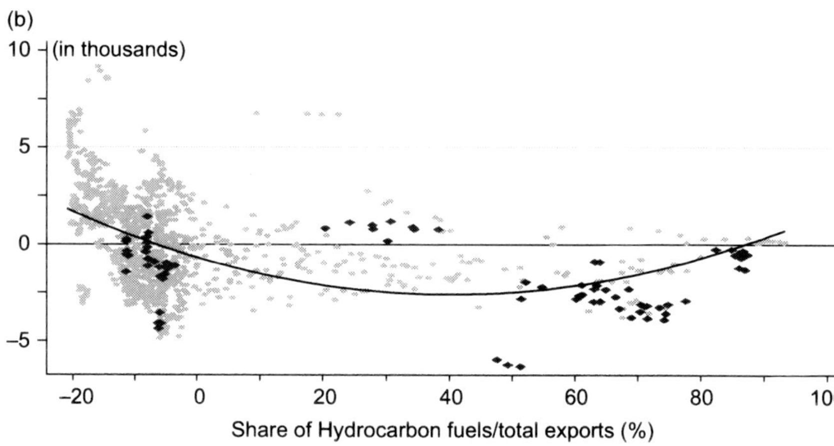

Figure 3.3b: Export Sophistication (EXPY) and the Share of Hydrocarbon Fuel Exports
Note:
1. Figure 2.3.b of Elbadawi and Gelb (2010).
2. The scatter is based on the fixed effects regression: Export Sophistication (EXPY) = 293.6556*** Per Capita GDP (PPP) + –3.2801*** Per Capita GDP^2 + –13.5386 Share of Hydrocarbon Fuels/Total Exports + 0.0912 Share of Hydrocarbon Fuels/Total Exports^2 + 5574.5618*** Const; where *** indicates significance at 1% level. No. of observations = 1214; and No. of countries = 142.
3. Data points in red (Figure is in gray scale) refer to oil and mineral-dependent countries.

3.3 An Empirical Model of the Real Exchange Rate

The most popular methodologies to determine the equilibrium RER are based on a single-equation, reduced-form model that attempts to account for current-account flow variables as well as factors influencing longer-run stock equilibrium.[9] Motivated by the theoretical models of Elbadawi and Soto (2008) and Elbadawi (1998), we estimate a version of such empirical models that emphasizes the *indirect* role of resource rent – through its impact on government expenditure and the accumulation of net foreign income – in influencing the long-run path of the RER as well as its short-run dynamics. Our model predicts the equilibrium RER to be more appreciated with higher terms of trade (TOT), larger productivity in the traded-goods sector relative to the non-traded sector (PROD), lesser trade openness (OPEN), higher government consumption (GOV), higher foreign aid (AID) and larger net foreign income (NFI), or less flexible exchange rate regimes (EXRregimes). Therefore, our specification is:

$$\log(RER)_{it} = \beta_{0i} + \beta_1 \log(TOT)_{it} + \beta_2 \log(PROD)_{it} + \beta_3 OPEN_{it} + \beta_4 \log(GOV)_{it} + \beta_5 NFI_{it} + \beta_6 AID_{it} + \beta_7' \text{EXRregimes}_{\text{it}} + \varepsilon_{it} \quad (3)$$

where subscripts i and t represent country and time indexes, respectively, and β_{0i} and ε_{it} are country-specific intercepts and disturbance terms.[10]

We estimate an error-correction model accounting for the aforementioned fundamentals for a world panel comprised of annual data for fifty countries for 1980–2011. We use three econometric estimation methods appropriate for an error-correction specification of Equation (3.3) applied to panel data. The pooled mean group (PMG) estimator – which imposes the restriction that all countries share the long-run coefficients; the more general mean group (MG) estimator – which assumes that the economies differ in their short- and long-run parameters; and the dynamic fixed effects (DFE) estimator, which assumes that all parameters are constant across countries, except for the intercept which is allowed to vary across countries. The choice between the three estimators entails a trade-off

[9] The underlying notion of equilibrium is essentially intertemporal as the path of the equilibrium RER is assumed to be influenced not only by the current value of the fundamentals, but also by anticipations regarding the future evolution of these variables [Edwards (1989) and Elbadawi (1994) provide pioneering analysis of theoretical and empirical equilibrium RER models, respectively.]

[10] See Appendix A for data definitions and sources and Appendix B for the country list.

between consistency and efficiency. The DFE estimator dominates the other two in terms of efficiency if the restrictions of equality of short- and long-run parameters are valid. If they are false, however, the DFE will generate inconsistent estimates. The MG estimator imposes no cross-country parameter restrictions and can be estimated on a country-by-country basis, provided that the time-series dimension of the data is sufficiently large. For our purposes, the PMG offers the best compromise between consistency and efficiency: we expect the long-run path of the RER to be driven by a similar process across countries, while the short-run dynamics around the long-run equilibrium path may differ from one country to another because it is likely to be driven by idiosyncratic news and shocks to the fundamentals.

Table 3.2 reports the results for the three estimation methods. The restriction of the PMG against the MG model can be tested using Hausman tests. The null hypothesis of equality of coefficients cannot be rejected at 10% level for six out of nine regressors, while it can be rejected at 1% or 5% levels for the remaining three regressors, namely terms of trade, government consumption and net foreign income. We, therefore, favor the PMG model against the MG and DFE estimators.

The results of the PMG regression are consistent with the theoretical and empirical literature for all of the RER fundamentals.[11] Save for the dummy for the free floating exchange rate regime, all long-run coefficient estimates are highly significant (at significance levels of 1% and 5% for all and less than 10% for the crawling peg regime). Moreover, all coefficients enter with the expected signs according to theory, except for the effect due to foreign aid, which was found to have promoted RER depreciation rather than appreciation.[12] In particular, these results show that fundamentals associated with natural resource rents (government consumption, net foreign income and to a lesser extent TOT) contribute significantly to RER appreciation. Regarding the short-term, the PMG results suggest that productivity, openness and government consumption have had significant effects that are also consistent with the direction of their long-run impact. Moreover, unlike its long-run effect, the crawling peg regime was found to

[11] Comparable findings in the literature include Chinn (1997) for productivity; Elbadawi and Soto (1997) and Drine and Rault (2004) for terms of trade; Maeso-Fernandez et al. (2002) for government consumption; and Elbadawi and Soto (2008) for the other variables.

[12] However, even when found to be associated with real exchange rate appreciation, the effect of foreign aid was minuscule and, hence, has no economic significance (Elbadawi et al. 2008).

promote real depreciation in the short run. The two pieces of evidence combined suggest that a crawling peg regime is not necessarily an effective monetary institution for the promotion of long-term real exchange rate competitiveness. The estimated average adjustment parameter is –0.165, about 80% of the absolute value of the one obtained by Edwards (1989) using a partial adjustment model for a group of twelve developing countries.

3.3.1 RER Undervaluation in Arab Oil Economies and Comparators

Using the estimation results of Table 3.2 and the methodology described in Appendix C, we construct indexes for the equilibrium real exchange rate (ERER) and real exchange rate misalignment (MIS). The ERER is obtained by feeding the estimated model with the permanent components of the fundamentals (estimated with the Hodrick–Prescott filter). These permanent components are characterized as sustainable levels and are therefore consistent with the concept of equilibrium. The ERER is normalized (through the country-specific intercept) so that the long-run misalignment for each country is set equal to zero. This imposes the plausible identification condition that no country can be overvalued (or undervalued) on a sustained basis for the full estimation period. The log of the actual RER is then subtracted from the log of the resulting normalized ERER to obtain the RER undervaluation (RERunderval) time-series measures for each country. The analysis can be developed using the three pivotal equations from Appendix C:

$$e_t^i = \widehat{\beta}' F_t^i + \widehat{\varepsilon}_t^i \tag{4}$$

$$\widetilde{e}_t^i = \overline{e}^i + \widehat{\beta}' \left(\widetilde{F}_t^i - \overline{\widetilde{F}}^i \right) \tag{5}$$

$$RERunderval_t^i = -\left(e_t^i - \widetilde{e}_t^i\right) = \widehat{\beta}' \left(\widetilde{F}_t^i - \overline{\widetilde{F}}^i \right) - \left(e_t^i - \overline{e}^i\right) \tag{6}$$

where e_t^i is the log of the real exchange rate for any given country i at time t; F_t^i and $\widetilde{F}_t^i$ are the vector of current and sustainable fundamentals, respectively; β is a vector of long-run coefficients; and a bar over a variable indicates the mean over time. Equation (3.4) expresses the log of the RER in terms of current fundamentals and a residual term, while Equation (3.5) specifies the log of the equilibrium RER that satisfies the aforementioned normalization condition. The equilibrium RER is

Table 3.2: *The Long- and Short-run Determinants of the Real Exchange Rate*

	Pooled Mean Group Estimation	Mean Group	Hausman Tests	Dynamic Fixed Effect
Long-run coefficients				
Terms of trade	0.089*	-0.730	-0.819	0.180***
(in logs)	1.760	-0.490	2.440	2.900
Productivity	0.391***	1.167*	0.776	0.452
(in logs)	24.230	1.790	1.070	12.600
Trade openness	-0.217***	-0.327	-0.062	-0.376***
	-10.800	-0.570	0.930	-7.510
Government consumption/GDP	0.283***	1.435	1.151	0.199***
(in logs)	6.050	1.130	2.090	3.130
Net foreign income/ GDP	1.128***	-0.095	-1.223	1.243***
	4.450	-0.020	6.230	2.900
Foreign aid (net of int'l reserve	0.000***	0.000	0.000**	0.000***
Accumulation)/GDP	-5.830	-1.970	0.000	-3.570
Exchange rate regime – crawl	0.044***	0.108	0.064	0.082
	1.720	0.410	0.420	1.570
Exchange rate regime – free fall	0.289***	0.056	-0.233	0.354***
	4.170	0.790	0.090	4.030
Exchange rate regime – floating	-0.128	0.060	0.188	0.155
	-1.550	1.000	0.050	1.000
Error correction coefficient	-0.165***	-0.471***		-0.198
	-5.540	-9.620		-14.010
Short-run coefficients				
D (terms of trade, logs)	0.0411	0.066		-0.005***
	0.940	1.260		-0.250
D (productivity, logs)	0.444***	0.463***		0.550
	18.320	15.260		30.590
D (trade openness)	-0.130***	-0.136***		-0.150***
	-5.170	-4.190		-7.810
D (government consumption/GDP, logs)	0.096***	0.062		0.142***
	3.370	1.530		6.340
D (net foreign income/ GDP)	0.218	-0.143		-0.001***

	Pooled Mean Group Estimation	Mean Group	Hausman Tests	Dynamic Fixed Effect
	0.560	-0.360		-0.010
D (foreign aid net of int'l reserve	0.000	0.000		0.000
Accumulation/GDP	1.080***	-0.670		-1.090
D (exchange rate regime - crawl)	-0.018	-0.001		-0.030***
	-2.830	-0.060		-2.920
D (exchange rate regime - free fall)	-0.016	-0.013		-0.081***
	-1.530	-0.960		-4.920
D (exchange rate regime - floating)	-0.004	-0.006		-0.068**
	-1.120	-1.100		-2.230
Intercept	1.836	4.929***		2.340***
	5.340	6.150		10.370
No. countries/no. observations	50/1647	50/1647	50/1647	50/1647

Source: Authors' calculations.
Numbers below coefficients correspond to *z* statistics.
***, **, * stand for significance at the 1%, 5% and 10%.

expressed as the sum of the mean of the observed RER and a term that depends on the difference between the sustainable fundamentals and their mean values ($\beta'(\widetilde{F}_t^i - \overline{\widetilde{F}}^i)$).

Equations (3.4) and (3.5) allow us to derive the expression for the RER undervaluation in Equation (3.6). Like the equilibrium RER index, the expression for undervaluation is also very intuitive. It suggests that, at any point in time, if the difference between the RER at time *t* and the average RER is in excess of the equilibrium depreciation component, the exchange rate is overvalued at time *t* and the extent of the overvaluation is given by the net difference. This expression also suggests that depending on the size of the equilibrium depreciation component, a higher than average real exchange rate is compatible with overvaluation (RERunderval<0), undervaluation (RERunderval>0) or equilibrium (MIS =0).

We present the evidence of the RER undervaluation for three types of country groupings: oil exporting vs. middle income non-oil (Figure 3.4); Arab oil economies vs. non-Arab oil (Figure 3.5); and, GCC vs. populous

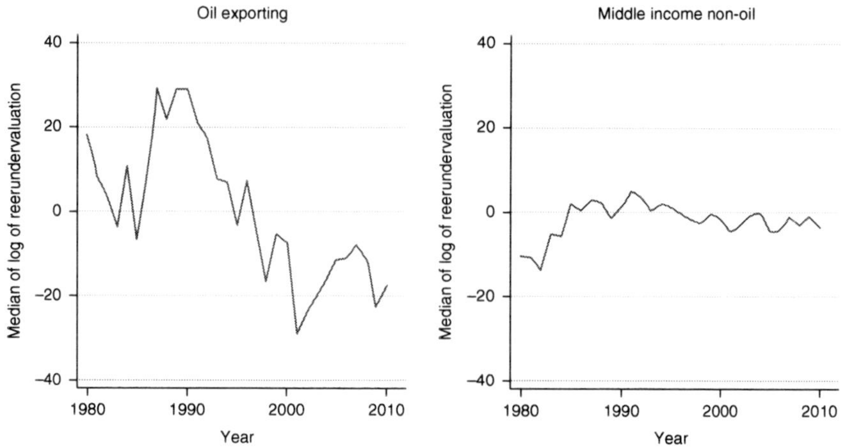

Figure 3.4: RER Undervaluation – Oil Exporting vs. Non-oil Middle Income

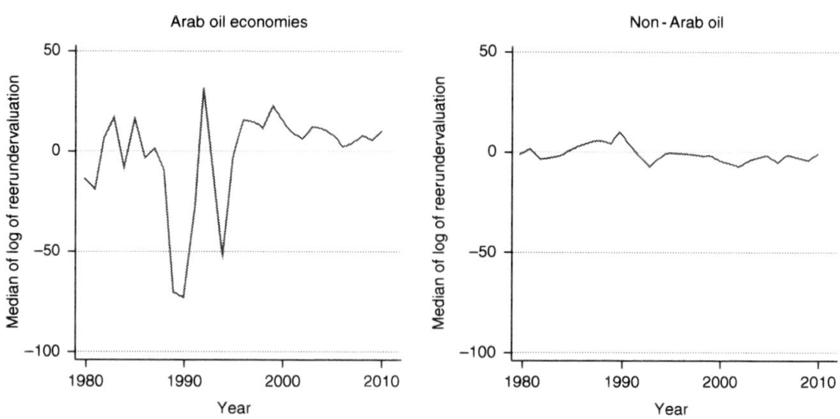

Figure 3.5: RER Undervaluation – Arab Oil Economies

Arab oil economies (Figure 3.6). The figures highlight that oil economies, both globally and in the Arab World, have experienced major RER volatilities, while the non-oil economies have been much more stable. Moreover, at the global level, non-oil middle income economies have managed to keep their RERs close to their equilibrium values, while the RERs in oil exporting economies have been overvalued since the mid-1990s (Figure 3.4). Although, like their non-Arab counterparts, the oil Arab countries failed to stabilize their RERs prior to the mid-1990s, they have, nevertheless, managed to keep their RERs more stable as well as

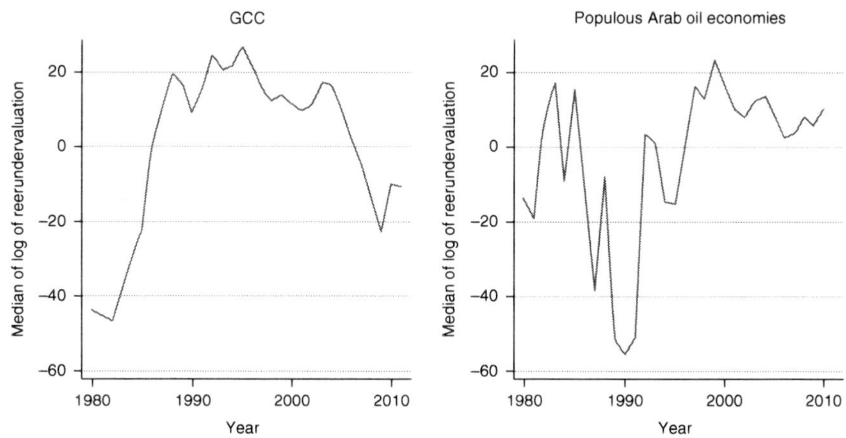

Figure 3.6: RER Undervaluation – GCC vs. Populous Arab Oil Economies
Notes:
1. Authors' elaboration using the regression results of Table 3.2 and Appendix C.
2. Oil exporting: all oil exporting countries in the sample (Appendix B).
3. Non-oil middle income: all non-oil middle income countries in the sample (Appendix B).
4. Oil Arab economies: GCC, Algeria, Sudan, Yemen.
5. Non-oil Arab: Morocco, Tunisia, Egypt, Lebanon, Jordan.
6. GCC: Bahrain, Kuwait, Oman, Qatar, Saudi Arabia (UAE is excluded for lack of data).
7. Populous Arab oil economies: Algeria, Sudan, Yemen.

undervalued for the entire period since 1995 (Figure 3.5). This suggests that the median oil exporting country in the Arab World has learned from the painful lessons of the mid-1980s, following the collapse of oil prices. Zooming into the evidence, however, it is possible to conclude that the GCC was responsible for the better performance in terms of RER stability and RER undervaluation (Figure 3.6). The evidence suggests that these countries' RERs were relatively stable and undervalued since prior to the turn of the 1990s decade until the recent global economic crisis. Instead, the populous Arab oil economies experienced very high RER volatility and only became competitive much later in the late 1990s. Nonetheless, compared to their non-Arab counterparts, it seems that oil exporting Arab countries have internalized the lessons of the 1980s much faster.

3.4 An Empirical Model of Export Performance

As discussed, export orientation has been credited as a successful development strategy leading to economic transformation. Furthermore, sustained export-oriented policies have been associated with significant export diversification, as countries initially limited to exploiting their endowments in natural resources have sought to avoid abrupt sector-specific shocks by moving into the production of non-traditional exports, such as manufacturing.

However, according to the evidence of Section 3.2, the Arab oil group exhibits the lowest level of export diversification, as suggested by the HHI for export concentration. Moreover, despite its relatively high income levels, it has relatively low shares of manufacturing exports, especially for knowledge- and skill-intensive types of manufacturing. Also, in addition to the failure to achieve technical upgrades in manufacturing, these countries were not able to produce and export more sophisticated exports, commensurate with their level of development. The export sophistication indexes for this group (EXPY) have been comparable or lower than those of much poorer emerging economies. Therefore, studying what fosters or hinders export diversification is crucial for a region like the Arab World. Moreover, the ability to study the determinants of various diversification measures that go beyond manufacturing exports allows us to capture the experiences of various resource-dependent countries such as Chile and Malaysia that have made substantial strides in diversifying their production and export structures by moving not just into manufacturing but also into high-value-added agricultural products. Unlike manufacturing exports per capita, one of our three export performance variables, the HHI and EXPY measures would allow us to capture such successful diversifications.

Previous literature studying the determinants of export diversification has focused on countries' factors of production (i.e., population, land per worker, natural resources) and/or geographic factors. Here we account for such factors, but we also focus on the role of resource rents and the real exchange rate.

Our main empirical results focus on a panel dataset that spans the period 1980 to 2011. The methodology employs the two-step generalized method of moments (GMM) estimator. The baseline model that we estimate is:

$$E_{it} = \beta_0 + \beta_1 E_{it-1} + \beta_2 LnPop_{it} + \beta_3 LnGDPPC_{it} + \beta_4 Credit_{it} + \beta_5 Agric_{it} + \beta_6 \mathrm{Rent}pc_{it} + \beta_7 Crisis_{it-1} + \mu_t + \varepsilon_{it} \tag{7}$$

where E is one of the three export performance indicators, namely manufacturing exports per capita, export sophistication (in turn, this is captured by three separate sophistication indices for goods, manufacturing and services exports) and concentration of exports; $LnPoP$ is the log of population size; *LnGDPPC* is the log of real per capita GDP; *Credit* is private credit to GDP; *Agric* is value added from agriculture as a share of GDP; *Rentpc* is the log of oil rents per capita; *Crisis* is the lagged indicator of currency crises as suggested by Freund and Pierola (2012) and μ stands for time effects and ε is the error term.

The baseline model is extended in several directions. First of all, given our intuition that oil rents per capita affect export performance, we test for the effectiveness of RER undervaluation (*RERunderval*) in stemming the negative impact of resource rents.

$$E_{it} = \beta_0 + \beta_1 E_{it-1} + \beta_2 LnPop_{it} + \beta_3 LnGDPPC_{it} + \beta_4 Credit_{it} + \beta_5 Agric_{it} + \beta_6 \text{Rentpc}_{it} + \beta_7 Crisis_{it-1} + \beta_8 RERund_{it} + \mu_t + \varepsilon_{it} \quad (8)$$

Subsequently, we augment our model to test for the role of RER undervaluation in stemming financial underdevelopment and institutional weaknesses. In addition, we also test whether or not RER undervaluation has a non-monotonic effect on the dependent variable by including the squared term of the RER undervaluation ($RERund^2$) in an expanded model laid out later in this chapter. The latter is motivated by the notion that RER undervaluation can only improve export performance up to a certain threshold, as no tradable exporting activity could be sustained without backward and forward linkages to non-tradable sectors.

The empirical results are presented in Tables 3.3–3.6, and they confirm the positive role of RER undervaluation on manufacturing exports and exports sophistication and its negative impact on export concentration. For all dependent variables, their past values are highly significant, confirming that all three measures tend to exhibit path dependency. The undervaluation variable exhibits the expected sign and is significant in all three regressions implying that an undervalued exchange rate would foster greater manufactured exports, lead to lower export concentration, and would support the venture into more sophisticated products. A number of other controls are also significant, namely oil rents undermine export performance and diversification and, to a lesser extent, sophistication, but also higher incomes per capita have the reverse effect. Other controls, including financial development, agriculture value added and crises, exhibit the expected signs but are not significant at conventional significance levels.

Alternative specifications suggest that the positive role of undervaluation is smaller in countries with sufficient financial development, at least for the case when this is proxied by the share of M2 in GDP. Given that the coefficient for undervaluation is smaller than in the baseline, this suggests

Table 3.3: *Resource Rents, RER Undervaluation and Manufacturing Exports (1980–2011)*

	Dependent Variable: Log Manuf Exports per Capita							
	[1]	[2]	[3]	[4]	[5]	[6]	[7]	[8]
Manuf exports per capita (–1) (in logs)	0.749***	0.759***	0.792***	0.792***	0.833***	0.695***	0.759***	0.824***
	(0.07)	(0.06)	(0.07)	(0.08)	(0.08)	(0.11)	(0.10)	(0.07)
Population (in logs)	0.029	0.064*	0.077*	0.106**	-0.034	-0.025	-0.036	0.097**
	(0.02)	(0.04)	(0.04)	(0.05)	(0.03)	(0.03)	(0.03)	(0.04)
GDP per capita (in logs)	0.248*	0.432***	0.472***	0.526***				0.468***
	(0.15)	(0.13)	(0.16)	(0.17)				(0.16)
Domestic credit to private sector	0.001		−0.00206					
	(0.00)		(0.00)					
Agricultural value added	−0.014							
	(0.02)							
Oil rent per capita (in logs)	−0.008**	−0.012*	−0.017**	−0.020**	0.006	0.009	0.008	−0.018**
	(0.00)	(0.01)	(0.01)	(0.01)	(0.01)	(0.01)	(0.01)	(0.01)
Crisis (−1)	−0.042	0.047	−0.042	−0.0346	−0.198	−0.079	−0.152	−0.020
	(0.05)	(0.06)	(0.05)	(0.05)	(0.12)	(0.16)	(0.13)	(0.04)
REER undervaluation(−)		0.003**	0.006***	0.008**	−0.014	−0.026	−0.021	0.012***
		(0.00)	(0.00)	(0.00)	(0.01)	(0.02)	(0.01)	(0.00)
Domestic credit to private sector* REER undervaluation			−0.00002					
			(0.00)					
M2				−0.00259*				
				(0.00)				
Government effectiveness* REER undervaluation					0.003			
					(0.00)			

Government effectiveness					−0.11*			
					(0.07)			
Regulatory quality* REER undervaluation						0.005*		
						−0.00283		
Regulatory quality						−0.290**		
						(0.12)		
Institutions * REER undervaluation							0.004*	
							(0.00)	
Institutions							−0.193**	
							(0.09)	
Dummy p50*REER undervaluation(1 if RER undervaluation > median)								−0.001
								(0.01)
Dummy p75* REER undervaluation (1 if REER undervaluation >75th percentile)								−0.007
								(0.01)
Constant	2.68	0.207	−0.857	−1.788	4.303**	7.465***	6.075**	−1.758
	(2.02)	(1.47)	(1.24)	(1.51)	(1.98)	(2.76)	(2.37)	(1.15)
No. of observations	1,954	2,147	2,131	2,027	1,081	1,081	1,081	2,147
No. of countries	118	118	118	118	113	113	113	118
First order serial correlation p-value	0.001	0.000	0.000	0.000				0
Second order serial correlation p-value	0.744	0.745	0.73	0.724				0.717
Hansen OID test p-value	0.834	0.949	0.692	0.568				0.786
No instruments	50	46	50	50				50

Standard errors in parentheses *** $p < 0.01$, ** $p < 0.05$, * $p < 0.1$

Table 3.4: *Resource Rents, RER Undervaluation and Exports Sophistication of Goods (1980–2011)*

Dependent Variable: Sophistication of Exports (in Logs)	(1)	(2)	(3)	(4)
Sophistication of goods exports (−1) (in logs)	0.875***	0.933***	0.857***	0.852***
	(0.216)	(0.116)	(0.264)	(0.074)
Population (in logs)	0.003	−0.007	−0.001	0.006
	(0.009)	(0.006)	(0.007)	(0.004)
GDP per capita (in logs)	0.016	−0.036		0.026
	(0.036)	(0.027)		(0.017)
Private credit to GDP		9.70e-05		
		(0.000311)		
Resource rents to GDP	−0.001	−0.002*	−0.001	−0.000
	(0.001)	(0.001)	(0.002)	(0.0007)
Reer undervaluation	0.000*	0.001*	−0.001	0.0004**
	(0.000)	(0.0007)	(0.002)	(0.000)
Private credit to GDP capita* REER undervaluation		−0.002*		
		(0.00129)		
Bad institutions * REER undervaluation			0.0001	
			(0.000)	
Bad institutions			0.011	
			(0.025)	
Resource rents to GDP* REER undervaluation				−0.002
				(0.001)
Constant	1.002	-0.212	1.324	1.070**
	(1.596)	(0.857)	(2.396)	(0.511)
No. of observations	1,087	1,000	593	1,087
No. of countries	72	71	72	72
First order serial correlation p-value	0.020	0.004	0.107	0.006
Second order serial correlation p-value	0.105	0.077	0.765	0.105
Hansen OID test p-value	0.333	0.896	0.891	0.636
No instruments	14	13	12	11

Standard errors in parentheses: *** $p < 0.01$, ** $p < 0.05$, * $p < 0.1$

that financial intermediation is one of the channels through which undervaluation contributes to higher manufacturing exports as it essentially constitutes a subsidy for the tradable sector. This finding corroborates similar evidence found for economic growth (e.g., Aghion et al. 2006). A similar intuition can be provided for the specifications where measures of poor institutional quality interacted with the undervaluation variable.

Table 3.5: *Resource Rents, RER Undervaluation and Exports Concentration (1980–2011)*

Dependent Variable: Export HHI Concentration Index	(1)	(2)	(3)	(4)
Concentration exports (−1)	0.779***	0.786***	0.738***	0.581***
	(0.058)	(0.058)	(0.086)	(0.082)
Population (logs)	−0.018	−0.128	−0.102	−0.574
	(0.175)	(0.183)	(0.197)	(0.447)
GDP per capita (logs)	0.428	0.396		0.282
	(0.414)	(0.371)		(1.114)
Private credit to GDP	−0.018	−0.018		−0.038**
	(0.013)	(0.011)		(0.017)
REER undervaluation	−0.023**	−0.022**	0.179*	0.025
	(0.011)	(0.011)	(0.099)	(0.025)
Resource rents to GDP	0.197***	0.191***	0.119**	0.158**
	(0.055)	(0.059)	(0.058)	(0.074)
Resource rents* REER undervaluation				−0.417*
				(0.251)
Private credit to GDP* REER undervaluation		0.010		
		(0.033)		
Bad institutions* REER undervaluation			−0.037*	
			(0.019)	
Bad institutions			−0.540	
			(0.755)	
Constant	−0.617	1.056	7.273	20.120
	(5.950)	(6.085)	(4.475)	(16.600)
No. of observations	1,604	1,684	891	1,684
No. of countries	91	96	92	96
First order serial correlation *p*-value	0.000	0.000	0.003	0.000
Second order serial correlation *p*-value	0.780	0.832	0.629	0.836
Hansen OID test *p*-value	0.490	0.104	0.509	0.705
No instruments	13	15	34	44

Standard errors in parentheses: *** $p < 0.01$, ** $p < 0.05$, * $p < 0.1$

The results seem to suggest that an undervalued exchange rate can make up for institutional weaknesses that would undermine manufacturing exports. This finding lends support to Rodrik's (2008) thesis that RER undervaluation can be a second-best approach to averting the negative role of bad institutions on growth performance. In view of their weak institutions, the oil-dependent Arab countries stand to gain the most by engineering real exchange rate undervaluation or at least avoiding extended overvaluation episodes (Figures 3.7 and 3.8).

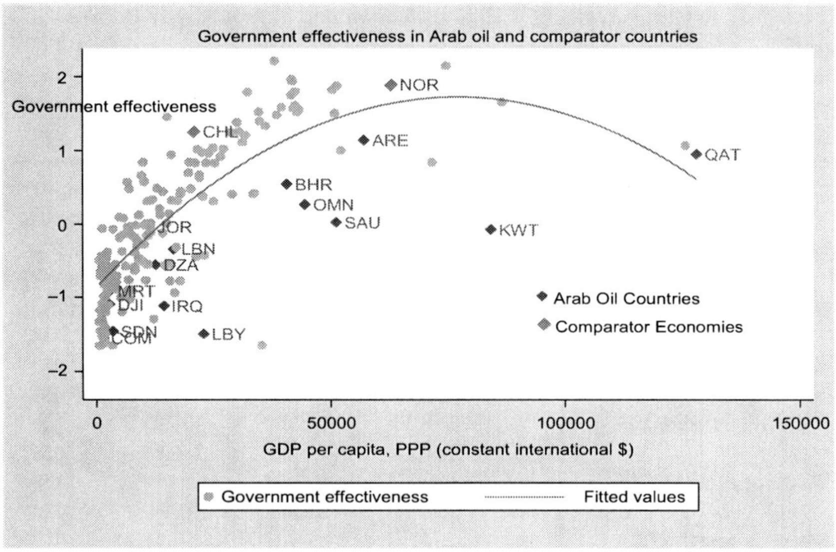

Figure 3.7: Government Effectiveness in Arab Oil and Comparator Countries

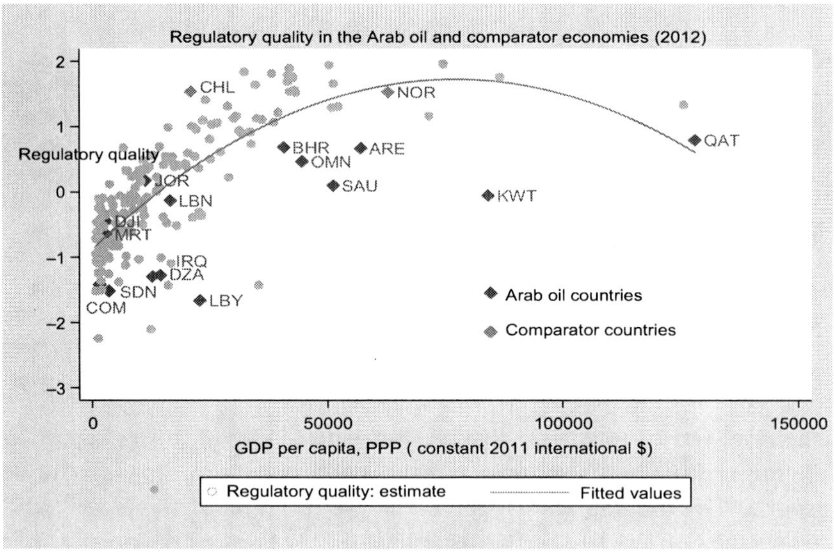

Figure 3.8: Regulatory Quality in the Arab Oil and Comparator Economies (2012)
Data source: Worldwide Governance Indicators, World Bank, 2014.

Table 3.6: *HP-gap RER Undervaluation and Export Performance (1980–2011)*

Variables	Manuf. Exports per Capita (in Logs)	Sophistication of Goods Exports (in Logs)	Concentration Exports
	(1)	(2)	(5)
Manuf. exports per capita (in logs) (−1)	0.756***		
	(0.0670)		
Sophistication of goods exports (in logs) (−1)		0.632***	
		(0.0897)	
Concentration exports (−1)			0.784***
			(0.077)
Population (in logs)	0.048**	0.013	−0.293
	(0.024)	(0.009)	(0.296)
GDP per capita (in logs)	0.381***	0.054***	−0.547*
	(0.121)	(0.017)	(0.317)
Crisis (−1)	−0.237		
	(0.172)		
Oil per capita (−1)	−0.009**		
	(0.004)		
Lgaphp_REER	0.008**		
	(0.003)		
Oil rents to GDP		−0.002	0.148***
		(0.001)	(0.048)
gaphp_REER		0.000***	−0.035**
		(0.000)	(0.014)
Constant	0.965	2.804***	14.85*
	(1.045)	(0.656)	(7.862)
No. of observations	2,147	1,087	1,816
No. of countries	118	72	99
First order serial correlation p-value	0.000	0.008	0.000
Second order serial correlation p-value	0.760	0.103	0.547
Hansen OID test p-value	0.949	0.642	0.095

In order to test for the robustness of our results, we repeated our key empirical specifications using an alternative, albeit much more rudimentary, measure of undervaluation. This measure was constructed by running an HP filter on the REER (and Log REER) series and then extracting its temporary component, which gives us two alternative measures of undervaluation: gaphp_REER (and Lgaphp_REER). The results (see Table 3.6) corroborate our findings that the RER undervaluation is conducive to greater manufacturing exports, higher export sophistication and lower export concentration and that it should be an effective policy response to the corrosive effects of oil rents on non-resource export performance.

3.5 Conclusions

The real exchange rate has played a central role in the development strategy of most countries, in particular when their ability to foster rapid productivity-driven growth is limited. This chapter develops RER misalignment series for a large panel of countries and studies the empirical link between RER undervaluation, export diversification and sophistication while also exploring possible interaction effects between RER undervaluation and the level of institutional and financial development. The chapter makes a number of important contributions. At the empirical level, though the evidence produced has wider applicability, we emphasize the implications of our findings for the resource-dependent economies in general, and Arab oil exporters in particular, given their diversity of exchange rate regimes and their high dependence on the resource rents.

This chapter builds model-consistent RER misalignment series for a large panel of countries. The RER misalignment series are generated from error-correction estimations for the equilibrium RER, based on structural determinants. The empirical results – based on a world sample of annual 1980–2011 data for fifty countries – show that long-run coefficients of all structural variables and short-run coefficients of some structural variables are significant and display expected signs according to theory. In particular, we find the long-run fundamentals that are highly important in resource-dependent economies – such as government expenditure, terms of trade, net foreign income and pegged exchange rate regimes – to have exerted a considerable appreciating impact on the RER. Subsequently, we used the long-run regression results to compute RER undervaluation series for each country.

The preliminary evidence on the tendency of oil economies to exhibit more volatile and overvalued (or less undervalued) RERs coheres with the estimated econometric effects of the RER fundamentals that are likely to be strongly associated with oil rents. However, it seems that the Arab oil countries, especially the GCC, have managed to keep these fundamentals closer to their sustainable long-term path for most of the post-1995 period until the onset of the recent global economic recession in 2008.

The chapter finds that those countries that have experienced a good measure of export diversification, increasing degrees of export sophistication and higher manufacturing per capita were also likely to have been able to avoid disequilibrium real exchange rate overvaluation. In fact, the evidence suggests an even stronger implication regarding exchange rate policy in that, not only is overvaluation bad for these export performance indicators, but that *undervaluation is good for all.* Therefore, the recent experiences of these countries should provide important lessons regarding the need to avoid high disequilibrium RER appreciation.

The additional and main contribution of this chapter is based on specifying several comprehensive export performance models that nest the variables within a standard specification, controlling for basic export performance fundamentals that are robustly identified in the empirical literature. The models are estimated with a dynamic system GMM estimator, using a global panel dataset that spans the period 1980–2011 and includes a large number of countries, ranging from 66 to 118 countries depending on the dependent variable. The results confirm the important positive role of RER undervaluation on export diversification and sophistication as well as in promoting a dynamic manufacturing sector. As expected, we find oil rent to have impeded export diversification, manufacturing sector growth and, to a lesser extent, export sophistication as well. Moreover, our results also show that RER undervaluation can be particularly effective in ameliorating the negative impact of oil rents on exports when financial markets are not sufficiently developed or when economies are characterized by underdeveloped institutions.

To the extent that oil and other resource-dependent economies are likely to be characterized by large institutional deficits and in some cases, underdeveloped financial markets, the evidence from this chapter suggests that a strategy to depreciate the real currency can be a viable industrial policy for promoting export diversification, technical upgrading and export sophistication. Moreover, this type of public policy can be used to minimize the need for excessive reliance on traditional vertical industrial policy, which usually requires high initial institutional capacity in order to succeed.

APPENDIX A: Definitions and Sources of Variables Used in Regression Analysis

Variable	Definition	Source
ERER Regressions		
Real effective exchange rate	An increase in the index reflects an appreciation (in logs).	International Monetary Fund International Financial Statistics
Government consumption expenditure	Ln(1+government consumption/GDP).	World Bank World Development Indicators
Official development assistance net of international reserves	Official development assistance – Change in gross international reserves.	World Bank World Development Indicators
Terms of trade	In logs.	World Bank World Development Indicators
Productivity	Ratio of per capita GDP over average per capita GDP industrial countries.	World Bank World Development Indicators
Openness	Residual of a regression of the log of the ratio of exports and imports as a share of GDP on the logs of area, population and dummies for oil exporting and landlocked countries.	World Bank World Development Indicators
Net foreign income	As a share of GDP.	World Bank World Development Indicators
Exchange rate regime	Three different variables that classify the exchange rate regime as pegged and crawling; free fall; and managed and free floating.	International Monetary Fund de facto classification of exchange rate regimes
Export Regressions		
Manufacturing exports	As a share of population (in logs).	World Bank World Development Indicators
Herfindahl–Hirschman index	Concentration index based on SITC 2–4 data (in logs).	UN COMTRADE
EXPY	Three indexes capturing export sophistication in goods, services and manufacturing. The variables are constructed as	From Anand et al. (2012)

Variable	Definition	Source
	the GDP per capita of countries exporting a particular good weighted by the value of exports summer over a country's export basket (in logs).	
REER undervalution	Linear transformation of the REER misalignment series estimated in the ERER regression.	Authors' calculation
GDP per capita	Real GDP as a share of population (in logs).	World Bank World Development Indicators
Population	In logs.	World Bank World Development Indicators
Currency crisis	Dummy variable indicating a currency crisis episode when it take a value of 1.	Laeven and Valencia (2008)
Resource rents	Rents (net of extraction costs) from hydrocarbon sources as a share of GDP.	World Bank World Development Indicators
Agricultural value added	As a share of GDP.	World Bank World Development Indicators
Private credit	As a share of GDP.	World Bank World Development Indicators
Money supply (M2)	As a share of GDP.	World Bank World Development Indicators
Institutional quality	Measures of government effectiveness and regulatory quality.	World Bank Worldwide Governance Indicators
Oil rent	As a share of population (in logs).	World Bank World Development Indicators

APPENDIX B: Country List

	Equilibrium RER Regression	Exports Regressions with Dependent Variable		
		Manufacturing Exports	HHI	EXPY
Albania		√	√	
Argentina	√	√	√	√
Armenia		√	√	
Australia	√	√	√	√
Austria	√	√	√	√
Azerbaijan		√	√	
Bahamas		√		
Bahrain		√		
Bangladesh	√	√	√	√
Barbados		√	√	
Belgium		√	√	
Belarus		√	√	
Belize	√	√	√	√
Benin		√		√
Bolivia	√	√		√
Bosnia Herzegovina		√		
Botswana	√	√	√	
Brazil	√	√	√	√
Bulgaria		√	√	
Burkina Faso	√	√	√	√
Burundi		√	√	√
Bhutan		√	√	
Cambodia		√	√	
Cameroon	√	√	√	√
Canada		√	√	√
Central African Republic		√		√
Chile		√	√	√
Colombia	√	√	√	√
Comoros		√	√	
Costa Rica	√	√	√	√
Croatia		√	√	
Cyprus	√	√	√	√
Czech Republic		√		
Denmark		√	√	√
Dominican Republic	√	√		

	Equilibrium RER Regression	Exports Regressions with Dependent Variable		
Ecuador	√	√	√	√
Egypt	√	√		√
El Salvador		√	√	√
Estonia		√	√	
Ethiopia	√	√	√	√
Fiji		√	√	
Finland	√	√	√	√
France	√	√	√	√
Germany	√	√	√	√
Ghana		√	√	√
Great Britain	√	√	√	√
Greece	√	√	√	√
Guatemala		√	√	√
Guinea		√	√	
Honduras	√	√	√	√
Hong Kong		√		√
Hungary		√	√	
Iceland	√	√	√	√
India	√	√	√	√
Ireland		√	√	√
Italy	√	√	√	√
Jamaica		√	√	√
Japan		√	√	√
Jordan	√	√	√	√
Kazakhstan		√	√	
Kenya	√	√	√	√
Korea	√	√		√
Kyrgyz Republic		√		
Laos				√
Latvia		√	√	
Lebanon		√	√	
Lithuania		√	√	
Luxembourg		√	√	
Macedonia		√		
Madagascar	√	√	√	√
Malaysia		√	√	√
Malawi		√	√	√
Maldives		√	√	
Mali	√	√	√	√
Malta	√	√	√	√
Mauritania		√	√	√

(*continued*)

(*continued*)

	Equilibrium RER Regression	Exports Regressions with Dependent Variable		
Mauritius		√	√	√
Mexico	√	√	√	√
Moldova		√	√	
Mongolia		√	√	√
Morocco	√	√	√	√
Namibia		√	√	
Nepal		√		√
Netherlands	√	√	√	
New Zealand	√	√	√	√
Norway	√			
Pakistan	√	√	√	√
Panama	√	√	√	√
Paraguay		√	√	√
Peru		√	√	√
Philippines	√	√	√	√
Poland		√	√	√
Portugal	√	√	√	√
Russia		√	√	
Senegal	√	√	√	√
Serbia		√	√	
Sierra Leone	√	√		√
Singapore		√	√	√
Slovakia		√		
Slovenia		√	√	
South Africa		√	√	√
Spain	√	√	√	√
Sri Lanka	√	√	√	√
Sudan	√	√		√
Swaziland		√	√	
Sweden	√	√	√	√
Switzerland	√	√	√	√
Syria		√		√
Tanzania		√		√
Thailand		√	√	√
Trinidad & Tobago		√	√	√
Tunisia	√	√	√	√
Turkey		√	√	√
Turkmenistan		√	√	
Uganda		√	√	√
Ukraine		√	√	
Uruguay		√	√	√
Venezuela	√	√		√

	Equilibrium RER Regression		Exports Regressions with Dependent Variable	
Vietnam		√		√
Yemen		√		√
Zambia	√	√	√	√

Note:
A check mark indicates that the country was included in the baseline regression estimation specified as the column heading under which the check mark is placed.

For those cases when a country does not get a check mark for the equilibrium RER estimation but is included in the growth or export regressions, this implies that its RER misalignment series was obtained by applying the out-of-sample coefficients.

APPENDIX C: Computing the Equilibrium Real Exchange Rate and RER Misalignment Indexes[13]

In order to determine the equilibrium RER, it is useful to collapse all of its determinants into a category we call *fundamentals*. Let e_{it} be the log of the observed real exchange rate for country i in time t. Then, we can write the equilibrium RER equation as:

$$e_t^i = \widehat{\delta}_0^i + \widehat{\beta}' F_t^i + \widehat{\varepsilon}_t^i \tag{9}$$

where i denotes a country and ε_t^i is a stochastic innovation or short-term fluctuation. Note that the intercept varies across countries. Let the equilibrium RER be as follows:

$$\widetilde{e}_t^i = \widetilde{\delta}_o^i + \widehat{\beta}' \widetilde{F}_t^i \tag{10}$$

where $\widetilde{F}_t^i$ refers to sustainable fundamentals, given by the permanent components of the fundamentals and $\widetilde{\delta}_0^i$ is a scaled country-specific intercept to be identified in the following text.

Under the assumption that the model is correctly specified, the real exchange rate undervaluation (RERund) is simply given by subtracting the observed RER from its equilibrium value:

$$RERund_{it} = \widetilde{e}_t^i - e_t^i = \left(\widetilde{\delta}_0^i - \widehat{\delta}_0^i\right) + \widehat{\beta}'\left(\widetilde{F}_t^i - F_t^i\right) - \widehat{\varepsilon}_t^i \tag{11}$$

The scaled intercept of the equilibrium RER ($\widetilde{\delta}_0^i$) must satisfy the following identification condition:

[13] This appendix is a modified version of the appendix in Elbadawi et al. (2012).

$$E_t(RERund_{it}) = E_t\left[\left(\widetilde{\delta}_0^i - \widehat{\delta}_0^i\right) + \widehat{\beta}'(\widetilde{F}_t^i - F_t^i) - \widehat{\varepsilon}_t^i\right] = 0 \qquad (12)$$

This condition requires that, for any given country, the expected value of the misalignment across time must be equal to zero. This is because eventually the RER must revert to its equilibrium level; otherwise, it will not be a "misalignment" but a permanent phenomenon. Though the expected value of the transitory components of the fundamentals (second right-hand side term) should be zero, we do not make that restriction to allow for potential misspecification of the decomposition procedure[14].

Noting that the first right-hand side term is time invariant, we have the following sample estimate for the equilibrium intercept term:

$$\widetilde{\delta}_0^i = \widehat{\delta}_0^i + \widehat{\beta}'\left[\frac{1}{n}\sum_t \left(F_t^i - \widetilde{F}_t^i\right)\right] + \overline{\widehat{\varepsilon}}_t^i \qquad (13)$$

Note that though the panel estimation requires that $E_{t,i}\left[\widehat{\varepsilon}_t^i\right] = 0$, $E_t\left[\widehat{\varepsilon}_t^i\right]$ is not, in general, equal to zero and can be estimated by the mean of the residuals $\overline{\widehat{\varepsilon}}^i = \frac{1}{n}\sum_t \widehat{\varepsilon}_t^i = \frac{1}{n}\sum_t e_t^i - \widehat{\delta}_0^i - \widehat{\beta}'\left(\frac{1}{n}\sum_t F_t^i\right)$ (from Equation (3.1)). Substituting for the mean residual in Equation (3.5), we have the final expression for the equilibrium intercept:

$$\widetilde{\delta}_0^i = \overline{e}^i + \widehat{\beta}'\left(\overline{F}^i - \overline{\widetilde{F}}^i\right) - \widehat{\beta}'\overline{F}^i = \overline{e}^i - \widehat{\beta}'\overline{\widetilde{F}}^i, \qquad (14)$$

where $\overline{e}^i, \overline{F}^i$ and $\overline{\widetilde{F}}^i$, respectively, denote the mean values (over time) of the actual RER, the fundamentals and their corresponding permanent components.

Equations (3.6) and (3.2) give us the ultimate expression for the equilibrium RER index:

$$\widetilde{e}_t^i = \overline{e}^i + \widehat{\beta}'\left(\widetilde{F}_t^i - \overline{\widetilde{F}}^i\right) \qquad (15)$$

This expression states that, for any given country *i*, the RER equilibrium index must be equal to the average of the observed RER over the estimation

[14] We show later that the expression for the equilibrium RER is the same whether or not we assume the expected values of the transitory fundamentals to be zero. Moreover, under the general case, the equation for misalignment generates the one with the expected value equal to zero as a special case.

period plus (minus) a component reflecting equilibrium appreciation (depreciation), where an equilibrium appreciation (depreciation) is required when the weighted permanent component of the fundamentals in time t is larger (smaller) than the corresponding average over the estimation period (second right-hand side term).

Subtracting the observed log RER from the above index gives the corresponding expression for misalignment[15]:

$$RERund_t^i = \widehat{\beta}' \left(\widetilde{F}_t^i - \overline{\overline{F}}^i \right) - \left(e_t^i - \overline{e}^i \right) \quad (8)$$

If the permanent components of the fundamentals are time invariant, the second term in the RHS of Equations (3.7) and (3.8) will be zero. The equilibrium RER will, therefore, be equal to the mean of the observed RER and the misalignment will be given by the deviation from the mean RER. This will be consistent with a variant of the PPP model. However, the PPP restriction is neither corroborated by theory nor the time-series characteristics of the fundamentals, especially for the case of developing countries.

References

Aghion, P., P. Bacchetta, R. Ranciere, K. R. (2006). Exchange Rate Volatility and Productivity Growth: The Role of Financial Development. *Journal of Monetary Economics* 56(4), 494–513.

Aguirre, A. and C. Calderon (2005). Real Exchange Rate Misalignments and Economic Performance. Working Paper No. 315. Central Bank of Chile.

Anand, R., S. Mishra and N. Spatafora (2012). Structural Transformation and the Sophistication of Production. IMF Working Paper WP/12/59. The International Monetary Fund, Washington DC, USA.

Carrere, C., V. Strauss-Kahn and O. Cadot (2007). Export Diversification: What's Behind the Hump? Clermont-Ferrand, Centre d'etudes et de recherché sur le development international (CERDI, Etudes et documents, E 2007.24).

Chinn, M. (1997). The Usual Suspects? Productivity and Demand Shocks and Asia-Pacific Real Exchange Rates. Pacific Basin Working Paper Series. Federal Reserve Bank of San Francisco.

[15] Under the assumption that $E_t\left[\widehat{\beta}'\left(F_t^i - \widetilde{F}_t^i\right)\right] = 0$, the corresponding expression for the equilibrium intercept is given by $\widetilde{\delta}_0^i = \overline{e}^i - \widehat{\beta}'\overline{F}^i$. However, since $\overline{F}^i = \overline{\overline{F}}^i + \frac{1}{n}\sum\left(F_t^i - \widetilde{F}_t^i\right)$ and the second RHS term are equal to zero by assumption/construction, the expression for the intercept is the same under both cases. Therefore, the expressions for the equilibrium RER (Equation (3.7)) as well as misalignment (Equation (3.8)) also remain the same.

Collier, P. and B. Goderis (2009). Structural Policies for Shock-Prone Developing Countries. CSAE Working Paper No. 2009-03, Department of Economics, University of Oxford.

Drine, I. and C. Rault (2004). On the Long-run Determinants of Real Exchange Rates for Developing Countries: Evidence from Africa, Latin America and Asia. Mimeo, Sorbonne University.

Edwards, S. (1989). *Real Exchange Rate, Devaluation and Adjustment: Exchange Rate Policy in Developing Countries*. Cambridge, MA: MIT Press.

Eichengreen, B. (2007). The Real Exchange Rate and Economic Growth. Paper prepared for the World Bank's Commission on Growth.

Elbadawi, I. (1994). Estimating Long-Run Equilibrium Real Exchange Rates. In *Estimating Equilibrium Exchange Rates*, J. Williamson, ed., Washington, D.C.: Institute for International Economics.

(1998). *Real Exchange Rate Policy and Non-traditional Exports in Developing Countries*. Helsinki: WIDER, the United Nations University.

(2002). Real Exchange Rate Policy and Non-traditional Exports in Developing Countries. In *Non-Traditional Export Promotion in Africa: Experiences and Issues*, G.K. Helleiner, ed., New York: Palgrave.

Elbadawi, I., L. Kaltani and R. Soto (2012). Aid, Real Exchange Rate Misalignment and Economic Growth in Sub-Saharan Africa. *World Development* 40(4), 681–700.

Elbadawi, I. and A. Gelb (2010). Oil and Economic Diversification in the Arab World. Approach paper. Economic Research Forum, Cairo, Egypt.

Elbadawi, I. and R. Soto (2008). Theory and Empirics of Real Exchange Rates in Sub-Saharan Africa and Other Developing Countries. World Bank Working Paper.

Elbadawi, I., L. Kaltani, and K. Schmidt-Hebbel (2008). Foreign Aid, the Real Exchange Rate, and Growth in the Aftermath of Civil Wars. *World Bank Economic Review* 22(1), 113–140.

Elbadawi, I. and G. Helleiner (2004). African Development in the Context of the New World Trade and Financial Regimes: The Role of the WTO and its Relationship to the World Bank and IMF, in *Africa and the World Trading System, Volume I: Selected Issues of the Doha Agenda*, A. Oyejide and W. Lyakurwa, eds., Trenton, NJ: Africa World Press, Inc., chapter 9.

(1997). Real Exchange Rates and Macroeconomic Adjustment in Sub-Saharan Africa and Other Developing Countries. *Journal of African Economies* 6(3), 74–120.

Freund, C. and M. Pierola (2012). Export Surges. *Journal of Development Economics* 97, 387–395.

Hausmann, R., J. Hwang, and D. Rodrik (2006). What You Export Matters. CEPR Discussion Paper No. 5444.

Hausmann, R., L. Prichett, and D. Rodrick (2005). Growth Accelerations. *Journal of Economic Growth* 10(4), 303–329.

Hausmann, R. and R. Rigobon (2003). An Alternative Interpretation of the 'Resource Curse': Theory and Policy Implications. In *Fiscal Policy Formulation and Implementation in Oil Producing Countries*, Davis, J. M., R. Ossowski and A. Fedelino, eds., Washington, DC: IMF.

Imbs, J. and R. Wacziarg (2003). Stages of Diversification. *American Economic Review* 93(1), 53–86.

IMF (2009). *Regional Economic Outlook: Middle East and Central Asia*. Washington DC: IMF.

Johnson, S., J. Ostry, and A. Subramanian (2007). Africa's Growth Prospects: Benchmarking the Constraints. IMF Working Paper WP/07/52, International Monetary Fund, Washington DC.

Klinger, B. And D. Lederman (2004). Discovery and Development: An Empirical Exploration of 'New' Products. World Bank Policy Research Working Paper No. 3450.

Maeso-Fernandez, F., C. Osbat, and B. Schnatz (2002). Determinants of the Euro Real Effective Exchange Rate: A BEER/FEER Approach. *Australian Economic Papers* 41 (4), 437–461.

Maloney, W. and D. Lederman (2007). Trade Structure and Growth. In *Natural Resources: Neither Curse nor Destiny*, W. Maloney and D. Lederman, eds., pp. 15–40. Stanford, California: Stanford University Press for the World Bank.

Rodrik, D. (2008). The Real Exchange and Economic Growth. *Brookings Papers on Economic Activities* 2, 365–412.

(2006). Industrial Development: Stylized Facts and Policies. Unpublished mimeo. Kennedy School of Government, Harvard University.

Sachs, J. and A. Warner (1997). Natural Resource Abundance and Economic Growth. Updated version of 1995 NBER working paper, 5398. Center for International Development and Harvard Institute for International Development, November.

Sala-i-Martin, X. (1997). I just Ran Four Million Regressions. Economics Working Papers 201, Department of Economics and Business, Universitat Pompeu Fabra.

Sekkat, K. and A. Varoudakis (1998). Exchange Rate Misalignment and Manufactured Exports in Sub-Saharan Africa. OECD Development Center, Technical Paper No. 134.

UNIDO. (2009). *Breaking In and Moving Up: New Industrial Challenges for the Bottom Billion and the Middle-Income Group*. Industrial Development Report 2009. Vienna, Austria: United Nations, Industrial Development Organization,

Williamson, J. (1997). Exchange Rate Policy and Development Strategy. In *Foreign Exchange Markets and Exchange Rate Policies in Sub-Saharan Africa, Journal of African Economies*, Elbadawi and Soto, eds., pp. 17–36. Supplement to Vol. 6(3).

4

Dutch Disease in the Services Sector

Evidence from Oil Exporters in the Arab Region

Ndiamé Diop and Jaime de Melo

ABSTRACT

This chapter shows that as per capita incomes increase, services sectors in resource-rich countries of the Arab World have declined as a share of gross domestic product (GDP) and of non-mining GDP. This negative relationship between the share of services in GDP and income per capita is opposite to global patterns and is linked to the large rents generated by natural resources in these countries. A large number of services sectors can now be contracted offshore or can be produced by temporary movement of service providers, implying that countries need to be competitive to maintain domestic production. Since rents from natural resources inflate wages and non-tradable prices, they tend to discourage domestic production of tradables, both goods and services. As the chapter highlights, the negative effect of rents are compounded by the negative impact of policy and regulatory restrictions on entry, and of business conduct on the development of services sectors. These restrictions create rents captured by "protected incumbents" or increase the real cost of producing services – in both cases inflating the price of services.

Keywords: Arab World, Dutch Disease, diversification, oil curse, rents

4.1 Introduction

For Arab resource-rich countries, finding ways to develop service sectors is crucial for widening the scope of job creation. First, the public sector, the largest employer in the region, has reached a saturation point in almost every single country. Going forward, rapid job creation in the public sector is likely to hold down aggregate labor productivity and deprive the private sector of the skilled labor it needs to grow. Second, rents from natural resources have been an obstacle to the development of manufacturing industries because of currency overvaluation leading to a low profitability for tradable manufactures (Rodrik 2008; Havrylyshyn 2010; Lopez-Calix et al. 2010). Third, agriculture has become a minor source of labor absorption and is expected to continue declining as a share of GDP, with income growth and technological advances. Thus, the services sector is the most promising source of job creation in Arab resource-rich countries.[1]

The fact that vibrant local services sectors can be developed in resource-rich countries has been taken for granted for a long time. This belief was underpinned by at least two combined theoretical considerations. First, Engel's law effects in consumption imply that demand for services tends to increase with income due to higher income elasticity of demand for services relative to agricultural products (Chenery and Syrquin 1975; Chenery et al. 1986). At the same time, services sectors, implicitly assumed to be largely non-tradable in earlier Dutch Disease models, would be positively affected by an appreciation of the real exchange rate subsequent to a natural resource boom (Corden and Neary 1982; Corden 1984).

While Engel's consumption effects are likely to operate, the "largely non-tradable" status of services implicit in the earlier Dutch Disease literature is no longer valid. The revolutions in technology, transportability and tradability of the past twenty years have significantly increased the tradability of services and have led to substantial cross-border "disembodied" trade in services (François and Hoekman 2010). A large number of services sectors are now able to be moved offshore, implying an imperative to be competitive to maintain domestic production. In addition to information and communications technology (ICT)-enabled services (back-office business process services, information technology services, software development and so forth) that can be delivered from a distance, many professional services (engineering, legal, accounting, auditing and the like) are now produced through temporary movement of professionals. In other words, firms located in a country such as Saudi Arabia are no longer dependent on domestic service providers and can import many of the services they need if local producers are not competitive.

[1] The proclivity for outmigration is limited for resource-rich countries. Instead, these countries are net receivers of migrant workers. Thus, outmigration is generally not a source of employment for nationals of these countries.

This chapter investigates whether resource-rich Arab countries have managed to diversify in services and, if not, why. The resource-rich Arab countries included in the analysis are: the high rent per capita group comprising the Gulf Cooperation Council (GCC) countries (Bahrain, Kuwait, Oman, Qatar, Saudi Arabia, United Arab Emirates) and the populous countries or low rent per capita countries (Algeria, Iran, Iraq, Libya, Syria, Yemen). For comparative purposes, a number of resource-poor labor-abundant countries (RPLA) are also included (Egypt, Jordan, Lebanon, Morocco, Tunisia). Section 4.2 presents the stylized facts about the size and performance of services sectors in the Arab region. It shows that contrary to the global pattern, the share of services in gross domestic product (GDP) has been declining with income in the resource-rich Arab countries. Using econometric analysis, Section 4.3 shows the negative role of rents in explaining this observed structural change. Section 4.4 discusses the role of microeconomic regulations and shows that rather than alleviating the negative effect of rents on the relative contraction of services, regulatory restrictions compound the difficulty of developing competitive services sectors in the Arab region. The chapter concludes by highlighting the contrast between resource-rich and resource-poor Arab countries. While rents hinder the development of services sectors and encourage import of services in resource-rich countries, they can indirectly widen the scope for diversifying exports and increasing job creation in resource-poor countries of the region by demonstrating the potential benefits of regional integration in services.

4.2 Services in the Arab Region: Stylized Facts

4.2.1 Recent Growth and Trade Performance

Although the services sectors in the Arab region have grown faster than average GDP in the past decade, they remain much less dynamic than the same sectors in East Asia, South Asia and, to a lesser degree, Europe and Central Asia. In the period 2000–2010, services sectors grew at an average annual rate of 9.5% in East Asia, 8.3% in South Asia and 5.4% in the Arab region. Within the region, they grew by 1% of GDP more in the Gulf Cooperation Council (GCC) countries than in the rest of the Arab region.

Figure 4.1 illustrates the importance of services sectors to growth. It compares growth in the real GDP per capita and growth in services value-added of all Arab countries. The positive relationship between the two variables is by no means causality, but it is a correlation implying that

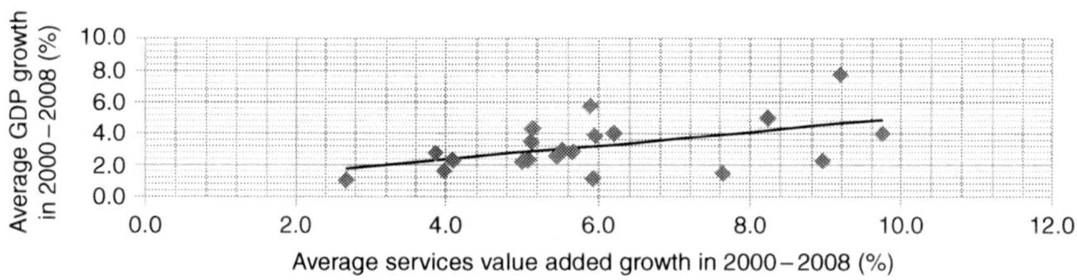

Figure 4.1: Services Value-added Growth and GDP Growth in the Arab Region
Source: World Development Indicators, World Bank.

Arab countries with high growth in services tend to have high GDP or, conversely, countries with high overall economic growth have high services growth. Ghani and Kharas (2010) have found this positive correlation in 136 countries.

The relative underperformance of the Arab region in growth in services is mirrored by its poor performance in services trade. In 2000–2008, trade in services grew 23% annually in South Asia, 16% in East Asia and 14% in Europe and Central Asia compared with 12.4% annually in the Arab region during the same period. Along with Latin America and Sub-Saharan Africa, the Arab region is a minor exporter in the global market for ICT, finance and other business services, which have exploded over the past decade. The Arab region's share in global trade of services stagnated at around 2.8% between 2000 and 2008.[2]

The lack of dynamism in the Arab World's services exports reflects its low value-added orientation. Exports of services remain largely dominated by travel services (tourism-related). In 2008, this category accounted for 53% of total services exports. Travel and transport together made up 78% of total exports. This is in sharp contrast with South Asia (driven by India), where ICT and finance are the leading export services, making up 55% of exports. Transport and travel services account for a mere 24% of total exports in that region (Figure 4.2).

4.2.2 Has the Arab Region Diversified Towards Services?

One of the key objectives that Arab resource-rich countries set for themselves in the early 1990s was to diversify away from natural resources toward services and manufacturing. Typically, as countries become richer, the share of agriculture declines, giving way to a rise in the share of manufacturing and services. This often happens because of technological

[2] Borchert et al. (2011).

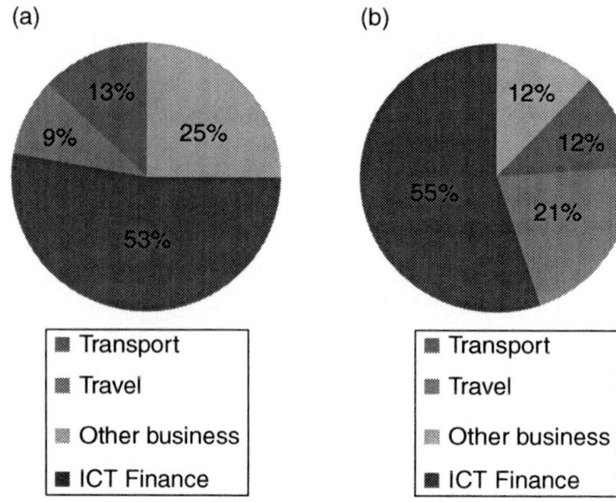

Figure 4.2: Composition of Exports in the Arab Region and South Asia, 2008. (a) Arab Region. (b) South Asia
Source: International Monetary Fund.

advances in agriculture, which increase agricultural productivity and drive resources out of agriculture toward manufacturing and services (Baumol 1967; Chenery and Syrquin 1975). At the same time, Engel's law stipulates that as households' income increases, the percentage of income spent on food decreases while the proportion spent on other goods and services increases. Thus, both supply and demand forces suggest that one should observe a decline in the share of agriculture in overall GDP and an increase in non-agricultural GDP as income increases.

Consistent with theoretical predictions, agriculture has shrunk over the past thirty years, to varying degrees, in all regions of the world. The share of agriculture in GDP contracted by more than 60% in East Asia, 58% in Europe and Central Asia, 46% in South Asia, 38% in Latin America and the Caribbean, 36% in the Arab region and 23% in Sub-Saharan Africa. Figure 4.3 shows how the economic structures of the different regions of the world have changed following the relative contraction of agriculture. In Asia, Europe and Central Asia, the contraction of agriculture has given way to a large services sector. In contrast with the rest of the world, the services sector has shrunk overall in the Arab world, but the share of industry (dominated by mining and oil) in GDP has increased. While this aggregate result certainly conceals large variations across Arab countries, it indicates that services have contracted in at least some countries of the region.

It is important to distinguish between different groups of countries within the Arab region, given the region's heterogeneity and the

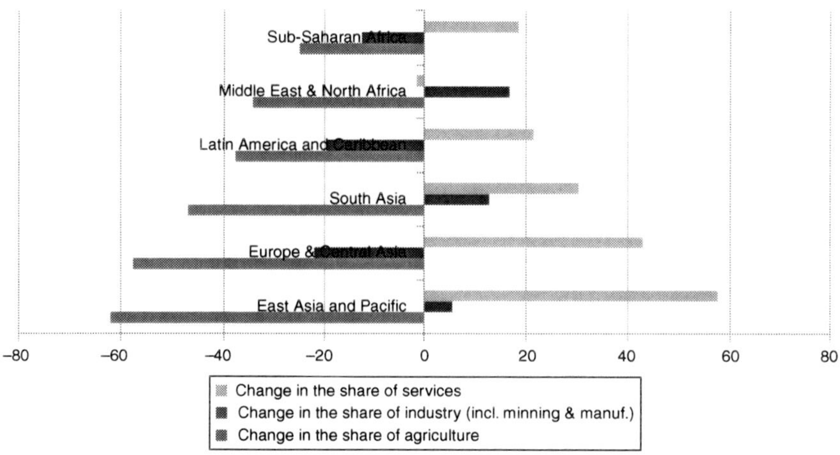

Figure 4.3: Changes in the Composition of GDP: 1980–1983 to 2007–2010
Source: World Development Indicators, World Bank.

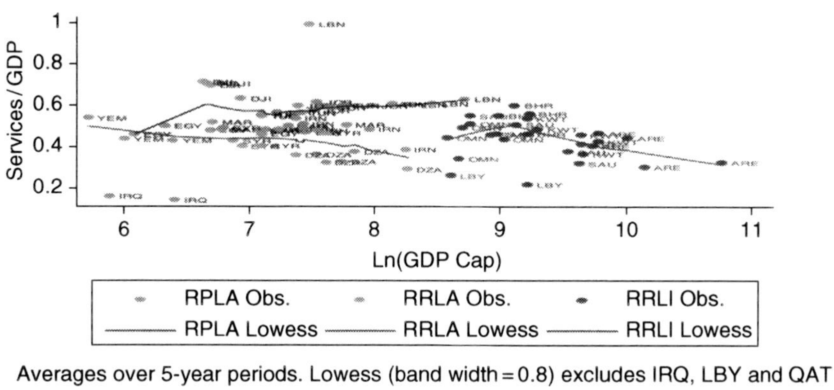

Figure 4.4: Services Share in GDP by Level of Income
Source: Authors.
Note: The figure shows averages over 5-year periods. RPLA = resource-poor labor-abundant; RRLA = resource-rich labor-abundant (populous resource-rich); RRLI = resource-rich labor-importing (GCC); Obs. = observations; Lowess = locally weighted scatter plot smoothing = fitting trend. Lowess (band width = 0.8) excludes Iraq, Libya and Qatar.

abundance of natural resources in many countries. Figure 4.4 shows a pooling (5-year averages for each observation) of the pattern of services in GDP for each of the three MENA groups: the GCC, the populous resource-rich economies and the resource-poor countries. The figure shows that the share of services in GDP is a decreasing function of per

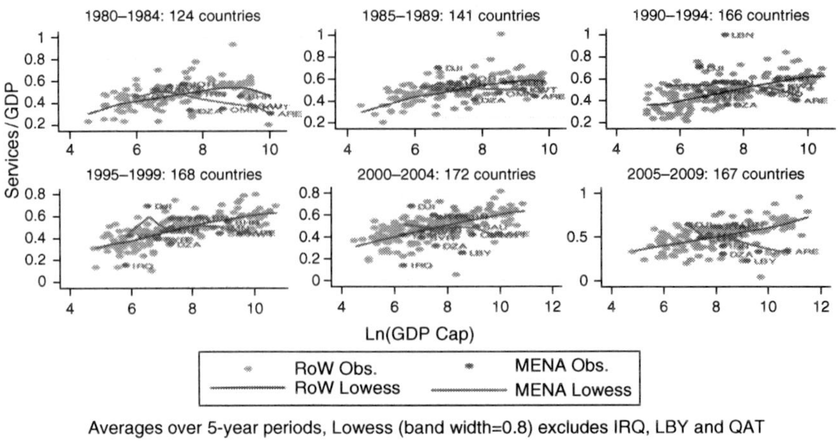

Figure 4.5: Services Share in GDP, Arab World vs. Rest of the World
Source: Authors.
Note: The figure shows averages over 5-year periods. RoW = rest of world; Obs. = observations; Lowess = locally weighted scatter plot smoothing = fitting trend. Lowess (band width = 0.8) excludes Iraq, Libya and Qatar.

capita GDP for the two resource-rich country groups while there is a clear positive correlation between these two variables for the resource-poor group. In other words, the decline in the share of services in GDP in the region is driven by the two resource-rich groups.

As Figure 4.5 shows, the contraction of services sectors in resource-rich Arab countries is at odds with global trends. There is indeed evidence of a positive correlation between the share of services in GDP and income per capita as shown by Hoekman and Mattoo (2008) who use a large cross-section sample of countries from around the world.

Before examining the correlates of the apparent relative shrinking of services in resource-rich Arab countries in the next section, it is important to examine whether the result obtained here is a statistical artifact, reflecting a rapid increase in the share of the mining sector (dominated by oil) in these economies. In other words, we need to control for the role of oil. One way to do this is to examine the changes in the composition of non-mining GDP over time and see how the share of services in non-mining GDP evolves.[3] Unfortunately, we do not have a complete series of non-mining value-added for many Arab countries. For the countries for which data is

[3] The mining sector in MENA is dominated by oil industries in resource-rich countries. Jordan, Morocco and Tunisia, all considered "resource-poor countries" in this volume, do

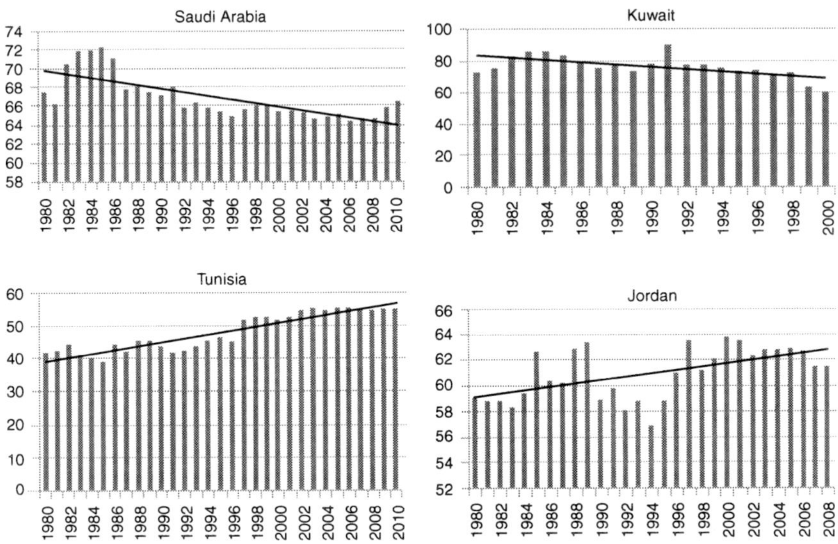

Figure 4.6: Share of Services in Non-mining GDP
Source: Authors.
Note: Data for Kuwait are not available after 2000.

available, however, the observation of a shrinking share of services over time is confirmed. This observation is in contrast with resource-poor countries, where the share of services in non-mining GDP is either increasing or stagnant. Figure 4.6, which shows the share of services in non-mining GDP declining in resource-rich Kuwait and Saudi Arabia while increasing in resource-poor Jordan and Tunisia, illustrates this point.

4.3 Relative Roles of Engel's Effects in Consumption and Rents

There are two traditional explanations for the positive correlation between the share of services and income. First is Engel's effects in consumption: as incomes rise, demand for services tends to rise owing to higher income elasticity of demand for services relative to agricultural products (Chenery and Syrquin 1975; Chenery et al. 1986). The second explanation is Baumol's cost "disease" effect: fast productivity growth in

have a sizable phosphates sector. Rents from phosphates, however, are limited compared with those from oil.

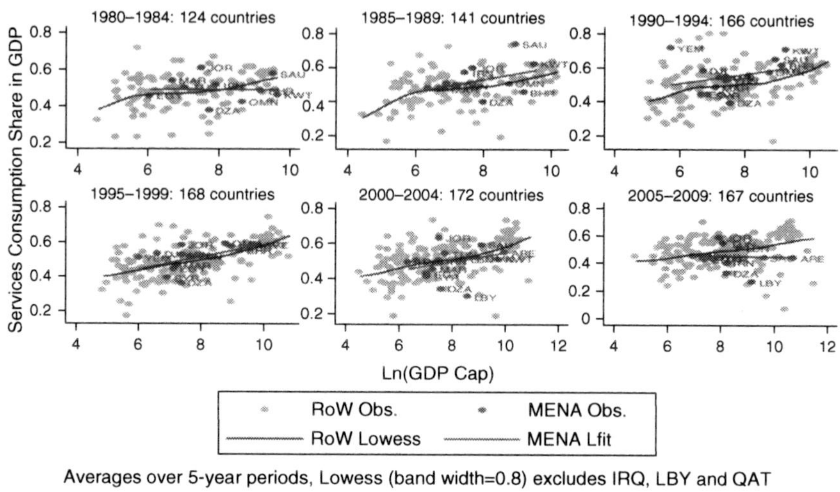

Figure 4.7: Share of Consumption of Services in GDP (%)
Source: Authors.
Note: The figure shows averages over 5-year periods. RoW = rest of world; Obs. = observations; Lowess = locally weighted scatter plot smoothing = fitting trend. Lowess (band width = 0.8) excludes Iraq, Libya and Qatar.

agriculture as a result of mechanization frees up resources allocated to manufacturing and services (Baumol 1967).

In addition to these effects, under the assumption that services are (largely) non-tradable, an appreciation of the real exchange rate subsequent to a natural resource boom would increase the relative profitability of the services sectors and promote their development. As noted, however, the revolutions in technology, transportability and tradability have made a large number of the services sectors tradable, weakening this argument. In any case, empirically, the positive correlation between the share of services in GDP and per capita income is well established for most countries in the world (Hoekman and Mattoo 2008).

4.3.1 Engel's Effects

We start by examining whether Engel's effects operate for services in the Arab region. Figure 4.6 confirms a rising share of consumption of services in GDP as GDP per capita rises. Thus, Engel's effects in consumption operate for services in the Arab World (Figure 4.7). However, a large and growing share of this consumption of services is satisfied by imports in the Arab

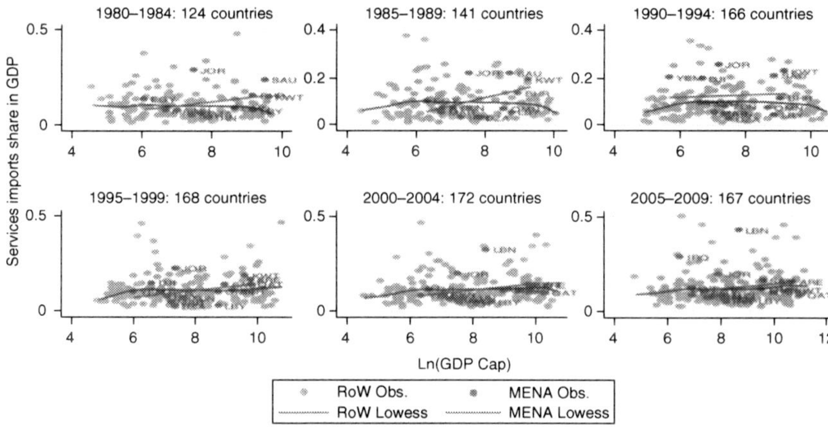

Figure 4.8: Share of Imported Services in GDP (%)
Source: Authors.
Note: The figure shows averages over 5-year periods. RoW = rest of world; Obs. = observations; Lowess = locally weighted scatter plot smoothing = fitting trend. Lowess (band width = 0.8) excludes Iraq, Libya and Qatar.

region in sharp contrast with the rest of the world. Figure 4.8 shows that imports of services have been growing in Arab countries, thereby compensating for the low domestic production of services in resource-rich countries and supplying the local consumption market.

4.3.2 The Role of Resource Rents

The pattern of production and imports of services observed earlier may be associated with a real exchange appreciation that accompanies the rents from exports of natural resources in the region. With an increasing share of services becoming tradable as transaction costs have fallen, rather than observing a general expansion of the services sector when the real exchange rate appreciates for the resource-rich group, one observes an increase in the imports of services that cannot be domestically produced competitively because of the depressed relative profitability of production in these activities. We now turn to the possible effects of rents and appreciation of the real exchange rate.

To explore the role of natural resources in the development of services sectors, we draw on data on rents from natural resources for a sample of 150 countries.[4] In Table 4.1, it is clear that the resource-rich populous and

[4] This data is obtained from the World Bank database on adjusted net savings (see details in Bolt et al. 2002). More specifically, rents are calculated as the difference between the

Table 4.1: *Share of Rents from Natural Resources in GDP%*

Region	1980–1984	1985–1989	1990–1994	1995–1999	2000–2004
Arab resource-poor	13	5	2	1	2
Arab RR populous	19	16	23	22	32
Arab GCC	55	35	34	30	41
East Asia and Pacific	4	4	3	3	4
Europe and Central Asia	3	1	1	3	3
Latin America and Caribbean	7	4	3	3	4
North America	6	3	2	2	3
South Asia	5	3	3	2	3
Sub-Saharan Africa	6	4	4	4	6

Source: Authors' computation from UNCTAD data.

the GCC countries have, by far, the largest share of rents in GDP in the world. Although data on rents in the period 2005–2010 are not available, one can safely assume that they have only grown larger as a result of the dramatic increase in oil price during that period.

To further explore the correlates of the share of services in GDP, Table 4.2 reports the results of the regressions of the share of value-added in services over GDP on country and year fixed effects, with the country fixed effects controlling for time-invariant omitted variables, and the time fixed effects for time-dependent common shocks. The regressions are run for a large number of countries including those in the Arab World (first three columns of Table 4.2), then for all countries except Arab countries, for Arab countries only and for the three subgroups within the region, namely resource-poor, populous and GCC. The first three columns confirm the positive correlation between the share of services in GDP and per capita GDP. A 10% increase in GDP per capita is associated with an increase of 0.2%–0.5% in the share of services in GDP. Column 5 shows that this correlation turns negative for the Arab regions, mostly because of the resource-rich countries (Columns 8 and 9).

As documented in Table 4.1, a large share of the GDP of resource-rich countries is composed of rents generated by their natural resources. To control for the association between rents and the share of services in GDP, Table 4.3 introduces the logarithm of the rents among the correlates. In all

market value of extracted materials and the average extraction cost and are expressed as a share of GDP. Rents from 150 natural resource-rich countries are compiled in this way from 1970 to 2004 (services data are available only since 1980).

Table 4.2: *Correlates of the Share of Services in GDP*

Services/GDP	(1) *All*	(2) *All*	(3) *All*	(4) *All but Arab*	(5) *Arab*	(6) *Arab*	(7) *RPLA*	(8) *Resource-rich populous*	(9) *GCC*
Ln(GDP Cap)	0.02***	0.02***	0.05***	0.03***	–0.07***	–0.02***	–0.01	–0.08***	–0.14**
	(0.00)	(0.00)	(0.00)	(0.00)	(0.02)	(0.01)	(0.01)	(0.02)	(0.06)
MENA Dummy		–0.13***	–0.02***						
		(0.03)	(0.01)						
Constant	0.19***	0.19***	0.12***	0.16***	0.97***	0.61***	0.73***	0.93***	1.66***
	(0.02)	(0.02)	(0.01)	(0.02)	(0.10)	(0.04)	(0.07)	(0.14)	(0.60)
Country fixed effects	Yes	Yes	No	Yes	Yes	No	Yes	Yes	Yes
Time fixed effects	Yes	Yes	No	Yes	Yes	No	Yes	Yes	Yes
Observations	3,644	3,644	3,644	3,311	333	333	127	88	118
R-squared	0.84	0.84	0.35	0.85	0.85	0.03	0.94	0.95	0.78

Source: Authors.

Notes: Robust standard errors in brackets. ***$p < 0.01$, **$p < 0.05$, *$p < 0.1$.

Table 4.3: *Determinants of the Services Share in GDP*

Services/GDP	(1) *All*	(2) *All*	(3) *All*	(4) *All but Arab*	(5) *Arab*	(6) *Arab*	(7) *RPLA*	(8) *Resource-rich populous*	(9) *GCC*
Ln (GDP Cap)	0.02***	0.02***	0.05***	0.03***	−0.07***	0	−0.01	−0.04**	0.08*
	(0.00)	(0.00)	(0.00)	(0.00)	(0.02)	(0.01)	(0.01)	(0.02)	(0.04)
Ln (Rents)	−0.002***	−0.002***	−0.005***	−0.005**	−0.01*	−0.01***	0	−0.07***	−0.17***
	(0.001)	(0.001)	(0.000)	(0.001)	(0.003)	(0.001)	(0.001)	(0.017)	(0.02)
MENA Dummy		−0.12***	0						
		(0.03)	(0.01)						
Constant	0.22***	0.22***	0.18***	0.19***	1.05***	0.72***	0.50***	2.13***	3.71***
	(0.02)	(0.02)	(0.01)	(0.03)	(0.11)	(0.03)	(0.08)	(0.35)	(0.40)
Country F.E.	Yes	Yes	No	Yes	Yes	No	Yes	Yes	Yes
Time F.E.	Yes	Yes	No	Yes	Yes	No	Yes	Yes	Yes
Observations	3,644	3,644	3,644	3,311	333	333	127	88	118
R-squared	0.84	0.84	0.43	0.85	0.85	0.4	0.94	0.96	0.85

Source: Authors.
Notes: Robust standard errors in brackets, ***$p < 0.01$, **$p < 0.05$, *$p < 0.1$.

the specifications, the share of rents in GDP is significantly negatively correlated with the share of services in GDP. Moreover, the relationship between the share of services in GDP and GDP per capita becomes positive for the GCC in regression 7. For the resource-rich populous countries (regression 6), the relationship is still negative, but much less significant. While these are only correlations because they do not account for the potential endogeneity of rents, nor does the inclusion of fixed effects only control for time-invariant omitted variables, it would appear that rents associated with natural resource abundance in Arab countries partially account for the negative correlation between services' production and per capita income.

4.4 The Role of Microeconomic Regulations

The preceding section shows that rents associated with natural resource abundance for Arab countries partially account for the negative correlation between services production and per capita income. Here we examine the role that policy and regulations may have played in the relative underperformance of their services sectors. Restrictions on entry and business conduct either create rents within the services sectors that are captured by "protected incumbents," or increase the real cost of producing services. In both cases, these restrictions inflate the price of services (Dee 2005). Opening up the services sectors to competition would therefore reduce the behind-the-border "tax equivalent" or the "productivity loss equivalent" of the restrictions, thereby enhancing the competitiveness of services.

The underdevelopment of the services sectors in resource-rich Arab countries has deep policy roots, in addition to the effects of rents. Indeed, regulatory barriers to market entry, licensing and business conduct remain significant in MENA. The World Bank has compiled a database of restrictions to trade in services in five services sectors of eleven countries of the Pan Arab Free Trade Agreement (PAFTA), five GCC countries, as well as Egypt, Jordan, Lebanon, Morocco, Tunisia and Yemen. Using this database, Borchert et al. (2011) shed light on the degree of restrictiveness in five key services sectors in MENA compared with the rest of the world.

Figure 4.9 shows the services trade restrictiveness average indexes (STRI) calculated by Borchert et al. (2011) using the World Bank database compiled from 102 countries. The GCC stands out as the region that is consistently more restrictive than the rest of the world in the five

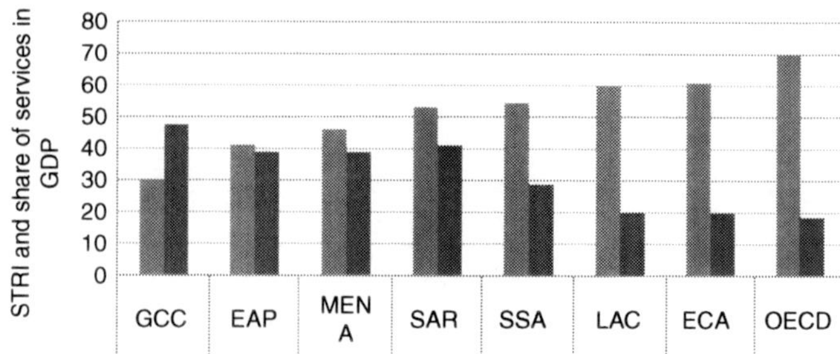

	GCC	EAP	MENA	SAR	SSA	LAC	ECA	OECD
Share services in GDP (%)	30	40.9	46.4	53.5	54.6	60.6	60.8	70
STRI	48.2	39.4	39.3	41.6	29.6	20.7	19.9	18.9

Figure 4.9: Restrictiveness of Services Trade Policies and Share of Services in GDP, Arab Region, GCC and Other Regions
Source: Borchert et al. (2011).

sectors surveyed.[5] When the share of services in GDP is overlaid on the same graph, there appears a net negative correlation between the magnitude of restrictions to trade in services and the share of services in GDP. Countries that have the most open services sectors are also those in which services make the largest contribution to overall GDP.

Furthermore, Borchert et al. (2011) identified the "desire on the part of government authorities to retain a considerable degree of regulatory discretion" as a distinctive feature of policies applied by PAFTA countries. As a result, even in areas free of explicit restrictions, de jure openness may not always imply a commensurate degree of de facto openness, and vice versa. In many instances, the restriction is on business conduct or related to foreign equity limits. Across different sectors, the granting of new licenses remains opaque and highly discretionary in many countries. This discretion creates uncertainty about the rules of the game and may discourage domestic and foreign investors in the services sectors.

The high restrictions found by Borchert et al. (2011) are also consistent with recent evidence pointing to a negative relationship between natural resources and the likelihood of undertaking microeconomic reforms. Drawing on the Doing Business data in a large sample of 133 countries, Amin and Djankov (2009) find that the proclivity to undertake microeconomic reforms that reduce regulation is much less in countries whose

[5] Note that the United Arab Emirates is not included in the sample. This country's economy is the most open one in the GCC.

exports are concentrated in abundant natural resources. In the same vein, Freund and Bolaky (2008) find that all 12 Arab countries are among the 50% most regulated economies among a large sample of 126 countries. Here, too, there may be a correlation with the presence of natural resources, given the multiple possibilities for regulatory capture in high-rent environments.

4.5 Export Diversification Opportunities for Resource-poor MENA Countries

If the increased tradability of services makes it challenging for resource-rich Arab countries to maintain domestic production of services, it offers formidable opportunities to resource-poor countries of the region. Indeed, given their cultural proximity and common language, these countries are well-placed to capture a share of the large and growing market of tradable services in resource-rich Arab countries. To take advantage of these opportunities, however, resource-poor countries will need to undertake autonomous reforms to improve their competitiveness and work with resource-rich countries to reduce barriers to labor mobility within the region[6]. The competition with "enclaves" within the resource-rich countries, where strategies are put in place to "force" diversification will remain stiff. Dubai, of the seven emirates of the United Arab Emirates, is a good example (Box 4.1).

Resource-poor countries will need to reduce the regulatory restrictions on business entry, licensing and business conduct in their own services sectors to foster competition. As noted, the restrictions are significant, including in sectors that have a high potential for capturing a growing market in the resource-rich countries of the region (such as professional services). In addition, improvements of the so-called backbone services (such as air transport and telecom services) will be needed. These services are core inputs into most economic activities in all sectors, including exporting service industries. To enhance their competitiveness, ICT-enabled and professional services firms will need to provide good quality and low-cost services.[7]

[6] See Marouani and Zaki (2011) for a comprehensive discussion of barriers to mode 4 (temporary movement of service providers) trade in MENA.

[7] For empirical evidence on the impact of backbone services on productivity and competitiveness, see, for instance, Arnold et al. (2011) and Arnold et al. (2008).

Box 4.1 Dubai's Successful Approach to Diversification

The United Arab Emirates (UAE) is a federation of seven emirates: Abu Dhabi, Ajman, Dubai, Fujairah, Ras Al-Khaimah, Sharjah and Umm al-Qaiwain. Abu Dhabi is the real "oil-rich" emirate of the UAE, accounting for about 90% of national oil production. Capitalizing on the financial support of Abu Dhabi, Dubai has positioned itself as a world center of finance, commerce, transportation and tourism. Dubai's model is based on attracting foreign direct investment, free movement of labor, capital and goods across borders, an efficient bureaucracy with no hassle to private firms and state-of-the-art infrastructure and backbone services. A key instrument was the creation of Free Trade Zones, which offer 100% foreign ownership with no taxation.

Dubai's market-oriented policies have led to significant diversification of the economy. New sectors have emerged, such as high-class tourism and international finance. The Dubai International Financial Centre offers 55.5% foreign ownership, no withholding tax, freehold land and office space, and a tailor-made financial regulatory system with laws similar to those governing leading financial centers in New York, London, Singapore and Zürich. Dubai has also developed internet and media free zones, offering 100% foreign ownership and untaxed office space for the world's leading ICT and media companies, with the latest communications infrastructure to service them. Recent liberalization of the property market, allowing non-citizens to buy freehold land has resulted in a major boom in the construction and real estate sectors.

While Dubai illustrates that diversification is possible in resource-rich countries, whether the Dubai model is sustainable is not clear. In 2008–2009, the UAE was hit hard by the global banking crisis and the collapse of the real estate bubble. The economy has slowly rebounded since then. Furthermore, the diversification of the economy did not fully solve the Emirates' employment problem. Despite the establishment of an "Emiratization" program in the early 2000s to boost employment of nationals in the public and private sectors, UAE citizens account for less than 1% of the labor force in the private sector.

A very positive aspect of the UAE's strategy for long-term diversification is the massive investments in education. In 2010, investments in education accounted for 22.5% ($2.6 billion) of the overall budget. Multiple governmental initiatives actively promote training of high school dropouts and graduates in a multitude of skills needed in the private sector. Beyond directly sponsoring educational initiatives, major research initiatives are funded by the Emirates Foundation for Philanthropy through competitive research grants, allowing universities such as the United Arab Emirates University or the Dubai School of Government to build and disseminate expertise in many areas.

Resource-poor governments also need to facilitate business-to-business contacts and work with their resource-rich partners to reduce barriers to mobility and promote "contract-based" movement of service suppliers.[8]

[8] See Hoekman and Ozden (2009) for details on the distinction between "contract-based" movement of services providers and employment-based movements of persons.

The main impediments to the movement of people arise from labor market laws in resource-rich countries, which rarely distinguish between temporary and permanent labor mobility. Restrictions in this category in the GCC (the main resource-rich services market) include burdensome and costly procedures for work permits, limitations on the length of stay, quantitative limits on work permits and sectoral bans, job nationalization, educational conditions and restrictions on foreign investment.[9]

Although a number of regulated professions feature nationality restrictions in both resource-poor and resource-rich countries, large niches exist. In Libya, for example, jobs such as accountants are reserved for Libyan nationals. In the United Arab Emirates, regulated professions in the areas of accounting, engineering, law, medicine, agriculture, economics, fishing, industry, management and technical consulting are open to practitioners and employees from any GCC member state and are accessible to non-GCC suppliers under certain conditions.

Bilateral agreements are a key vehicle for fostering trade in services among neighboring countries. For instance, Iraq and Lebanon have concluded an agreement aimed at increasing economic cooperation between them, including exchange of expertise, specialists and trainers (Article 3 of the Iraq-Lebanon bilateral agreement). A similar agreement was concluded between Kuwait and Lebanon, stipulating that the two parties shall facilitate the procedures for granting entry visas to businessmen in both countries (Article 6). Other agreements have been signed by Lebanon and the Syrian Arab Republic that promote labor mobility between the two countries. Yet even though these agreements mention some provisions related to exchanging expertise or facilitating visa procedures, they do not include direct provisions organizing the temporary movement of people. Resource-poor countries interested in exporting services to the GCC and other resource-rich Arab countries should aim at signing comprehensive bilateral agreements that address the specific barriers their firms are facing in these markets.

In brief, the services sectors of Arab countries present interesting complementarities and a huge potential for trade, but all countries need to implement relevant policies to exploit these potential benefits.

9 An elaborate administrative mechanism exists to regulate the inflow and residence of non-GCC migrant workers to GCC countries. All foreign workers and their dependents entering a GCC country are issued a resident visa for the number of years stipulated in the work contract. All such visas are issued under the authority of a sponsor that wishes to hire the foreign worker.

4.6 Concluding Remarks

This chapter has shown that despite recent good growth performance, services sectors in resource-rich Arab countries have been declining as a share of GDP (and as a share of non-mining GDP) over time, even as income per capita has increased. This negative relationship between the share of services in GDP and income per capita is opposite to observed global patterns. The analysis here suggests that this result is linked to the large rents generated by natural resources in these Arab countries. A large number of services sectors can now be stationed offshore or produced by temporary movement of service providers, implying that countries need to be competitive to maintain local production. Rents from natural resources tend to inflate wages and non-tradable prices in resource-rich countries, thereby appreciating the real exchange rate and discouraging domestic production of tradable goods and services. This explains why resource-rich Arab countries have become large importers of tradable services and why only domestic production of non-tradable services (such as real estate, retail trade, hotels and restaurants) has really developed.[10]

Microeconomic regulations on business have tended to compound the problem, rather than compensating for it. Restrictions on business entry, licensing and business conduct are indeed significant and correlate negatively with the share of services in GDP. Arab countries, in particular resource-rich ones, stand out for heavy and discretionary restrictions of services sectors compared with the rest of the world. These restrictions either create rents within the services sector that are captured by "protected incumbents" or increase the real cost of producing services – in both cases inflating the price of services and further reducing competitiveness of tradable services sectors.

Regardless of the relative weight of different explanatory factors, the underdevelopment of tradable services in Arab resource-rich countries creates a big challenge for these countries. Indeed, manufacturing is difficult to develop competitively in these countries for similar reasons (rents and Dutch Disease) – oil-related industries are capital-intensive rather than labor-intensive, and public sectors are bloated. The scope for creating jobs for nationals in the productive sectors of these countries is limited. This situation is particularly problematic for resource-rich countries with

[10] The rapid development of a non-tradable service, namely, the real estate sector, in conjunction with the underdevelopment of tradable services is an illustration of this point.

abundant labor supply, such as Algeria, Saudi Arabia, Syria and Yemen. It is crucial for these countries to reduce regulatory restrictions in services sectors and invest heavily in education to compensate for the negative effect of rent on competitiveness.

Resource-rich economies should strive to reduce production costs, thus offsetting the negative effect of rents on production in the non-resource tradable sectors. This can be done by reducing regulatory restrictions on entry and competition in these sectors. Experience from resource-rich countries around the world also shows the importance of investing in human capital and strengthening institutions (see Gelb 2011 for a summary). Finland, the Republic of Korea and Norway are examples of countries that have invested in building a high-quality human capital base and have successfully diversified into high-tech manufacturing and services. Similarly, there is strong evidence that institutions matter for diversification (Gelb 2011; Acemoglu et al. 2005). Gelb (2011, 67) argues that manufacturing sectors are "heavily dependent on strong contract enforcement, a rule of law and generally strong business environment." These arguments equally apply to services, if not more strongly so. Institutions that prevent or reduce rent-seeking are also important, as the example of Botswana shows (see Acemoglu et al. 2005).

If the increased tradability of services makes it challenging for resource-rich Arab countries to maintain domestic production of services, it offers formidable opportunities to resource-poor countries of the region. Indeed, given their cultural proximity and common language, these countries are well placed to capture a share of the large and growing market for tradable services in resource-rich Arab countries. To capture these opportunities, however, resource-poor countries will need to undertake autonomous reforms to improve their competitiveness, and work with resource-rich countries to reduce barriers to labor mobility within the region. More specifically, they will need to reduce their own restrictions to entry and competition in professional services, improve their backbone services (such as telecom and transport) and proactively engage resource-rich countries in reducing barriers to trade and mobility in resource-rich countries through specific bilateral agreements.

References

Acemoglu, D., S. Johnson, and J. Robinson. (2005). Institutions as a Fundamental Cause of Long-run Growth. In *Handbook of Economic Growth,* vol. 2, P. Aghion and S. Durlauf, eds., pp. 385–472. Amsterdam: North-Holland.

Amin, M., and S. Djankov. (2008). Democracy and Reforms. Policy Research Working Paper 4835, World Bank, Washington, DC.

Arnold, J. M., B. S. Javorcik, and A. Mattoo. (2011). Does Services Liberalization Benefit Manufacturing Firms? Evidence from the Czech Republic. *Journal of International Economics* 85 (1): 136–46.

Arnold, J. M., A. Mattoo, and G. Narciso. (2008). Services Inputs and Firm Productivity in Sub-Saharan Africa: Evidence from Firm-level Data. *Journal of African Economics* 17 (4): 578–99.

Baumol, W. J. (1967). Macroeconomics of Unbalanced Growth: The Anatomy of Urban Crisis. *American Economic Review* 62 (June): 415–26.

Bolt, K., M. Matete, and M. Clemens. (2002). *Manual for Calculating Adjusted Net Savings.* Washington DC: Environment Department, World Bank.

Borchert, I., B. Gootiiz, and A. Mattoo. Forthcoming. (2011) *Policy Barriers To International Trade in Services: New Empirical Evidence.* Washington, DC: World Bank.

Chenery, H., S. Robinson, and M. Syrquin, eds. (1986). *Industrialization and Growth: A Comparative Study.* Oxford, U.S.: Oxford University Press for the World Bank.

Chenery, H., and M. Syrquin. (1975). *Patterns of Development, 1950–1970.* Oxford, U.K.: Oxford University Press for the World Bank.

Corden, W. M. (1984). Booming Sector and Dutch Disease Economics: Consolidation and Survey. *Oxford Economic Papers*, 36: 359–80.

Corden, W. M. and J. P. Neary. (1982). Booming Sector and De-industrialisation in a Small Open Economy. *Economic Journal*, 92(368, December): 825–48.

Dee, P. (2005). A Compendium of Barriers to Services Trade. Paper prepared for the World Bank, www.crawford.anu.edu.au/pdf/Combined_report.pdf.

Francois, J., and B. Hoekman. (2010). Services Trade and Policy, *Journal of Economic Literature, American Economic Association* 48(3, September): 642–92.

Freund, C., and B. Bolaky. (2008). Trade, Regulations, and Income. *Journal of Development Economics* 87 (2, October): 309–21.

Gelb, A. (2011). Economic Diversification in Resource Rich Countries. Center for Global Development, Washington, DC, www.imf.org/external/np/seminars/eng/2010/afrfin/pdf/Gelb2.pdf.

Ghani, E., and H. Kharas. (2010). Service-led Growth in South Asia: An Overview. In *The Service Revolution in South Asia,* E. Ghani, ed. New York: Oxford University Press.

Havrylyshyn, O. (2010). Does the Global Crisis Mean the End of Export-led Open-economy Strategies? Paper prepared for World Bank, Washington, DC.

Hoekman, B., and A. Mattoo. (2008). Services Trade and Growth, Policy Research Working Paper Series 4461, World Bank. Washington, DC.

Hoekman, B., and C. Ozden. (2009). The Euro-Mediterranean Partnership: Trade in Services as an Alternative to Migration? Policy Research Working Paper Series 5049, World Bank. Washington, DC.

Lopez-Calix, J., P. Walkenhorst, and N. Diop. (2010). *Trade Competitiveness of the Middle East and North Africa.* Washington, DC: World Bank, Directions in Development.

Marouani, M., and C. Zaki. (2011). Trade in Mode 4 in MENA and Barriers to Labor Mobility. Draft paper.

Rodrik, D. (2008). The Real Exchange Rate and Economic Growth. *Brookings Papers on Economic Activity* 2008: 2.

World Economic Forum and OECD. (2005). Arab World Competitiveness Report. Geneva: World Economic Forum.

5

The Political Economy of Public Sector Employment in Resource-dependent Countries[1]

Omer Ali[2] and Ibrahim Elbadawi[3]

ABSTRACT

In this chapter, the authors propose a political economic explanation for the well-documented difference in labor market institutions between high natural resource per capita countries and those that are natural resource dependent but whose populations are large. This dichotomy is widespread but particularly apparent in the oil-rich countries of the Arab World. They argue that the natural resource endowment influences the policy that a dictatorship chooses. It is optimal for governments in countries of the first type to employ a large proportion of the population in the public sector, while it is optimal to set up a repressive security apparatus and employ a smaller proportion of the population when the natural resource endowment is small relative to the size of the population. Empirical support is provided for this theory using global data on public wages, rents per capita and other correlates.

Keywords: Arab World, GCC, political economy, oil curse, game theory

[1] This chapter is a product of the Economic Research Forum's Research Initiative on Arab Development (RIAD). The views expressed in this chapter do not necessarily reflect the official positions of the Dubai Economic Council or the Economic Research Forum. An earlier version of the chapter was presented at the research workshop on "Understanding and avoiding the oil curse in the Arab world", Cairo, October 7–8, 2011. We would like to thank Hasan Ersel, our lead discussant, for his insightful comments as well as other participants at the Cairo workshop. We thank Kabbashi Suliman for a careful reading and valuable comments, including a title suggestion with which we agreed. Ibrahim Elbadawi would like to acknowledge the excellent research support by Ms. Dhuha Fadhel of the Dubai Economic Council. Omer Ali would like to thank George Stefanidis and Guillaume Sublet for useful comments and suggestions.

[2] omerali@ucla.edu [3] ielbadawi@dec.org.ae

5.1 Introduction

Relative to the size of their populations, or scale of their economies, oil and other mineral endowed countries (hereafter referred to as resource-rich countries) tend to have large governments, characterized by generous public sector wages and other compensations. For example, during 2000–2007, the top fifteen oil exporting countries received a net annual average per capita income from oil of around (real ppp) US$ 20,000 and spent an equivalent of (real ppp) US$ 3,655 per citizen in the form of public sector wage bill per year. This compares with about US$ 586 real ppp dollars of per capita public sector wage bill for other non-resource developing countries. However, the phenomenon is not confined to resource-rich countries, because advanced industrial economies also have large governments. For example, during the same period, the OECD countries spent more than US$ 2,600 real ppp dollars per citizen on the public sector wage bill (Table 5.1). In these and other open economies, it has been argued that large government expenditures, including on wages and transfers, provide a social insurance function against the risks associated with high economic openness (e.g., Rodrik 1998).

Though resource-rich, especially oil-rich, countries also have large trade/GDP ratios due to the dominance of the resource exports in their economies, we argue in this chapter that the mechanism driving large public sector employment in these countries is entirely different from the one associated with the open but mostly-democratic countries analyzed in the received literature. Instead, in the resource-rich and largely-authoritarian countries, we would argue, redistributive political economy considerations are likely to be the main culprit. The ruling elites in these countries would attempt to avert a revolt by the public that might lead to regime change by using the labor market to transfer a fraction of the resource rents to the latter. However, the elites are not likely to adopt a pure public sector employment strategy, as they might also rely on political repression for pre-empting or quelling incipient or unfolding revolt. In high resource but population-scarce countries, the elites are likely to rely more on expanding public employment and less on political repression. The opposite is likely to happen in moderately endowed but populous countries. The Gulf Cooperation Council (GCC) member countries provide the most notable example of the former, while the other oil-rich but populous Arab economies epitomize the latter. This is clearly borne out by the evidence of Table 5.2, which compares the public sector wage bill and the extent of political repression between the two groups. The median oil

Table 5.1: *Public Wage and Correlates in Resource-rich and other Countries*

Country Grouping	Real Rent Per Capita (PPP $)	Real Wage Per Capita (PPP $)	Physical Integrity	Democracy Gap	Dummy for Regime Change	Polity
Top 15 oil exporting	19,992	3,655	3.9	0.7	0.03	-3.9
Other oil exporting ($\geq$ 150,000 bpd)	3,000	790	3.4	0.5	0.02	0.4
Top mineral exporting ($x \geq 50\%$ of exports)	115	298	4.9	0.3	0.03	4.5
Other mineral exporting ($15 \leq x \leq 50\%$ of exports)	281	343	4.7	0.4	0.04	1.4
OECD	586	2,607	6.6	0.0	0.0	9.3
Other developing countries	170	642	5.1	0.3	0.0	3.4

Notes: Definition of variables: For rent per capita, political repression and real public wage per capita, see notes to Table 5.2; and for the definitions of polity, democracy gap and regime change dummy see Table A.1.

Top 15 oil exporting: GCC , Algeria, Angola, Iran, Iraq, Kazakhstan, Libya, Nigeria, Russia, Venezuela.

Other oil exporting ($\geq$ 150,000 ppd): Argentina, Azerbaijan, Belarus, Brunei, Chad, China, Colombia, Congo, Republic of, Ecuador, Egypt, Equatorial Guinea, Gabon, India, Indonesia, Malaysia, Sudan, Syria, Thailand, Trinidad & Tobago, Vietnam, Yemen.

Top minerals exporting ($x \geq 50$ % of exports): Botswana, Congo, Dem. Rep., Guinea, Jamaica, Namibia, Niger, Sierra Leone, Zambia.

Other minerals exporting ($15 \leq x \leq 50$ % of exports): Armenia, Bolivia, Central African Republic, Cuba, Ghana, Guyana, Jordan, Kyrgyzstan, Mali, Mauritania, Mongolia, Morocco, Papua New Guinea, Peru, South Africa, Suriname, Tanzania, Togo, Ukraine, Uzbekistan.

OECD: Australia, Austria, Belgium, Canada, Chile, Czech, Republic, Denmark, Estonia, Finland, France, Germany, Greece, Hungary, Iceland, Ireland, Israel, Italy, Japan, Korea, Luxembourg, Mexico, Netherlands, New Zealand, Norway, Poland, Portugal, Slovak Republic, Slovenia, Spain, Sweden, Switzerland, Turkey, United Kingdom, United States.

Table 5.2: *Public Sector Employment vs. Political Repression (Annual Average: 2000–2007)*

Country	Rent per Capita	Public Wage Bill per Capita	Index of Political Repression
GCC Countries			
Qatar	40,446	11.9	6.8
Kuwait	25,007	8.1	5.9
United Arab Emirates	15,556	9.2	6.4
Bahrain	8,584	3.5	5.9
Saudi Arabia	8,239	3.9	4.5
Oman	6,975	2.1	7.3
GCC Mean	11,898	6	6.1
Populous Arab Oil Exporters			
Iraq	1,775	0.5	0.3
Algeria	1,563	0.4	3.3
Syria	532	0.1	2.9
Egypt	313	1.9	2.8
Sudan	288	0.1	0.6
Yemen	–	0.1	2.9
Non-GCC Median	532	0.2	2.8

Units: Rent per capita is in thousands of current US dollars.
Public wage bill per capita is in thousands of real PPP dollars.
Notes:
We extend the resource rents database of Collier and Hoeffler (2009), who use data from the World Bank's adjusted savings project. We calculated the rents for each commodity by subtracting the cost from the commodity price. Rents per unit were then multiplied by the amount extracted and summed across the different commodities. Rent per capita was then calculated by dividing rents in current US dollars by population; while real rent per capita was obtained by first converting current rents into real (PPP) values and then dividing by population. Natural resources for which rent data were available are: oil, gas, coal, lignite, bauxite, copper, iron, lead, nickel, phosphate, tin, zinc, silver and gold. The data are described in Hamilton and Clemens (1998) and available from http://lnweb18.worldbank.org/ESSD/envext.nsf/44ByDocName/GreenAccountingAdjusted Net Savings
Political repression is measured by the index of physical integrity rights, which ranges from 0 (most repressive) to 8 (least repressive) and accounts for the incidence of torture, extrajudicial killing, political imprisonment and disappearances that are attributable to the government (Cingranelli–Richards 2008).
Public wage bill per capita is given by total government expenses on compensation of employees (from the IMF World Economic Outlook database) divided by population.

rent per capita in the GCC in 2000–2007 stood at an annual average of US$ 11,898 (in real PPP dollars), which is more than twenty times the average rent for the median country from the populous Arab oil-exporting economies. In a similar vein, the median annual public sector wage bill per capita in the GCC, around US$ 6,000 (in real PPP dollars), was fifteen times the wage for the populous Arab oil-exporting economies. On the other hand, compared to the other group, the GCC countries were found to be far less likely to resort to political repression. On a scale from 0 (most politically repressive) to 8 (least repressive), the median GCC country scored 6.1, while the median populous oil-exporting Arab country scored 2.8.[4]

There is a dearth of empirical work on Middle Eastern labor markets, precipitated by a general unavailability of data. The work that does exist, however, corroborates our observations that: public sector employment is dominant in the labor markets of per capita resource rich countries and is significant but relatively less important in those oil-dependent countries in the Middle East with large populations. Said (1996) documents the relative largess of the public sectors in several Arab countries. In Kuwait, for example, between 1985 and 1992, 91% of nationals were employed by the government, whereas this figure was significantly lower in non-GCC economies: 56.6% in Algeria in 1990, 34% in Egypt in 1992 and 47% in Jordan in 1991 (Table 5.1; Said 1996, pp. 5–6). Shaban et al. (1993) note that for the Gulf countries in general, the government sector remains the only sector that the majority of new national entrants to the labor force would consider and where employment is perceived as a right of the citizen. This perception demonstrates how almost all the national workforce is in some form employed by the government.

In terms of the received theoretical literature, we discuss a few papers that model public sector employment as a means through which a dictatorship redistributes resources in an effort to remain in power. This is by no means an exhaustive review of the literature and we continue to introduce studies that are relevant to specific parts of the models we develop in the following text. Perhaps the study that is closest in spirit to ours, Robinson et al. (2006) present a model of clientelism, in which politicians use natural resource rents to buy support through public sector employment in order to win political contests. In a similar approach to ours, they model patronage (redistribution to supporters) as the offer of employment in the public sector (Robinson et al. 2006, 449). In defense of this assumption, the authors invoke Auty's (2001) claim that "public employment can be a politically appealing way to redistribute rents." Their results agree with our analysis. More specifically, their Proposition 5.3 (i), which states that "a permanent resource boom increases public sector employment and decreases private sector employment," is precisely what we have proposed in this chapter. Gelb et al. (1991) show

[4] For definitions of the rent and public wage per capita variables and a description of the index of political repression, see the notes to Table 5.2.

how a government absorbs workers into the public sector as a result of the political pressure it faces. In their study, however, there is no role for natural resource wealth. While we propose that as natural resource wealth increases, the propensity of a dictatorship to repress decreases, there are also arguments in the literature that propose the opposite relationship between natural resource wealth and repression. Verwimp (2003), for example, outlines a general model of dictatorship, along similar lines to Wintrobe (1990), the outcome of which is a dictatorship that increases repression as its budget constraint is relaxed. However, the focus of his study, Rwanda, falls well below our threshold for "natural resource rich" and the comparative statics of repression with respect to natural resource wealth in this case can be explained by the non-monotonic effects we derive in Section 5.4.

Section 5.2 presents the basic game theoretic model, which is extended in section 5.4. We solve for the game theoretic equilibrium conditions, which will be shown to crucially depend on size of the per capita resource rents. Section 5.3 empirically tests our predictions regarding the role of rents as a determinant of the size of public sector employment, conditional on other controls motivated by the model. Section 5.5 concludes.

5.2 Theory

5.2.1 Basic model

The following model attempts to explain the differences in labor market institutions between GCC economies – those with high natural resource wealth per capita – and other Middle Eastern economies with some natural resource wealth but large populations. We take a political economic approach, arguing that the attempt of non-democratic governments to remain in office is an influential factor in the way labor market institutions develop. More specifically, we focus on the size of the public sector as a metaphor for the extent of redistribution that takes place from state to citizen. Alesina et al. (2000) provide a theoretical model as well as empirical support for why politicians use disguised redistributive policies as opposed to blatant transfers. We argue that governments in the Middle East consider public sector employment as a means to redistribute resources to their citizens, thereby "buying" their political allegiance.

We begin with a situation in which the elite and citizens play a two-stage game. The population of the elite has mass 1 while that of the citizens has mass N (where $N > 1$). There is a public sector in which L of the citizens are employed. They receive a constant, exogenous wage of w

(where $w > 0$).[5] Public sector employment is preferred to employment in the private sector. We capture this feature by assuming that unless a citizen is employed in the public sector, they receive a zero payoff.

The elite determine the size of the public sector – L. Crucially, citizens are randomly assigned public sector jobs. Moreover, they observe the elite's choice of L, so each citizen's expected utility is $\frac{wL}{N}$. Once the elite have chosen L, citizens decide whether or not to revolt. The collective action problem does not arise in this situation because of the symmetry of citizens' preferences. Each citizen privately decides whether or not to revolt. Since citizens are identical, if it makes sense for one citizen to revolt, it makes sense for all. We model the cost and uncertainty associated with challenging the status quo by introducing a probability p. This can be considered as the probability that an attempted revolt succeeds. Other authors have included a "punishment" for citizens in the event of a revolt's failure (Acemoglu and Robinson (2001), for example). Essentially, the role of punishment is to diminish the expected utility from revolt. This same effect can be achieved by decreasing the value of p. For the purposes of this model, it suffices to assume that citizens receive a zero payoff in the event of a revolt's failure. The value of p is fixed in order to restrict the policy instruments through which the elite influence citizens' decision-making to the redistribution of rent in the form of public sector jobs. Moreover, although endogenizing p (for example, making it inversely related to the size of the public sector) may result in a more realistic model, qualitatively, it will say the same as the model does now: namely that higher investments in the size of the public sector reduce the probability of a successful revolt. With p endogenous, the probability of a successful revolt will be reduced continuously over some interval in $[0,1]$, while with a fixed p, the probability of a successful revolt jumps to zero if the elite are able to bribe the citizens and avert a revolt altogether, and is otherwise p. See Section 5.4.2 for an extension in which p is endogenous.

This approach is one way to analyze a phenomenon that has been often documented, but seldom modeled explicitly. For example, Robinson et al. (2006) identify public sector employment as a tool through which politicians gain patronage, noting that "there is a large degree of consensus in the political science literature that patronage refers to the way in which party politicians distribute public jobs or special favors in exchange for electoral support." Alesina et al. (1999) provide empirical evidence of politicians using public sector employment as a redistributive device in

[5] We justify this seemingly restrictive assumption at the end of this section.

US cities. These studies, among others, point to the importance of public sector employment in rent distribution. Yet we have encountered few attempts to model this role explicitly.

Although this is a two-stage model, there is essentially one period in which the interaction between the elite and citizens takes place. At the beginning of the period, the elite receive a natural resource rent of R. This resource must finance the public sector wage bill, wL, and the remaining amount can be consumed by the elite. The timing of employment is not important for our argument. Indeed, we only adopt this framework because of its tractability. A more accurate description of our idea is an infinitely repeated game where a new group of citizens is offered public sector jobs each period (those offered jobs in previous periods keep them). Each citizen then has a positive probability of getting a public sector job at some point in the future and hence his or her decision, whether or not to revolt, will depend wholly on the size of this probability. Framing the problem in this way addresses the issue of why unemployed citizens do not revolt after public employment has been distributed, but it simultaneously complicates the model.

We are now ready to fully describe and solve the game being played by the elite and the representative citizen. In the first stage, the elite choose a value for L. Having observed this choice, citizens decide whether or not to revolt. The payoffs are as follows: in the case of a revolt, the elite get an expected payoff of $(1-p)(R-wL)$, while the citizens get $\frac{pR}{N}$ each, where p is the probability that the revolt succeeds. Notice that there is an assumption that if the revolt is attempted and is successful, then the elite can expect to receive a zero payoff. Moreover, if there is a revolt that is unsuccessful, the citizens can expect a zero profit. In addition, in the latter case, the elite still lose an amount equivalent to the public sector wage bill. This serves to characterize the destabilizing effect of social upheaval. The act of a revolt, per se, results in a loss of income to the elite. This serves as a justification for the elite's preference to avoid a revolt. If there is no revolt, the elite get the balance of the government budget less the public sector wage bill, $R-wL$, and the citizens each get an expected payoff of $\frac{wL}{N}$. The game is described in Figure 5.1.

To solve the game, we look for a subgame perfect equilibrium. Proceeding by backward induction, we first analyze the decision of the citizen whether or not to revolt. It is clear to see that a citizen chooses not to revolt if the following condition holds:

$$EU^c(No\ revolt) = \frac{wL}{N} \geq \frac{pR}{N} = EU^c(Revolt) \quad (1)$$

Although the N term seems redundant at first, its presence on either side of the inequality is important. On the left-hand side, $\frac{L}{N}$ describes the

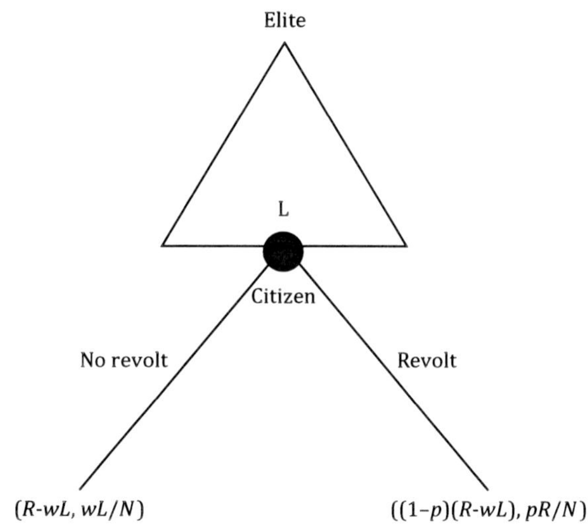

Figure 5.1: The Extensive Form of the Game Between the Elite and Citizens

proportion of the population in the public sector, namely, the importance of the public sector in the economy. On the right-hand side, $\frac{R}{N}$ describes natural resources per capita – the variable we have chosen to distinguish between GCC and non-GCC economies. Through the analysis of the game, we will aim to show how higher natural resources per capita precipitate a swelling of the public sector through the effort of the elite to remain in power and preempt a revolt.

Having established the conditions under which citizens prefer not to revolt (Equation 5.1), we now consider the elite's decision problem. The elite choose the size of the public sector, which subsequently influences the citizens' decision. If citizens do revolt, the elite receive an expected payoff of $(1-p)(R-wL)$ (i.e., the probability that the revolt fails multiplied by their expected consumption). If citizens do not revolt, the elite receive a payoff of $R - wL$, the natural resource rent less the public sector wage bill. The choice of L determines whether or not citizens revolt. Namely, there is a critical level L^* such that Equation (5.1) holds for all $L \geq L^*$. The following identity follows trivially:

$$L^* = \frac{pR}{w} \tag{2}$$

Clearly, if the elite set up a public sector at all, they will choose its size to be L^*. This is because citizens will revolt for all values $L < L^*$. But in that case, since the elite expect $(1-p)(R-wL)$, they might as well choose $L = 0$.

In addition, citizens choose not to revolt for all values $L \geq L^*$, however, since the elite then get a payoff of $R - wL$, which they try to maximize subject to a "no revolution constraint" (Equation 5.1), they will choose $L = L^*$.

Now, the relevant question is whether the elite will set up a public sector at all. It is beneficial for the elite to set one up if the following condition holds:

$$EU^e(L = L^*) = R - wL^* \geq (1 - p)(R) = EU^e(L = 0) \qquad (3)$$

This is true whenever $pR \geq wL^*$. Substituting for the value of L^* from (Equation 5.2), the condition becomes $pR \geq w\frac{pR}{w}$. This inequality always holds weakly and hence the elite weakly prefer to avert a revolt by creating a public sector of size L^*.[6]

Proposition 5.1 *If Equation (5.3) holds, there is a subgame perfect equilibrium in which the elite choose L = L* and citizens do not revolt.*[7] *If Equation (5.3) does not hold, the subgame perfect equilibrium is one in which the elite do not set up a public sector (L = 0) and citizens revolt.*

Having solved the model, we are now in a position to investigate the comparative statics of equilibrium quantities. More specifically, how does the relative size of the public sector change with per capita natural resources? Since Equation (5.3) always holds, the elite always attempt to avert a revolt by choosing $L = L^*$. Using Equation (5.2) and dividing both sides by N, we can get an expression for the relative size of the public sector on the left-hand side and per capita natural resources on the right. It is then straight forward to note that the two quantities are indeed proportional. As the value of natural resources per capita increases, so does the relative size of the public sector in equilibrium. This agrees with our initial observation that public sector employment, and the extent of redistribution more generally, in GCC economies (high $\frac{R}{N}$) is substantially more important than in resource-rich Middle Eastern economies with large populations (low $\frac{R}{N}$).

$$\frac{L^*}{N} = \frac{p}{w}\frac{R}{N} \qquad (4)$$

We need w to remain constant in order to avoid multiplicity of equilibria. The elite would be able to choose between a number of different wage and

6 Introducing risk aversion would make the elite strictly prefer to avert a revolt.

7 Off the equilibrium path, the elite choose L<L*, then citizens revolt, whereas if they choose L>L*, citizens do not revolt.

public sector size pairs that all dissuade citizens from attempting a revolt (increasing one and decreasing the other would keep the citizens' utility unchanged). To motivate this assumption we argue that the public sector wage provides for a minimum standard of living (and hence is bounded below) and if we assume that the specific public sector jobs we are concerned with are predominantly entry level government jobs, then *w* is bounded above. Choosing a fixed level for *w* would then seem less arbitrary.

5.2.2 A Model with Perfect State Repression

A noticeable feature of authoritarian regimes in general, and Middle Eastern dictatorships in particular, is their reliance on a state security apparatus to stifle dissent and repress the opposition. Many economists have modeled dictatorships' use of repression for this purpose (See Wintrobe 1990 and Ali 2009). Nurmikko (2008), for example, studies the repeated interaction between a leader and the army (or state security) in the presence of an opposition that seeks to replace the leader through electoral competition. Similarly to the model presented here, the leader can stay in power through two ways: they could either bribe the opposition (which then redistributes the resources among the population) or they could pay the military to repress the opposition. It is shown that in a collusive equilibrium, the leader pays the army to repress the opposition. The paper differs from our chapter in its comparative statics, which say that the larger the available resources, the more likely it is that the leader uses the army since they would be more willing to remain in power. Although this line of reasoning is convincing for relatively small amounts of natural resources, it simply fails to explain the institutional arrangement in Middle Eastern countries with substantial hydrocarbon reserves. As the value of a dictatorship's natural resource wealth grows, the redistributive policy becomes more attractive as the relative efficiency of repression diminishes in comparison (see Ali 2009). We will argue in the following model that the formation of a security apparatus enables governments to fund smaller public employment programs, reducing the amount of redistribution from state to citizens required for the dictator to remain in power.

In the following model, the broad structure of the game outlined earlier remains unchanged. One minor assumption that is introduced is on the payoff citizens expect to get following a successful revolt.

Whereas previously it was assumed that citizens expect to get a share of the natural resource wealth ($\frac{R}{N}$), we now replace this with the fixed quantity D denoting the payoff in "democracy." There are several reasons why this is a reasonable assumption. First, in the previous case, it was assumed that each citizen knew the amount of natural resource rents the government received and, in addition, knew the population with whom these rents would be shared. In reality, we would not expect the average citizen to be mindful of this information. In the case of Saudi Arabia, for example, the exact amount of natural resource wealth has been treated as a "state secret." D represents how much citizens value democracy. This could mean how much they value the political and civil liberties per se, or how much they expect their income to be in a democracy, or a composite of the two. In either case, it represents an expectation of what democracy will be like for each citizen. Indeed it is this expectation, rather than any objective quantity, that is likely to be the most influential factor in the decision whether or not to revolt. In Burnell and Youngs (2010), empirical evidence is presented showing that although support for democracy is high among citizens of new democracies, satisfaction with the performance of these democracies is low among the same group of citizens. This is clear evidence that citizens' valuation of democracy is distinct from its immediate monetary benefits and justifies the use of an exogenous parameter to describe the expected utility from democracy instead of the payoff each citizen receives when a revolt succeeds.[8]

In the game shown in Figure 5.2, the elite decide whether or not to set up a state security apparatus. If they do set one up, there is no threat of a revolt from citizens and the elite simply consume the remainder of the government budget after having paid for the cost of the set up. We model the cost of setting up the state security apparatus as a fraction, θ, of the natural resource rents.

We justify this modeling strategy as follows. The state security apparatus is given some power by the elite. They are likely to employ a large number of people, to be armed and well organized. Indeed, left alone, they may be a potential threat to the elite themselves. To insure against a coup, the elite essentially bribe the security apparatus. This bribe, however, would need to be proportional to the amount of resource rents that appear each period,

[8] This does not mean that citizens do not value monetary rewards. It could be the case, for example, that citizens value discounted future monetary rewards, which are perceived to be higher under democratic rule.

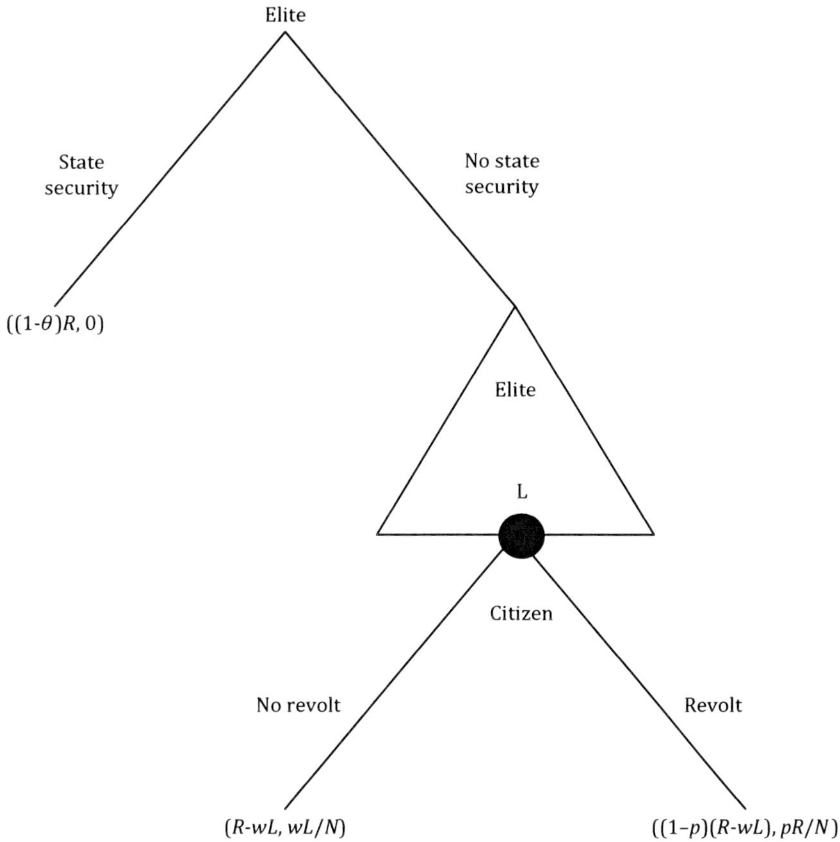

Figure 5.2: The Extensive Form of the Game Between the Elite and Citizens When the Elite Have an Option to Form a State Security Apparatus that Removes the Option of Citizens to Revolt

since the higher the potential booty, the higher the temptation to dislodge the elite. This is why the amount paid to the state security organ is the fraction θ of natural resource rents, R.

If the elite decide not to set up a security apparatus, they are again faced with the option of creating a public sector in which to employ the citizens. They choose the size of this public sector before citizens decide whether or not to revolt. Similar to the model described earlier, the size of the public sector influences the citizen's probability of getting a public sector job and hence, their expected payoff from not revolting.

We proceed to find the subgame perfect equilibria of this game by backward induction. First, we consider the citizens' decision in the final stage of the game. They decide not to mount a revolt when the following condition holds:

$$EU^c(No\ revolt|NSS) = \frac{wL}{N} \geq pD = EU^c(Revolt|NSS) \quad (5)$$

The size of the public sector making citizens weakly prefer no revolt is $L^* = \frac{pDN}{w}$. The elite will never choose $L > L^*$ since their payoff in the event of no revolt is $R - wL < R - wL^*$ for all such L. Otherwise, they will choose $L = 0$ since for all other $L^* > L > 0$, citizens revolt and the elite get a payoff of $(1 - p)(R - wL) < (1 - p)R$, where the right-hand side is the elite's payoff when $L = 0$. Exactly like our previous model, the elite's choice reduces to the following dilemma: either they set up a public sector of size L^* or they do not set up a public sector at all. The elite will choose to set up a public sector when the following expression is true:

$$EU^e(L = L^*|NSS) = R - wL^* \geq (1 - p)(R) = EU^e(L = 0|NSS) \quad (6)$$

Rearranging this equation, we see that the elite will choose to set up a public sector whenever $\frac{R}{N} \geq D$. This expression is more likely to hold as the natural resources per capita becomes higher.

Finally, we study whether the elite decide to set up a state security apparatus or not. First, we will introduce the following parametric assumption:

$$\theta < p \quad (7)$$

This assumption ensures that in the event that Equation (5.6) does not hold and that it is optimal for the elite not to set up public sector employment, they would prefer to avert a revolt by setting up a state security apparatus than simply face a revolt in the hope that it would fail. The only remaining issue to consider is whether the elite will choose the state security route when they can prevent a revolt by redistribution. The elite choose redistribution over the state security route when the following inequality holds:

$$EU^e(L = L^*|NSS) = R - wL^* \geq (1 - \theta)(R) = EU^e(SS) \quad (8)$$

This simplifies into the condition $\frac{R}{N} \geq \frac{pD}{\theta}$. Notice that this condition is more likely to hold the higher are natural resource rents per capita, which is consistent with our explanation and agrees with the observation that GCC governments, those with the highest natural resource rents per capita, consistently choose the redistribution strategy over the state security one.

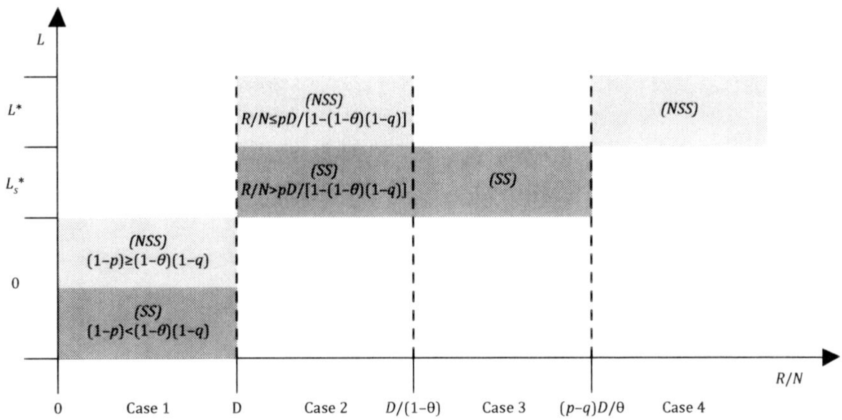

Figure 5.3: The Optimal Policy of the Elite as R/N Increases

Proposition 5.2 *If Equation (5.6) holds, there is a subgame perfect equilibrium in which the elite choose NSS and* $L = L^*$, *and citizens do not revolt.*[9] *If Equation (5.6) does not hold, the subgame perfect equilibrium is such that the elite choose SS.*[10]

We also explore the implications of a model where state repression is imperfect. Namely, state repression fails to extinguish a revolt with some probability. The details of this model can be found in Section 5.4. We derive the main results in Proposition 5.3, which is summarized in Figure 5.3. Below a certain level of natural resources per capita ($\frac{R}{N}$), there is some ambiguity about the optimal policy of the elite. However, under some parametric assumptions, the model is consistent with our theory of how the elite choose their policies in order to remain in power. In case 1, we can assume that $(1-p) < (1-\theta)(1-q)$ and that the elite do set up a state security apparatus but do not employ any citizens in the public sector. In case 2, we can assume that for all levels of natural resources per capita in this region, $\frac{R}{N} > \frac{pD}{1-(1-\theta)(1-q)}$, and that the elite continue to maintain the state security apparatus, however, now they employ some of the citizens in the public sector. In case 3, there is no longer any ambiguity about the policy of the elite and they always set up a state security apparatus and employ L_s^* citizens in the public sector. In case 4, the elite always choose

9 Off the equilibrium path, if the elite choose L<L*, citizens revolt, whereas if they choose L>L*, citizens do not revolt.

10 Off the equilibrium path, the elite choose L=0 and citizens revolt. Moreover, citizens revolt for all values of L<L* and do not revolt for all values of L>L*.

not to set up a state security apparatus and employ a larger proportion of the citizens in the public sector ($L^* > L_s^*$). If these parametric assumptions hold, we can see that below a certain level of natural resources per capita ($\frac{(p-q)D}{\theta}$), the elite choose to deal with a threat of revolution from the citizens by setting up a state security apparatus, whereas above that level, they choose, instead, to redistribute natural resource wealth to the citizens through public sector employment. Moreover, we can see that for very low levels of natural resources per capita ($\frac{R}{N} < D$), the elite do not employ any citizens in the public sector and are forced to face a revolt.

These parametric assumptions enable us to neatly apply the model to Middle Eastern labor markets. We argue that the populous, natural resource-dependent countries fall in cases 1, 2 and 3, whereas the GCC economies fall in case 4. For governments in the former group, it is then optimal to set up a state security apparatus and employ a (relatively) small proportion of citizens in the public sector. This is confirmed by the figures of the proportion of workers in the public sector for Algeria, Egypt, Jordan and Kuwait in Said (1996). Around the early 1990s, the proportion of workers in the public sector in Kuwait, a member of the GCC, was between almost double (compared to Algeria) to almost three times as large (compared to Egypt) as the same proportion in non-GCC economies.

5.3 Data

The theoretical model suggests that elites in resource-rich countries would, under broad conditions, prefer to expand public sector employment in order to avert a revolution. However, they would also prefer to combine carrots with sticks by building a coercive capacity to minimize the potential of successful regime change. Figure 5.4 contains the profile of rents per capita for the major oil and mineral producing regions, which shows the oil-rich, population-scarce GCC group leading the way. Moreover, out of the twenty-six leading oil and mineral exporters, there are fourteen Arab and African countries. These countries, as the model predicts, have some of the largest public sector employment per citizen in the world.[11] However, the advanced OECD countries also tend to have even larger public sector employment. As discussed in Section 5.1, the large public sectors in these

[11] For the GCC countries, rents and wages per capita are based on national populations, which exclude the sizable migrant population in these countries. In 2004, for example, the share of nationals in total population accounts for 80% in Oman, 73% in Saudi Arabia, 62% in Bahrain, 36% in Kuwait, 30% in Qatar and only 19% in the UAE.

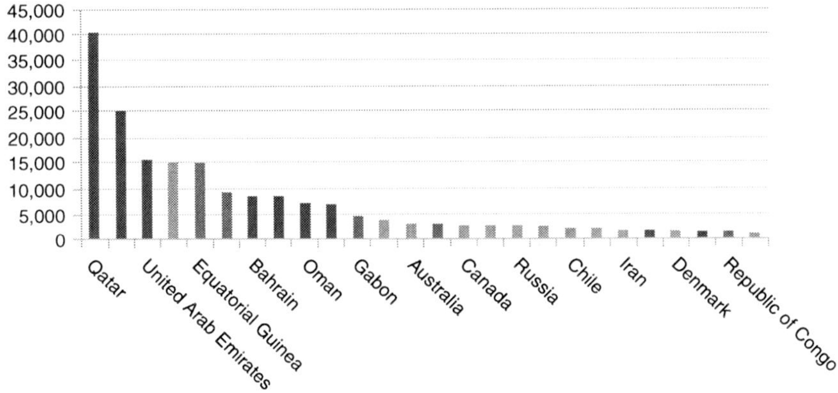

Figure 5.4: Resource Rent Per Capita Across the Globe (In Current US Dollars: Average 2000–2007)
Source: see notes to Table 5.2.

countries might be attributed to their high degree of openness (Rodrik 1998). Moreover, to the extent that public sector wages and other social outlays are influenced by private sector wages, democracy might also be of relevance to this phenomenon, if indirectly. In a follow-up paper, Rodrik (1999) finds that controlling for productivity and the level of development, among other factors, democracies tend to pay higher wages in the manufacturing sector. We will probe further into these issues later in this section in the context of the formal empirical regression analysis.

Though the model is developed in terms of the share of public sector labor force to national population, data on public sector employment is rather limited and is not available for many countries, especially resource-rich ones. Instead, we use on public sector wages and compensation per capita (w_pc). In fact this variable, we would argue, is a more appropriate measure of resource commitment to public sector employment. To see this, we write $w\_pc = \frac{w}{N} = \frac{w}{L_p}\frac{L_p}{N}$, which decomposes public sector wage per capita into two multiplicative components: the average public sector wage and the share of public sector labor to total national population. This accounts for both the wage rate as well as the size of public sector employment relative to total national population. As such, it is a more encompassing measure of how much a share of the rents on public sector employment the elites are willing to spend.

As regards the model's prediction that elites in resource-rich countries also rely on building state security for repressing potential revolts, there is also some supporting evidence to this effect. There is ample evidence on the large shares of the military and security budgets in authoritarian oil and other resource-rich countries on military and security. However, most of this evidence might also reflect other considerations, such as external threats (e.g., Ross 2009). We avoid this potential endogeneity by using a direct measure of government repression, available in the recently developed Cingranelli–Richards dataset (2008). This measure, called *physical integrity rights*, constructs an annual variable that ranges from 0 (most repressive) to 8 (least repressive) and accounts for the incidence of torture, extrajudicial killing, political imprisonment and disappearances that are attributable to the government. According to the physical integrity (PI) variable, the resource-rich economies appear to be among the leading repressive countries (Table 5.1), again consistent with the theory developed in this chapter.

Moreover, most, but certainly not all, resource-rich countries do not extend the franchise, as reflected by their deplorable median polity (Table 5.1). The polity IV index is based on two concepts: "institutionalized democracy" (DEM) and "institutionalized autocracy" (AUT). The DEM score is coded according to four measures of regime characteristics: competitiveness of executive recruitment, openness of executive recruitment, constraints on the chief executive and competitiveness of political participation. These measures, along with regulation of participation, contribute to the AUT score. The polity score (POL) is computed by subtracting the AUT score from the DEM score, resulting in a score from –10 (strongly autocratic) to 10 (strongly democratic).

Furthermore, non-democratic resource-rich countries are among some of the most extreme examples of authoritarianism. To measure this, we construct the democracy gap index (DemocGap), which is a standardized distance from the democracy frontier for country *c* in period *t*:

$$DemocGap_{ct} = \frac{\{Max(Polity_{it}) - Polity_{ct}\}}{\{Max(Polity_{it}) - Min(Polity_{it})\}} \tag{9}$$

This variable ranges between 0 (when a country is exactly at the frontier of polity with the maximum value of 10) and 1 (when a country is exactly at the bottom of the polity scale at the minimum of –10). This index is aimed at capturing the democratic aspirations of the public in the society, which also reflects the extent of sacrifice the elites had to incur in resource-rich countries, in terms of forgone rents in order to satisfy the public without extending the franchise. As the model suggests (Proposition 5.3), a high enough augmented democracy gap (adjusting for the model parameters) that exceeds available rents per capita would lead to a choice of low public employment by the elites, because even if they decide to spend the entire rents on public employment, it still would not avert a revolt. As such, the model predicts a negative association between a wide democracy gap and public sector employment.

For the elites, avoiding successful regime change is the key objective of the game. We measure regime change by a binary variable, which assigns a value of 1 for the regime change year, where regime change in year t is defined by an absolute change of 3 or more on the polity scale relative to the previous year $(t - 1)$. The data suggests that regimes in resource-rich countries do not seem to have faced significantly higher risk of regime change than other non-resource developing countries (Table 5.1). We estimate the probability of regime change, p, as a function of resource rents (R_pc) and the degree of political repression (PolRep: as measured by physical integrity), among other controls:

$$P = P\left(R\_pc, (R\_pc)^2, PolRep, Controls\right) \tag{10}$$

The estimates from this probability model assess whether resource-rich countries are different from other countries in terms of the hazard of regime change due to their being resource-rich (i.e., the marginal impact of R_pc). Moreover, as suggested by the theory in the previous section, the aforementioned model allows for testing for the presence of a non-monotonic effect of resource rents. Specifically, the testable implication of this would be to check if there exists a per capita resource rent threshold below which the ruling elites will prefer not to share the rents in order to avoid a revolt and possible loss of power. Also, the model allows testing whether or not political repression reduces the likelihood of successful regime change. On the other controls, we follow Morrison (2009), who analyzes the impact of non-tax revenue per capita on regime change, conditional on a host of controls deemed to be important additional explanatory variables. These controls include past regime instability,[12] initial polity, initial level and growth of per capita income, ethno-linguistic fractionalization[13] and the share of urban population.

We estimate the aforementioned probability model using a panel data set of 3,249 country years in 1970–2007 and covering 187 countries.

[12] This is accounted for by the number of times the country experiences regime changes in the past up to the year/period in question.

[13] The index of ethnic fractionalization is a measure of latent conflict in a society and is given by, where is the share of group in country. This index gives the probability that two randomly selected individuals from a population belong to two different groups.

Appendix Table A.1 provides the definitions of the variables used in the estimation of the probability model and the subsequent one for the public wage; Table A.3 contains country-by-country averages for public wages and resource rents; and Table A.2 presents the overall summary statistics for the regression variables. In applying the fixed-effects estimator to models with qualitative dependent variables based on panel data, the conditional logit model seems to be the preferred choice. However, it requires strict exogeneity of the regressors, and stationarity over time. Because these conditions are frequently violated in economic data, the random-effects estimator is an attractive alternative. In the panel data context, the probit model is computationally tractable, while the logit model is not. The only limitation of probit models is that they require normal distributions for all unobserved components, a feature that may characterize most unobserved, random components, but that is notoriously absent in cases where variables are truncated (e.g., prices must be positive).[14] In light of this, we choose the discrete choice random-effects probit for estimating the probability of regime change.

The results are contained in Table 5.3. We find that both resource rents and political repression are robustly and negatively associated with the likelihood of regime change, thus confirming the two key predictions of our hypothesized regime change model. Moreover, the resource rents effect was actually accounted for by those countries belonging to the top quartile of the resource-rich countries in the sample (regressions 3 and 4 of Table 5.3). According to this finding, only those ruling elites in countries generating real (PPP) per capita rents at US$ 317 or more will be willing to use rents as an instrument for fending off an impending revolt. This threshold is slightly above the real rents per capita of Egypt in 2007; hence, even populous oil-exporting economies like Syria, Algeria and Iraq are well above this threshold (Table 5.2). We also find that regimes with a history of political instability are likely to experience higher probability of regime change, while initial democracy tends to reduce the likelihood of regime change. These aforementioned four pivotal findings are robustly significant across the four regressions of the table. However, there is only weak evidence that economic growth tends to reduce the likelihood of regime change (regressions 1 of Table 5.3). Moreover, we also find the other controls analyzed by Morrison (2009), including the share of urban population,

[14] See Elbadawi, Schmidt-Hebbel and Soto (2011) for a more detailed discussion of the econometric properties of logit and probit estimators in panel data estimation for the case of qualitative dependent variables.

Table 5.3: *Probability of Regime Change (Full Sample)*

	REP [1]		REP [2]		REP [3]		REP [4]	
Variable	Coefficient	Z	Coefficient	Z	Coefficient	Z	Coefficient	Z
Real GDP_ pc growth rate	−0.68	−1.11	-	-	-	-	-	-
Log Rent_pc	−0.09	−2.75	−0.1	−3.42	-	-	-	-
Past instability	0.11	3.88	0.11	3.75	0.11	3.63	0.11	3.61
Polity (−1)	−0.04	−5.2	−0.05	−5.96	−0.05	−5.87	−0.05	−5.91
Political repression	−0.04	−1.76	−0.04	−1.96	−0.05	−2.05	−0.05	−2.02
Share of urban POP	0	−0.54	-	-	-	-	-	-
ELF	−0.18	−0.85	-	-	-	-	-	-
POP density	0	−0.53	-	-	-	-	-	-
Dum_25*log Rent_pc	-	-	-	-	0.01	0.13	-	-
Dum_25 −75*log Rent_pc	-	-	-	-	−0.04	−0.64	-	-
Dum_75*log Rent_pc	-	-	-	-	−0.07	−1.87	−0.06	−3.21
Constant	−1.26	−4.99	−1.39	−7.52	−1.63	−6.16	−1.73	−12.84
Observations	3218	-	3249	-	3249	-	3249	-
LR statistic	3.45	-	3.9	-	4.16	-	4.65	-
Value	0.032	-	0.024	-	0.021	-	0.015	-
Log Likelihood	−492.15	-	−499.04	-	−498.15	-	−499.62	-

Note: ELF = Ethno-linguistic fractionalization, REP=Random effect probit.

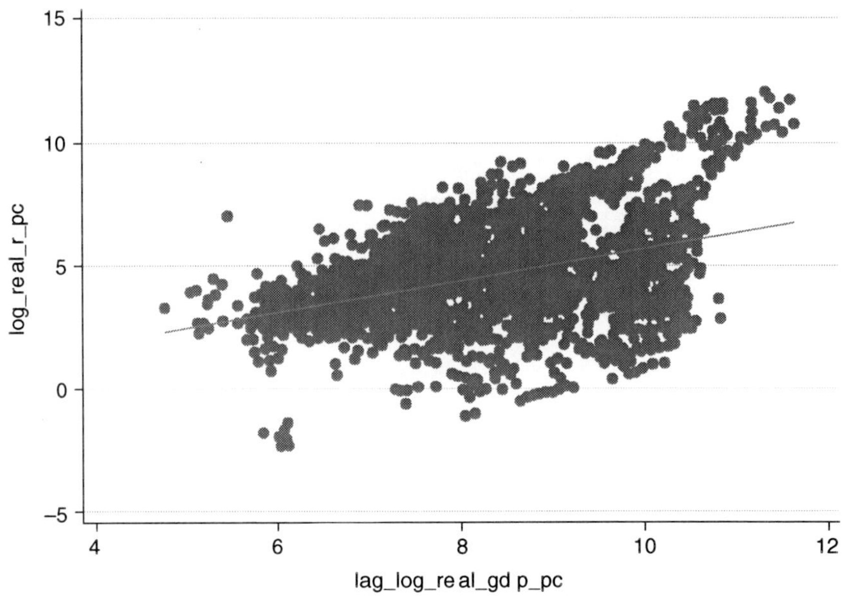

Figure 5.5: The Income and Resource Rents Scatter

population density and ethno-linguistic fractionalization to be uniformly insignificant (regression 1); hence, we dropped them from subsequent regressions. We also dropped the lagged income per capita variable because it is very highly correlated with rents (Figure 5.5).

The random probit results of Table 5.3 hold a remarkable degree of similarity when the model is estimated using simple pooled probit regressions (reported in Table A.4). A more important robustness check is provided by the results of Table 5.4, which is based on a restricted sample of only those countries with an initial polity score of 6 or less. This results in a smaller sample of only 1,942 country years derived from 87 countries over 1970–2007. Despite the much smaller sample and the consequent reduced variability for key variables, the results of Table 5.4 broadly corroborate the findings of the full sample regressions. However, when per capita growth is included, rents become only weakly significant (regression 1), but regain significance when the former is removed from subsequent regressions.[15] However, the political repression variable becomes less significant in all four regressions of Table 5.4. Though the restricted sample is more

[15] It appears that, like lagged per capita income, growth and rents might be multi-collinear as well.

Table 5.4: *Probability of Regime Change (Restricted Sample: Initial Polity ≤6)*

	REP [1]		REP [2]		REP [3]		REP [4]	
Variable	Coefficient	*Z*	Coefficient	*Z*	Coefficient	*Z*	Coefficient	*Z*
Real GDP_ pc Growth Rate	-1.41	-1.96	-	-	-	-	-	-
Log Rent_pc	-0.05	-1.18	-0.08	-2.36	-	-	-	-
Past instability	0.1	2.53	0.09	2.31	0.09	2.25	0.08	2.17
Polity (-1)	-0.04	-3.51	-0.04	-3.98	-0.04	-3.97	-0.04	-3.93
Political repression	-0.02	-0.79	-0.03	-1.32	-0.03	-1.36	-0.03	-1.35
Share of urban POP	0	-0.54	-	-	-	-	-	-
ELF	-0.07	-0.32	-	-	-	-	-	-
POP density	0	1.11	-	-	-	-	-	-
Dum_25*log Rent_pc	-	-	-	-	-0.02	-0.19	-	-
Dum_25 -75*log Rent_pc	-	-	-	-	-0.05	-0.71	-	-
Dum_75*log Rent_pc	-	-	-	-	-0.06	-1.43	-0.04	-2.1
Constant	-1.55	-5.16	-1.4	-6.2	-1.53	-4.71	-1.69	-10.38
Observations	1923	-	1942	-	1942	-	1942	-
LR statistic	0.81	-	1.83	-	1.95	-	2.34	-
Value	0.185	-	0.088	-	0.082	-	0.063	-
Log likelihood	-388.18	-	-396.62	-	-396.45	-	-397.23	-

Note: ELF = Ethno-linguistic fractionalization, REP=Random effect probit.

consistent with the theoretical model since it is premised on the absence of full functioning democracy, we prefer the results of the full sample. This is because the results from the latter are broadly similar to those of the restricted one. In addition, it has the advantage of accounting for the theoretically important political repression effect. Therefore, we use the full sample regressions to construct the predicted probability of regime change (p^H). We interpret this predicted probability as an empirical proxy for the theoretical term $(p - q)$ in the equilibrium solution of the game. This constructed variable will be among the controls in the empirical estimation of the rent per capita effect on public sector wages and compensation per capita.

5.3.1 Hypothesis Testing

As the theoretical model suggests, assuming that the elites in resource-rich societies decide to set up a public employment sector $(L > 0)$ in order to avert a revolution, the condition that they also build a state security system (SS) and employ a proportion of the citizen $L_S = L_S^*$ (compared to the alternative of an expanded public sector employment system without state security $(L = L^*)$) is given by: $\frac{(p-q)D}{c} - \frac{R}{N} \geq 0$ (from Equation 5.10), where c is the cost of setting up state security. Hence, we can state the following probability function:

$$\Pr(L \geq L_S*) = 1 - G\left\{\frac{(p-q)D}{c} - \frac{R}{N}\right\} \tag{11}$$

Equation (5.11) motivates the following empirical model:

$$ln(w\_pc) = a_1 + a_2 ln(R\_pc) - a_3 p^H - a_4 DemocGap + a_5 ln(GDP\_pc) \tag{12}$$

In Equation (5.12), we proxy the share of public sector labor to total population $(\frac{L}{N})$ by public wage bill per capita (w_pc); $p - q$ by the predicted probability of regime change conditional on political repression (P^H); the democratic aspiration of the public (D) by the democracy gap index (DemocGap); and the cost of setting up a state security apparatus (c) by real GDP per capita. Finally, we enter per capita variables in logarithmic form to bring their scales to closer conformity with those of P^H and DemocGap, which also allows for direct estimation of elasticities.

Moreover, Proposition 5.3 of Section 5.4 suggests a non-monotonic effect for R_pc, where there exists a threshold level of rents per capita, beyond which R_pc is positively associated with w_pc. To account for this, we also estimate the following extended version of Equation (5.12).

$$ln(w\_pc) = a_1 - a_2 ln(R\_pc) + a_3 ln(R\_pc)^2 - a_4 p^H - a_5 DemocGap \\ + a_6 ln(GDP\_pc) \quad (13)$$

We estimate Equations (5.12) and (5.13) using a global panel dataset composed of 1,109 country years covering 87 countries during 1970–2007, including most of the oil- and mineral-exporting countries.[16] The results of the fixed-effects regressions strongly corroborate the theoretical model predictions (Table 5.5). Starting with the estimation of Equation (5.12) (regression 1 of Table 5.5), we find rents per capita to be positively associated with public wages per capita, where an increase of one dollar (in real ppp terms) of rents per national would lead to an increase of about 10 cents in the public sector wage bill per national. Moreover, other model predictions are also borne out by the evidence, where high conditional probability of regime change or democracy gap exerts a negative influence on the public wage rate. As discussed earlier, to the extent that private wage setting is likely to influence wages in the public sector, especially in advanced industrial democracies, our results on the negative effect of the democracy gap index also coheres with other evidence in the literature, most notably that of Rodrik (1998). He finds that, controlling for productivity, income levels and other controls, there is a robust and statistically significant association between democracy and wages in manufacturing. He argues that this is attributed to the fact that democracy entails greater freedom of association and collective bargaining. Moreover, the process of political participation, competition and contestation may increase the bargaining power of workers.

However, regressions 2 and 3 of the table suggest that there is no direct non-monotonic effect for rent per capita. In regression 2, the square of rents is highly insignificant, while the estimated coefficients due to the first quartile, middle 50% and top quartile of the rents variable are all significant and have similar orders of magnitude. Moreover, these estimates are not statistically significantly different from the estimated linear effect of regression 1. This clearly suggests that the non-monotonic effect predicted by the model is rather indirect and operates through the probability of regime change channel. The results of the full sample regressions are strongly

[16] However, due to multi-collinearity between rents and lagged GDP per capita, the results worsen considerably when both variables are included in the same regression; hence, we dropped the latter from all of the wage regressions.

Table 5.5: *Public Sector Wages Per Capita (Log_pw_pc) (Full Sample)*

	FE [1]		FE [2]		FE [3]		FE [4]		FE [5]	
Variable	Coeff	*t*	Coeff	*t*	Coeff	*t*	Coeff	*t*	Coeff	*t*
Log Rent_ pc	0.1	4.61	0.07	1.34	-	-	0.06	2.86	0.08	3.24
Sqr Log Rent_ pc	-	-	0	0.48	-	-	-	-	-	-
Prob regime change	-2.93	-3.76	-2.98	-3.79	-3.02	-3.85	-3.09	-4.01	-3.16	-3.63
DemocGap	-0.22	-2.42	-0.22	-2.37	-0.21	-2.26	-0.13	-1.34	-0.18	-1.73
Trade openness (OPEN)	-	-	-	-	-	-	0	4.74	-	-
Adjusted trade openness (*Adj_OPEN*)	-	-	-	-	-	-	-	-	-0.08	-2.13
Dum_25*log Rent_pc	-	-	-	-	0.1	2.95	-	-	-	-
Dum_25 -75*log Rent_pc	-	-	-	-	0.09	3.49	-	-	-	-
Dum_75*log Rent_pc	-	-	-	-	0.1	4.43	-	-	-	-
Constant	5.69	49.19	5.74	36.22	5.7	44.09	5.51	45.67	5.83	44.38
No of observations	1109	-	1109	-	1109	-	1109	-	935	-
R-Squared	0.0739	-	0.0741	-	0.0748	-	0.0941	-	0.067	-
Adj *R*-Squared	-	-	-	-	-	-	-	-	-	-
F	26.73	-	20.09	-	16.22	-	26.08	-	14.95	-

Note: FE = Fixed effect, Coeff = Coefficient.

corroborated by those of the restricted sample (Table 5.6), with only one exception regarding the democracy gap (DemoGap), which became insignificant. As in the case of the probability of regime change estimation, this is due to the reduced variability and the smaller sample, which dropped to 701 country years covering only 65 countries, for which initial polity was equal to or less than 6.

5.3.2 Robustness Checks

Resource-rich economies, especially oil-rich ones, tend to be highly open economies due to their heavy reliance on resource exports. In a widely quoted paper, Dani Rodrik (1999) finds strong empirical evidence linking large governments, including large public sector employment, to exposure to trade openness. He also provides a theoretical explanation for this phenomenon in that large government expenditures in open economies act as a social insurance against the ensuing external risks associated with openness. Therefore, to the extent that resource-rich countries tend to be open economies, the observed large public sector employment in these countries might be the product of openness, rather than their being resource rich. In this case, the theoretical underpinnings behind the large public sectors in resource-rich economies will be very different from that of this chapter's model. We check the robustness of our results against this competing theory by adding openness to the fixed-effects regressions of Tables 5.5 and 5.6. We use two measures of openness: one is total trade ratio to GDP (OPEN); and the other is an adjusted measure (*Adj_OPEN*), which is derived as a residual from the OLS regression of $(Exports + Imports)/GDP$ on log of population, log of area, landlocked and oil exporter dummies. The results are reported in regressions 4 and 5 of the two tables. For all four regressions, the rent per capita effect remains highly significant and positively associated with public wages. Moreover, the direct trade openness variable (OPEN) was also positively and highly significantly associated with public wage as predicted by Rodrik's theory. However, this measure is problematic because it does not adjust for country characteristics. Instead, the adjusted openness measure (*Adj_OPEN*) was significant but implausibly negative. In view of the fact that the adjusted openness measure is more appropriate for assessing the policy impact of openness, the aforementioned results make clear that the observed phenomena of large public sectors in resource-rich economies appears to be better explained by the "authoritarian bargain" as modeled in

Table 5.6: *Public Sector Wages per Capita (Log_pw_pc) (Restricted Sample: Initial Polity ≤6)*

	FE [1]		FE [2]		FE [3]		FE [4]		FE [5]	
Variable	Coeff	*t*	Coeff	*t*	Coeff	*t*	Coeff	*t*	Coeff	*t*
Log Rent_ pc	0.09	3.46	0.13	1.76	-	-	0.05	1.85	0.08	2.71
Sqr Log Rent_ pc	-	-	0	-0.59	-	-	-	-	-	-
Prob regime change	-3.16	-3.52	-3.11	-3.44	-3.23	-3.57	-3.49	-3.93	-3.45	-3.54
DemocGap	-0.16	-1.39	-0.17	-1.45	-0.15	-1.26	-0.04	-0.35	-0.09	-0.7
Trade openness (OPEN)	-	-	-	-	-	-	0.01	4.63	-	-
Adjusted trade openness (*Adj_OPEN*)	-	-	-	-	-	-	-	-	-0.13	-2.83
Dum_25*log Rent_pc	-	-	-	-	0.07	1.57	-	-	-	-
Dum_25 -75*log Rent_pc	-	-	-	-	0.07	2.19	-	-	-	-
Dum_75*log Rent_pc	-	-	-	-	0.09	3.04	-	-	-	-
Constant	5.43	37.05	5.34	25.28	5.49	32.69	5.18	33.77	5.49	34.29
No of observations	701	-	701	-	701	-	701	-	632	-
R-Squared	0.0682	-	0.0687	-	0.0697	-	0.0987	-	0.0736	-
Adj *R*-Squared	-	-	-	-	-	-	-	-	-	-
F	15.49	-	11.69	-	9.48	-	17.35	-	11.24	-

Note: FE = Fixed effect, Coeff = Coefficient.

this chapter than by the other competing argument of social insurance against the risks associated with trade openness.

5.4 Extensions

5.4.1 A Model with Imperfect State Repression

In the model discussed earlier in this chapter, we assumed that state repression (setting up the state security apparatus) was perfectly effective. Namely, that once a state security apparatus is set up, citizens simply cannot revolt. Governments, as a result, do not institute public sector employment. However, this is far from the reality. Public sector employment in the Middle East, amongst non-GCC economies (those with the most active state security apparatuses), is substantial. In Algeria, Egypt and Jordan, public sector employment as a percentage of total employment was 56.6% (1990), 34% (1992) and 47% (1991) respectively (Said 1996).

In the following framework, we consider the case in which the state security does not repress citizens perfectly. Instead, setting up the state security apparatus makes it more difficult for citizens to effect regime change. We model the effect of the state security apparatus as a reduction in the probability that a revolt succeeds. Letting $q < p$, we now have the situation in Figure 5.6.

The setup is analogous to the one in Section 5.2. However, now when the elite set up a state security apparatus, they must again choose a level of public employment, L_s, in order to avert a revolt by citizens. The level of public employment necessary to dissuade citizens from revolt in the case in which the elite do not set up a state security apparatus is the same as before: $L^* = \frac{pDN}{w}$. The elite prefer to set up a public sector with this level of employment when Equation (5.6) holds. If the elite do set up a security apparatus, the minimum level of public sector employment necessary to avert a revolt is $L_s^* = \frac{qDN}{w}$. Their expected utility from choosing the state security route and employing L_s^* citizens in the public sector is then: $EU^e(L_s = L_s^* | SS) = (1 - \theta)R - qDN$. They prefer this course of action to the one where they set up a state security apparatus but do not avert a revolt ($L_s = 0$) whenever the following condition, analogous to Equation (5.6), holds.

$$\frac{R}{N} \geq \frac{D}{(1 - \theta)} \tag{14}$$

Notice that whenever Equation (5.9) holds, so does Equation (5.6). Suppose Equation (5.9) holds (i.e., the elite always choose to avoid a revolt

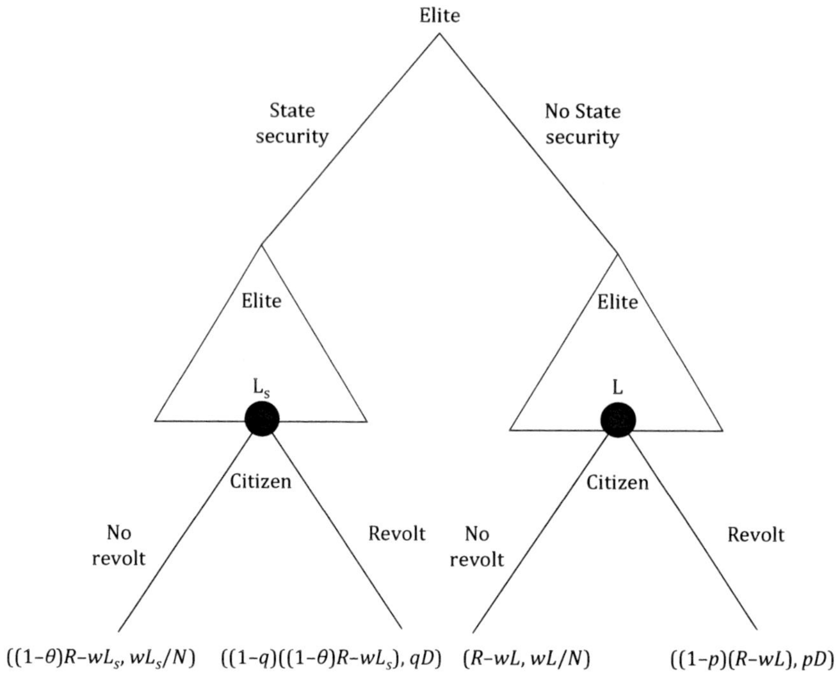

Figure 5.6: The Extensive Form of the Game Between the Elite and Citizens with Imperfect Repression

by distributing public employment). The question is now whether the elite will set up a state security apparatus or not. They will choose the state security path whenever the $EU^e(L_s = L_s^*|SS) \geq EU^e(L = L^*|NSS)$, which simplifies to the following condition:

$$\frac{(p-q)D}{\theta} \geq \frac{R}{N} \tag{15}$$

Clearly, this condition is more likely to hold when θ is smaller. Indeed, θ is the proportion of natural resources that the elite sacrifice in order to set up the state security apparatus. The smaller it is, the cheaper is the cost of repression. More interesting, however, is the fact that, ceteris paribus, the condition is more likely to hold when the ratio $\frac{R}{N}$ is smaller. This condition, together with Equation (5.9), represents the constraint that the ratio $\frac{R}{N}$ needs to satisfy for the elite to prefer to set up a state security apparatus and employ a proportion of the citizens in the public sector.

$$\frac{(p-q)}{\theta}D \geq \frac{R}{N} \geq \frac{1}{1-\theta}D \tag{16}$$

This range of values for $\frac{R}{N}$ is non-empty whenever $\frac{1-\theta}{\theta} > p - q$. Whenever $\frac{(p-q)}{\theta}D < \frac{R}{N}$ and $\frac{R}{N} \geq D$, the elite prefer to set up a large public sector in the absence of a state security apparatus. The equilibria of this game crucially depend on the value that the ratio $\frac{R}{N}$ takes in addition to a number of parametric relationships. The following proposition summarizes these conditions.

Proposition 5.3 *The subgame perfect equilibria of this game depend on the size of* $\frac{R}{N}$.

$$\text{Case 1}: 0 < \frac{R}{N} < D:$$

The elite choose (SS, $L_s = 0$*) whenever* $(1-\theta)(1-q) > (1-p)$. *Citizens choose to revolt.*

$$\text{Case 2}: D \leq \frac{R}{N} < \frac{D}{1-\theta}:$$

The elite choose (SS, $L_s = 0$*) whenever* $\frac{R}{N} \leq \frac{pD}{1-(1-\theta)(1-q)}$. *Citizens then choose to revolt. Otherwise, the elite choose (NSS,* $L = L^*$*) and citizens choose not to revolt.*

$$\text{Case 3}: \frac{D}{1-\theta} \leq \frac{R}{N} < \frac{(p-q)D}{\theta}:$$

The elite choose (SS, $L_s = L_s^*$*) and citizens choose not to revolt.*

$$\text{Case 4}: \frac{R}{N} > \frac{(p-q)D}{\theta}:$$

The elite choose (NSS, $L = L^*$*) and citizens choose not to revolt.*

5.4.2 A Model With Endogenous Probability Of Revolt Success

5.4.2.1 No Repression

The setup is the same as before. The elite choose L, how many citizens to employ in the public sector at the beginning of the game. After observing this, citizens decide whether or not to revolt. If no revolt takes place, citizens are randomly picked to fill the vacant government jobs created by the elite, who have to finance the wage bill. If citizens revolt, they succeed with probability[17] $p(L) = 1 - \frac{L}{N}$. This probability is decreasing in the number of government jobs created by the elite and, in the limit when $L = N$ (everyone is employed), it is zero. This simple specification captures the idea that the larger the public sector, the more citizens will be employed, the higher the probability of employment in the public sector and hence, the less likely it will be that individual citizens will decide to participate in anti-regime activity and, therefore, the less likely it will be that agitation would reach the critical mass necessary to effect regime change. On the other hand, if the public sector is small, more citizens would participate in anti-regime activities (since their expected benefit from acquiescence is now much smaller) and that would make regime change more likely to emerge. In effect, this formulation captures the idea that there is a collective action problem[18] to be solved amongst citizens. The higher is the expected gain from keeping the regime intact, the less likely it will be that the collective action problem is solved successfully.

An alternative interpretation of this model answers the question of why unemployed citizens do not organize amongst themselves and revolt against both employed citizens and the elite.[19] If we assume that all unemployed citizens always choose to revolt against the regime, then it would be plausible to assume that the probability that their revolt succeeds depends on their numbers relative to the number of employed citizens. In that case, the functional form for p specified earlier captures the idea that the larger the pool of unemployed citizens, the higher the probability of revolt success.

Let D be the citizens' expected utility from democracy. If we normalize the payoff to citizens in the event of a failed revolt to 0, we have the following quantities:

[17] This section was born from the insightful comments we received on a previous draft from Hasan Ersel.

[18] Olson (1965) proposed that a group of individuals that attempts a collective action (one that requires input from many individuals), whose benefits are non-excludable, must contend with the tendency of individual members to "free ride" on the contribution of others. The example of agitating for regime change is an almost ideal illustration of this tension. Since it is practically impossible to verify which citizens contributed to the effort to topple a government, all citizens benefit from a change in the political system. However, participating in anti-regime activities is costly since it involves a certain risk of imprisonment and injury. However, the participation of many individuals is required for regime change to be successful.

[19] We are, again, thankful to Hasan Ersel for raising this point.

$$EU^C(Revolt|L) = p(L)D = \frac{(N-L)}{N}D$$
$$EU^C(No\ \ Revolt|L) = \frac{L}{N}w$$
$$EU^E(Revolt|L) = (1-p(L))(R-wL) = \frac{L}{N}(R-wL)$$
$$EU^E(No\ \ Revolt|L) = R - wL$$

Citizens do not revolt if $EU^C(No\ \ Revolt|L) \geq EU^C(Revolt|L)$. The following expressions show this condition with a general D and when $D = \frac{R}{N}$.

$$\frac{L}{(N-L)} \geq \frac{D}{w}$$
$$\frac{L}{(N-L)} \geq \frac{R}{N}\frac{1}{w}, \qquad if\ D = R/N$$

These conditions have some interesting properties. First, for any values of w, R and N, there always exists an L that ensures that the condition holds. Namely, if the government offers employment to a sufficiently large number of citizens, it would be able to avert a revolt. Second, for higher ratios of natural resources per capita, the number of citizens that the government needs to employ in the public sector in order to avert a revolt is higher. If we let L^* be the size of the public sector that only just averts a revolt, then we see that this quantity is increasing in R/N. Crucially, the minimum fraction of citizens employed in the public sector required to avert a revolt, L^*/N, is increasing in natural resources per capita whenever we assume that $D = \frac{R}{N}$.

$$\frac{L^*}{N} = \frac{D}{D+w} = \frac{\frac{R}{N}}{\frac{R}{N}+w}$$

Next, we ask whether or not the elite will choose this minimum public sector size that ensures no revolt. The elite's payoff when they choose to avert a revolt is shown in Equation (5.17), while Equation (5.18) shows their payoff when they choose another $L < L^*$ and citizens revolt in response. We then find the value of L that maximizes their payoff, call it L^R and compare $EU^E(L^*)$ with $EU^E(L^R)$.

$$EU^E(L^*) = (L^*|No\ Revolt) = R - wL^* = R - \frac{wDN}{D+w} = R - \frac{wR}{\frac{R}{N}+w} = \frac{R^2}{R+wN} \tag{17}$$

$$EU^E(L|Revolt) = (R - wL)\frac{L}{N} \tag{18}$$

$$\frac{d}{dL}EU^E(L|Revolt)\bigg|_{L=L^R} = \frac{R}{N} - 2\frac{wL}{N}\bigg|_{L=L^R} = 0$$

$$L^R = \frac{R}{2w}$$

Assumption 1 $\frac{R}{N} < w$

Notice that D does not enter this expression. If we let $D = \frac{R}{N}$, however, then we can see that $L^R < L^*$ if and only if $\frac{R}{N} < w$, which is the interesting case. If, on the other hand, $L^R > L^*$, a government can always choose L^* and consume the balance of natural resource rents less the wage bill, since choosing L^R will always result in less consumption: the wage bill is higher, and it is uncertain whether or not the elites will manage to hold on to power. Suppose $L^R < L^*$ and consider the elite's decision whether to implement L^* or L^R,

$$EU^E(Revolt|L^R) = \left(R - wL^R\right)\frac{L^R}{N} = \left(R - w\frac{R}{2w}\right)\frac{R}{2wN} = \frac{R}{2}\frac{R}{2wN} = \frac{R^2}{4wN}$$

Then, the elite choose L^* whenever $\frac{R^2}{R+wN} \geq \frac{R^2}{4wN}$, which is the same as $\frac{R}{N} \leq 3w$. But since we have already assumed that $\frac{R}{N} < w$, this is always satisfied. This gives rise to the following proposition.

Proposition 5.4 *Whenever $\frac{R}{N} < w$, the elite prefer to choose $L = L^*$ and citizens choose not to revolt.*

Proposition 5.4 shows that for any values of the exogenous parameters w, R and N that satisfy assumption 1, the elite always prefer to avert a revolt and the relative size of the public sector they choose to set up with respect to the total population (L^*/N) is increasing in natural resources per capita (R/N).

5.4.2.2 Perfect Repression

Now consider the case in which the elite have an option to effectively repress demands for democratization at some cost. Formally, there is an action O that the elite can choose that costs a fixed amount ϕ and results in their remaining in power and consuming the remainder of the natural resources $(R - \phi)$. To solve the model, we now need to consider whether the elite would prefer to choose O instead of employing citizens in the public sector. The elite's utility from choosing to repress the citizens is:

$$EU^E(O) = R - \varphi$$

Suppose $\frac{R}{N} < w$, then we know, by Equation (5.4), that the elite would choose $L = L^*$ if they do decide to employ citizens in the public sector. The relevant quantities for comparison to solve for the elite's optimal choice are then $EU^E(O)$ and $EU^E(L^*)$. The following condition illustrates when the former quantity is greater than the latter.

$$EU^E(O) > EU^E(L^*) \quad \Leftrightarrow R - \varphi > \frac{R^2}{R + wN} \quad \Leftrightarrow R - \frac{R^2}{R + wN} > \varphi$$

First, we assume that ϕ is positive. Let $\phi^*(R) = R - \frac{R^2}{R+wN}$ be the threshold value of ϕ below which it always pays for the elite to repress a revolt by citizens. The larger the value of $\phi^*(R)$, the more likely it is that the elite choose repression. The cost of repression per capita, $\phi^*(R)/N$, can be expressed as follows:

$$\frac{\varphi^*(R)}{N} = \frac{R}{N} - \frac{\left(\frac{R}{N}\right)^2}{\left(\frac{R}{N}\right) + w}$$

Using this expression, we can study how the threshold value of the cost of repression per capita, $\phi^*(R)/N$, changes with per capita resources R/N.

$$\frac{d\frac{\varphi(R)}{N}}{d\frac{R}{N}} = 1 + \frac{\frac{R}{N}}{\frac{R}{N} + w}\left(\frac{\frac{R}{N}}{\frac{R}{N} + w} - 2\right)$$

It can be shown that $\frac{d\frac{\phi(R)}{N}}{d\frac{R}{N}}$ is always positive, which means that higher resources per capita make it more likely that the government chooses repression.

5.5 Summary and Conclusion

In the preceding text, we proposed an explanation for the widely documented difference in labor market institutions between resource-dependent countries with high per capita rents (as a result of small populations) and those with large populations. With reference to the Middle East's authoritarian regimes, we identified the strategic interaction between government and citizens as the driving factor behind this difference. We argued that authoritarian regimes that rule over small populations with access to substantial natural resources have a policy tool that is simply unavailable to other authoritarian governments. In their effort to remain in power, these governments have the wherewithal to redistribute enough resources to their populations to effectively remove the incentive to revolt. In addition, we argue that the public sector is the mechanism of choice for governments to effect this redistribution. Public sector jobs essentially become funnels, channeling income to the citizens of the country.

In Section 5.2, we set up the basic model – a game theoretic environment where the government (elite) and citizens play a two-period sequential game. The only available policy instrument the elite have is the size of the public sector labor force. We assume there is no private sector employment. After observing this choice, citizens can stage a revolt, which succeeds with a certain probability p. The elite always prefer to set up public sector employment (Equation 5.3). What emerges is that the relative size of the public sector workforce set up by the elite is proportional to the value of natural resources per capita in the economy. In Section 5.2.2, we enriched the modeling environment by introducing another policy alternative for the elite – they can repress the population of citizens by paying a proportion θ of the natural resource rents they receive. We also slightly alter the setup by making the citizens' expected value in democracy an exogenous parameter. In this new model, we again derive the required relationship between natural resources per capita and the size of the public sector labor force. Specifically, there is a level, D, of natural resources per capita such that the elite set up public sector employment for all levels of natural resources per capita above D and choose to repress the population of citizens for all levels of natural resources per capita below it.

Finally, we relax the assumption that repression is perfectly effective. Now, choosing repression does not guarantee that the elite are unchallenged but, instead, decreases the size of the public sector needed to be set up by the elite in order to avoid a revolt. It is this final setup that yields our most interesting results. We find that only for some parameter values do we get a monotonic relationship between the size of the public sector workforce and natural resources per capita. For low levels of natural resources per capita, the government sets up a state security organ and public sector employment is kept low. Beyond a threshold level, the optimal policy for the elite becomes setting up a large public sector work force in the absence of a state security organ.

In Section 5.3 of the chapter, we empirically investigate some of the claims made in the preceding sections. We use public sector wages and compensation per capita as a measure of the extent of redistribution, denoted in the model by the relative size of the public sector work force.

This latter measure is unavailable for most countries of interest and, in addition, we argue that public sector wages and compensation per capita are a more appropriate measure for what we are attempting to quantify which is the share of the rents that the elites are willing to spend on public sector employment. We first construct the predicted probability of regime change, which, according to our theory, should be negatively associated with public wage bill. This variable is based on a behavioral probit regression estimation that accounts for rents and the degree of political repression. We make use of the recently developed physical integrity variable (PI) due to the Cingranelli–Richards dataset (2008) to measure government repression. According to this measure, the populous resource-rich countries are among the leading repressors. We find that both resource rents and political repression are negatively and robustly associated with the probability of regime change. However, the resource rents effect is non-monotonic, where ruling elites will only use rents to avert a potential revolt when the rents are in excess of about (real PPP) US$ 317 per capita. Instead, in societies with lower rents than this threshold, elites are likely to exclusively resort to political repression as a one track strategy. This is consistent with the evidence that populous resource-rich countries tend to be highly repressive.

More importantly, our main theoretical result regarding the positive impact of rents on public sector wages was very strongly corroborated by the empirical findings. Using a global panel dataset composed of 1,109 country years covering 128 countries, we estimate fixed-effects regressions and find that a one dollar increase in rents per capita leads to around a 10 cents rise in the public wage per capita. Moreover, as the model predicts, we find that regime change and democracy gap have had negative and significant impacts on public wages, suggesting that, other things equal, public wages are likely to be smaller in resource-rich countries with extreme autocracy or when the ruling elites are faced with a high probability of regime change. However, we do not find support for a direct non-monotonic rent effect on wages, though there exists an indirect non-monotonic rent effect through the probability of regime change channel.

Although the results of the theoretical model hold with only natural resources in the government budget, there may be an argument for including an explicit private sector in the model, with the elite's income becoming $R + Y$, where Y is the elite's income from production. First, the private sector is more important than the natural resource sector in non-GCC economies. Moreover, the "non-lootable" nature of the private sector may act as a disincentive to revolt as well as a disincentive for using repression.

Whereas presently, the payoff from private sector employment (zero) is the same as the payoff following a failed revolt, in the presence of an explicit private sector, the payoff in the former case could be positive, making revolt more costly and perhaps explaining the lack of an association between regime change and resource rents. It is uncertain, at this stage, how much of the analysis this will change. So far, the model is very simple, which lends to empirical testing and this extension could be the work of future research.

APPENDIX A: Data Description and Summary

Table A.1: *Definition of Variables*

Public Wage Bill per Capita	General Government Expense, Compensation of Employees (ppp) per Capita
Incidence of regime change	If the absolute value of [polity (t) – polity (t – 1)] is equal to 3 or more, the variable regime change in year t = 1, while if it is less than 3, the variable regime change in year t = 0
(Predicted) probability of regime change	Predicted using probit model in table 3
Past instability	Past regime instability in year t = Sum of the variables regime change for all years T that are less than t
Rent_pc	Resource rent per capita (ppp)
Index of political repression	Political repression is measured by the index of physical integrity rights, which ranges from 0 (most repressive) to 8 (least repressive) and accounts for the incidence of torture, extrajudicial killing, political imprisonment and disappearances that are attributable to the government (Cingranelli–Richards dataset 2008).
Real GDP_pc growth rate	log Real GDP per capita ppp (t) – log Real GDP per capita ppp (t - 1)
Real GDP_pc(-1)	Open-period lagged Real GDP per capita ppp
Initial polity	Polity in year 1970; the scale ranges from +10 (strongly democratic) to –10 (strongly autocratic); See: *Polity IV Project: Dataset Users' Manual*
Democracy gap	Constructed index based on polity score = (10 –(polity)/20)
Trade openness (OPEN)	Openness in current prices, % in current prices
Adjusted trade openness (Adj_OPEN)	Constructed as the residuals of reg (exports + imports)/ GDP on lnapop (log of average population),

Public Wage Bill per Capita	General Government Expense, Compensation of Employees (ppp) per Capita
	landlock, lnland (log of area), oilx (dummy for oil exporters)
Share of urban POP	People living in urban areas as defined by national statistical offices, see: World Bank database
POP density	People per sq. km of land area, see: World Bank database
ELF (ethno-linguistic fractionalization)	ELF indices, Philip Roeder (we used ELF 85), see Ethnolinguistic Fractionalization.doc

Table A2: *Data Summary (1970–2007)*

	Mean	Median	Minimum	Maximum
Public wage bill per capita	1197	469	0.16	14141
Incidence of regime change	0.04	0.03	0	0.29
(Predicted) probability of regime change	0.04	0.03	0.003	0.17
Past instability	1.04	0.92	0	7.08
Rent_pc	1687.37	100.42	1.02	62623.83
Index of political repression	5.14	5.31	0.32	8
Real GDP_pc growth rate	0.003	0.01	-0.1	0.12
Real GDP_pc(-1)	9030	4941	381	62386
Initial polity	-1.29	-5	-10	10
Democracy gap	0.45	0.47	0	1
Trade openness (OPEN)	0.83	0.76	0.17	3.49
Adjusted trade openness (Adj_OPEN)	0.01	-0.09	-0.84	3.2
Share of urban POP	48.7	48.6	6	100
POP density	228.3	53	1.3	12597
ELF (ethno-linguistic fractionalization)	0.46	0.47	0	0.98

Table A3: *Public Wage and Resource Rents by Country (1970–2007): In Real PPP Dollars*

Country	Rent per Capita	Public Wage Bill per Capita	Country	Rent per Capita	Public Wage Bill per Capita	Country	Rent per Capita	Public Wage Bill per Capita
Afghanistan	32	44	Cote d'Ivoire		149	Kazakhstan	1,884	413
Angola	1,512	316	Cameroon	169	127	Kenya	20	147
Albania	140	233	Congo, Republic of	708	197	Kyrgyzstan	16	207
United Arab Emirates	62,624	11,151	Colombia	247	482	Cambodia	269	-
Argentina	475	1,129	Comoros	14	149	Kiribati	-	-
Armenia	19	433	Cape Verde	5	769	St. Kitts & Nevis	-	1,775
Antigua and Barbuda	-	1,746	Costa Rica	104	1,037	Korea, Republic of	7	-
Australia	908	-	Cuba	126	-	Kuwait	37,279	10,518
Austria	144	3,239	Cyprus	5	4,981	Laos	95	99
Azerbaijan	1,836	239	Czech Republic	65	1,280	Lebanon	2	808
Burundi	23	56	Djibouti	13	724	Liberia	256	-
Belgium	10	-	Dominica	4	705	Libya	4,849	1,335
Benin	44	71	Denmark	284	-	St. Lucia	-	1,071
Burkina Faso	30	-	Dominican Republic	83	342	Sri Lanka	27	-
Bangladesh	25	-	Algeria	993	366	Lesotho	35	279
Bulgaria	113	378	Ecuador	801	-	Lithuania	184	-
Bahrain	12,246	4,461	Egypt	332	1,632	Luxembourg	-	4,966

Bahamas	35	1,839	Eritrea	18	97	Latvia	216	-
Bosnia and Herzegovina	51	598	Spain	34	2,617	Macao	-	-
Belarus	202	1,719	Estonia	193	-	Morocco	31	-
Belize	62	832	Ethiopia	27	50	Moldova	3	265
Bermuda	-	-	Finland	320	-	Madagascar	19	42
Bolivia	381	-	Fiji	62	-	Maldives	-	456
Brazil	240	906	France	41	3,652	Mexico	722	-
Barbados	100	2,602	Micronesia, Fed. Sts.	-	-	Marshall Islands	-	-
Brunei	25,166	7,094	Gabon	3,767	596	Macedonia	16	-
Bhutan	401	-	United Kingdom	470	-	Mali	17	56
Botswana	121	907	Georgia	15	228	Malta	-	-
Central African Republic	37	42	Germany	56	2,288	Montenegro	-	-
Canada	1,408	3,620	Ghana	51	101	Mongolia	173	-
Switzerland	21	-	Guinea	100	122	Mozambique	27	69
Chile	661	934	Gambia, The	16	63	Mauritania	174	204
China	75	-	Guinea-Bissau	25	65	Mauritius	1	1,131
Malawi	16	61	Equatorial Guinea	2,468	113	Panama	21	445
Malaysia	840	-	Greece	40	-	Peru	218	356
Namibia	-	790	Grenada	-	1,497	Philippines	56	234

(*continued*)

Table A3 (*continued*)

Country	Rent per Capita	Public Wage Bill per Capita	Country	Rent per Capita	Public Wage Bill per Capita	Country	Rent per Capita	Public Wage Bill per Capita
Niger	16	35	Guatemala	118	210	Palau	-	-
Nigeria	331	-	Guyana	285	260	Papua New Guinea	563	-
Nicaragua	60	110	Hong Kong	-	-	Poland	99	1,138
Netherlands	578	3,030	Honduras	141	291	Puerto Rico	-	-
Norway	3,527	-	Croatia	178	1,414	Portugal	54	2,257
Nepal	46	53	Haiti	22	56	Paraguay	109	-
New Zealand	440	-	Hungary	254	-	Qatar	61,819	14,141
Oman	5,572	1,737	Indonesia	295	192	Slovenia	50	2,083
Pakistan	56	-	India	47	-	Sweden	244	-
Romania	323	1,289	Ireland	77	2,958	Swaziland	171	783
Russia	3,637	664	Iran	1,904	669	Seychelles	-	1,799
Rwanda	31	50	Iraq	1,897	533	Syria	455	104
Saudi Arabia	7,331	3,473	Iceland	-	-	Chad	102	73
Sudan	136	53	Italy	53	2,881	Togo	60	48
Senegal	23	101	Jamaica	433	-	Thailand	103	551
Singapore	-	-	Jordan	21	271	Tajikistan	7	85
Solomon Islands	154	-	Japan	17	-	Turkmenistan	5,343	0
Sierra Leone	47	92	Serbia	78	753	Timor-Leste	-	-
El Salvador	50	-	Sao Tome and Principe	-	351	Tonga	7	-
Somalia	21	-	Suriname	646	1,085	Trinidad &Tobago	3,965	1,129

Vanuatu	50	-	Slovak Republic	48	37	Tunisia	271	875
Samoa	95	-	Uzbekistan	891	-	Turkey	27	573
Yemen	-	105	St.Vincent & Grenadines	-	577	Tanzania	34	29
South Africa	176	854	Venezuela	1,874	-	Uganda	43	49
Congo, Dem. Rep.	41	15	Vietnam	155	-	Ukraine	225	760
Zambia	229	108	-	-	-	Uruguay	46	573
Zimbabwe	9	8	-	-	-	United States	549	4,165

Table A4: *Probability of Regime Change (Full sample)*

	PP [1]		PP [2]		PP [3]		PP [4]	
Variable	Coefficient	Z	Coefficient	Z	Coefficient	Z	Coefficient	Z
Real GDP_pc growth rate	−0.7	−1.2	-	-	-	-	-	-
Log rent_pc	−0.08	−2.92	−0.09	−3.68	-	-	-	-
Past instability	0.13	5.61	0.13	5.6	0.13	5.53	0.12	5.47
Polity (−1)	−0.04	−5.31	−0.04	−6.21	−0.04	−6.09	−0.04	−6.13
Political repression	−0.03	−1.68	−0.04	−1.83	−0.04	−1.92	−0.04	−1.89
Share of urban POP	0	−0.6	-	-	-	-	-	-
ELF (ethno-linguistic fractionalization)	−0.18	−1.01	-	-	-	-	-	-
POP density	0	−0.65	-	-	-	-	-	-
Dum_25*Log Rent_pc	-	-	-	-	0.01	0.06	-	-
Dum_25 – 75*Log Rent_pc	-	-	-	-	−0.04	−0.75	-	-
Dum_75*Log Rent_pc	-	-	-	-	−0.06	−1.97	−0.05	−3.36
Constant	−1.3	-6.02	−1.43	-9.05	−1.64	-6.94	−1.74	-14.82
Observations	3218		3249		3249		3249	
LR χ^2	94.91	-	95.93	-	97.44	-	94	-
Prob $> \chi^2$	0	-	0	-	0	-	0	-
Pseudo R^2	0.088	-	0.087	-	0.089	-	0.086	-

Note: PP = pooled probit.

References

Acemoglu, D. (2006). A Simple Model of Inefficient Institutions. *Scandinavian Journal of Economics* 108(4): 515–546.

Acemogu, D., and J. Robinson. (2001). A Theory of Political Transitions. *The American Economic Review* 91(4): 938–963.

Alesina, A., R. Baqir, and W. Easterly. (2000). Redistributive Public Employment. *Journal of Urban Economics* 48(2): 219–241.

Ali, O. A Theory of Dictatorship. Unpublished MSc dissertation, Department of Economics, Coventry, UK: University of Warwick.

Auty, R. (2001). *Resource Abundance and Economic Development*. Oxford, UK: Oxford University Press.

Burnell, P., and R. Youngs. (2010). *New Challenges to Democratization*. Florence: Taylor and Francis.

Cingranelli, D. L., and D. L. Richards. (2008). Cingranelli-Richards (CIRI) Human Rights Dataset 2008 (Available from www.humanrightsdata.org.)

Collier, P., and A. Hoeffler. (2009). Testing the Neocon Agenda: Democracy in Resource-Rich Societies. *European Economic Review* 53(3): 293–308.

Elbadawi, I., K. Schmidt-Hebbel, and R. Soto. (2011). Why do Countries have Fiscal Rules? Paper presented at the Catholic University of Chile Conference on "Economic Policy in Emerging Economies", Santiago, Chile, October 27–28, 2011.

Gelb, A., J. B., Knight, and R. H. Sabot. (1991). Public Sector Employment, Rent Seeking and Economic Growth. *The Economic Journal* 101(408): 1186–1199.

Hamilton, K. and M. Clemens. (1998). Genuine Saving Rates in Developing Countries. World Bank, Washington D.C. (available from http://lnweb18.worldbank.org/ESSD/envext.nsf/44ByDocName/ GreenAccountingAdjusted Net Savings).

Morrison, K. (2009). Oil, Nontax Revenue, and Regime Stability. *International Organization* 63(1): 107–138.

Nurmikko, S. (2008). Survival of Political Leadership. Economics Discussion Papers, Department of Economics, University of Essex.

Olson, M. (1965). *The Logic of Collective Action*, Massachusetts, USA, Harvard University Press.

Rodrik, D. (1998). Why Do More Open Economies Have Bigger Governments? *Journal of Political Economy* 106(5): 997–1032.

Rodrik, D. (1999). Democracies Pay Higher Wages. *Quarterly Journal of Economics* 114 (3): 707–738

Robinson, J. A., R. Torvik, and T. Verdier. (2006). Political Foundations of the Resource Curse. *Journal of Development Economics* 79(2): 447–468

Ross, M. (2009). Oil and Democracy Revisited. Unpublished mimeo, UCLA Department of Political Science, Los Angeles, CA

Said, M. (1996). Public Sector Employment and Labor Markets in Arab Countries: Recent Developments and Policy Implications. Working Papers 9630, Economic Research Forum

Shaban, R., R. Assaad, and S. Qudsi. (1993). Labour Markets in Arab Countries: A Survey. First Annual Conference on Development Economics, Initiative to Encourage Economic Research in the MENA, Cairo June 4–6

Verwimp, P. (2003). The Political Economy of Coffee, Dictatorship and Genocide. *European Journal of Political Economy* 19(2): 161–181.

Wintrobe, R. (1990). The Tinpot and the Totalitarian: An Economic Theory of Dictatorship. *American Political Science Review* 84(3): 849–872

6

The Oil Curse and Labor Markets

The Case of Saudi Arabia

Hend Al-Sheikh* and S. Nuri Erbas*

ABSTRACT

Oil revenues enable the government to afford generous wages and expansive employment policies in Saudi Arabia. The availability of low-cost foreign labor combined with a rapidly growing working age population result in a large disparity between public and private sector salaries. This segmentation skews Saudi worker preferences for public sector employment and increases their reservation wages for private sector employment, resulting in high unemployment. Government initiatives such as those entailed by the 2011 royal decrees aiming at increasing public sector employment and compensations would further exacerbate unemployment in the long run. The main manifestation of the oil curse in Saudi Arabia is thus through labor market segmentation and the persistently high unemployment rate.

JEL Classification: J11, J24, J31, J38, J82

Keywords: Oil curse, rentier state, Saudi Arabia, labor markets, segmentation, reservation wage, public sector employment, unemployment

6.1 Introduction

Oil endowment facilitated unprecedented economic and social development in Saudi Arabia and the Gulf region. Since the 1970s, oil income has

* The views expressed in this chapter are the personal views of the authors and do not represent the views or the official positions of IPA or IMF. We thank Dr. Ibrahim Elbadawi for his comprehensive comments. The authors are solely responsible for any remaining errors.

enabled economic prosperity through large spending on infrastructure, investment in human capital and provision of a wide range of social services. It has also generated positive externalities for growth across the region by increasing trade and cooperation, and contributed to an impressive rise in human development indicators.[1]

However, abundant oil wealth also has a negative impact on economic development. The conjecture of a negative correlation between oil endowment and economic growth is broadly called the oil curse. This correlation has been researched widely in the literature, with particular focus on two main sources of the curse. One source is the loss of competitiveness due to the pegging of domestic currencies to the US dollar, the reserve currency on which oil is priced globally. Although the expediency of pegging enables stable inflation and interest rates, it reduces non-oil export competitiveness. The Gulf Cooperation Council (GCC) countries' currencies are pegged to the dollar. In those countries, loss of non-oil export competitiveness curbs non-oil sector growth and diversification, which cannot be compensated for by other means (technology, know-how, export sophistication) (Elbadawi and Gelb 2010). Low economic diversification limits private sector investment opportunities and causes private investment to become dependent on the stimulus of government expenditure. Moreover, oil price volatility reflects on overall macroeconomic instability, while pegging disables active monetary policy, a fundamental macroeconomic management tool. In addition, foreign capital inflows from oil exports create inflationary pressures.[2] Even when domestic currency is not pegged, large demand for oil, and therefore for the oil-exporting country's currency, results in currency appreciation (the Dutch Disease).

[1] According to the World Bank, from 1980 to 2009, Saudi Arabia's overall human development index rose from 65 to 76; life expectancy increased from 62 to 74 years; infant mortality declined from 66 per 1,000 live births to 16 and education expenditure rose from 3% of GNI to 7%.

[2] An oil-rich country and the country whose currency is used for the peg may also have conflicting monetary policy requirements in different historical episodes. For example, when oil prices rose to very high levels in 2007–2008, the GCC countries faced unprecedented inflationary pressures and needed to follow a contractionary monetary policy, whereas the United States was compelled to follow an expansionary monetary policy in the face of the breaking financial crisis. It has been argued that more flexible exchange rate regimes can enhance GCC trade competitiveness and stimulate economic diversification and growth, especially if the intended GCC monetary union becomes a reality (Abed, et al. 2005; Khan 2009; Nandwa 2011).

The other main source of the oil curse is the ill effects of rent seeking and the political largesse oil income enables resulting in inefficient use of the resource, particularly in the absence of strong institutions and checks and balances for the management of oil income (Elbadawi and Soto this volume; Collier and Hoeffler 2008). In that context, a fundamental impact of the oil curse is on labor markets, which is the main focus of this chapter in the case of Saudi Arabia. The source of the oil curse on Saudi Arabia's labor markets is the politically expedient largesse that oil income enables by making affordable financing of high public sector employment and wages. Combined with availability of low-cost foreign labor and a rapidly growing working age population, large disparity between government and private sector salaries skews Saudi worker preferences for government sector employment, increases their reservation wages for private sector employment and results in high unemployment.

The chapter is divided as follows. In Section 6.2, we make some observations on the oil curse in relation to institutional weaknesses. In Section 6.3, we review the underlying causes of labor market distortions in Saudi Arabia that can be traced to the oil curse, including past policies and new initiatives. In Section 6.4, we document the salient facts about Saudi Arabia's labor markets that result in high unemployment among citizens. In the light of this, in Section 6.5, we evaluate the possible impact of the 2011 Royal Decrees on labor markets. Those decrees include measures directly related to persistent and high Saudi unemployment. We compare the new measures to past labor market policies and argue that they are not likely to decrease long-term unemployment and may even exacerbate the existing labor market distortions. In Section 6.6, we conclude with some observations on labor market policies that can correct labor market distortions and increase employment of citizens.

6.2 The Oil Curse and Institutions

Institutional weaknesses may have a greater impact on long-term economic performance of oil exporters than the Dutch Disease. Many studies have found a negative correlation between oil endowment and growth.[3] A common finding of most studies is that resource curse cannot be attributed

[3] Recent literature includes Ali and Elbadawi (2012), Anshasy and Bradley (2009), Arezki and van der Ploeg (2010), Elbadawi and Gelb (2010), Elbadawi and Soto (2011),

to resource availability but rather to the way resources are exploited and managed. The point is that governments should be concerned with the institutional arrangements that ensure sound policies for resource exploitation, revenue management and investment, in order to prevent political largesse, rent seeking, waste and corruption. Great benefits can be reaped through developing institutions that ensure firm rules for resource management, including transparency and decentralized decision-making independent of political interference. With strong institutional arrangements for efficient resource use, revenue management and investment, oil endowment need not have a negative impact on economic diversification and growth.[4]

6.2.1 Institutional Weaknesses

The institutional approach to the oil curse falls in the purview of the broad and well-documented hypothesis that economic development is positively correlated with strong institutions.[5] Some advanced industrial countries, including United Kingdom, Norway and Holland, have become oil exporters by utilizing advanced technologies for oil extraction (deep sea drilling, shale mining). They already had strong institutions that had evolved through their history of industrialization without disproportionate dependency on resource endowments. Established democratic regimes, good governance and public accountability have enabled those countries to manage their oil resources efficiently and without creating rent-dependent societies. However, in countries where abundant oil resources were much more easily accessible with less advanced and less costly technologies, including Saudi Arabia and the other GCC countries, democratic regimes and public accountability had not taken root historically before oil became a globally demanded commodity. Those countries were initially less developed, less skilled and much poorer, with no appreciable industry and industrial work discipline, but they quite suddenly became major exporters that dominate global oil production and exports at present.

Frankel (2010–2011), Hausmann and Rigobon (2002), OECD (2010), Schmidt-Hebbel (this volume), van der Ploeg (2007, 2009, 2010), van der Ploeg and Poelhekke (2009).

4 An OECD report (2010) underlines: "Weather resources are a blessing or a curse depends on several factors: Is resource income consumed or saved? Is it invested in high or low return enterprises? What percentage of the population benefitted? To what extent the government has been transparent and responsible in managing the resources?"

5 A seminal reference from new institutional economics is by North (1994). Rodrik et al. (2002) provide extensive evidence that the role of institutions in growth performance is more significant than geography and trade.

If a country that has already established good governance comes upon oil, then the natural resource endowment is managed well and evidently, it is not a curse on economic performance. However, if a country that has historically less developed institutions suddenly faces extraordinary demand for its resource endowment with the emergence of oil-based energy production, then oil wealth becomes a curse that has a negative impact on economic growth. That outcome does not come as a surprise, especially if all that the country has to do is take a pick to the sand and pump the oil that is there as a result of the country's primordial geology, its history that has placed it in that particular geography and the initial state of its institutions and governance quality whose historical evolution does not have much to do with oil wealth. Ironically, before the combustion engine, oil might have been a curse because it contaminated precious water resources.

It appears that developing countries without appreciable resource endowments have made greater strides not only in growth performance, but also in democratization, openness and public accountability during the last half century.[6] Among other things, this phenomenon reflects the emergence of competitive working and middle classes that inevitably vie for political power and obtain it in varying degrees. Without resource rent, those classes cannot be pacified and co-opted, primarily because they pay the taxes that support political power. As the maxim goes, there is no taxation without representation and no representation without taxation. However, along with the oil windfall, rent distribution falls in the hands of existing political elites that run it. Oil income enables the inherited *ancien régimes* to perpetuate themselves and maintain the *status quo* by pacifying emerging working and middle classes through rent distribution, as expedient for the elite (Ali and Elbadawi in this volume). As much as oil is a curse on development when indigenous institutions are weak, oil may also be a curse on indigenous institutional development, not to mention external support for *ancien régimes* motivated by secure access to oil.

6.2.2 Weak and Strong Institutions and Disincentives

In the case of Nigeria, Sala-i Martin and Subramanian (2003) propose redistribution of oil income through per capita transfers to citizens in order

6 Collier and Hoeffler (2008) offer some evidence that democracy appears to enhance economic growth except in the presence of substantial natural resources. They underline that democracy (elections) and resource rents can have a deleterious impact on economic performance because such rents undermine the development of checks and balances.

to prevent corruption, waste and political patronage.[7] This proposal is similar to the arrangement that exists in Alaska where each US citizen and resident of the state for at least one year receives a dividend check out of the interest earned on oil royalties deposited into an account called the Alaska Permanent Fund, created in 1976 after oil was discovered (Hartzok 2004).[8]

Variants of such funds already exist in most GCC countries, notably including the legally established national oil funds in the United Arab Emirates and Kuwait, although returns from fund assets are not distributed to citizens, but reinvested for future generations. However, operations of those funds remain opaque, an issue addressed internationally by the Santiago Summit in 2008. In contrast to Alaska, Norway and Chile, opaqueness (including closely guarded size of accumulated wealth) reflects institutional weaknesses, which points to a large measure of political arbitrariness – or at least the possibility of it – in using accumulated national wealth, including soft budget constraints and general lack of well-defined long-term budgeting rules.[9] Social and political pressures on budgets typically become manifest in the form of ad hoc wage increases granted to government sector workers, more government sector hiring and increased expenditure on various subsidies and transfers (we will discuss this point in detail in the case of 2011 Royal Decrees in Saudi Arabia).[10]

However, it should be noted that in Kuwait, strictly binding and explicit legal restrictions apply to the oil fund: A legally mandated share of oil export receipts must be transferred to the fund and, in order to make a transfer to the government, an act of Parliament is necessary (a large transfer was made only once in the history of Kuwait's fund to repair the damage of the first

[7] The classical argument on the taxation side is the efficiency and simplicity of lump sum or head tax, especially when more complex income tax schemes are difficult to administer and open doors to evasion and corruption. Head tax is regressive and creates inequalities. However, such concerns may be addressed through redistributive schemes financed by the portion of oil income retained by the government, as in Alaska.

[8] The law mandates that 25% of oil income be invested in the Fund and the rest be spent on public goods and services (education, social services and infrastructure).

[9] See Schmidt-Hebbel's (this volume) comprehensive review of the wealth funds of Chile and Norway and the evolution of transparent rules and oversight that apply to those funds.

[10] Such pressures are not confined to oil funds, but they also emerge in other countries where government revenue is short and politicians placate social aversion to increased taxation by raiding accumulated social security funds on the expedient presumption that one day they can be replenished. But votes are counted before that day comes, if it ever does. The result is passing on the burden of taxation to future generations or debasing social security obligations by inflation tax. For the same reasons, bankrupt pension plans in local governments are not uncommon.

Gulf war). In Saudi Arabia, there is no national oil fund. The Saudi Arabia Monetary Agency holds reserves accumulated from balance of payments surpluses and independent organizations also hold large reserves (Table 6.1). The operations of the latter remain rather opaque.

Oil funds can minimize waste and corruption, provided they are managed transparently and independently of political influence. Nevertheless,

Table 6.1: *Saudi Arabia: Summary Economic Indicators*

	2008	2009	Provisional 2010	2003–2010 Averages
	Oil Sector			
Proven reserves (billion barrels)	264	265	–	–
Oil export price (US$/barrel)	95	59	76	61
	Real Sector			
	(Percent growth)			
GDP	4.2	0.2	3.8	4.0
Oil	4.3	-7.8	2.2	3.2
Non-oil	4.3	3.5	4.9	4.5
Private sector	4.8	2.7	4.2	4.8
Government sector	2.4	5.2	6.3	3.2
	(Percent share in total)			
GDP	100	100	100	100
Oil	30	28	27	31
Non-oil	70	72	73	69
Private sector	47	48	48	46
Government sector	23	24	24	23
	External Sector			
	(In US$ billions)			
Merchandise exports (CIF)	313	192	240	200
Oil	281	163	207	175
Percent share in total	*90*	*85*	*86*	*88*
Non-oil	32	29	33	24
Percent share in total	*10*	*15*	*14*	*12*
Official foreign assets (net)	503	474	520	318
Saudi Arabia Monetary Agency	438	405	440	263
Government institutions and independent organizations	66	69	79	56

Sources: Saudi Arabian authorities, IMF, British Petroleum and SIMIRAD, and authors' calculations.

Table 6.2: *Saudi Arabia: Fiscal Indicators (2003-2010 Averages)*

	In percent of					
	GDP	Non-oil GDP	Total Revenue	Total Expenditure	Current Expenditure	Oil Exports
Total revenue	46	94	100			
Oil	40	82	87			77
Non-oil	6	12	13			
Total expenditure	34	69		100		69
Current	26	53		78	100	54
of which:						
Wages and salaries	14	28	33	41	53	29
Subsidies	1	2	2	3	4	2
Capital	7	16	19	22		16

Sources: Saudi Arabian authorities, IMF, British Petroleum and SIMIRAD, and authors' calculations.

they can also impel creation of rent-based societies and curb work effort, even in countries with well-developed institutions and legally binding checks and balances. In Alaska, it is acknowledged that annual dividends have created resistance to taxation even for the purposes of maintaining a balanced budget. Moreover, as we document here in the case of Saudi Arabia, rent transfers may create an endowment effect with a deleterious impact on employment and may serve as a disincentive for workers to acquire marketable skills (Ali and Elbadawi, this volume).

6.2.3 Conduits of Oil Rent Transfers: Budgets and Public Sector Employment

Without an active monetary policy because of currency pegging, the remaining tool for macroeconomic management in the GCC countries is fiscal policy. Heavy reliance on oil revenue for budgetary financing is where the oil curse becomes much more visible because budget is the main conduit of transferring oil rents to citizens and the main channel of transfer is the government sector wage bill.[11]

[11] In Saudi Arabia, the budgetary wage bill absorbs more than 50% of current expenditure and around 30% of oil exports (Table 6.2).

Reliance on oil income for budgetary revenue tends to soften budget constraints and bloat public expenditure. In the absence of significant taxation to raise non-oil revenues, citizens' claims for more expenditure (increased public sector salaries, transfers and subsidies) acquire a greater measure of legitimacy and forcefulness when oil prices rise. Evidently, without taxation, there is a large constituency for more expenditure financed by higher oil rent but a small constituency for levying taxes to finance more expenditure. Budgetary expenditure tends to rise when oil income is high, resulting in procyclical fiscal policy. But much of the increased expenditure becomes vested as a permanent part of the budget (in particular, the government sector wage bill) and cannot be decreased when oil income is low. Ali and Elbadawi (this volume) present strong evidence that per capita oil income is positively correlated with wage income, with a quarter of a one-dollar increase in oil income going to the public sector wage bill.

Furthermore, using public sector wages and benefits, transfers and subsidies (below-market energy prices and subsidized housing loans and utilities) as conduits of transferring oil income to citizens creates price distortions and attendant market inefficiencies that amount to inefficient utilization of oil income with a negative impact on growth performance. Such fiscal eventualities underline the importance of robust institutional arrangements and checks and balances for the management of oil revenue (credible budget rules, hard expenditure constraints, greater reliance on taxation and other non-oil revenue, and rationalization of subsidies and transfers).

6.2.4 Saudi Arabia: The Oil Curse and Income Inequality

The impacts of the oil curse on intragenerational and intergenerational income inequality should also be highlighted. The first concerns the income disparity between workers of the same generation and the second concerns the sustainability of income across generations over time. For assessing oil income sustainability in the actuarial sense, the fundamental criterion is the sustainability of per capita oil income from the existing endowment into the foreseeable future. On that basis, with its vast oil endowment, Saudi Arabia easily passes the actuarial sustainability test, even with its fast population growth (Table 6.1).[12] However, this is a

[12] In the actuarial evaluation of *exhaustible* resource sustainability, the discounted present value of estimated oil endowment under the ground is hypothetically converted to financial capital at an initial period. As a minimum requirement of sustainability, growth

narrow definition of sustainability because it ignores labor market equilibrium. A broader and more substantive sustainability assessment needs to take into account the sustainability of *total* per capita income, that is, the imputed per capita oil income *plus* wage income (Al-Sheikh and Erbas 2013).[13] Sustainability of wage income is much more relevant for policy formulation when actuarial oil income sustainability is easily at hand.

There is a rising disparity between increasing public and declining private sector wages of Saudis, which indicates that oil income is not distributed equitably among workers of the same generations because public sector workers – accounting for more than 50% of total Saudi employment – receive substantially higher salaries and benefits than private sector workers.[14] Furthermore, if the average wage of future generations is lower than the average wage of present generations, then intergenerational total per capita income sustainability is not viable, even if per capita oil income is sustainable. For full employment of future generations, private sector wages need to decline in the long run to a level below the present level; therefore, intergenerational total income is *not* sustainable. Thus, with declining wages, the oil curse also creates intragenerational and intergenerational income inequality.

6.3 The Oil Curse on Saudi Arabia's Labor Markets: General Observations

6.3.1 Demography and Labor Market Distortions

Many of the aforementioned symptoms of oil dependency have been observed in the GCC countries, including Saudi Arabia, the largest

rate of estimated income from financial capital should match population growth rate so as to leave per capita income (transfer) constant over time. Saudi Arabia's proven oil reserves in 2009 were estimated at 265 billion barrels, projected to last about eighty years at the current rate of production (3.2 billion barrels per year) (Table 6.1).

13 We think oil endowment is in a special category that compels immediate right policy decisions for the benefit of future generations. This is because oil endowment is *exhaustible* and it is not private but *public* wealth to which all generations are entitled. As such, its sustainable utilization for improving labor and other income of future generations is a fundamental pillar of public policies to be followed by the *custodian*, but not the owner of the natural resource, that is, the government.

14 It is difficult to argue that, on average, public sector Saudi workers are so much more productive than private sector workers and so they are paid higher wages. Sfakianakis et al. (2011) provide evidence for sharply declining government sector productivity (output per worker) since 2004; they also report a declining trend in private sector productivity since 2005.

economy in the GCC.[15] In Saudi Arabia, in common with the other GCC countries, there are a number of tension points (distortions) in labor markets whose sources can be traced to oil income. Those tension points primarily reflect the underlying demographics and government wage and employment policies.

Demographic evidence indicates that Saudi Arabia has a young indigenous population, reflecting the high population growth rate. Its indigenous labor force is expanding alongside rapid growth of the working age population and increasing labor force participation, as more males and females seek higher education and females join the labor force in increasing numbers. In countries with such demographic characteristics, unless investment and capital accumulation keep up with rapid population growth, the typical result is a decline in the average wage. However, Saudi Arabia's resource endowment affords it high public sector employment and wages. This employment policy is the major conduit through which oil income is transferred to citizens, but creates significant tension points in labor markets. Oil endowment has enabled economic prosperity and social development but at the same time, it has created a huge welfare state. It has engendered a society accustomed to generous transfers and subsidies with a preference for public sector employment. Relative to private sector employment, this preference reflects higher wages, generous pension plans and other benefits, shorter working hours and, importantly, a secure source of lifetime income irrespective of work effort and productivity. However, labor markets have become increasingly strained in the face of a rapidly growing indigenous population.

In Al-Sheikh and Erbas (2013), the following tension points in Saudi Arabia's labor markets were identified.[16] Various transfers and subsidies extended to citizens as a minimum guaranteed income (importantly, free education and healthcare) create an *endowment effect*, so some citizens may choose not to work, relying only on the endowment income and family support. Additionally, the endowment effect may serve as an incentive for

[15] Individual country circumstances vary, and the structure of the GCC economies and the composition of their exports are changing. For example, in the UAE and Bahrain, the non-hydrocarbon sector accounted for 66–75% of GDP; the GCC average of non-hydrocarbon sector share is estimated at about 50% (IMF 2010). In Saudi Arabia, the same share averaged at about 70% of GDP in 2003–2010. Our comparative observations suggest that the smaller the oil endowment per capita, the smaller the oil curse on labor markets.

[16] By and large, the following tension points are representative of labor markets of the other GCC countries in varying degrees.

delaying labor force participation, for example, by extending the duration of university education because education is provided free of charge. Average salary (even excluding benefits) in the government sector is much higher than the average salary that Saudi citizens can earn in the private sector. The higher public sector wage sets a benchmark for a *reservation wage* for citizens and serves as a disincentive for them to seek employment in the private sector at the going wages. Those who delay employment in the private sector in anticipation of public sector employment register as unemployed. Availability of *cheap foreign labor* at nearly all skill levels at wages significantly lower than citizens' wages decreases citizens' competitiveness in the private sector. On average, the *skill level* of foreign workers is higher and their work effort and productivity exceeds those of citizens. Moreover, it appears that there is a social stigma attached to citizens accepting low-skill jobs, perhaps attributable to an implicit endowment effect since citizens feel they are entitled to oil wealth and income. Easy availability and large-scale employment of cheap low-skilled foreign labor might compound that stigma. Saudi quotas established for selective job categories under the rubric of Saudization policies (mostly clerical jobs, especially in the banking and retail sectors) might be compelling evidence of such social attitudes. *Foreign labor can be easily hired and fired* both in the public and private sectors, whereas laying off Saudi workers in both sectors is difficult and costly, even for the fundamental economic reason of low productivity relative to wages in the private sector. While the number of foreign workers can easily be adjusted to demand reflecting cyclical market conditions and long-term growth, the *Saudi population and labor force are growing rapidly* and high growth will persist in the longer run according to various demographic projections. Saudi students' *educational preferences* lean towards majors that are more inclined for the government sector rather than private sector jobs (Arts, Humanities and Islamic Studies). Overall, some 40% of Saudi students prefer majors that do not have competitive marketability in the private sector relative to foreign skilled labor.

As a result, there has been a *persistent and large decline in wages* that citizens can earn in the private sector and *persistent and high unemployment* in the last decade.

6.3.2 The Oil Curse and Past Labor Market Policies

A menu of employment policies has long been on the table to correct labor market distortions, including, most significantly, public sector hiring and wage and education policies to improve citizens' marketable skills relative to foreign workers. In reality, however, public sector hiring has been increasing steadily in Saudi Arabia. Furthermore, while private sector wages have been declining sharply, public sector wages are rising, reflecting both regular yearly scale raises and generous discretionary raises granted by the government typically when oil income is high. The growing disparity between public and private sector wages has exacerbated labor market distortions in the private sector by skewing Saudi preferences for government sector employment and increasing their reservation wages for private sector employment, thus resulting in high unemployment among citizens (especially young educated females and, to a smaller extent, young males).

In view of Saudi Arabia's vast oil wealth and citizens' entitlement to it as ultimate owners, the political difficulty of radically changing public sector employment and wage policies is easy to appreciate. It is a particularly difficult political choice to allow public sector wages to decline over time to levels more compatible with private sector wages in order to stimulate a preference for private sector employment among citizens. Consequently, instead of active market-based wage and employment policies, the government has followed passive policies that aim at increasing Saudi employment, establishing Saudi employment quotas in certain sectors by fiat.

6.3.3 The Oil Curse, Employment and the Royal Decrees of 2011

A recent initiative mandated by the Royal Decrees of 2011 (February 23 and March 18) created a large economic policy package to address rising social needs. The Royal Decrees provide for more public sector hiring, further government salary increases and higher transfers. They also promulgate unemployment benefits and institute more aggressive guidelines for Saudization. We assess this recent package's possible impact on labor markets, Saudi private sector employment and wages, in the light of the lessons we draw from the impact of past policies.

It is Saudi Arabia's large oil income that enables the financing of the large and generous public expenditure package the Royal Decrees mandate. The measures relating to labor markets and unemployment amount to a continuation of past policies that have created tension points in labor markets, with a significant negative impact, but without a significant positive impact on Saudi private sector employment and wages. In the longer run, those measures might even further exacerbate Saudi unemployment and are unlikely to rein in the persistent decline in Saudi private sector wages. Similar policies in the past have contributed to unemployment in the face of the rapidly growing Saudi labor force. If such policies are continued, the disparity between public and private sector wages will increase into the foreseeable future. That disparity is also the main source of income inequality between present and future generations that are equally entitled to the exhaustible resources and the income they yield.

6.4 Saudi Labor Markets and Unemployment

In this section, we examine some salient characteristics of Saudi demographics and labor markets in relation to the tension points discussed earlier.

6.4.1 Demography and Labor Force

6.4.1.1 *Total Population*

Saudi population growth rate is high, averaging 2.4% per year during 2007–2009. This population is young and 30% of the population is below fifteen years of age, compared to the industrial countries' average of 18% (Table 6.3). According to the US Census Bureau (USCB) projections, total Saudi and non-Saudi population will grow by a cumulative 23% during 2010–2030.

6.4.1.2 *Saudi 15–34 Age Group Population*

The USCB projections for the share of the Saudi and non-Saudi 15–34 age group population is 41% of the total population during 2010–2020. The same source projects that the cumulative growth (7%) and population share (38%) of this age group will remain high for the next two decades (2010–2030) (Table 6.3). Compared to industrial countries, the share of this age group is much higher.

In recent years, the population growth rates of the 25–29 and 30–34 age groups have been significantly higher than those of the other age groups. The 20–34 age group makes up more than half of the labor force and the bulk of entry and junior level employment (Table 6.6).

6.4.1.3 *Labor Force Composition and Growth*

Saudi workers account for around half of the total Saudi and non-Saudi labor force (Table 6.4). Saudi males have a 48% share in total male labor force and Saudi females have a 54% share in total female labor force. Males dominate the Saudi labor force (83%).

Table 6.3: *Saudi Arabia: Demographic Characteristics*

	(In percent)	
2007–2009 averages	Population growth rate	Population share
Total	3.4	100
Saudi	2.4	70
Male	2.4	35
Female	2.4	35
Non-Saudi	5.8	30

	(In percent)		
		15–34 age group	
Projected cumulative (Saudi and non-Saudi)	Total population growth rate	Growth rate	Average share in total pop.
2010–2020	8	10	40
2010–2025	16	9	39
2010–2030	23	7	38

	(Percent share in total)			2007–2009 average pop. growth rate
	2009–2010 averages			
		Saudi Arabia		
Age group	Total	Male	Female	(in percent)
<15	30	28	33	-0.7
15–65+	70	72	67	2.8
15–34	41	41	40	2.3
15–19	10	9	11	0.9
20–24	10	10	10	2.0
25–29	10	11	10	3.2
30–34	10	10	9	3.0
20–29	21	21	20	2.6
20–34	31	31	30	2.7
35–65+	29	31	27	3.7
		2010		

	US, Canada and Australia average		
Age group	Total	Male	Female
<15	18	19	18
15–65	82	81	82
15–34	27	28	27

(continued)

Table 6.3 (*continued*)

	US, Canada and Australia average		
Age group	Total	Male	Female
15–19	7	7	7
20–24	7	7	7
25–29	7	7	7
30–34	7	7	7
20–29	14	14	13
20–34	21	21	20
35–65+	55	53	56

Sources: Saudi Arabia CDIS, US Census Bureau, authors' calculations.

In recent years, the average Saudi labor force growth was 3.2%, with the male labor force growing by 3.5%. Those rates are significantly higher than the overall population growth rate (2.4%). High labor force growth rates reflect high growth rates in the 20–25 and 30–34 age groups.

6.4.1.4 *Employment*

The Saudi employment-to-population ratio is very low (20%) compared to the industrial countries' average (58 percent) (Table 6.4).[17] The female ratio is especially low (3%). This may be explained by the following facts: (a) the share of Saudi population below 15 years of age in total is large (30%); (b) most Saudis continue education until the age of 20; and (c) to some extent, the high unemployment rate that labor force entrants face may delay labor force entry until the age of 20.

In recent years, the overall Saudi employment grew at the rate of 2.7%; female employment growth rate was more than 4% (Table 6.5). Overall employment growth rate was significantly below the population growth rate in the 25–29 and 30–34 age groups (more than 3%) (Table 6.3).

6.4.1.5 *Participation*

The Saudi labor force participation ratio is also very low (23%), around one-third of the industrial countries' average (Table 6.3).[18] The female

[17] Employment-to-population ratio is the proportion of civilian non-institutional population aged 15–16 years of age and older who are employed (The US Bureau of Labor Statistics – BLS – definition).

[18] Labor force participation ratio is the total labor force as a percent of the non-institutional civilian population aged 15–16 years of age and older (BLS definition).

Table 6.4: *Saudi Arabia: Labor Force Characteristics*

(2007–2009 Averages Unless Otherwise Indicated)			
	(Percent Share in Total)		
	Total Labor Force	Total Male Labor Force	Total Female Labor Force
Total labor force	100	100	100
Saudi[1]	49	–	–
Male	41	48	–
Female	8	–	54
Non-Saudi	51	–	–
Male	44	52	–
Female	7	–	46
	(In percent of)		
	Total male population		Total female population
Saudi labor force[1]	–		–
Male	38		–
Female	–		8
	(Percent growth)		
Labor force[1]		2.4	
Saudi		3.2	
Male		3.5	
Female		1.8	
Non-Saudi		1.6	
(2007–2009 averages unless otherwise indicated)			
	(In percent)		
	Saudi Arabia (Saudi only)		Industrial countries[3]
	Age group		
	15+		15/16+
Employment-to-population ratio[2]			
Total	20		58
Male	18		65
Female	3		52

(continued)

Table 6.4 (*continued*)

(2007–2009 Averages Unless Otherwise Indicated)		
Labor force participation ratio[4]		
Total	23	62
Male	20	69
Female	4	55
Labor force composition		
Male	83	54
Female	17	46
Labor force participation ratio[4]	Age group[5]	
	15–34	15/16–34
Total	41	27
Male	22	14
Female	18	13

Sources: Saudi Arabia CDSI, U.S. Bureaus of Labor Statistics (BLS), US Census Bureau (UCB), authors' calculations.

[1] Data reflect household surveys of a sample of 23,000, including civilian as well as informal sector participants who are considered private sector employees. Data compilation methodology changed in 2009. The reference for the detailed explanation of the methodology used is provided in the reference of this paper.

[2] The proportion of the civilian non-institutional population aged 15/16 years of age and older who are employed (US Bureau of Labor Statistics – BLS – definition).

[3] BLS data; simple averages for United States, Canada, Australia, Japan, France, Germany, Italy, Netherlands, Sweden and United Kingdom; 2007–2010.

[4] Total labor force as a percent of the civilian non-institutional population aged 15/16 years of age and older (BLS definition).

[5] US Census Bureau data; United States, Canada and Australia only; 2010.

ratio is especially low (4%). Low participation may be partly explained by: (a) the large share of the population less than 15 years of age in total and (b) low 15–19 age group participation (in 2009, that age group made up only 1% of total Saudi labor force) (Table 6.6).

6.4.1.6 *Saudi Unemployment*

The overall Saudi unemployment rate averaged 10.5% during 2007–2009. The unemployment rate for males was nearly 7%. In contrast, the unemployment rate for females was very high, at 27% (Table 6.5). However, during the same period, overall unemployment registered a significant decline, mainly reflecting the sharp decline in unemployment rate for females.

Table 6.5: *Saudi Arabia: Saudi and Non-Saudi Employment and Unemployment, 2007–2009*

	2007–2009 Average (Percent Change)
Employed	
Saudi	2.7
Male	3.8
Female	4.3
Non-Saudi	0.8
Unemployed	
Saudi	−2.2
Male	−1.0
Female	−5.1
Non-Saudi	5.3
Unemployment rate	
Saudi	−2.2
Male	−1.0
Female	−5.1
Non-Saudi	5.3
	2007–2009 average (in percent of respective labor forces)
Unemployment rate	
Saudi	10.5
Male	7.3
Female	26.7
Non-Saudi	0.5

Sources: Saudi Arabia CDSI, authors' calculations.

In 2009, Saudi workers in the 15–34 age group made up 52% of the total Saudi labor force, around the same as the 20–34 age group whose population increased rapidly (Tables 6.3 and 6.6). Therefore, it is not surprising that the unemployment rate for Saudis in the 15–34 age group (20%) is nearly double the overall Saudi average. The unemployment rate (40%) for the 15–19 age group is dramatically higher, nearly four times the Saudi average; however, that group accounts for only 1% of the Saudi labor force. The unemployment rate for the 25–29 age group (20%) is also twice as large as the overall Saudi average. This is in line with the large population share (30%) and high population growth (2.7%) of the 20–34 age group (Table 6.3). However, the unemployment rate for the 30–34 age group (6%) is significantly lower than the overall rate, although the female unemployment rate in that group is very high (around 16%) (Table 6.6). The unemployment rate is negligible for the 35–65+ age group (less than 1%).

Table 6.6: *Saudi Arabia: Unemployment Rates and Shares of Saudi and Non-Saudi Workers in Saudi, Non-Saudi and Total Labor Force by Age Group, 2009*

	(In percent)								
	Unemployment rate in respective labor force[2]			Share in respective labor force[3]			Share in total labor force[4]		
Age group[1]	Male	Female	Total	Male	Female	Total	Male	Female	Total
	Saudi								
15–65+	6.9	28.4	10.5	100	100	100	42	8	50
15–34	13.6	41.1	19.5	49	68	52	44	12	56
15–19	43.4	45.9	43.5	1	0	1	53	2	55
20–24	30.3	71.2	39.3	11	16	11	59	17	76
25–29	12.7	45.5	20.3	19	29	21	48	14	62
30–34	3.7	15.5	6.0	18	23	19	36	9	45
20–29	19.1	54.5	27.1	30	44	32	63	10	73
20–34	13.2	41.1	19.2	48	67	51	44	12	56
35–65+	0.6	2.1	0.8	51	32	48	39	5	44
	Non-Saudi								
15–65+	0.3	0.7	0.3	100	100	100	43	7	50
15–34	0.3	0.7	0.3	95	81	62	36	8	44
15–19	0.0	0.0	0.0	1	0	1	40	5	45
20–24	0.1	0.2	0.1	4	6	4	20	5	24
25–29	0.1	0.2	0.1	12	16	13	32	7	38
30–34	0.1	0.3	0.1	22	29	23	46	9	55
20–29	1.2	1.7	1.3	16	22	16	22	34	24
20–34	0.2	0.6	0.3	38	4	13	36	8	44
35–65+	0.0	0.1	0.0	62	0	0	50	6	56

Sources: Saudi Arabia Ministry of Economy and Planning, authors' calculations.
[1] Age group 15–65+ is defined as the total labor force.
[2] Saudi unemployment rate is measured with respect to Saudi labor force only. Non-Saudi unemployment rate is measured with respect to non-Saudi labor force only.
[3] Saudi share is measured with respect to Saudi labor force only. Non-Saudi share is measured with respect to non-Saudi labor force only.
[4] Saudi share and non-Saudi share are measured with respect to total Saudi and non-Saudi labor force.

The very high unemployment rates for Saudi females reflect the high population share of the 20–34 age group in general, as well as the increasing female education levels and labor force participation. The younger workers in the 15–29 age group, especially females, bear the brunt of unemployment. In view of the projections that the share of the 15–34 age group will stay high in the next two decades, it is likely that high unemployment will persist in the long run.

6.4.1.7 Non-Saudi Unemployment and Competition

The unemployment rate for non-Saudis is nearly zero, which is not surprising because the foreign labor force is in Saudi Arabia to work (Table 6.6). Notice that non-Saudi concentration is high in the 15–34 age group labor force (95% of males). On average, non-Saudis make up 44% and Saudis make up 56% of the labor force in that age group.

Nearly nine out of ten workers in the private sector are non-Saudis (see next section). Given foreign workers' more marketable skills and work effort, they pose stiff competition for Saudi workers in that sector.

6.4.2 Private Sector Employment and Salaries

6.4.2.1 Employment

Non-Saudi workers make up the bulk of private sector employment with an 88% share; Saudi share is only 12% (Table 6.7).

In 2007–2009, the overall Saudi employment in the private sector declined by around 1% but female employment growth was impressive (7.6%) (Table 6.7). Nevertheless, the female Saudi share in total private sector employment is low (1% of total employment and 7% of Saudi employment).

In the period 2007–2010, the overall Saudi private sector employment gain is 1% and female employment growth is even higher (9.4%). Notwithstanding the impact on data of methodology change in 2009, the overall Saudi employment gain in the private sector in 2010 is impressive (6.3%). This apparent upswing in private sector Saudi employment is promising. It is possible that, facing large unemployment, Saudi workers are accepting lower wages in the private sector and employers are responding to the large decline in wages. The gains in private sector Saudi employment may also reflect the impact of the Saudization efforts in certain sectors.

The bulk of Saudi workers in the private sector are employed in three job categories: clerical, sales and services (60%) (Table 6.8). Saudi females make up 97–98% of total Saudi and non-Saudi employees in clerical and sales job categories. This points to Saudi worker (especially female) preferences for such jobs and the impact of the Saudization efforts in those sectors.

Those ratios are reversed for non-Saudi workers; 84–98% of foreign workers are employed in occupations other than clerical, sales and services.

Table 6.7: *Saudi Arabia: Private Sector and Government Sector Employment Growth and Shares*

	Saudis			Non-Saudis			Saudi and Non-Saudi
Period	Male	Female	Total	Male	Female	Total	
	(Percent change)						
	2007–2009 averages						
Total	1.8	5.3	2.5	8.6	3.8	8.5	7.1
Private sector	−1.2	7.6	−0.7	8.7	1.9	8.6	7.3
Government sector	6.0	5.1	5.7	−3.0	9.1	2.5	5.5
	2007–2010 averages						
Total	2.1	4.5	2.6	6.7	2.7	6.6	5.6
Private sector	0.5	9.4	1.0	6.8	1.3	6.7	5.9
Government sector 1	4.5	3.8	4.3	−2.3	6.8	1.9	4.1
	2009–2010 averages[1]						
Private sector	5.7	14.9	6.3	0.9	−0.7	0.8	1.4
	2007–2009 averages[2]						
	(Percent share)						
Total	100	100	100	100	100	100	100
Private sector	56	16	48	99	72	99	88
Government sector	44	84	52	1	28	1	12
	(Percent share in total employment)						
Private sector	11	1	12	86	1	88	100
Government sector	62	30	92	4	4	8	100
	(Shares in total Saudi and total non-Saudi employees)						
Private sector	93	7	100	98	2	100	–
Government sector	67	33	100	50	50	100	–

Sources: Saudi Arabia Ministry of Civil Service, authors' calculations.

1 The reported change in government sector employment during 2009–2010 is zero.

2 Only the 2007–2009 averages are presented because there is no appreciable difference between the averages for 2007–2009 and the averages for 2007–2010. The methodology change in 2009 creates discontinuity between the period 2009–2010 and the previous years.

Saudi male workers make up 4% of employment in the basic engineering support category, while foreign male workers make up 98%. Similarly, foreign workers dominate in the agriculture, professional/specialist, technical and industrial categories.

Table 6.8: *Saudi Arabia: Composition of Private Sector Employment by Occupation, 2008–2009 Average*[1]

	In Percent of Total Workers in Each Occupation Category			
	Saudi Workers		Non-Saudi Workers	
	Male	Female	Male	Female
Occupations				
Directors and managers	8	4	0	0
Professionals/specialists	9	15	6	12
Technicians	8	25	5	44
Clerical	22	26	0	0
Sales workers	10	9	3	0
Services	28	18	35	33
Agriculture[2]	2	0	9	0
Industries[3]	3	2	4	10
Basic engineering support	11	1	38	1
Total	100	100	100	100
Memorandum items:	(Percent shares)			
Clerical, sales, services	60	53	38	33
Other occupations	40	47	62	67
	(In percent of total Saudi and Non-Saudi workers in each occupation category)			
Occupations				
Directors and managers	89	98	11	2
Professionals/specialists	16	41	84	59
Technicians	16	24	84	76
Clerical	94	97	6	3
Sales workers	29	98	71	2
Services	9	23	91	77
Agriculture[2]	2	25	98	75
Industries[3]	7	11	93	89
Basic engineering support	4	48	96	52

Sources: Saudi Arabia Ministry of Labor, authors' calculations.

[1] Methodology change in 2009 does not appreciably affect averages from 2008 to 2009.

[2] Including animal husbandry and fisheries.

[3] Including transportation equipment workers and other related workers.

6.4.2.2 Salaries

Table 6.9 shows the large nominal decline in private sector monthly salaries both for Saudi and non-Saudi workers in recent decades. In 1996–2010, the decline in the weighted average Saudi salary is more than 50%. Considering that the cumulative CPI inflation rate in 1996–2010 was about 27%, the decline in real salaries is much higher.[19]

It appears that the decline in private sector Saudi salaries has decreased both in the periods 2007–2009 and 2007–2010; salaries may have risen by 12% in 2010. Data on percent changes in salaries by education level in 2008–2010 indicate that the decline in salaries in 2008–2010 was mainly for the less educated group (Group A), while for the more educated group (Group B), salaries did not appreciably change.[20] Keeping in mind that the unemployment rate for the Saudi workers in the 15–34 age group is high (20%) (Table 6.6), it is reasonable to surmise that Group A is largely made up of the 15–34 age group workers, so it is likely that this is the age group whose salaries declined the most (data on age groups and education levels are not available).

While Saudi private sector salaries have declined sharply in the last decade, so have non-Saudi salaries by a larger margin. In 2007–2010, the Saudi weighted average salary was four times the non-Saudi salary (Table 6.9). Although this large disparity reflects the concentration of foreign workers in low-pay and low-skill jobs, even excluding agricultural, industrial support and basic engineering support workers, Saudi weighted average wage is about three times the non-Saudi weighted average wage. Saudi private sector wages are not competitive with non-Saudi wages. With much lower wages and better skills and work effort on average, non-Saudi workers are highly competitive relative to Saudi workers in the private sector, which explains the 88% share of non-Saudi workers in that sector.

6.4.3 Government Sector Employment and Salaries

6.4.3.1 Employment

Government sector employment grew steadily for both male and female Saudi workers in 2007–2010, by 4–6% (Table 6.7). While 52% of Saudi

[19] In 2008–2009, occupation-wise, the largest decline in wages was in the clerical, sales and services job categories where Saudi workers concentrate in the private sector; the decline in the weighted average salary in those categories was 14% (the caveat applies that data compilation methodology changed in 2009).

[20] Group A's weighted average salary has risen sharply in 2010 (10–13%); however, for the same group, data indicates a 4.3% decline in male workers' salaries in 2008–2010 (the caveat applies that data compilation methodology changed in 2009).

Table 6.9: *Saudi Arabia: Weighted Average Monthly Salaries in Private Sector, 1996–2010*

	Weighted Average Salary per Month (in SRls)					
	Saudi			Non-Saudi		
Year	Male and Female	Male	Female	Male and Female	Male	Female
1996	7,210	n.a.	n.a.	2,261	n.a.	n.a.
2000	7,043	n.a.	n.a.	2,354	n.a.	n.a.
2007	3,624	3,722	2,260	1,011	1,000	1,573
2008	3,601	3,680	2,418	1,008	998	1,578
2009 [1]	3,125	3,192	2,247	765	757	1,343
2010 [1]	3,500	3,590	2,418	912	902	1,639
Ratio of Saudi salary to non-Saudi salary	(2007–2010 average)					
	3.8	3.9	1.5	–	–	–
	(Percent change)					
	Private sector salary					
Period						
1996–2010[1]	-51.5	n.a.	n.a.	-59.7	n.a.	n.a.
1996–2008	-50.1	n.a.	n.a.	-55.4	n.a.	n.a.
2009–2010[1]	12.0	12.5	7.6	19.2	19.1	22.0
	Weighted average salaries by education groups					
2008–2009						
Group A[1,2]	–	-15.1	-6.9	–	-28.0	-17.7
Group B[1,3]	–	0.0	0.0	–	0.0	-0.2
2009–2010						
Group A[1,2]	–	13.0	10.1	–	47.8	28.3
Group B[1,3]	–	0.0	0.0	–	0.0	-0.3
2008–2010						
Group A[1,2]	–	-4.3	2.3	–	4.7	4.9
Group B[1,3]	–	0.1	0.0	–	0.0	0.2

Sources: Saudi Arabia authorities, authors' calculations.

[1] The figures for 2009, 2010 reflect change in methodology in 2009.

[2] Group A comprises: Illiterate, Read and Write, Elementary School, Secondary School, High School, Diploma and Bachelor's degree.

[3] Group B comprises: Masters Degree, Ph.D., Higher Diploma, Fellowship.

workers are employed in the government sector, only 1% of non-Saudi workers are employed in that sector. While the Saudi male concentration in the government sector is 56%, Saudi female concentration is 84%, indicating a strong female preference for government sector employment.

6.4.3.2 Salaries

In contrast to Saudi private sector salaries, government sector salaries have been steadily increasing, reflecting regular yearly merit raises (about 4%) and extraordinary salary increases granted at government discretion (Table 6.10). The weighted average Saudi salary (including transportation allowance) in the government sector is three times higher than the weighted average Saudi salary in the private sector. Unless private sector Saudi wages make unprecedented large gains, the 2011 Royal Decrees' impact will increase this ratio. It is not surprising that Saudi workers have a strong preference for government sector employment.

6.4.4 Educational Profile of Saudis

Education has been found to be a significant determinant of growth.[21] In terms of investment in education, Saudi Arabia measures up rather well to high-income developed countries.[22] However, the type of education also matters. For example, the attainment of technical education has more

[21] Elbadawi and Soto (this volume) provide a review and some supporting results. Gylfason (2011) argues that resource extraction industries are less labor intensive and do not require high-skilled labor. Workers released from traditional types of employment (e.g., agriculture) are not skilled enough to find employment in other industries and unless their education is an active policy goal, such workers do not enhance their skills, especially when a good part of their sustenance comes from oil rent transfers. To those observations, we should add the presence of cheap foreign labor in Saudi Arabia at all skill levels.

[22] According to World Bank Development Indicators, the following facts emerge. Share of expenditure on education in total government spending averaged at around 23% and the share of education in GDP averaged at 6.5% in 2000–2009, compared to high-income countries' average of around 13% and 5%. The literacy rate for youth (ages 15–24) in 2009 was 98% compared to high-income countries 100%. Adult literacy rate (ages 15 and above) in 2009 was 86%, compared to nearly 100% in high-income countries (this may reflect the impact of low-skilled expatriate workers). Also in 2009, expenditure per student (in percent of per capita GDP) in primary and secondary education averaged around 18%, compared to high-income countries' average of 23%. Saudi Arabia fares better in terms of spending on education than the GCC average.

Table 6.10: *Saudi Arabia: Weighted Average Government Sector Compensation, 2005–2010*

	Salary/Month	Average Transportation Allowance/month	Total Salary/Month
		(In SRls)	
2005–2007[1]	9,789	38	9,827
2008[2]	10,278	38	10,316
2009[2]	10,792	38	10,830
2010[2]	11,332	38	11,370

Government sector total Saudi salary in multiples of private sector Saudi salary	
2007	2.7
2008	2.9
2009	3.5
2010	3.2
2007–2010 average	3.1

Sources: Saudi Arabia Ministry of Civil Services, Al-Sheikh and Erbas (2013), authors' calculations.

[1] An extraordinary raise on scale salaries was granted in 2005 and remained in effect through 2007. Extraordinary raises apply only to scale salaries and transportation allowance. The annual merit raise in SRLs is not affected by extraordinary raises.

[2] A 5% extraordinary raise in scale salaries was granted as of 2008 to apply in 2008, 2009 and 2010.

immediate marketability in industry and management, as opposed to and without prejudice, say, a degree in divinity or medieval literature. Market-oriented education enhances a worker's labor market competitiveness and educational preferences strongly reflect market demand. The main market incentive to choose one type of education over another is the potential wage a worker anticipates to earn. Labor market distortions that affect wages in a given profession can therefore have a significant effect on educational preferences. The fundamental labor market distortion that has an impact on educational preferences in Saudi Arabia is the much higher wages in the government sector relative to the private sector, along with government sector jobs being almost exclusively reserved for Saudis.

Table 6.11 shows the numbers and shares of higher education degrees obtained by Saudi students and their educational preferences. Saudi females have a significantly larger share among total graduates and around 7% of females in the 20–25 age group are obtaining higher education degrees. This points to increasing female labor force participation, which may further strain the decline in both male and female private sector salaries. Furthermore, given the strong female preference for government sector jobs, female private sector unemployment may rise further in the coming years.

The educational preferences of a significant proportion of Saudi students lean towards majors that appear to be more suitable for government sector employment than for private sector employment in competition with skilled foreign workers. On average, close to 40% of Saudi students major in areas more oriented toward government sector employment.[23] This is especially true in the case of students already in higher education (registered) and those seeking graduate degrees. To a significant extent, this preference might reflect the market response of students to higher wages in the government sector. It also likely reflects the Saudization quotas favoring Saudis in clerical, sales and services job categories reflecting Saudi occupational preferences.

On the other hand, notice that more than 80% of Saudi entrants to higher education prefer more private sector oriented majors, business administration having the largest weight (27%). This is a promising trend for Saudi employment in the private sector. If that trend persists in the coming years, we may expect Saudi employment to increase in the private sector, especially if, after graduation, entrants to the labor force accept lower wages relative to the government sector in the face of high unemployment in the 20–29 age group (Table 6.5).

6.5 The 2011 Royal Decrees: What Not to Do

The political aspects of this initiative are beyond the purview of this chapter, but its impact on Saudi labor markets and employment calls for assessment. The measures undertaken with the Royal Decrees are summarized in Box 6.1 (Explanatory Remarks). The cost of the initiatives is estimated at SAR 400 billion (US$ 110 billion). The intention was to spend about SAR 117 billion (US$ 31 billion) in 2011. According to our classification, the components of the package that have a direct bearing on labor markets are under the categories of government sector employment,

[23] Our classification of majors that appear to be more oriented toward government sector employment is in line with the estimate by Sfakianakis et al. (2011); they estimate that 41% of Saudi students prefer majors in Humanities and Arts.

Table 6.11: *Saudi Arabia: Higher Education Degrees Obtained by Saudi Males and Females, 2007–2009, and, Saudi Educational Preferences, 2010*

	2007	2008	2009	2007–2009 average
	Number of students			
Total	95,353	99,908	101,341	–
Male	37,544	39,337	40,047	–
Female	57,809	60,571	61,294	–
	(In percent total)			
Total	100	100	100	100
Male	39	39	40	39
Female	61	61	60	61
	(In percent of 20–29 age group population[1])			
Total	2.7	2.8	2.8	2.7
Male	2.1	2.2	2.2	2.1
Female	3.3	3.4	3.4	3.3
	(In percent of 20–24 age group population[1])			
Total	5.3	5.5	5.6	5.5
Male	4.2	4.3	4.4	4.3
Female	6.5	6.7	6.7	6.6
	Student Educational Preferences, 2010			
	Graduate students	Registered students	New entrants	Total
	(In percent of total)			
Government sector oriented[2]	64	60	17	38
Islamic Studies	13	9	4	7
Education	15	13	3	8
Humanities	16	19	4	11
Other[3]	20	19	6	13
Private sector oriented[2]	36	40	83	62
Business administration	8	14	27	20
Information	7	7	14	11
Life Sciences	5	5	10	7
Physics	4	4	9	6

(continued)

Table 6.11 (*continued*)

	2007	2008	2009	2007–2009 average
Engineering	4	4	9	7
Other[4]	8	7	15	11
Total	100	100	100	100

Sources: Saudi Arabia Ministry of Education, authors' calculations.

[1] Calculated by prorating US Census Bureau estimates for Saudi and non-Saudi population for the age group using the figures for Saudi (male and female) share in total Saudi and non-Saudi population, reported by Saudi Arabia authorities.

[2] Authors' classification.

[3] Includes: teacher training; arts; social and behavioral sciences; health; social services; others.

[4] Includes: law; mathematics and statistics; media and journalism; manufacturing and productive industries; architecture.

transfers, welfare payments and subsidies. Those components amount to 30% of the package, with the rest allocated to longer-term capital expenditure.[24] The package shown in Box 6.1 does not include two important additional measures. One measure is the promulgation of unemployment benefits and the other is the new guidelines to accelerate Saudization of employment in the private sector.

Large one-time transfers to the Real Estate Development Fund (SAR 40 billion) and the Saudi Credit and Savings Bank (SAR 20 billion) include subsidy elements. The Real Estate Development Fund is the entity that pays the SAR 500,000 transfer, increased from SAR 300,000 by the 2011 Royal Decrees, for Saudis to build their own homes. The Saudi Credit and Savings Bank gives interest free loans. Loans approved for social needs (marriage, family and building conservation) made up 99% of total approved loans in 2008. It is also noteworthy that the two-month's salary bonus granted to government sector employees also applies to stipends, salaries and scholarships paid to students in public institutes and universities. No doubt such transfers and subsidies meet important social needs and provide relief for lower income families. However, they also create a significant endowment effect that deters or delays Saudi labor force participation; they also enhance Saudi workers' reservation wages for private sector employment.

[24] Importantly, the SAR 250 billion allocated to housing construction (more than 60% of the total package) will go to good use in meeting housing shortages in the country.

In the government sector employment category, salaries and benefits of government sector employees are increased further, widening the gap between government and private sector Saudi salaries.[25] Such measures result in higher Saudi reservation wages for private sector jobs and a strong preference for government sector jobs, as they have in the past. Granted that while the intended 60,000 increase in government sector employment will significantly reduce Saudi unemployment at present, unless the government continues enlarging government sector employment in the future, this measure cannot contain unemployment in the face of the rapidly growing working age population in the long run. The government sector cannot indefinitely be a repository of employment for those who are unemployed in the private sector. Increasing private sector employment of Saudi workers is imperative for achieving any long-term decline in unemployment.

The sharp increase in government sector minimum wage to from SAR 1,750 to SAR 3,000 per month will increase the existing disincentive for Saudi workers to improve their qualifications and work effort to seek private sector employment and render them less competitive with non-Saudi workers who overwhelmingly dominate the private sector. In recent years, the weighted average non-Saudi salary was less than SAR 1,000 per month or one-third of the new Saudi minimum wage in the government sector. Moreover, the promulgation of an unemployment allowance (to come into effect in 2012, according to the IMF) in the amount of SAR 2,000 per month is double the average of non-Saudi salaries in the private sector and is a relatively attractive match to the new minimum government sector wage of SAR 3,000 per month.[26] Although to be paid up to the maximum of 12 months, the generous unemployment benefit relative to

[25] Our estimates indicate that, at the average merit raise of 4%, the weighted average government sector salary excluding transportation allowance rises to SAR 11,785 per month. With the two-month's salary granted to government employees as a bonus under the 2011 Royal Decrees, the average salary rises to more than SAR 13,000 in 2010. This corresponds to a more than 16% raise in 2011, which may be compared to the reported average increase in private sector salaries by 12% in 2010 (Table 6.9). Although the salary bonus in 2011 is a one-time thing, it is also likely that the sharp increase in government sector minimum wage will increase the weighted average government sector salary even further in the coming years.

[26] In 2011, Saudi Arabian Monetary Agency (SAMA) opened 1.2 million accounts for unemployment benefits recipients, reflecting the expected number of applicants. The number of Saudi unemployed was estimated at about 0.9 million in 2010. It appears that along with growing number of unemployed in 2011, the authorities anticipate a large number of applications from those workers who did not count among the unemployed previously.

government sector minimum wage may effectively delay alternative employment for Saudi workers at the lower skill and lower wage rungs in the private sector. Recalling that the Saudi private sector weighted average salary is estimated at SAR 3,500 per month in 2010 or about one-third of the weighted average Saudi government sector salary, the new measures will increase that disparity further. For a low-skilled worker, the minimum wage of SAR 3,000 per month may be more attractive than the average wage of SAR 3,500 per month in the private sector (which includes the wages of more skilled and more senior workers), especially if the low-skilled worker is covered for a year by unemployment payments of SAR 2,000 per month, while he or she delays private sector employment with an eye on government sector employment. Such a worker's preference for government sector employment is compounded by less stressful working conditions, better benefits and longer vacations, guaranteed life-time employment, better pensions and perhaps by the higher social status attached to a lower-level government job than to a lower-level private sector job by the prevailing social values. Of course, in the meantime, such a worker will register as unemployed.

Finally, it is likely that the new and on the surface more aggressive measures for Saudization by fiat (the "traffic light" rating of private sector firms by the degree of Saudi employment) will not find a wide political constituency in the private sector, as in the past. This simply reflects the economic reality of the profit-maximizing firms seeking cheaper, more skilled and more dedicated labor, which, at this time, non-Saudi workers supply amply and that supply is not likely to diminish in the foreseeable future[27]. Recall that, at present, non-Saudi workers earn one-third to one-fourth of the salaries Saudi workers earn and make up 88% of private sector employment.

Thus, in our view, the measures taken in the government sector employment category, and some elements of welfare payments and subsidies, do not address the documented tension points in the Saudi labor markets that result in high unemployment among Saudis. To the contrary, those measures appear to fall in the "*what not to do*" category because they may further exacerbate the existing labor market distortions that result in high unemployment.

[27] Sfakianakis et al. (2011) report that close to a million visas were issued for foreign workers in 2009, double the number issued in 2005. In 2009–2010, two million visas were issued.

6.6 Conclusions: The Oil Curse on Labor Markets and What to Do

Saudi Arabia's labor market conditions can be traced to its vast oil endowment. The source of the oil curse on Saudi Arabia's labor markets is the politically expedient largesse that oil income enables by making affordable financing of high public sector employment and wages, transfers and subsidies that serve to deter adoption of market-based policies to increase Saudi employment. In the long run, the cumulative impact of those policies may even elevate the existing tension points to a breaking point.

In contrast, if oil income did not exist, the public sector could not afford to be as large a repository of employment for citizens at wages far above private sector wages so as to have a significant impact on Saudi's reservation wages for private sector employment. Saudi workers would have a compelling incentive to seek private sector jobs at the going market wages and to improve their marketable skills, productivity and work effort. Thus, Saudi unemployment would decline, employment and wages would follow a natural path reflecting capital accumulation and population growth and labor markets would be on the long run full-employment path.

In our view, labor market policies to reduce unemployment are in stark contrast to the recent initiatives and have been on the table for a long time.[28] Those policies fall into the category of "*What to do*" and include, the hardening of budget constraints which disable discretionary salary raises for government sector workers. As noted in the introduction, there is a consensus in the literature that what matters when it comes to avoiding the oil curse is not natural resource availability but rather the way that resources are managed and invested. The wide disparity between Saudi government sector wages and private sector wages needs to be reduced in order to incentivize job seeking in the private sector at the going wages, acquisition of better marketable skills, higher productivity and better work habits. Unlimited access to cheap foreign labor at all skill levels needs to be restricted. Although the availability of low-wage foreign labor is a boon to Saudi employment and economic growth, labor markets as open as those in Saudi Arabia are not a reality in industrial countries. Foreign entry to labor markets in industrial countries is restricted to

[28] For a more detailed discussion, see Al-Sheikh and Erbas (2013).

protect employment of nationals. Such protection results in higher labor costs and lower growth. However, the technological endowment of industrial countries and their workers' overall high skill levels relative to foreign workers compensate for efficiency losses. This is not the case in Saudi Arabia. Therefore, labor market policies need to face the reality that there is a trade-off between cheap foreign labor and higher employment for Saudis. The recent measures also include some disincentives for private sector firms with low Saudi employment in the form of visa application restrictions to hire foreign workers. Whether such measures will be more successful than the quota system remains to be seen. However, as in the past, such measures are unlikely to find a strong constituency behind them in the private sector where the profit motive dominates hiring decisions.

There are no income or consumption taxes in Saudi Arabia apart from some user fees and charges, such as foreign worker license fees paid by employers. Thus, implicit benefits from the use of public goods and subsidized goods and services are not recouped by the government budget to a significant extent. This setup, in effect, is a substantial implicit subsidy to foreign workers and their private sector employers that remains untaxed. Taxation of consumption and income may increase the relative cost of employing foreign workers by increasing their reservation wages and it can provide much stronger policy tools to increase Saudi private sector employment through tax incentives. Moreover, taxation may provide additional revenue to finance employment measures and welfare and transfer schemes and, thus, it may reduce intragenerational and intergenerational inequalities in the present distribution of oil income.

Market mechanisms that motivate educational preferences of students toward higher paying jobs appear to be perversely in effect in Saudi Arabia. Indeed, a large proportion of Saudi students opt for majors that are more suitable for government sector employment. There is a need to reorient the education system to endow Saudi students with more technical and marketable skills so that they become more competitive with skilled foreign workers in the private sector. However, "*this will demand drastic shifts in cultural perceptions of both work and entitlement*," as Sfakianakis et al. (2011) underline. A promising trend is that more than 80% of the majors that entrants to higher education prefer are private sector oriented. This is likely to increase the Saudi employment in the private sector in the coming years.

These policies are not easy political choices, at least in the short run. On the other hand, pecuniary placation of potentially restive segments of the society might do more harm than good in addressing unemployment pressures in the long run.

EXPLANATORY REMARKS ON SELECTIVE INITIATIVES TAKEN UNDER THE 2001 ROYAL DECREES

Box 6.1 shows the various expenditures mandated by the 2011 Royal Decrees. We classify those measures under three categories: measures that have an impact on government sector wages and employment; measures that include transfers, welfare payments and subsidies; and, capital expenditures.

1. *Two-month's salary bonus to all government sector employees (SRLs 32 billion)*: This is a one-time bonus. All military personnel were moved up by one step in the salary scale; there are virtually no female employees in the military, excluding those employed in small numbers at border checkpoints. Two-month's bonus also applied to student stipends, salaries and scholarships paid in public institutes and universities.
2. *Sixty thousand new employees to be hired by the Interior Ministry (SRLs 2.2 billion)*: The kinds of jobs that will be created are yet to be announced.
3. *Social Security (SRLs 6.5 billion)*: Reportedly allocated mainly to the Ministry of Social Affairs toward subsidies and transfers for poverty relief and related social programs.
4. *Student scholarships (SRLs 1.1 billion)*: Allocated for study abroad.
5. *Real Estate Development Fund (SRLs 40 billion)*: This fund assists Saudis to build their own homes by paying a subsidy, which was increased from SRLs 300,000 to SRLs 500,000 in 2011. The SRLs 40 billion is a one-time transfer to the Fund. The transfer may lower long waiting period for such assistance.
6. *Credit and Savings Bank (SRLs 20 billion)*: This bank extends interest-free credit for trade and industry and other production activities, and, for social needs (building conservation, marriage and family loans). In 2008, loans for social needs made up 99% of total approved loans; marriage and family loans together made up 96% of total approved loans. Total outstanding loans in 2010 were SRLs 14.6 billion.

Box 6.1: Saudi Arabia: Measures Adopted Under the 2011 Royal Decree

Measure	Estimated cost (In SRL billion)
Government sector employment[1]	**48.7**
Two-month's salary bonus to public sector civilian employees. Military personnel moved up by one step in base salary scale.	32.0
Cumulative 15% cost-of-living allowance (5% per year) in 2008–2010 to be treated as a part of base salary toward increased pension upon retirement.[2]	4.5
Increase in minimum wage for government employees (from SRL 1,750/mo. to SRL 3,000/mo.)	10.0
Hiring 60,000 new employees at Ministry of Interior[3]	2.2
Transfers, welfare payments and subsidies[3]	**69.5**
Social security (part of welfare payments and subsidies)	6.5
Real Estate Development Fund	40.0
Credit and Savings Bank	20.0
Other[4]	3.0
Capital expenditures[3]	**282.0**
Building of 500,000 housing units	250.0
Transfer to Housing Authority	15.0
Transfers for building and expansion of hospitals	16.0
Other[4]	1.0
Total	**400.2**
	In percent of 2010 GDP
Total	**24.8**
Government sector employment	3.0
Transfers, welfare payments and subsidies	4.3
Capital expenditures	17.5

Sources: Saudi Arabian authorities, IMF, authors' classification.

1 Expenditure intended for 2011 unless otherwise indicated. Government sector employment category excluded intended unemployment benefits.

2 Government sector retirees also benefit from the same allowance.

3 Some expenditure to be spread over several years.

4 Includes various transfers.

References

Abed, G., S. N. Erbas and B. Guerami. (2005). The GCC Monetary Union, Some considerations for the Exchange Rate. In *Exchange Regimes and Policies*, Proceedings of a Conference Held in Abu Dhabi, United Arab Emirates, December 16–17, 2002, Arab Monetary Fund.

Al-Sheikh, H. and S. N. Erbas. (2013). Labor Income and Oil Wealth: A Broader Sustainability Framework for Saudi Arabia. *Middle East Development Journal*, 5 (2): 13–31.

Anshasy, A. and M. Bradley. (2009) Oil Price Volatility and Fiscal Response in Oil-Exporting Countries. OxCarre Conference Paper. Oxford University.

Arezki, R. and F. van der Ploeg. (2010). Do Natural Resources Depress Income Per Capita? Research Paper. Oxford Centre for the Analysis of Resource Rich Economies (OxCarre), Department of Economics, Oxford University.

Beidas-Strom, S. et al. (2011). *Gulf Cooperation Council Countries: Enhancing Economic Outcome in an Uncertain Global Economy*. Washington. D.C.: Middle East and Central Asia Department, International Monetary Fund, (www.imf.org/external/pubs/ft/dp/2011/1101mcd.pdf, accessed January 14, 2012).

Collier, P. and A. Hoeffler. (2008). Testing the Neocon Agenda: Democracy in Resource-Rich Societies. Research Paper. 2008–13, Oxford Centre for the Analysis of Resource Rich Economies (OxCarre), Department of Economics, Oxford University.

Elbadawi, I. and A. Gelb. (2010). Oil, Economic Diversification and Development in the Arab World. ERF Research Report No. 35.

Elbadawi, I. and O. Ali. (2016).

Elbadawi, I. and S. Raimundo. (2016).

Frankel, J. (2010–2011). The Natural Resource Curse: A Survey. Discussion Paper, Cambridge, Mass.: Harvard Environmental Economics Program, September, 2010 (www.hks.harvard.edu/m-rcbg/heep/papers/FrankelHEEPDP21.pdf, accessed May 10, 2012).

Gylfason, T. (2011) Natural Resource Endowment: A Mixed Blessing? CESifo Working Paper Series 3353, CESifo Group Munich, accessed April 8, 2011.

Hartzok, A. (2004). Citizen Dividends and Oil Resource Rents: A Focus on Alaska, Norway and Nigeria. Earth Rights Institute (www.earthrights.net/docs/oilrent.html, accessed May 14, 2012).

International Monetary Fund, Saudi Arabia (2011). Article IV Consultation, Staff Report 2011 (www.imf.org/external/pubs/ft/scr/2011/cr11292.pdf, accessed January 11 2012).

Hausmann, R. and R. Rigobon. (2002). An Alternative Interpretation of the Resource Curse: Theory and Policy Implications. NBER Working Paper No. 9424, Cambridge, Massachusetts.

Khan, M. (2009) The GCC Monetary Union: Choice of Exchange Rate Regime. Peterson Institute for International Economics, Working Paper 09–1, April.

Nandwa, B. (2011). Exchange Rate Regimes in the Arab Oil Economies: A Time to Rethink the Paradigm? Paper prepared for the ERF Project on Oil and Economic Diversification in the Arab World as part of the Research Initiative in Arab Development (RIAD), October, 7–8, Cairo.

North, D. (1994). Economic Performance Through Time. *American Economic Review* 84(3): 359–368.

OCED Report, 2010, A Good Practice Paper: Natural Resources and Pro-Poor Growth, DAC Guidelines and Reference Series.

Rodrik, D., A. Subramanian and F. Trebbi. (2002). Institutions Rule: The Primacy of Institutions over Integration and Geography in Economic Development. Working Paper 02/189 (Washington, D.C.: International Monetary Fund).

Sala-i-Martin, X. and A. Subramanian. (2003). NBER Working Paper Series No. 9804 (June).

Saudi Arabian Monetary Agency. (2010) Annual Report (www.sama.gov.sa/Pages/Home.aspx, accessed December 9, 2011).

Saudi Arabian Ministry of Civil Services. Civil Employees Statistics (www.mcs.gov.sa/pages/default.aspx, accessed, January 11, 2011).

Saudi Arabian Ministry of Higher Education. Higher Education Statistics (www.mohe.gov.sa/en/studyinside/universitiesStatistics/Pages/default.aspx, accessed January 14, 2011).

Saudi Arabian Ministry of Labor Portal. Statistical Yearbooks 2008, 2009, 2010 (http://portal.mol.gov.sa/Sites/default.aspx, accessed January 20, 2011).

Saudi Arabia Central Department of Statistics and Information (www.cdsi.gov.sa, accessed January 10, 2011).

Manpower Survey Methodology (www.cdsi.gov.sa/english/index.php?option=com_content&view=article&id=50&Itemid=130, accessed January 9, 2011).

Schmidt-Hebbel, K. (2016).

Sfakianakis, J., D. Merzaban and T. Al Hugail. (2011). Quota Counting: New Saudi Employment Rules to Shake Up Private Sector. *Bank Saudi Fransi Research Paper* (June 14).

Van der Ploeg, F. (2007). Africa and Natural Resources: Managing Natural Resources for Sustainable Growth. Background Paper for the African Development Bank, Tunis-Belvedère.

(2009). Global Crisis and Developing Countries: Financial, Environmental, Resource and Food Perspectives, Policy Paper 5,Oxford Centre for the Analysis of Resource Rich Economies (OxCarre), Department of Economics, Oxford University.

Van der Ploeg, F. and Steven P. (2008). Volatility and the Natural Recourse, *Research Paper* 2008-03, Oxford Centre for the Analysis of Resource Rich Economies (OxCarre), Department of Economics, Oxford University.

(2010). Natural Resources: Curse or Blessing, CESifo *Working Paper 31225*, Munich.

7

Resource Rents, Political Institutions and Economic Growth*

Ibrahim Elbadawi and Raimundo Soto

ABSTRACT

This paper contributes to the empirical literature on oil and other point-source resource curse. We find that the curse does exist but conditional on bad political governance. Unlike previous studies, we estimate a flexible econometric growth model that accounts for long-term country heterogeneity and cross-dependency and retains the virtues of the recent literature, including short-run flexibility, cointegration and error-correction mechanisms. We unpack political institutions into those reflecting the degree of inclusiveness (Polity) and credibility of intertemporal commitments (political check and balances) and find that resource-rich countries with low levels on both scores are likely to experience the curse, while those with high enough levels may turn resource rents into a driver of growth. Countries with high scores on only one dimension may avoid the curse but are not likely to effectively use resource rents to promote growth. This suggests that for the oil-rich Arab World to achieve sustained growth, not only democracy is a necessary condition but also the creation of a strong system of political checks and balances.

JEL Classification: O13, P16, O43

Keywords: Oil and natural resource curse, economic growth, democracy, political checks and balances

* We are grateful of generous comments received from our discussant Mahmoud El-Gamal.

7.1 Introduction

> "Projects of mining, instead of replacing capital employed in them, together with ordinary profits of stock, commonly absorb both capital and stock. They are the projects, therefore, to which of all others a prudent law-giver, who desired to increase the capital of his nation, would least choose to give any extraordinary encouragement..."
>
> Adam Smith (1776)[1]

As the aforementioned quote suggests, the curse associated with oil and other point-source rents is real, though contemporary scholarship also suggests that it is not predestined. In particular, the recent empirical growth literature finds that the existence of the curse is conditional on bad governance. However, a minority opinion still questions its existence when resource abundance measures (such as net resource exports per capita) are used instead of the resource dependency measures (such as resource exports/GDP), which has been the staple of the cross-country econometric literature on the development impact of oil.[2] Indeed, many, though not all, oil-rich countries have managed to successfully transform the rents from oil into substantial gains in terms of higher consumption and other aspects of social welfare for their citizens and investments in useful public goods for their economies (Sachs 2007). A comparison of oil-rich to non-oil developing countries makes clear that the former do in fact outperform the latter in terms of the levels of a wide range of economic and social indicators. Also, as far as the overall development indicators are concerned, the major oil-rich Arab member countries of the GCC obviously belong to a much higher league compared to other countries (Elbadawi and Gelb 2010).

However, other contributions to the literature find that even when measures of resource abundance are used, the curse seems to hold against a variety of robustness checks.[3] Oil income is intrinsically temporary because it is derived from non-renewable, depletable stocks. It is also unreliable because oil prices are highly volatile (Collier et al. 2009) and adequate risk coverage is not always available. Therefore, the consequences of failure to properly manage the volatility of oil income or to effectively

1 Quoted in Lederman and Maloney (2008).

2 See for example Lederman and Maloney (2007), who argue that, unlike the resource dependency measure, the abundance measure is consistent with theory; and that when used as a proxy for resource rents, no oil curse is found. Instead, they find a robust positive association between resource wealth and long-term growth.

3 See, for example, Arezki and van der Ploeg (2008).

use it to accumulate large and sufficiently diverse stocks of tangible (e.g., infrastructure) and intangible (e.g., human and knowledge capital and good institutions) types of capital are likely to be extremely dire for oil-rich countries.

More recently, Collier and Goderis (2009) use a panel cointegration empirical growth model to analyze two issues that, in our view, constitute the point of departure for further research on the oil curse question. Firstly, they find robust evidence that the *change* in non-agricultural export prices is positively associated with economic growth, but the price *level* has a strong negative effect. This suggests that commodity booms have positive short-term effects on output but adverse long-term effects. Therefore, the curse operates in the long run. Secondly, these authors also find that, conditional on bad governance, controlling for the real exchange rate, public consumption and private consumption as channels of the resource curse, higher commodity prices no longer have a negative long-run effect. The empirical significance of these channels corroborates the recent political economy literature, which predicts that permanent resource booms when government accountability is lacking allow politicians to expand public sector employment or to directly boost private consumption to enhance their popularity (e.g., Robinson and Torvik 2005; Robinson et al. 2006). In addition to these distributional aspects in resource economies, another strand of the literature suggests that bad governance also discourages overall savings and promotes excessive spending, which is reflected in appreciated real exchange rates (e.g., Matsen and Torvik 2005). Moreover, the significance of the real exchange rate channel also coheres with the recent literature on its role as an instrument in the development strategy for economic diversification; sophistication and growth (e.g., Elbadawi and Gelb, 2010).

This chapter contributes to this strand of the growth literature along three dimensions. First, following on Collier and Hoeffler (2009), we unbundle the role of institutions in the resource curse literature by accounting for both democracy and checks and balances. For the latter, however, we use a recently developed index of political checks and balances by Henisz and Zelner (2010). As argued by Barma et al. (2011), democracy accounts for political inclusiveness, which is critical for the long-term stability and growth in rent-dependent societies. However, the electoral competition aspect of the democratic process is likely to be at odds with the "credibility of inter-temporal commitment," another critical ingredient for success in resource-rich societies. These two opposing channels of democracy (inclusiveness versus the short-run politics of electoral competition) might explain the lack of robust association between democracy and long-term growth in the literature. We argue that the presence of a strong system of political checks and balances should augment democracy, because it is likely to be difficult for politicians to tamper with inter-temporal commitments in order to achieve short-term electoral gains.

The second contribution of this chapter is in the measurement of the impact of resource rents on economic growth. Our growth model properly accounts for most theoretically and empirically relevant growth fundamentals. Conditional on this encompassing set of growth correlates, we test for the significance of resource rents, in addition to democracy and checks and balances, as well as the interaction of the latter with rents. We embed such a growth model in a novel econometric estimation technique (called second generation panel data models) which allows for the derivation of the country-specific growth impact of rents. Our econometric model retains much of the virtues of recent studies (e.g., Collier and Goderis 2009; Arezki et al. 2011), such as short-run flexibility, cointegration and error-correction mechanisms. However, it adds controls for the crucial dimensions of country heterogeneity (differing short- and long-run parameters for each economy) and cross-section dependency (arising from common factors). This, we argue, should provide insights toward explaining country-specific growth experiences, such as those of the oil-rich countries of the Arab world.

The cross-country, time series literature on economic growth has been criticized for its reliance on questionable identification assumptions and lack of proper account of endogeneity. Our models belong to a new brand of panel models which account for identification by cointegration and are thus quite resilient to endogeneity biases. By allowing for country-specific estimation, it does not force short- or long-run structures to be the same across countries. Additionally, by allowing different speeds of adjustment, it does not force the dynamics to be the same across economies. We acknowledge, nevertheless, that macroeconomic cross-section time series models based on annual data may be limited when it comes to capturing the subtleties of the impact of oil exports on economic activity. However, by pooling the experiences of several economies, we aim to unveil general trends and identify issues across the board that would otherwise be too difficult to glean through the individual scrutiny of a large number of stories.

Section 7.2 contains a selective review of the literature to motivate the subsequent empirical growth analysis. This review will be focused mainly on the institutional aspects of the resource curse, though we also discuss the economic correlates of the curse from the perspective of their role as channels through which the latter impact growth, rather than being the true underlying causes. Section 7.3 presents the standard growth model and replicates the main findings of recent literature on the growth impact of rents conditional on political institutions. Section 7.4 undertakes a critical econometric assessment of the received empirical literature and estimates the resource curse effects in a more flexible econometric model. Section 7.5 concludes.

7.2 Understanding the Oil Curse[4]

Explanation of the oil curse has evolved from purely economic theories into the realm of political economy. Economic explanations, most notably the Dutch Disease (e.g., Corden 1982; Corden and Neary 1982; Sachs 2007) and the more recent volatility view (e.g., Gelb and Grasmann 2008), have been questioned in the recent literature (Elbadawi and Gelb 2010). On the contrary, the political economy view of the curse has received ample support in the recent empirical growth literature, which finds that the existence of the curse is conditional on bad governance (e.g., Collier and Goderis 2009; Arezki et al. 2011).

To motivate the empirical framework for analyzing the growth impact of oil and other point-source resource rents, the discussion in this section will be substantially focused on the institutional strand of the curse literature. This, nevertheless, does not imply that Dutch Disease and volatility effects cannot be present in badly managed economies as channels through which the curse might operate (e.g., Collier and Goderis 2009). Hence, we start with a brief review of these approaches before taking up institutional issues.

Despite the appreciation of the RER and the subsequent squeeze on the non-resource tradable sectors, the pure Dutch Disease effect cannot explain the overall collapse of the economy, as evidenced in the collapse of the non-oil GDP usually associated with the oil curse phenomenon. This is because, at the theoretical level, it is not clear as to why the declining share of the non-resource tradable sectors could not be compensated for by the rising share of the non-traded goods. However, the phenomenon might become a "true" disease (i.e., an oil curse) under two very adverse conditions. First, if the squeezed activities assume some special qualities that

[4] This section draws heavily from Elbadawi and Gelb (2010).

could not be compensated for by the expanding resource and non-traded goods sectors. Therefore, an extended version of the Dutch Disease thesis also assumes that exporting activities in general entail some dynamic properties, such as learning by exporting, and that manufacturing in particular generates increasing returns to scale (e.g., Matsuyama 1992; Bigsten et al. 2002; Rodrik 2007). Second, it also assumes that the RER appreciation is too extreme, relative to the long-term "sustainable" exchange rate. This, however, might be a plausible assumption because the marginal impact of oil booms on the RER can be quite substantial, with elasticity estimates ranging from 40% to 50% (Korhonen and Juurikkala 2007). Though both premises might be perfectly plausible, there is no systematic evidence that real exchange rate overvaluation, for example, might be a cause of an unconditional resource curse regardless of the quality of political governance.

The volatility view of the curse argues that because oil prices (and hence revenues) tend to be very volatile, oil-dependent economies tend to experience frequent and oftentimes extreme volatility in economic activity. As noted by Hausmann and Rigobon (2003), a one standard deviation shock to the price of oil (estimated at 30–35%) can generate an income shock as high as 6% of GDP in an economy where oil accounts for 20% of GDP. This is a very high volatility effect compared with the median shock for industrial countries (about 2%) or even developing countries (at 4%). The failure to cope with this extreme volatility, it is argued, has been the main factor behind the post-boom economic collapse experienced by most oil-dependent economies. Country experiences show how destructive the oil cycles can be when not properly managed. Gelb and Grasmann (2008) cite three such examples, including the Mexican debt crisis precipitated by the debt-financed spending spree against expectations of a longer oil boom that failed to materialize after 1981. Even more glaring examples are those of Nigeria (Budina and Wijnbergen 2008) and Venezuela, which saw its per capita output reduced to half its real value following the end of the 1974–1980 oil boom (Hausmann 2001).

In an attempt to provide a stronger explanation for the curse, Hausmann and Rigobon (2003) develop a model to motivate their "inefficient specialization" hypothesis, which augments the basic Dutch Disease model by accounting for the impact of the RER volatility on the non-resource tradable sector. They argue that for a resource curse to happen, the non-resource tradable sector must be substantially squeezed beyond a critical size or almost disappear and that the economy must be characterized by financial imperfections. They show that the near disappearance of the tradable sector will be associated with a highly volatile real exchange rate, because with a very squeezed or no non-resource tradable sector, only expenditure switching forces will be at play in response to the shocks from the oil sector. In turn, financial imperfection would cause interest rates to rise with increased volatility of the real exchange rate. In their model, a vicious circle between greater volatility and interest rates, on the one hand, and lower investment in the tradable sector, on the other, will lead to a steady squeeze of the latter until it eventually disappears and the economy specializes "inefficiently" in non-tradables. They show that this economy will exhibit higher interest rates, lower capital and wages and more depreciated exchange rates.

However, as the received empirical literature makes clear, the net growth impact of the Dutch Disease, volatility or inefficient specialization do not seem to have impacted long-term growth in well managed economies. Instead, oil-rich but poorly governed countries are not likely to acquire the knowledge and the institutional capacity to take the right decisions regarding extraction, savings and investments; to adopt the most appropriate macroeconomic framework for avoiding the Dutch Disease; or to better manage the rampant volatility that plague their economies. For example, the recent literature on managing oil economies has called for a strong role for the state in oil-rich countries with regard to dealing with oil corporations (e.g., Stiglitz 2007). To maximize the returns for their oil resources, these countries need to deal effectively with the vastly more knowledgeable multinational oil corporations on relatively complex issues, such as the overall terms of agreements; properly evaluate fiscal terms of contracts; or how best to auction oil rights.[5] However, as important as they may be, the most serious drag on these countries' ability to effectively manage the oil rents for development is likely not of knowledge or institutional capabilities. Instead, the most devastating aspect of the governance deficits in these countries is probably the lack of accountability associated with the scale and nature of the rents or the lack of inclusive institutions for distributing rents or managing the risks associated with volatility.

On the latter issue, Rodrik (1999) links the volatility story to the governance one and argues that the effect of external shocks on growth,

5 In their book on escaping the resource curse, Humphreys et al. (2007) devote considerable attention to these issues. See also Collier et al. (2008) and Collier and Goderis (2009), who strongly advocate auctioning as an instrument for selling oil rights by countries emerging out of conflicts, which tend to lack capable institutions and knowledge and are highly susceptible to corruption.

and economic performance in general, is not just the outcome of the failure of adjustment policies in the technical sense; it also reflects the interaction of these shocks with "latent" social conflicts in society on the one hand and institutions for conflict management, on the other. In societies with deep social conflicts (for example, societies fractionalized or polarized along ethnic, religious and cultural lines; or economic class) and weak social and political institutions for mediating conflicts among social groups, Rodrik argues that the economic costs of external shocks are magnified by the ensuing growth-retarding distributional conflicts triggered by these shocks.[6] Using a simple empirical proxy to test this hypothesis in a global cross-sectional database, Rodrik finds that the interaction term (shock × social conflict × lack of economic and institutional capacity) is robustly associated with the collapse of growth experienced by many oil-importing developing countries following the oil price hikes in the 1970s. More recently, Elbadawi (2005) analyses the collapse of growth in the oil-dependent Arab world in 1985–1994 (relative to the boom era of 1975–1984) and finds that it can also be explained by Rodrik's framework. Like Rodrik's analysis for the case of growth volatility and long-term sustainability, most of the received literature suggests that political institutions that promote inclusiveness and checks and balances also tend to promote long-term growth and other aspects of development in oil-rich societies. For example, in a theoretical model of patronage politics in the context of resource wealth, Robinson et al. (2006) show that good institutions may restrain this dysfunctional behavior. Also Mehlum et al. (2006) present empirical evidence in support of the critical role of good institutions in resource-rich economies.

In a more recent contribution to the growth literature in resource-rich economies, Collier and Hoeffler (2009) draw a distinction between democracy as an instrument for political inclusiveness and also electoral competition, and political checks and balances as an instrument for accountability. In this context, they analyze an empirical growth model using global data from 1970–2001 and a new measure of resource rents.[7]

[6] Social conflicts and lack of effective institutions for mediating them could affect the response to external shocks in many ways, for example, by delaying adjustment in fiscal policy and in key relative prices (most notably the real exchange rate), by generating increased uncertainty in the economic environment and by diverting resources from productive to distributive activities.

[7] They built a new series of resource rents using the following methodology. First, they defined rents as the difference between natural resource price and extraction costs. Second, for each point-source natural resource, they multiplied the estimated rents per

They find that high natural resource rents and open democracy (as measured by Polity IV) interact badly as determinants of growth, but checks and balances offset this adverse effect. Based on their results, they conclude that democratization in resource-rich economies needs to emphasize strong checks and balances.

Their empirical framework is motivated by a simple intuitive model that generates uncertain outcome for the growth impact of democracy in resource-rich economies, depending on whether or not democracy entails enough checks and balances that limit embezzlement of funds and hence maximize provisions for growth-enhancing public goods. Thus, the key insight of their study is modeling the determination of checks and balances. They assume that politicians would like to heavily tax in order to generate patronage but they will be deterred from doing so because high taxation provokes scrutiny. They assume that patronage expenditure, P, is determined by the product of the tax rate, t, and disposable income, Y, and the proportion of revenue which can be embezzled for patronage, e, which is simply given by a linear negative function of the tax rate: $e = \alpha(1 - t)$. The decision problem for the corrupt politician is thus given by:

$$P^{\max} = \max_{\text{wrt } t} \alpha(1 - t) \cdot t \cdot Y \quad (1)$$

which solves for the patronage maximizing level of the tax rate: $t^* = 0.5$ and the maximum resources available for patronage (P^*) and public goods (G^*):

$$P^* = Y\alpha/4 \quad (2)$$

$$G^* = (2 - \alpha)Y/4 \quad (3)$$

Hence, in this simple model, Equations (7.2) and (7.3) describe the outcome of electoral competition with the restraint of scrutiny endogenized. Now when resource rents are introduced at a rate, r, available revenue becomes:

$$[t(1 - r) + r]Y \quad (4)$$

Note that the resource rents are not subject to scrutiny. The new optimization problem now becomes:

$$P^{r^{Max}} = \max_{wrt\ t} \alpha(1 - t) \cdot [t(1 - r) + r] \cdot Y \quad (5)$$

The patronage maximizing tax rate is given by:

unit of output by the volume extracted; then they add up the outcome for such resources (e.g., oil, gas, coal, iron, copper, etc.). Finally, for each year, they divide total rents by GDP for the country in question.

$$t^{**} = (1 - 2r)/(2 - 2r) \tag{6}$$

This result suggests that the higher the revenue from resource rent, r, the lower the tax rate. However, though the resource discovery augments income, it is not clear whether or not it will increase the provision of public goods, depending on the scale of resource discovery, r, and the rate of embezzlement α. They show this by analyzing the interesting case of a scale of resource discovery that exactly doubles national income, hence $r = 0.5$, which (by Equation 7.6) will also be consistent with $t^{**} = 0$. At this rate, both income and government revenue are doubled, because the state previously received half of the national income ($t^{*} \cdot Y = 0.5Y$) and now it receives all of the previous national income before the resource discovery (because $r \cdot 2Y = 0.5.2Y = Y$). Also now with the tax rate driven to zero dampening the effect of scrutiny, the rate of embezzlement (relative to Y) increases from $\alpha/2$ (Equation 7.2) to α. Therefore, the provision for public goods in the presence of resource rents is given by:

$$G^{**} = (1 - \alpha)2Y \tag{7}$$

Comparing G^{**} with G^{*} (in Equation 7.3) suggests that the provision of public goods will worsen in the presence of resource rents if:

$$\alpha > 0.857 \tag{8}$$

Equation (7.8) is the upshot of these authors' model, which they use to explain some specific country experiences. For example, they argue that in a democracy with strong checks and balances, α will be small, hence provision for public goods is likely to increase. This will be the case of Norway, where strong checks and balances limit embezzlements and increases provisions for public goods even with small resource discovery, r. However, for Nigeria, the likely high rate of embezzlement (as should be inferred from its ranking across a range of governance indicators) has probably worsened the provision of public goods in the context of its moderate size rate of resource discovery. Finally, Saudi Arabia would be the polar opposite to Norway, though with the same outcome, in a qualitative sense. Though the rate of embezzlements is likely to be very high in this country, again judging by the low governance ranking and the zero rates of taxation, the sheer size of the resource discovery is likely to improve provision of public goods.

Therefore, this simple model provides a nice empirical framework for testing not only the impact of contestable democratic politics on the development of resource-rich societies, but the specific features that might make democracy effective in these countries.

7.3 Revisiting the Impact of Resource Rents on Economic Growth

The empirical literature on the determinants of economic growth is vast. A survey by Durlauf et al. (2005) indicates that around 150 variables have been found to be statistically correlated to economic growth in some countries across periods of time. We draw from this extensive empirical literature and estimate an encompassing model which seeks to link a country's economic growth rate to economic, political and social variables using a large sample of countries and time periods. This framework allows studying the natural resources curse, its transmission mechanisms and the role that political/institutional variables such as democracy and checks and balances can play to ameliorate or inhibit the curse altogether.

We estimate the following variation of a traditional growth regression in levels (for reasons we explain later):

$$y_{it} = \alpha y_{it-1} + \beta' X_{it} + \mu_i + \lambda_t + \varepsilon_{it} \tag{9}$$

where y_{it} is the log of per capita output, X_{it} is a set of variables postulated as growth determinants, λ_t is a period-specific effect, μ_i represents unobserved country-specific factors and ε_{it} is the regression residual. The subscripts i and t refer to country and time period, respectively. On the right-hand side of Equation (7.9), the regression model includes the level of per capita output at the start of the period (to account for transitional convergence or slow adjustment) and a set of explanatory variables measured during the same period. The time-specific effect, λ_t, allows us to control for international conditions that change over time and affect the growth performance of all countries in the sample (e.g., a global recession). The term μ_i accounts for unobserved country-specific factors that both drive growth and are potentially correlated with the explanatory variables.

7.3.1 Growth Determinants

We focus on variables that have received the most attention in academic literature and in policy circles. Following Loayza and Soto (2002), these variables are divided into four groups: transitional convergence, structural policies and institutions, stabilization policies and external conditions (see Appendix A for details on the definitions and sources of the data).

Transitional Convergence: One implication of the modern models is that the growth rate depends on the initial position of the economy.

The conditional convergence hypothesis maintains that, ceteris paribus, poor countries should grow faster than rich ones because of decreasing returns to scale in production. We control for the initial position of the economy by including the *lagged level of real per capita GDP* in the set of explanatory variables.

Structural Policies and Institutions: There is a general consensus that economic growth can be affected by public policies and institutions. We consider explanatory variables representing major categories of public policies. The first area of structural policies is *education and human capital formation* in general. Human capital can counteract the forces of diminishing returns in other factors of production – such as physical capital – to sustain long-run growth. We measure the outcome of policies directed toward increasing human capital with educational attainment obtained from Barro and Lee's (2011) database. The second policy area is related to *financial integration in world markets*. Well-functioning financial systems promote long-run growth as they facilitate risk diversification, help identify profitable investment projects and mobilize savings to them. Our measure is of an institutional nature – as opposed to the observed outcomes of such financial integration – as we use the index of capital account openness proposed and calculated by Chinn and Ito (2010). The third area is *international trade openness*. There are several channels through which trade affects economic growth: (a) inducing higher total factor productivity as a result of specialization and the exploitation of comparative advantages; (b) producing market expansion and use of scale economies; (c) helping diffuse technological innovations and improved managerial practices; (d) lessening anti-competitive practices of domestic firms and (e) reducing incentives for firms to conduct rent-seeking activities that are mostly unproductive. Our measure of openness is the volume of trade (real exports plus imports) over GDP, adjusted for the size (area and population) of the country, for whether it is landlocked and for whether it is an oil exporter. The fourth area is related to the *government burden*. Although governments can play a beneficial role for the economy, they can be a heavy burden if they impose high taxes, use this revenue to maintain ineffective public programs and a bloated bureaucracy, distort market incentives and interfere negatively in the economy by assuming roles most appropriate for the private sector. We account for the burden of the government through the ratio of government consumption to GDP. The fifth important area of policy involves the availability of *public services and infrastructure*. Whether they are treated as classic public goods or as subject to congestion, public services and infrastructure can affect growth by entering directly as inputs of the production function, by serving to improve total factor productivity and by encouraging private investment as they help protect property rights. There are a few alternative measures of public services and infrastructure. Among these, the variable with the largest cross-country and time series coverage is telecommunications capacity, measured by the number of telephone lines per capita.

Stabilization Policies: We include stabilization policies as determinants of economic growth for two reasons. From an economic perspective, stabilization policies affect not only cyclical fluctuations, but also long-run growth. Fiscal, monetary and financial policies that contribute to a stable macroeconomic environment and avoid financial and balance-of-payments crises are important for long-run growth. By reducing uncertainty, they encourage firm investment, reduce disputes for the distribution of ex-post rents and allow economic agents to concentrate on productive activities (rather than trying to manage high risk). From an econometric viewpoint, including stabilization policies improves the regression's fit and increases forecasting power over horizons that are relevant to economic policy. The first area in this category is related to the *lack of price stability*, which we measure by the average inflation rate. The second area is related to *external imbalances and the risk of balance-of-payments crises*. The occurrence of *systemic banking crises* accounts for the deleterious effect of financial turmoil on economic activity, particularly over short and medium horizons. The occurrence of banking crises is measured by a simple discrete dummy variable.

External Conditions: Economic growth is shaped not only by internal factors, but also by external conditions that influence the domestic economy in both the short and long runs. We include two additional variables in the growth regression: the *terms-of-trade shocks* affecting each country individually and a *period-specific shift* affecting all countries in the sample. Terms-of-trade shocks capture changes in both the international demand for a country's exports and the cost of production and consumption inputs. The period-specific shifts (or time dummy variables) summarize the prevalent global conditions at a given period of time and reflect worldwide recessions and booms.

7.3.2 Sample and Estimation Methodology

We use dynamic panel data models to study the evolution of real per capita GDP growth. Contrary to the standard literature, we do not conduct our analysis using averages of five-year periods; as discussed earlier and demonstrated now, the time series properties of the variables suggest the presence of non-stationarity and also of cross-country correlation in errors and non-observable variables. New econometric techniques developed by Pesaran (2006) and extended by Kapetanios et al. (2011) allow for consistent estimation in these less restrictive setups.

Our sample is dictated by data availability, particularly that for oil-dependent economies, which have a tradition of poor statistics and information secrecy. It contains ninety countries representing all major world regions (see Appendix B for the complete list of countries) and the time span for the econometric estimation is the period 1975–2009. Although we have been very careful in assembling our data, we acknowledge that some variables are weak in that available quantitative measures are not adjusted for quality. For example, secondary education achievement in developed economies may be of very different quality from what can be achieved in low-income countries. There is a classic trade-off between increasing the generality of the model at the cost of losing fidelity.

Much of the literature focuses on estimating growth models using the generalized method-of moments (GMM) estimator for dynamic models of panel data, which was introduced by Arellano and Bond (1991) and Arellano and Bover (1995). These estimators deal effectively with the three important challenges posed by the growth model. First, the regression equation is dynamic in the sense that it includes a lagged-dependent variable model to account for slow adjustment to long-run equilibrium. Second, the regression equation includes an unobserved country-specific effect, which cannot be accounted for by regular methods (such as the within estimator) given the dynamic nature of the model. Third, the set of explanatory variables includes some that are likely to be jointly endogenously determined with the growth rate. Moreover, the GMM estimator is best suited for the case of panel data models with a large number of cross-section units and relatively short time periods.

7.3.3 Estimation Results for the Standard Model

Table 7.1 presents the results obtained when estimating the empirical model by GMM techniques. Column (1) in the table corresponds to the basic specification while columns (2) to (5) extend the basic model to test in a candid way the oil curse and the role of political and institutional variables.

Table 7.1: *Econometric Results: Long-run Growth Determinants Dependent Variable: Real GDP per Capita*

Variable	(1)	(2)	(3)	(4)	(5)
			Standard Controls		
Education	0.062***	0.062***	0.052***	0.057***	0.051***
(attainment, in logs)	(0.011)	(0.011)	(0.011)	(0.010)	(0.011)
Trade Openness	0.035***	0.036***	0.035***	0.035***	0.035***
(% of GDP, in logs)	(0.006)	(0.006)	(0.006)	(0.006)	(0.006)
Capital Account Openness	0.002	0.002	0.001	0.00*	0.002*
(index)	(0.002)	(0.001)	(0.002)	(0.002)	(0.002)
Government Burden	-0.044***	-0.045***	-0.045***	-0.045***	-0.045***
(gov. consumption % of GDP, logs)	(0.006)	(0.006)	(0.006)	(0.006)	(0.006)
Inflation	-0.029***	-0.029***	-0.027***	-0.029***	-0.028***
(log (1+inflation rate))	(0.003)	(0.003)	(0.003)	(0.003)	(0.003)
Infrastructure	0.002	0.001	0.001	-0.001	0.002
(telephones per capita, in logs)	(0.004)	(0.004)	(0.004)	(0.004)	(0.004)
Workers' Remittances	0.014***	0.014***	0.013***	0.014***	0.013***
(% of GDP)	(0.004)	(0.004)	(0.004)	(0.004)	(0.004)
Systemic Banking Crisis	-0.010**	-0.009*	-0.009*	-0.009*	-0.009*
(dummy)	(0.005)	(0.005)	(0.005)	(0.005)	(0.005)
Terms-of-trade Shocks	0.034***	0.036***	0.035***	0.035***	0.035***
(dev. from HP trend)	(0.009)	(0.010)	(0.009)	(0.010)	(0.009)
Lagged real GDP per capita	0.986***	0.985***	0.986***	0.987***	0.986***
	(0.006)	(0.006)	(0.006)	(0.006)	(0.006)
			Resource Rents and Political Institutions		
Resource Rents	-	-0.049*	-0.032	-0.024	-0.026
(as % of GDP)		(0.029)	(0.031)	(0.030)	(0.031)
Checks and Balances	-	-	0.028***	-	0.025***
(Polity index)			(0.008)		(0.008)
Democracy	-	-	-	0.010**	0.004
(index)				(0.004)	(0.004)
Constant	0.116***	0.126***	0.129***	0.120***	0.127***
	(0.041)	(0.041)	(0.041)	(0.041)	(0.041)

Note: Number of countries = 90, number of observations = 2,743, maximum number of instruments = 605, time and country-fixed effects included. (*,**,***) indicate statistical significance at the 90%, 95% and 99% respectively.

Transitional Convergence. The coefficient on the lagged level of per capita GDP is statistically very significant and close to one, suggesting the possible presence of unit roots. In the latter case, the GMM estimator could be averaging over a set of possibly spurious regressions. Residuals-based test, however, reject the presence of spurious correlations.

Structural Policies and Institutions. All variables related to structural policies present coefficients with expected signs and statistical significance. Economic growth increases with improvements in education, trade openness and infrastructure. It decreases when governments apply an excessive burden on the private sector. Financial openness has the expected positive association with economic growth, but it is imprecisely estimated. These results are broadly consistent with vast empirical literature on endogenous growth, including Barro (1991) on the role of education, trade and government burden; Dollar (1992) on trade openness; and Levine et al. (2000) on financial depth. Infrastructure – proxied as usual by telephone lines per 1,000 inhabitants – has a positive effect on growth as found by Canning, Fay and Perotti (1994), but is very imprecisely estimated, probably because of the small year-to-year changes when using annual data (it is also highly correlated with education levels in our sample).

Stabilization Policies. All estimated coefficients for these variables carry the expected signs and are statistically significant. Economic growth generally decreases when governments do not carry out policies conducive to macroeconomic stability, including the absence of financial and external crises. Like Fischer (1993), we find that an increase in the inflation rate leads to a reduction in economic growth. The frequency of systemic banking crises has also a particularly negative effect on economic growth.

External Conditions. Positive terms-of-trade shocks foster growth. As noted by Easterly and others (1993), good luck (for example in the form of favorable commodity price shocks) can be as important as good policies in explaining growth performance over medium-term horizons.

We extend the basic regression to include resource rents and the results are presented in column (2) of Table 7.1. We use World Bank data on resource rents as share of GDP. The unit measure of rents is the difference between world prices and the average unit cost of extraction. Total rents are simply the unit measure of rents times the quantity extracted or harvested. The measure is simple and relatively easy to collect for a large number of countries. It is obviously very limited in that it excludes the non-negligible exploration and development costs, as well as financial, intermediation and marketing costs. As can be seen, the estimated coefficients for the standard controls are not affected by the inclusion of the new variable. Moreover, we obtain a negative, significant estimated, parameter for the resource rent variable, consistent with the "natural-resource curse hypothesis" as discussed in Sachs and Warner (1995). According to Gylfason (2011), four main channels of transmission, from natural resource abundance or intensity to slow economic growth, have been suggested in the literature. First, through Dutch Disease-linked phenomena, where the abundance of natural resources lead to currency overvaluation and economic cycles (the latter as a result of fluctuations in the international price of commodities and volatility in the exchange rate). Second, natural resource rents in conjunction with ill-defined property rights, imperfect or missing markets and lax legal structures may lead to rampant rent-seeking, thus diverting resources away from more socially fruitful economic activity (Gelb 1988). Third, natural resource abundance may reduce private and public incentives to accumulate human capital due to a high level of non-wage income – for example, dividends, social spending and low taxes. Finally, natural resource abundance may blunt private and public incentives to save and invest and harm the development of the financial sector, thereby impeding economic growth.

Note that we control for an eventual "contagion channel" of the resource curse, whereby non-resource exporting countries are indirectly harmed by the effect of the resource boom of a neighboring country via labor markets. An example is Yemen, which has had significant economic volatility as a result of oil-cycles because of the fluctuations in remittances of Yemeni workers employed in oil-rich countries such as Saudi Arabia and the UAE. In Table 7.1, we use unrequited worker remittances to control for this phenomenon, but we hasten to acknowledge that other forms of contagion are possible (e.g., via asset prices). We also acknowledge that the resource curse may operate only after a certain threshold is surpassed, but our model is linear and, even if country specific, it excludes threshold effects. It is nevertheless difficult to identify a criterion to set such country-specific thresholds when working with a large sample of countries. It remains a challenge for future work to develop an econometric technique capable of dealing with threshold effects in a dynamic panel data context.

Note that the resource curse implicitly requires countries to have institutional and political impediments to enact efficient policies to mitigate the possible negative externalities of natural resource exploitation. Resource rents can be properly allocated to education, research and innovation, building institutions and the provision of public goods, resulting in higher economic growth. This in turn would require, at least, the coordination of two elements. First, a mechanism capable of aggregating individual preferences in society and of channeling them through the political structure so that the population has an adequate representation in policy decision making. Second, a structure of checks and balances capable of monitoring that those preferences are respected and that those responsible of enacting economic policies actually perform their duties.

We, thus, include in our estimation, two additional variables. First, since there is some consensus that democratic regimes tend to allow for more participatory decision making than other political systems, we include the variable Polity 2 from Project Polity IV: regime authority characteristics and transitions datasets which takes the value −10 if the country is non-democratic and 10 when there is a high degree of democracy. Second, while democracy and stability are certainly important for economic development, there is also evidence to suggest that rules are primarily the manifestation of an implicit contract with the electorate, a public signal of the commitment to maintain mutually agreed standards of fiscal discipline (see Debrun and Kumar 2007). Therefore, we include a measure of political risk and checks and balances: we use the political constraint index (POLCON-V) developed originally by Henisz (2000) and later refined and extended by Henisz and Zelner (2010). This index is a quantitative measure of the institutional constraints faced by authorities and evaluates the extent to which any one political actor or the replacement for any one actor (e.g., the executive or a chamber of the legislature) is constrained in his or her choice of future policies. It considers elements such as the number of independent branches of government, veto power over policy changes, the party composition of the executive and legislative branches and the preference heterogeneity within each legislative branch.

When these additional variables are included in our econometric model, the natural resource curse, however, seems to be unsupported by the data. Columns (3) and (4) present the results of the estimation of the base model when including each institutional/political variable by itself, while column (5) presents the results of the inclusion of both variables in the model. Again, estimated coefficients for the other regressors are not affected by the change in specification, with the only exception being resource rents that are now statistically insignificant in all models. In conclusion, we confirm, in an encompassing growth model and relatively large and updated dataset, the recent findings from the received literature on the natural resource curse, which suggests that the curse exists, but is conditional on bad political governance such as lack of democracy and political checks and balances.

7.4 Toward a More Flexible Econometric Model

Standard long-run growth models, like the one described in the previous section, suffer from limitations and, in general, do not guarantee an adequate representation of the determinants of economic growth. Consider, again, Equation (7.9). The first important limitation of the aforementioned model lies in the implicit restriction that growth rates in all countries are but realizations of the same stochastic process. By pooling all countries in the same regression model, one implicitly assumes that all countries would react in the same way to changes in the fundamental variables (common parameters α, β, μ and λ), independent of their state of development, existing conditions or previous history. This, of course, seems an implausible restriction.

For this reason, different econometric techniques have been developed to permit more flexible representations of the data. The inclusion of individual or country effects allows for some heterogeneity in initial conditions, but continues to restrict the other parameters from being the same across countries. Time dummies can be used to capture cross-country shocks affecting all economies in a similar fashion. Embedding the model in Equation (7.9) in an error correction structure (see Equation 7.10) allows for country-specific short-run dynamics and adjustment to equilibrium while restricting only the parameters in the long-run equilibrium model to be the same for all countries. Estimation via pooled-mean group estimators (PMG) provides consistent estimation of the parameters when N and T are large enough.

$$\Delta y_{it} = -\phi_i\left(y_{it-1} - \beta_i' X_{it-1} - \mu_i - \lambda_{t-1}\right) + \sum_{j=0}^{q} \theta_{ij} \Delta X_{it-j} + \sum_{j=0}^{q} \delta_{ij} \Delta y_{it-j} + \eta_{it} \tag{10}$$

As noted by Islam (1995), among others, the growth regression in Equation (7.10) is in fact a levels regression where y_{it} has been subtracted from both sides to investigate out-of-steady state behavior. This approximation is, thus, only valid in a proximity to the steady state.

While estimating the growth model using PMG techniques allows for significant heterogeneity in the short-run dynamics, it still forces all countries to react identically (and linearly) to long-run shocks. Economic intuition indicates that the growth impact of, say, education or opening an economy to foreign trade, ought to have diminishing returns; yet, the model in Equation (7.10) treats all countries alike in the long run.

A second limitation of the standard growth model is the restriction placed among the regressors themselves, for which the model in Equation (7.10) assumes to be complete and uncorrelated. However, the multiplicity of studies on growth, each testing new or additional regressors, indicates that most specifications are bound to be incomplete and vulnerable to omitted variable problems. Chief among the omitted variables is, of course, total factor productivity (TFP).[8] Economic theory provides a number of alternative mechanisms through which TFP in different economies can be correlated, ranging from technology transfer, economic integration, access to information, trade agreements. Westerlund and Edgerton (2008) note that when studying macroeconomic and financial data, cross-sectional dependencies are likely to be the rule rather than the exception because of strong inter-economy linkages. This type of heterogeneity introduces cross-section correlation, or dependence between the regression error terms, which can lead to inconsistency and incorrect inference in standard panel econometric approaches.

Based on this discussion, a general specification for the growth model is proposed by Eberhardt and Teal (2010) and summarily described in the following set of equations:

$$y_{it} = \beta_i x_{it} + \mu_{it} \tag{11}$$

$$\mu_{it} = \alpha_i + \lambda_i f_t + \varepsilon_{it} \tag{12}$$

$$f_t = \rho f_{t-1} + \omega_t \tag{13}$$

$$x_{it} = \pi_{mi} + \delta_{mi} g_m + \rho_{mi} f_{mi} + \upsilon_{it} \tag{14}$$

[8] Though some aspects of TFP could be accounted for by institutions, human capital and public policy, others, such as technological change are difficult to control for and remain as unobservable variables.

$$g_t = \kappa g_{t-1} + \eta_t \tag{15}$$

Equation (7.11) indicates that the observed output (y_{it}) – in our case, per-capita GDP – relates to a set of observed (x_{it}) and unobserved fundamentals (μ_{it}). The unobserved fundamentals μ_{it} in Equation (7.12) are represented by a combination of country-specific fixed-variables (α_i) representing initial conditions and country specificities (e.g., being landlocked); a set of common unobservable factors (f_t) with heterogeneous country impacts or factor loadings (λ_i); and a random perturbation. Equation (7.13) allows for persistence in these common factors. Equation (7.14) introduces an empirical representation of the observed inputs in order to indicate the possibility for endogeneity: the input variables (x_{it}) are driven by a set of common factors (g_{mt}) as well as an additional set of factors (f_{mi}), which may also drive output y_{it}. Equation (4.6) indicates that the common factors are persistent over time, which allows for the setup to accommodate non-stationarity in the factors ($\rho = 1, \kappa = 1$) and thus the observables. It further allows for various combinations of cointegration: between output y and inputs x, between output y, inputs x and (some of) the unobserved factors f_t, or non-cointegration.

The proposed econometric model for long-run growth – based on the so-called "second generation" panel data techniques – retains much of the virtues of the standard literature (short-run flexibility, cointegration and error-correction mechanisms and so on), but it also adds the crucial dimensions of country heterogeneity (itself reflected in differing short- and long-run parameters for each economy) and cross-dependency (arising from common factors). We thus first test for common factors and later estimate the full model.

7.4.1 Testing for Cross-country Correlations

We use Pesaran's (2006) CD test for cross-dependency. This test employs the correlation coefficients between the time series for each panel member. In our dataset of N = 90 countries, for instance, this would be the 90 × 89 correlation between country i and all other countries, for i = 1 to N – 1. Referring to these estimated correlation coefficient between the time series for country i and j as ρ_{ij}, the Pesaran CD statistic is then computed as

$$CD = \sqrt{2/N(N-1)} \sum_{i=1}^{N-1} \sum_{j=1}^{N} \sqrt{T_{ij} \rho_{ij}} \tag{16}$$

Table 7.2: *Econometric Results: Testing for Cross-country Correlations*

Variable	CD-Test	*p*-value	Correlation	Absolute Correlation
Education	308.50	0.000	0.898	0.906
Trade Openness	64.46	0.000	0.190	0.361
Capital Account Openness	n.a.			
Government Burden	7.88	0.000	0.020	0.380
Inflation	70.48	0.000	0.206	0.304
Infrastructure	268.08	0.000	0.782	0.823
Workers' Remittances	117.91	0.000	0.330	0.487
Systemic Banking Crisis	n.a.			
Terms-of-trade Shocks	11.81	0.000	0.034	0.239
Resource Rents	58.46	0.000	0.161	0.370

Note: n.a. = not available because there is no within-unit, time variation.

where T_{ij} is the number of observations for which the correlation coefficient was computed. Under the null hypothesis of cross-section independence, the aforementioned statistics are distributed standard normal for $T_{ij} > 3$ and N sufficiently large. The test is robust to non-stationarity (the spuriousness would show up in the averaging), parameter heterogeneity or structural breaks and was shown by its author to perform well even in small samples.

In Table 7.2 it can be seen that for all variables where the index can be computed, it is possible to reject the null hypothesis of no cross-country correlation at 99% confidence. Correlation, not surprisingly, is quite high in education (as most countries have engaged in massive human capital formation programs) and in infrastructure (for similar reasons). These results call for the use of common correlated effects estimators.

7.4.2 Econometric Estimation of the Common Correlated Effects Model

We estimate the empirical models using Pesaran's (2006) common correlated effects (CCE) estimators. This estimator avoids obtaining explicit estimates for the unobserved common factors (f_t)(as in Bai et al. 2009) and accounts for their presence implicitly by adding cross-section averages for the dependent and independent variables to the regression equation. The specification assures that coefficients on the implied common factors are allowed to differ across countries (equivalent to λ_i differing across i).

The Pesaran (2006) common correlated effects mean group estimator (CCEMG) allows for the empirical setup as laid out in Equations (7.11) to (7.15), which considers cross-section dependence, time-variant unobservable variables with heterogeneous impact across panel members, and problems of identification (note that β_i is unidentified if the regressor contains a non-null λ_i)). The CCEMG solves this problem with a simple but powerful augmentation of the group-specific regression equation: apart from the regressors (x_t) and an intercept, this equation now includes the cross-section/panel averages (for the entire panel $i = 1, \ldots, N$) of the dependent and independent variables: $\overline{y}_t$ and $\overline{x}_t$. Together these can account for the unobserved common factor, (f_t) and given the group-specific estimation, the heterogeneous impact (λ_i) is also given. The coefficients (β_i) are again averaged across panel members, where different weights may be applied.

In empirical applications, the estimated coefficients on the cross-section averaged variables as well as their average estimates are not interpretable in a meaningful way: they are merely present to blend out the biasing impact of the unobservable common factor. The focus of the estimator is on obtaining consistent estimates of the parameters related to the observable variables. The CCEMG approach is robust to the presence of a limited number of "strong" factors as well as an infinite number of "weak" factors – the latter can be associated with local spillover effects, whereas the former represent global shocks. Furthermore, as shown by Kapetanios, Pesaran and Yamagata (2011), these factors may be nonstationary.

The Pesaran (2006) estimator yields consistent and efficient estimates of the parameters and is robust to structural breaks in the data, a feature that is important in the context of long-run growth models for economies that may have engaged in significant reforms. Alternative estimation approaches (Bai 2009; Bai et al. 2009) involve estimation of the number of "relevant" factors $\boldsymbol{f}_t$ first and then the factors themselves, which hinges crucially on the assumption that all factors in the data generating process (DGP) are of the "strong" type, thus excluding "weak factors," (e.g., spatial correlations such as neighborhood effects) (Pesaran 2009). The CCE estimators can account for the presence of strong factors as well as an *infinite* number of weak factors, while no prior knowledge of the cointegrating properties of the observable and/or the unobservable variables is required, since the method is robust to all these scenarios (Eberhardt and Teal 2010).

Table 7.3: *Econometric Results: Long-run Growth Determinants Allowing for Common Correlated Effects*

Dependent Variable: log of Real GDP per Capita	Full Sample	Full Sample	Advanced Countries	Advanced Countries	Emerging Countries	Emerging Countries
	(1)	(2)	(3)	(4)	(5)	(6)
Education	0.121	0.237*	0.403	0.206	0.184	0.162
(years of attainment, in logs)	(0.112)	(0.133)	(0.287)	(0.372)	(0.137)	(0.202)
Trade Openness	0.043***	0.035*	0.049*	0.051**	0.045***	0.029*
(% of GDP, in logs)	(0.015)	(0.018)	(0.027)	(0.022)	(0.013)	(0.017)
Capital Account Openness	-0.001	0.009**	0.009*	0.002	0.005	0.004
(index)	(0.004)	(0.004)	(0.005)	(0.011)	(0.004)	(0.006)
Government Burden	-0.113***	-0.119***	-0.357***	-0.406***	-0.040**	-0.056***
(gov. consumption % of GDP, logs)	(0.029)	(0.03)	(0.068)	(0.084)	(0.017)	(0.022)
Inflation	-0.038*	-0.095***	-0.051	-0.006	-0.057***	-0.119***
(log (1+inflation rate))	(0.024)	(0.028)	(0.059)	(0.129)	(0.019)	(0.031)
Infrastructure	0.121***	0.093***	0.072	0.007	0.133***	0.079***
(telephones per capita, in logs)	(0.020)	(0.028)	(0.082)	(0.100)	(0.021)	(0.030)
Workers Remittances	0.0002	-0.035**	-0.005	-0.003	-0.008	-0.028*
(% of GDP, in logs)	(0.013)	(0.014)	(0.015)	(0.029)	(0.015)	(0.017)
Terms-of-trade Shocks	-0.001	-0.002	-0.043	-0.081	-0.010	-0.0002
(log dev. from HP trend)	(0.020)	(0.024)	(0.050)	(0.060)	(0.017)	(0.026)
Systemic Banking Crisis (dummy)	–	–	–	–	–	–
Resource Rents	-0.306*	-0.111	0.562	1.247	-0.301*	-0.010
(as % of GDP)	(0.188)	(0.215)	(0.954)	(1.661)	(0.186)	(0.179)
Checks and Balances	–	0.039**	–	-0.007	–	0.021
(index)		(0.019)		(0.404)		(0.015)
Democracy	–	-0.002**	–	0.002	–	-0.002
(Polity IV index)		(0.001)		(0.002)		(0.002)
Constant	6.278***	6.749***	1.287	2.342**	4.810***	6.057***
	(0.666)	(0.829)	(1.015)	(1.169)	(0.674)	(0.92)
Countries	*90*	*90*	*21*	*21*	*69*	*69*
Observations	*2,742*	*2,728*	*632*	*632*	*2,110*	*2,096*
RMSR	*0.0171*	*0.0120*	*0.0066*	*0.0047*	*0.0186*	*0.0134*

Note: Estimated using the common correlated effects mean group estimator proposed by Pesaran (2006). Robust regression methods are used to control for outliers.(*, **, ***) indicate statistical significance at the 90%, 95% and 99%, respectively.

Table 7.3 presents the results of the estimation of the economic growth model using the CCE technique for the full sample and also splits the countries between advanced and emerging economies, because, as will be discussed here, there is evidence of heterogeneous behavior among different countries. Column (1) collects the results for the growth model without controlling for political factors and compares directly with column (2) of Table 7.1. It can be seen that the country-averaged CCE coefficients partially confirm the results obtained using GMM models. Only in the cases of trade openness, inflation and capital account openness do the new estimates fall within two standard deviations of the GMM estimate. On the contrary, the estimated parameters for government burden and infrastructure are significantly larger than GMM estimates, while those for education, worker remittances and terms-of-trade shocks are imprecisely estimated. With respect to our variables of interest, the CCE estimation finds evidence of a resource curse that disappears once political variables are included in the estimation as shown in column (2) of Table 7.3.

Further inspection of the individual country results suggests that emerging economies behave very differently from more advanced economies and that pooling both types of countries can be potentially misleading when studying the effects of resource rents on economic growth. We split the econometric analysis into two subsamples: one including the twenty-one advanced economies and the other comprising the sixty-nine emerging countries. In columns (3) and (4), it can be seen that there is no evidence of a resource curse for advanced economies, even if political factors are excluded from the analysis. On the contrary, in emerging countries, the resource curse shows in the significant parameter found in column (5) of Table 7.3 which, nevertheless, disappears once political factors are taken into account. The estimated parameters for these political variables appear to be statistically insignificant due to the high co-linearity between them (the sample correlation is 70%); however, the notorious reduction in the root mean square error (RMSE) of around 30% (from 0.0186 to 0.0134) justifies their inclusion.

Note also that the asymmetric behavior of advanced and emerging countries does not refer to the resource curse only. When comparing columns (4) and (6), we found that growth in emerging economies seems to be less affected by the government burden (government consumption in emerging economies is 12.9% in our sample vis-à-vis 19% for advanced economies) but far more affected by inflation and infrastructure levels.

7.4.3 On the Heterogeneity of the Resource Curse

Close examination of the data on democracy and checks and balances reveals the presence of clusters. Figure 7.1 presents a scatter diagram of the two variables. It can be seen that there is a clear, positive correlation between democracy and political checks and balances. This is not surprising since in achieving higher levels of democracy, societies increasingly value the importance of enacting efficient mechanisms of checks and balances. Note that the majority of Arab countries are ranked as non-democratic and non-accountable economies. Nevertheless, while no Arab economies are deemed as democratic, some countries display average levels of political self-control as reflected in the average checks and balances levels achieved by Egypt, Kuwait and Morocco.

The positive association between democracy and checks and balances is subject to important deviations. First and foremost, several countries enjoy high levels of democracy but suffer from relative low levels of political checks and balances, as is the case in Greece, Jamaica and Colombia. On the other hand, some economies – typically in East Asia – enjoy high levels

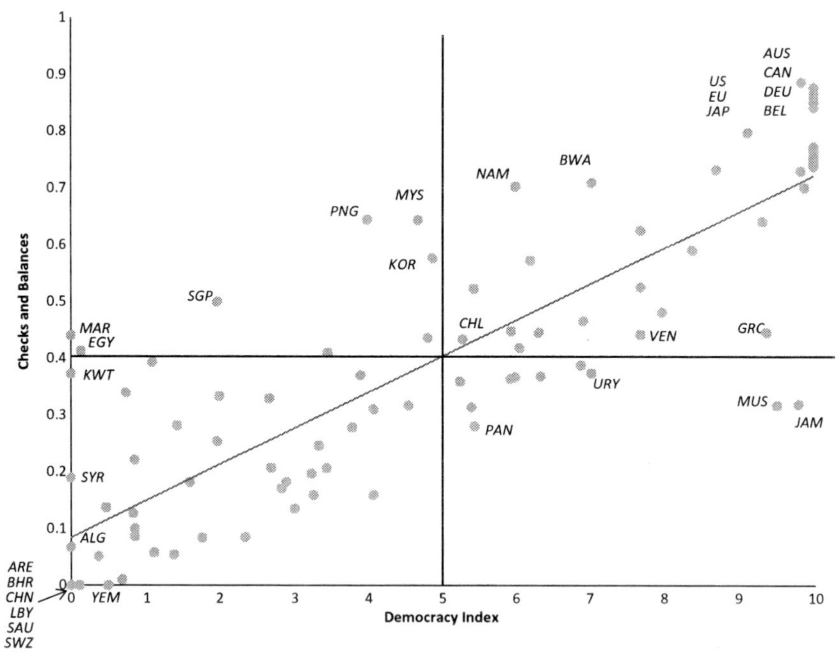

Figure 7.1: Scatter of Political Variables Average 1975–2009

of political responsibility, but restrictive levels of democracy. The latter would include Singapore, Hong Kong and Malaysia and, at times, Korea. Finally, the positive correlation in Figure 7.1 should be taken with care, as averages over long periods of time may darken the role of these variables in economic life. For example, it can be seen that Chile and Uruguay rank in a relatively mediocre position, which is explained by their move from an initial period of draconian military regimes in the 1970s and 1980s towards highly democratic regimes in the 1990s and 2000s.

After repeated testing, we decided to split our sample according to the following cluster criteria:

- Countries with *low level of democracy,* if the index is equal or below 6
- Countries with *high level of democracy,* if the index is above 6
- Countries with *low degrees of checks and balances,* if the index is lower than 0.35
- Countries with *high degree of checks and balances,* if the index is above 0.35

We ran a restricted version of our model in Table 7.3 for each of these subsamples, retaining all fundamentals except for democracy and checks and balances. In Table 7.4, we report the results, focusing only on the "natural resource curse" parameter. One should bear in mind that these subsamples are of smaller size and might, therefore, be subject to the influence of outliers; we trim the results from outliers using a robust regression technique that uses an initial screening based on Cook's distance >1 to eliminate gross outliers before calculating starting values and then perform Huber iterations followed by bi-weight iterations, as suggested by Li (1985).

Table 7.4: *Estimated Long-run Effect of Resource Rents on Economic Growth Allowing for Common Correlated Effects*

	High Checks and Balances	Low Checks and Balances
High Democracy	4.369***	-5.125
	(2.291)	(5.125)
Low Democracy	−0.46	−1.657**
	(0.281)	(0.882)

Note: The estimated parameters of standard controls used in the estimation are excluded to save space. (*,**,***) indicate statistical significance at the 90%, 95% and 99%, respectively.

Three main conclusions can be derived from Table 7.4. First, countries failing to achieve high enough standards of democracy and checks and balances will most likely fail to prevent the resource curse (as implied by the negative and significant rents effect found). Second, countries with above average democratic standards and in-place checks and balances can avoid the resource curse. It is noteworthy that countries such as Botswana, Chile and Peru (where natural resources comprise the majority of exports) have been able to use their resource availability to sustain long-run growth. Third, countries achieving high enough standards of checks and balances, but that are not democratic, will likely be able to nullify the resource curse (i.e., a statistically insignificant rents effect). However, they will not be able to turn the resource rents into a driver for higher growth, as would have been the case if the estimated parameter were positive. Finally, there are only four countries to conform to the group of economies with high democracy levels and low checks and balances to derive statistically and economically significant conclusions. The results clearly indicate that the resource curse operates in countries with low levels of democracy and poor mechanisms of checks and balances.

To assess the political–economy trajectory of specific resource-dependent economies, it would be more appropriate to combine the implications of the aforementioned econometric results with other more nuanced approaches. For example, Barma et al. (2011) introduce the concepts of the "credibility of intertemporal commitment" and "political inclusiveness":

- **The credibility of intertemporal commitment** – or the degree to which policy stability and bargains over time can be enforced and deviations from such agreements are subject to sanction; and
- **The overall political inclusiveness of the prevailing state-society compact** – or the extent to which diverse social, economic and political viewpoints are incorporated into decision making, and a sense of either collectivist or clientelist welfare is privileged over purely elite interests.

Using checks and balances (C&B) and Polity as approximate proxies for the credibility of intertemporal commitment and political inclusiveness respectively, we construct a typology of four distinct country settings

Table 7.5: *Democracy, Checks and Balances as Indicators of Credibility and Intertemporal Commitments*

Polity (Political inclusiveness)	Checks and Balances (C&B) (Credibility of intertemporal commitment)	
	Low C&B: Less credible/weaker enforcement	**High C&B**: More credible/ stronger enforcement
Low Polity: Less inclusive/ less collectively oriented	**Patrimonial Rule** Individualized political authority, built on a hierarchy of cronyism; emphasis on private (elite) goods; exploitation of public resources for private gain	**Hegemonic Government** Institutionalized one-party regime; either predatory or benevolent; emphasis on private (elite) goods with particular and public goods. More inclusive /more collectively oriented
High Polity: More inclusive/ More collectively oriented	**Clientelist Pluralism** Political competition based on extensive use of clientelism; provision of particular goods; low horizontal accountability	**Programmatic Pluralism** Electoral competition based on programs geared toward collective welfare enhancement; provision of public goods; democratic accountability

Note: adapted from table 2 of Ross et al. (2011).

(Table 7.5). The usefulness of this typology hinges on the fact that, though Polity and C&B are not perfectly orthogonal, they are not perfectly correlated (Figure 7.1).

If we strictly adhere to the measured Polity and C&B, all oil-rich Arab countries, for example, will be placed under the "patrimonial rule," which is the most vulnerable group to the oil curse. However, in reality, the conservative and long-reigning monarchies of the GCC (as well as Morocco and Jordan) are more likely to achieve higher level of intertemporal commitment than would be suggested by the numerical checks and balances index. Using a broad assessment of regime characteristics, Ross et al. (2011) argues that the monarchies in the Arab region appear to be more credible than the so-called "republican" oil-dependent Arab states, such as Algeria, Iraq, Sudan, Syria and Yemen. Therefore, it seems plausible that the GCC countries might make the "hegemonic group," while the latter remain in the most vulnerable group. Such a typology would augur well with the observed growth experience of the two groups.

Finally, having established that bad political governance is the underlining cause of the resource curse, we now assess the role of certain economic variables as channels for these institutional effects. Confining the analysis to the case of low democracy and low checks and balances (the last column of Table 7.4), we test for the role of real exchange rate overvaluation and current account, reflecting the Dutch Disease view of the oil curse; and stock of debt to account for the debt overhang influence on growth.[9] To establish such causative links of these variables with growth, we must satisfy three conditions:

- First, that these variables are linked to the presence of natural resource rents, which is, in fact, the essence of the Dutch Disease literature for the case of the real exchange rate overvaluation or the unsustainable domestic absorption (high current account deficit or high external debt).
- Second, these intervening variables are also likely to be strongly correlated with growth – again these three variables were found to be among the growth fundamentals in the recent empirical growth literature.
- Third, the addition of an intervening variable to the growth regression reduces both the size and/or the statistical significance of the estimated effect of the resource rents.

Table 7.6 reports the results, focusing only on the "natural resource curse" parameters and the three intervening economic variables; results for other controls are not critical for this analysis, and hence, are not reported. The results suggest that the current account and stock of debt individually render the resource rents insignificant, while both are negative and highly significant (regressions 2 and 3). However, when both variables are included in the same regression, only the latter remains significant, though the resource rents estimator continues to be insignificant (regression 6). On the other hand, real exchange rate overvaluation enters insignificantly, though it also renders the resource rent insignificant (regression 1). Perhaps this result is due to the fact that the RER overvaluation index used in this regression is based on price comparisons (Rodrik 2008) and is,

[9] These are the same set of variables used in Collier and Goderis (2009), except that we use the more appropriate measure of real exchange rate overvaluation, while they used the real exchange rate variable.

Table 7.6: *Econometric Results: Long-run Growth Determinants Allowing for Transmission Channels Low Democracy and Low Checks and Balances*

Variable	(1)	(2)	(3)	(4)	(5)	(6)	(7)
Resource Rents	–0.098	–0.139	–0.041	–0.015	0.230	–0.088	0.091
(as % of GDP)	(0.250)	(0.213)	(0.445)	(0.149)	(0.244)	(0.267)	(0.070)
RER Overvaluation	-0.002	–	–	–0.105**	–0.112**	–	–0.166***
(logs)	(0.060)			(0.070)	(0.029)		(0.051)
External Debt Stock	–	–0.184***	–	-0.165**	–	–0.257**	–0.281***
(% of GNI)		(0.090)	–	(0.072)		(0.100)	(0.096)
Current Account Balance	–	–	–0.191*	–	–0.321**	–0.230	–0.335***
(as % of GDP)			(0.105)		(0.133)	(0.154)	(0.149)
Observations	898	772	767	762	753	679	628
Countries	47	41	43	41	42	39	34

Note: The estimated parameters of traditional controls used in the estimation are excluded to save space.

therefore, more basic than the model-based approach that accounts for the non-traded goods equilibrium and the intertemporal external balance of an economy.[10] However, when added to the stock of debt (regression 4), to the current account (regression 5), or when all three variables are included (regression 7), the three variables were all negatively and highly significantly associated with growth, while the resource rents become insignificant. In comparing the results of these regressions, it appears that accounting for all three channels produces the best results in terms of the precision of estimates, degree of significance, as well as the size of the estimated coefficients.

7.5 Conclusions

There is now a near consensus among scholars and development practitioners alike that under certain conditions, oil and mineral resource rents can be harmful to development, hence generating the so-called natural resource curse. Moreover, there is consensus that the curse is a long-term phenomenon that is not necessarily incompatible with short-run growth

[10] See, for example, Aguirre and Calderon (2005), Elbadawi and Soto (2008) and Elbadawi, Kaltani and Soto (2011), among others.

spells during the boom cycle. The recent debate about whether or not and how a curse might happen has been waged in context of the endogenous growth model. The application of modern panel data econometrics allows testing of the various theories that have been discussed in the literature for explaining the curse. Reasonably robust evidence now exists on the central role of political institutions, showing that the curse is more likely to happen in non-democratic countries or, especially, those that do not have strong enough systems of political checks and balances.

However, these findings are obtained in highly restrictive econometric models that assume no country heterogeneity, as in panel cointegration; or at best allow for heterogeneity, but only in the short run. We argue, in this chapter, that due to the strong cross correlations in growth fundamentals, the received literature might produce biased results. Instead, we apply second generation panel estimators that account for country heterogeneity and correlations in both the short and long runs.

Moreover, we unpack political institutions into those reflecting the degree of inclusiveness (Polity) and credibility of intertemporal commitments (political checks and balances). We find that resource-rich countries with low levels on both scores are likely to experience the curse, while those with high enough levels on both managed to turn the resource rents into a driver of growth. However, those countries with low Polity but high checks and balances or those with high Polity but low checks and balances might be able to avoid the curse, but are likely to not effectively use the resource rents to promote growth. These findings suggest that for the oil-rich Arab World to achieve sustained growth, the Arab Spring should not only bring democracy, as badly needed as it is in this region, but should also lay the foundations for strong systems of political checks and balances.

Finally, a cautionary note as well as a challenge for future research is in order. While the advantages of second generation panel data estimators are promising, one should acknowledge that their limitations are yet to be explored. Consequently, we take our results as indicative of a need for further research. In particular, we need to better understand the properties of the new second generation panel data econometrics, in terms of the power of the test and the robustness against potential model misspecification. Likewise, at this nascent stage of the econometric literature on common-correlated effects, threshold effects cannot be tested for, although they were found to be present when using first generation panel data econometrics (fixed-effects, GMM, etc.).

APPENDIX A: Definitions and Sources of Variables Used in Regression Analysis

Variable	Definition and Construction	Source
Real Per Capita GDP	Ratio of total GDP to total population. GDP in US dollars of 2000.	World Bank (2011)
Education and Human Capital Formation	Educational attainment, computed as completion rates in five-year age intervals for 146 countries from 1950 to 2010. Exponential interpolation used to obtain annual data.	Barro and Lee (2011)
Trade Openness (% of GDP)	Residual of a regression of the log of the ratio of merchandise trade (% of GDP), on the logs of area and population, as well as dummies for oil-exporting and landlocked countries.	Author's calculations, based on data from World Bank (2011)
Index of Capital Account Openness	The index is based on the binary dummy variables that codify the tabulation of restrictions on cross-border financial transactions reported in the IMF's Annual Report on Exchange Arrangements and Exchange Restrictions (AREAER).	Chinn and Ito (2011)
Government Burden	Ratio of government consumption to total GDP.	World Bank (2011)
Inflation	Measured by annual log-change of the consumer price index.	World Bank (2011)
Main Telephone Lines per 1,000 Workers	Telephone lines connecting a customer's equipment to the public switched telephone network.	World Bank (2011)
Remittances	Workers' remittances and compensation of employees received (% of GDP).	World Bank (2011)
Systemic banking crises	Number of years in which a country underwent a systemic banking crisis, as a fraction of the number of years in the corresponding period.	World Bank (2011)
Terms of Trade	Ratio of export unit values to import unit values.	World Bank (2011)
Resource Rents	Total natural resources rents (% of GDP) computed as the difference between world prices and the average unit cost of extraction times the quantity extracted or harvested.	World Bank (2011)
Democracy	Regime authority characteristics and transitions datasets.	Polity IV Project (2011)
Checks and Balances	Political constraint index (POLCON-V)	Henisz and Zelner (2010)
Real Exchange Rate Overvaluation	Balassa–Samuelson adjusted measure of the domestic price level.	Rodrik (2008)
Foreign debt	External debt stocks, total (DOD, current US$) as share of GNI.	World Bank (2011)
Current account balance		World Bank (2011)
Period-specific shift	Time dummy variable.	Authors' construction

APPENDIX B: Sample of Countries

United Arab Emirates	Gabon	Norway
Argentina	United Kingdom	New Zealand
Australia	Ghana	Pakistan
Austria	Gambia, The	Panama
Burundi	Greece	Peru
Belgium	Guatemala	Philippines
Benin	Honduras	Papua New Guinea
Burkina Faso	Hungary	Portugal
Bangladesh	Indonesia	Paraguay
Bahrain	India	Rwanda
Bolivia	Ireland	Saudi Arabia
Brazil	Iran, Islamic Rep.	Sudan
Botswana	Iceland	Senegal
Central African Republic	Israel	Singapore
Canada	Italy	Sierra Leone
Switzerland	Jamaica	El Salvador
Chile	Jordan	Sweden
China	Japan	Swaziland
Cote d'Ivoire	Kenya	Syrian Arab Republic
Cameroon	Korea, Rep.	Togo
Congo, Rep.	Sri Lanka	Thailand
Colombia	Lesotho	Trinidad and Tobago
Costa Rica	Morocco	Tunisia
Germany	Madagascar	Turkey
Denmark	Mexico	Uganda
Dominican Republic	Mongolia	Uruguay
Algeria	Mauritius	United States
Ecuador	Malawi	Venezuela, RB
Egypt, Arab Rep.	Malaysia	South Africa
Spain	Namibia	Congo, Dem. Rep.
Ethiopia	Niger	Zambia
Finland	Nicaragua	Zimbabwe
France	Netherlands	

References

Aguirre, A. and C. Calderon (2005). Real Exchange Rate Misalignments and Economic Performance. Central Bank of Chile Working Paper No. 315.

Arellano, M. and S. Bond (1991). Some Tests of Specification for Panel Data: Monte Carlo Evidence and an Application to Employment Equations. *The Review of Economic Studies* 58(2): 277–297.

Arellano, M. and O. Bover (1995). Another Look at the Instrumental Variable Estimation of Error-components Models. *Journal of Econometrics* 68(1): 29–51.

Arezki, R., K. Hamilton and K. Kazimov. (2011). Resource Windfalls, Macroeconomic Stability and Growth: The Role of Political Institutions. IMF Working paper WP/ 11/142, International Monetary Fund, Washington DC.

Arezki, R. and F. Van der Ploeg (2008). Can the Natural Resource Curse be Turned into a Blessing? Role of Trade Policies and Institutions. IMF Working Paper 07/55, International Monetary Fund, Washington, DC.

Bai, J. (2009). Panel Data Models With Interactive Fixed Effects. *Econometrica* 77: 1229–1279.

Bai, J., Kao, C. and Ng, S. (2009). Panel Cointegration with Global Stochastic Trends. *Journal of Econometrics* 149(1): 82–99.

Barma, N., K. Kaiser, T. M. Le and L. Viñuela (2011). *Rents to Riches? The Political Economy of Natural Resource Led Development*. Washington, DC: World Bank Press.

Barro, R. (1991). Economic Growth in a Cross Section of Countries. *Quarterly Journal of Economics* 56(2): 407–443.

Barro, R. J. and J. W. Lee (2011). A New Data Set of Educational Attainment in the World 1950–2010. NBER Working Paper No. 15902.

Bhattacharyya, S. and R. Hodler (2010). Natural Resources, Democracy and Corruption. *European Economic Review* 54(4): 608–621.

Bigsten, A., P. Collier, S. Dercon, M. Fafchamps, B. Gauthier, J. W. Gunning, A. Oduro, R. Oostendorp, C. Pattillo, M. Soderbom, F. Teal and A. Zeufack (2002). Do African Manufacturing Firms Learn from Exporting? Working Paper WPS/ 2002-09. Centre for the Study of African Economies, Oxford, UK.

Budina N. and S. van Wijnbergen (2008). Managing Oil Revenue Volatility in Nigeria: The Role of Fiscal Policy. Chapter 10 in *Africa at Turning Point?* D. Go and J. Page (eds). Washington. D.C.: World Bank.

Canning D., M. Fay and R. Perotti (1994). Infrastructure and Growth. In *International Differences in Growth Rates*, M. Baldassarri, L. Paganetto and E. Phelps, (eds). Rome: St. Martin's Press.

Chinn, M. and H. Ito (2008). A New Measure of Financial Openness. *Journal of Comparative Policy Analysis* 10(3):307–320.

Collier, P. and A. Hoeffler (2009). Testing the Neocon Agenda: Democracy in Resource-rich Societies. *European Economic Review* 53(3): 293–308.

Collier, P., F. van der Ploeg; M. Apence and A.J. Venables (2009). Managing Resource Revenues in Developing Economies. OxCarre Research Paper 15, University of Oxford.

Corden, W. M. and J. P. Neary (1982). Booming Sector and De-Industrialization in a Small Open Economy. *The Economic Journal* 92 (368): 825–848.

Collier, P. and B. Goderis. (2009). Structural Policies for Shock-prone Developing Countries. CSAE Working Paper No. 2009-03, Department of Economics, University of Oxford.

Corden, W. M. (1982). Exchange Rate Policy and the Resource Boom. *Economic Record* 58(160): 18–31.

Debrun X. and M. S. Kumar (2007). The Discipline-Enhancing Role of Fiscal Institutions: Theory and Empirical Evidence. IMF Working Papers 07/171, International Monetary Fund.

Dollar, D. (1992). Outward-oriented Developing Economies Really Do Grow More Rapidly: Evidence from 95 LDCs, 1976–1985. *Economic Development and Cultural Change* 40(3): 523–44.

Durlauf, S. N., P. A Johnson, and Temple, J. (2005). Growth Econometrics. In, *Handbook of Economic Growth*, volume 1, P. Aghion and S. Durlauf (ed.), chapter 8, pp. 555–677. Amsterdam, The Netherlands: Elsevier.

Easterly, W., et al. (1993). Good Policy or Good Luck? Country Growth Performance and Temporary Shocks. *Journal of Monetary Economics* 32(3): 459–83.

Eberhardt, M. and F. Teal (2011). Econometrics for Grumblers: A New Look at the Literature on Cross-Country Growth Empirics. *Journal of Economic Surveys* 25(1): 109–155.

Elbadawi, I., L. Kaltani and R. Soto (2012). Aid, Real Exchange Rate Misalignment and Growth in Sub-Saharan Africa. World development (forthcoming).

Elbadawi, I. (2005). The Politics of Sustaining Growth in the Arab World: Getting Democracy Right. Working Paper Series No. 2, 2004, American University of Beirut Institute of Financial Economics.

Elbadawi, I. and A. Gelb. (2010). Oil and Economic Diversification in the Arab World. Policy Research Report No. 35, December 2010, Economic Research Forum, Cairo, Egypt.

Elbadawi, I. and R. Soto (2008). Theory and Empirics of Real Exchange Rates in Developing Countries. Unpublished mimeo, Development Economic Research Group, World Bank, November.

Fischer, S. (1993). The Role of Macroeconomic Factors in Growth. *Journal of Monetary Economics* 32(3): 485–511.

Gelb, A. (1988). *Oil Windfalls: Blessing or Curse?* Washington DC.: Oxford University Press.

Gelb, A. and S. Grasmann (2008). Confronting the Oil Curse. Unpublished mimeo, The World Bank, Washington DC.

Gylfason, T. (2011). Natural Resources Endowment: A Mixed Blessing? CESifo Working Paper No. 3353.

Gylfason, T., T. T. Herbertsson and G. Zoega (1999). A Mixed Blessing. *Macroeconomic Dynamics* 3(2): 204–225.

Hausmann, R. (2001). Venezuela's Growth Implosion: A Neo-classical Story? In *In Search of Prosperity: Analytic Narratives on Economic Growth*, D. Rodrik (ed.). Princeton, NJ: Princeton University Press.

Hausmann, R. and R. Rigobon (2003). An Alternative Interpretation of the 'Resource Curse: Theory and Policy Implications. NBER Working Paper No. 9424.

Humphreys, M., J.D. Sachs and J.E. Stiglitz (2007). *Escaping the Resource Curse*. New York: Columbia University Press.

Henisz, W. and B. A. Zelner (2010). *Measures of Political Risk Database*. Georgetown University: The McDonough School of Business.

Henisz, W. (2010). Political Constraint Index (POLCON) Dataset, accessed from www-management.wharton.upenn.edu/henisz/. on April 6, 2013.

Islam, N. (1995). Growth Empirics: A Panel Data Approach. *Quarterly Journal of Economics* 110(4): 1127–1170.

Kapetanios, G., M. H. Pesaran and T. Yamagata (2011). Panels with Non-Stationary Multifactor Error Structures. *Journal of Econometrics* 160(2): 326–348.

Korhonen, I. and T. Juurikkala (2007). Equilibrium Exchange Rates in Oil-Dependent Countries. BOFIT Discussion Papers 8/2007, Bank of Finland, Institute for Economies in Transition.

Lane, P. R. and A. Tornell (1996). Power, Growth, and the Voracity Effect. *Journal of Economic Growth* 1: 213–241.

Levine, R., N. Loayza and T. Beck (2000). Financial Intermediation and Growth: Causality and Causes. *Journal of Monetary Economics* 46(1): 31–77.

Li, G. (1985). Robust Regression. In *Exploring Data Tables, Trends, and Shapes*, D. C. Hoaglin, F. Mosteller, and J. W. Tukey (eds), 281–340. New York: Wiley.

Loayza, N. and R. Soto (2002). *Economic Growth. Sources, Trends, and Cycles*. Banco Central de Chile: Santiago de Chile.

Maloney, W. and D. Lederman (2008). In Search of the Missing Resource Curse. Unpublished mimeo, The World Bank, Washington DC.

(2007). Trade Structure and Growth. In *Natural Resources: Neither Curse nor Destiny*, W. Maloney and D. Lederman (eds.) pp. 15–40. Stanford, California: Stanford University Press for the World Bank.

Matsen, E. and R. Torvik (2005). Optimal Dutch Disease. *Journal of Development Economics* 78(2): 494–515.

Matsuyama, K. (1992). Agricultural Productivity, Comparative Advantage, and Economic Growth. *Journal of Economic Theory* 58(2): 317–334.

Mehlum, H., K. Moene and R. Torvik (2006). Institutions and the Resource Curse. *The Economic Journal* 116: 1–20.

Neary, J. P. and S. J. G. van Wijnbergen (1986). *Natural Resources and the Macroeconomy*. Cambridge, MA: MIT Press.

Pesaran, M. H. (2006). Estimation and Inference in Large Heterogeneous Panels with a Multifactor Error Structure. *Econometrica* 74(4): 967–1012.

Robinson, J. A., R. Torvik and T. Verdier (2006). Political Foundations of the Resource Curse. *Journal of Development Economics* 79: 447–468.

Robinson, J. A. and R. Torvik (2005). White Elephants. *Journal of Public Economics* 89: 157–566.

Rodrik, D. (1999). Where Did All The Growth Go? External Shocks, Social Conflict and Growth Collapses. *Journal of Economic Growth* 4(4): 385–412.

(2007). Why Does the Real Exchange Rate Matter to Growth? Unpublished mimeo, Kennedy School of Government, Harvard University.

(2008). The Real Exchange Rate and Economic Growth. Mimeo, Kennedy School of Government, Harvard University.

Ross, M., K. Kaiser and N. Mazaheri. (2011).The Resource Curse in MENA: Political Transitions, Resource Wealth, Economic Shocks and Conflict Risks. Policy Research Working paper 5742, World Bank, Washington DC.

Sachs, D. J. (2007). How to Handle the Macroeconomics of Oil Wealth. In *Escaping the Resource Curse*, M. Humphreys, J. D. Sachs and J. E. Stiglitz (eds.), pp. 173–193. New York: Columbia University Press.

Sachs, J. and A. Warner (1995). Natural Resource Abundance and Economic Growth. NBER Working Paper 5398. Cambridge, MA.

Stiglitz, J. E. (2007). What is the Role of the State? In *Escaping the Resource Curse*, M. Humphreys, J. D. Sachs and J. E. Stiglitz (eds.), pp. 23–52. New York : Columbia University Press.

Westerlund, J. and Edgerton, D.L. (2008). A Simple Test for Cointegration in Dependent Panels with Structural Breaks. *Oxford Bulletin of Economics and Statistics* 70(5): 665–704.

8

Fiscal Institutions in Resource-rich Economies

*Lessons from Chile and Norway**

Klaus Schmidt-Hebbel**

ABSTRACT

Resource-rich economies in general, and Arab oil exporters in particular, are at a critical juncture, facing the challenge of revamping their fiscal policy institutions and conduct to strengthen macroeconomic and financial stability, raise growth, and improve intra/inter-generational equity. This paper starts by reviewing the international evidence on fiscal policies and outcomes in resource-rich economies at large and Arab oil-exporting countries in particular, which suggests that strong fiscal (and political) institutions can turn the resource curse into a blessing. Then the paper provides comparative reviews of Chile's and Norway's decade-long experience in setting up new fiscal institutions and rules to manage their resource rents, aiming at and, in fact, attaining more macroeconomic and financial stability, higher growth and improved equity. Specific reform lessons to strengthen fiscal institutions and policies are drawn for resource-rich economies and Arab oil exporters.

JEL Classification: E61, E62, E63, H61

Keywords: Fiscal institutions, fiscal rules, resource-rich economies, Chile, Norway

* Financial support from the Economic Research Fund is gratefully acknowledged. I thank Hasan Ersel, Raimundo Soto and, in particular, Ibrahim Elbadawi for excellent comments to previous drafts. I also thank Francisco Muñoz for very proficient assistance.

** Catholic University of Chile. email: kschmidt-hebbel@uc.cl

8.1 Introduction

Resource-rich economies (RREs) in general, and Arab oil-exporting countries in particular, are at a critical juncture, facing the challenge of revamping their fiscal policy institutions and conduct to strengthen macroeconomic and financial stability, raise growth and improve intra/intergenerational equity. The ongoing world price boom that benefits exporters of oil and other commodities offers RREs a very favorable window of opportunity to adopt deep fiscal reforms.

Section 8.2 reviews the recent international evidence on fiscal policies and outcomes in REEs in general and Arab oil exporters in particular. Among RREs, Chile and Norway revamped their fiscal policy frameworks a decade ago, setting up new fiscal institutions and rules to manage their resource rents, aiming at more macroeconomic and financial stability, higher growth and improved inter-generational equity. Sections 8.3 and 8.4 describe and assess the conduct of policy under the new fiscal institutions and rules in Chile and Norway and relate them to fiscal and macroeconomic performance. Section 8.3 analyzes Chile's case, focusing on the country's fiscal institutions and outcomes, fiscal rule, sovereign wealth funds (SWFs) and the shortcomings of its current institutional framework and proposals for reform. Section 8.4, on Norway, reviews the country's institutional setup, fiscal rule, budget performance and SWFs. The subsequent section derives specific policy lessons from Chile and Norway for RREs and Arab oil exporters. Section 8.6 concludes.

8.2 Evidence on Fiscal Policies and Outcomes in Resource-rich Economies and Arab Oil-producing Countries

Economic history abounds with negative examples of inconsistent and inefficient design and execution of fiscal policy in RREs, which lead to macroeconomic and financial crises, debt defaults, financial underdevelopment, low growth and inequity.

The following is a brief summarization of the evidence on fiscal policies and their outcomes in RREs in general. Following that is a review of the corresponding evidence available for oil-producing Arab countries in particular.

8.2.1 Resource-rich Economies

8.2.1.1 Natural Resources: Curse or Blessing?

In countries where the relative size of the resource sector is similar, a resource boom is more likely to benefit those economies with better initial institutions (Robinson et al. 2003). A more refined, subsequent analysis shows that institutions are endogenous to both resource rents and the size of tax revenue from the non-resource sector. When the latter taxes are low and the resource sector is large, resource rents corrupt institutions and lead to low growth; the opposite – a resource blessing – is observed where non-resource taxes are large (Salti 2008). Complementary evidence shows that RREs typically raise less revenue from non-resource taxation, as a result of high resource rents. This causes governance problems as lower domestic taxation reduces the incentive for public scrutiny by governments (Bornhorst et al. 2008).

8.2.1.2 Fiscal Pro-cyclicality

A large body of empirical international research shows that fiscal (and monetary) pro-cyclicality is more likely in countries with poor fiscal governance, high levels of corruption, low fiscal credibility, domestic financial under-development and weak integration into world financial markets (Ilzetzki and Végh 2008; Calderón and Schmidt-Hebbel 2008; Abdih et al. 2010; Calderón et al. 2016). This evidence is particularly relevant for many RREs, where fiscal policy is pro-cyclical. For example, non-oil primary government balances worsened substantially in oil-producing countries during the 2003–2008 oil-price boom and improved in 2009 – hence exhibiting strong policy pro-cyclicality (Villafuerte and López-Murphy 2010).

8.2.1.3 Weak Fiscal Sustainability

Fiscal policy sustainability is weak in many resource-rich economies. Long-term fiscal sustainability positions (measured by fiscal policy positions relative to estimated sustainable fiscal benchmarks) have worsened in oil-producing countries during the 2003–2008 boom (Villafuerte and López-Murphy 2010).

8.2.1.4 Fiscal Vulnerability to Resource Price Reversals

Fiscal positions are often vulnerable to reversals after price booms in resource-rich economies. In oil-producing countries, a relatively small decline in oil prices leads to a substantial increase in net financing needs, as a result of absent fiscal and exchange rate adjustment (Villafuerte and López-Murphy 2010).

8.2.1.5 Macroeconomic Volatility, Dutch Disease and Lower Growth

Fiscal pro-cyclicality and lack of strong fiscal fundamentals lead to macroeconomic volatility, reflected particularly in large volatility of the real exchange rate and activity. Fiscal policy pro-cyclicality has amplified business cycles in oil-producing countries (Villafuerte and López-Murphy 2010; Abdih et al. 2010). Pro-cyclical government spending leads to Dutch Disease during revenue booms, reflected in exchange rate misalignment, loss of export competitiveness and "large non-resource current account deficits"?.[1] Moreover, contrary to the expectation that resource rents are used to finance much-needed public investment, public capital is lowered by resource rents worldwide – a perverse effect that is mitigated by better institutions and lower macroeconomic volatility (Bhattacharyya and Collier 2011).

8.2.1.6 Fiscal and Political Institutions

Growing cross-country evidence and country studies show that political institutions generally and fiscal institutions in particular, help turn a natural resource curse into a blessing. Weak fiscal policy institutions have adverse effects that go well beyond policy pro-cyclicality and Dutch Disease. Opaque budgetary management and external control, lack of transparent fiscal policies and budgets and poor budgetary accountability lead to ineffective and inefficient government spending, misuse of government resources and corruption.

At the heart of good fiscal and political institutions is good governance, based on adequate political checks and balances, transparency and accountability, and well-designed and strongly-enforced incentives faced by politicians – ultimately democracy (Torvik 2012). RREs that have been blessed by natural resource rents are those that have put in place strong fiscal and political institutions, including Botswana, Chile, Malaysia, Mauritius and Norway (Gylfason 2012). Strong systematic empirical evidence for the world on the links between resource rents, political institutions and growth are provided by Elbadawi and Soto (this volume). They show that RREs with both a high degree of inclusiveness (a measure of democracy) and strong political checks and balances turn resource rents into a blessing, exhibiting larger growth than with the rents alone. On the contrary, RREs

[1] Recent evidence shows that the size of the fiscal balance coefficient as a determinant of the current account balance is much larger in OPCs than in other industrial and emerging economies (Morsy 2009).

that lack democracy and/or strong checks and balances, suffer from the ultimate manifestation of a resource curse: low growth.

8.2.2 Arab Oil-exporting Countries

As discussed earlier in the volume, the Arab region is highly dependent on hydrocarbons. Currently, the twenty-two-country Arab League includes eleven hydrocarbon (oil and gas)-exporting countries, which are included in the group of RREs discussed earlier. On average, the dependence of Arab oil-exporting countries on oil (and gas) is larger than the dependence of the average RRE in the world on its resource exports. The whole region accounts for close to half of the oil reserves and one quarter of the natural gas reserves in the world, and their hydrocarbon sector represents 50% of GDP and at least two-thirds of government revenue.

Like RREs in general, many Arab oil exporters show evidence of a natural resource curse reflected in fiscal pro-cyclicality, macroeconomic volatility, RER misalignment and lack of productive diversification. In oil- and gas-exporting Gulf Cooperation Council (GCC) countries, government spending follows oil revenue, confirming expenditure pro-cyclicality (Fasano and Wang 2002).

Of the eleven oil-exporting Arab countries, seven countries are rich labor-importing economies and four are less rich labor-abundant economies. Fiscal policy institutions tend to be more developed in the labor-importing, richer Arab countries. For example, among the world's oldest, largest and best-managed sovereign wealth funds are those belonging to UAE (Abu Dhabi), Kuwait and Saudi Arabia.

However, many manifestations of Dutch Disease are not found to be correlated with the divide between the GCC and populous Arab oil exporters. Fiscal policy (reflected by the fiscal balance) is found to be pro-cyclical in all Arab resource-rich countries except Kuwait (Selim and Zaki, this volume). While there is no evidence of very large real exchange rate appreciation in response to higher oil rents, three Arab countries (Algeria, Kuwait and Saudi Arabia), exhibit severe Dutch Disease reflected by exchange rate appreciation and either a stagnation or a sharp reduction in their manufacturing sector (Selim and Zaki, this volume).

Recent case studies provide in-depth evidence on the links between oil rents, fiscal institutions and macroeconomic performance in Arab oil exporters. Saudi Arabia's large government sector pays wages that are significantly higher than those paid by the private sector, crowding out private business and contributing to high unemployment (Al-Sheikh and

Erbas, this volume). Yemen exhibits some of the standard oil curse symptoms derived from its new oil and gas wealth. In Yemen, weak fiscal and political institutions have led to pro-cyclical fiscal policy, Dutch Disease, a sub-optimal use of oil wealth and political instability and conflict (Al-Iriani 2012). Sudan's experience has been similar to Yemen's: oil rents and weak fiscal and political institutions have led to fiscal policy pro-cyclicality, real exchange rate misalignment, negative TFP growth and an exacerbation of political conflict (Suliman, this volume).

8.3 Chile's Fiscal Institutions and Outcomes

8.3.1 Chile's Institutional Framework for Fiscal Policy

A modern institutional framework for the conduct of fiscal policy and financial management should aim at addressing the principal agent problems that arise between voters and political authorities due to government impatience, lack of representation of future generations, electoral competition, sensitivity to special interest lobbies, corruption and use of asymmetric and biased information (von Hagen 2005; Wren-Lewis 2010). To overcome these distortions and negative externalities, the academic literature and international experience suggests adopting an institutional framework for fiscal policy based on the following components (Ter-Minassian 2010; IMF 2009, 2010; Debrun et al. 2009): a fiscal responsibility law, modern financial management, a planning horizon that exceeds one year, a fiscal rule for the budget, rules for government asset and liability management, requirements on accountability and public information on the government's financial management, effective external control and auditing, and establishment of a fiscal council and/or fiscal committees.

The following is a description of Chile's development of the eight latter components of a modern institutional framework.

8.3.1.1 Fiscal Responsibility Law

Several countries have adopted fiscal responsibility laws, specifying budgetary procedures and rules to strengthen fiscal policy transparency, accountability and stability. Fiscal responsibility laws establish requirements on the executive branch to state transparently its short- and medium-term policy objectives, to set short- and medium-term targets for fiscal flows and stocks, to provide transparent information on budget planning and execution to the legislature, to implement a fiscal policy conducive to fiscal stability and solvency and to account ex-post for its fiscal policy execution and attainment of policy objectives (Lienert 2010).

However, fiscal responsibility laws do not always spell out an explicit fiscal rule. Conversely, fiscal rules are often adopted in the absence (or outside) a fiscal responsibility law. For example, in 2008, Germany adopted a fiscal rule by constitutional amendment, not as part of a fiscal responsibility law.

Fiscal responsibility laws are often enacted as legal tools in support of fiscal consolidation, which sometimes occurs under a stabilization program agreed with the IMF, so that its emphasis is on budgetary control rather than effective fiscal policy. However, under different conditions, such laws are adopted with the aim of strengthening the institutional framework for fiscal policy, focusing on the aforementioned components of a modern institutional framework – like in Chile's case.

Chile enacted Law No. 20128 on Fiscal Responsibility in 2006 to strengthen the institutional framework for its fiscal rule. The rule had been adopted as an administrative decision five years earlier by a previous administration, as a way to tie its own hands and possibly its successors', but it lacked legal support. Therefore, the 2006 Law was not adopted to enforce fiscal adjustment but to develop the institutional framework required to strengthen the links between the fiscal rule, the use of government savings and the establishment of two sovereign wealth funds. Its major provisions are the following (Rodríguez et al. 2006):

(i) Legal requirement for a new government administration to define and publish the fiscal policy framework for its four-year term and its implications for the government's "structural" (i.e., cyclically adjusted) balance.
(ii) Legal requirement for the administration to issue an annual report on the financial state of the government, its fiscal sustainability and the macroeconomic and financial implications of fiscal policy, as well as an estimation of its cyclically adjusted balance.
(iii) Requirement of annual estimation of government contingent liabilities (required to determine the cyclically adjusted balance target).
(iv) Establishment of a Pension Reserve Fund (PRF) to finance future pension liabilities of the government. During its first ten years of existence, the PRF accumulates funds at a yearly floor equivalent to 0.2% of GDP and a ceiling of 0.5%, where the actual figure within the aforementioned range is determined by the government.

The funding sources for the floor are government revenues or government savings.

(v) Establishment of an Economic and Social Stabilization Fund (ESSF), starting with the foreign resources accumulated in the former Copper Revenue Compensation Fund (which is abolished) and other foreign asset holdings of the government at the end of 2006. The budget surplus (after payment into the PRF) is transferred into the ESSF and resources are withdrawn from the ESSF to finance budget deficits (including payments into the PRF).

(vi) Government option to engage in annual capital transfers to the Central Bank of Chile (CBC) for the latter's capitalization during five years (2007–2011), on an amount bound by a ceiling of 0.5% of GDP and subject to availability of a government surplus after payment into the PRF. The government holds the right to decide how to allocate the budget surplus to the ESSF and to CBC capitalization, subject to the aforementioned restrictions.

(vii) International investment of the resources held in the two sovereign wealth funds – PRF and ESSF – can be executed directly by the Treasury (under the Ministry of Finance, MoF) or can be outsourced to the CBC or private fund managers hired by the CBC. The law also establishes a new Advisory Financial Committee for Fiscal Responsibility Funds of the MoF to advise the minister on investment regulations and decisions related to the two sovereign wealth funds.

Note that the latter law does not impose any specific fiscal rule on the government. It rather implies a requirement to adopt and implement a fiscal policy framework aiming at fiscal sustainability, based on a distinction between actual and cyclically adjusted government balances. This provides sufficient flexibility to new governments to define the explicit formula for the fiscal rule they commit to at the start of their administration. In addition, the law defines new institutions – adopted in 2006–2007 – that strengthen operation of the fiscal rule.

8.3.1.2 Financial Management of the Budget

International best practice of budgetary management is based on the adoption of the following core elements for efficient financial management by the MoF (Ter-Minassian 2010):

(a) Strong constitutional and legal powers granted to the MoF for legal initiatives with a budgetary impact,

(b) Strong political power of the MoF to adopt and implement the budgets pertaining to other ministries, regional and municipal governments, independent government agencies and public enterprises,
(c) High technical capacity for budgetary planning and ensuring its consistency with the government's fiscal policy objectives,
(d) Adoption of modern accounting principles,
(e) Efficient legislative approval of the budget, limiting the possibility of legislative amendments that contradict government fiscal policy and
(f) Significant capacity in monitoring and obtaining early information on budget and in implementing budgetary corrections of observed deviations.

International evidence shows that countries that meet the latter requirements attain better performance in budget execution, stability and sustainability of fiscal policy (Alesina and Perotti 1996; Stein et al. 1998; Alesina et al. 1999; Ter-Minassian 2010).

Chile's strong presidential system empowers the government with budgetary initiative (not shared with congress) and strong control over budget design and negotiation with congress. The executive holds the sole power of tax initiative. Congress lacks line-item veto and is only entitled to vote for or against the government's proposed budget bill. It also lacks in-house capabilities to assess and evaluate in depth current fiscal policy, projections in the budget bill and underlying macroeconomic assumptions.

However, significant budget negotiations take place between the government represented by the MoF and congress before the budget comes to a vote. The latter negotiations tend to focus on budget (i.e., spending) composition rather than on taxation or the overall resource constraint, not the least because of the constraints imposed by the fiscal rule on the budget's overall resource envelope.

8.3.1.3 Budget Planning Horizon

Many countries are reforming their budget planning, shifting from one-year to multi-year horizons. This change is aimed at different objectives: (i) achieving greater transparency, consistency and intertemporal sustainability of fiscal policy, (ii) reducing government short-term bias in public spending decisions and (iii) committing to a path of gradual correction of initially unsustainable fiscal positions.

On the one hand, there are governments that lack fiscal rules but adopt multi-annual budgeting to enhance sustainability and credibility of fiscal policy. On the other hand, there are governments with one-year budgeting, but that have also adopted fiscal rules based on cyclical adjustments. The latter governments have an implicit multi-annual budget horizon by distinguishing between actual and cyclically adjusted government accounts and balances. However, in the latter case, adopting explicit multi-year budgetary and fiscal targets contributes to additional potential gains in fiscal policy sustainability and credibility, as well as in macroeconomic stability.

Chile has a one-year budgeting framework in place. However, the aforementioned 2006 Fiscal Responsibility Law includes two provisions that require governments to adopt a fiscal policy that is, at least in principle, consistent with a horizon that exceeds one year: (i) the requirement imposed on new administrations to present their four-year fiscal policy framework and its implications for the cyclically adjusted government balance, (ii) the requirement on the MoF to submit annually, together with the draft budget law, a medium-term budget projection.

Moreover, by adhering to a cyclically adjusted balance rule, governments are required to base next year's budget proposal on medium-term projections for GDP and the price of copper.

With one exception, Chile's governments are not required to submit long-term projections for government accounts and budgets, based on a systematic assessment of fiscal sustainability and regular evaluation of long-term consequences of fiscal decisions with budgetary implications, like those conducted by the US Congressional Budget Office and other governments in the world. The one exception is the undertaking of a technical study, required by the 2006 Law, to assess the long-term funding needs of the Pension Reserve Fund (one of Chile's sovereign wealth funds) to meet the government liabilities derived from the government pension subsidy (*Pensión Básica Solidaria*).

8.3.1.4 Fiscal Rules

The number of countries with fiscal rules at the national level has grown steadily over the past two decades, from a universe of ten countries in 1990 to thirty in 2001 (the year of adoption of the Chilean fiscal rule) and fifty-one in 2009 (IMF 2009). Countries with fiscal rules had in place on average 2.5 active fiscal rules in 2009.

Fiscal rules are adopted to attain one or more of three objectives: sustainability of public debt, control of government size and contribution to cyclical stability. Rules differ widely across countries in how they are defined. A major category of rules are those that define numerical targets (targets, ceilings or floors) for government balances, overall revenue or expenditure levels that are independent of the business cycle. A paramount example of the latter rules is the Stability and Growth Pact ceilings on government deficits (3% of GDP) and debt levels (60% of GDP) for euro zone member countries – which were systematically violated before 2008 and almost universally thereafter. A second category are fiscal rules that aim at stabilizing cyclically adjusted balances, allowing for cyclical changes in actual government balances, with the aim of implementing a counter-cyclical fiscal policy or at least avoiding pro-cyclical policy bias. According to the IMF (2009), some ten countries (including Chile) had such rules in place in 2009. Among them, only Germany, the UK and Sweden had a fiscal rule in place that defines a numerical target for the actual budget balance throughout the economic cycle.

The cited study (IMF 2009) reviews the relative advantages of different fiscal rules, their design and complementarity with other institutional changes, the institutional of deviations. It also assesses positively the fiscal and macroeconomic effects of fiscal rules in the world.

A recent world-wide panel data study by Elbadawi, Schmidt-Hebbel and Soto (2015)provides empirical evidence on why countries adopt and maintain rule-based fiscal policy regimes. Their reported evidence shows that fiscal conditions, government stability, inflation targets, fixed exchange rates, global financial integration, democracy, political checks and balances and overall development raise countries' likelihood to have fiscal rules in place.

Chile is among the ten countries with a fiscal rule that aims at stabilizing cyclically adjusted government balances. Together with Norway, and in contrast to the other eight cases, Chile corrects not only for the cyclical influence of the business cycle as reflected by GDP deviations from trend but also for the cyclical deviations of the price of the main commodity export good from its trend level. Hence, Chile's and Norway's experiences are unique in the world and are exemplary for commodity-exporting countries, as highlighted by IMF (2009).

However, unlike other countries with fiscal rules in place, Chile has neither established ex-ante escape clauses from the rule that would apply under pre-specified conditions, nor ex-post sanctions for violating the rule and ex-post corrections of deviations from the rule. More on Chile's rule is discussed later.

8.3.1.5 Rules for Management of Government Assets and Liabilities

As the commodity-exporting countries benefited from significant fiscal surpluses during the extended boom in commodity prices over the last decade, many of them have started to save their surpluses in sovereign wealth funds (SWFs) invested in internationally diversified portfolios. This is observed in countries with and without fiscal rules. Until 2009, thirty-seven countries had SWFs in place, with outstanding aggregate assets of US$ 3.8 trillion. In the case of Chile, the outstanding investments of its two SWFs added up to US$ 16.6 billion at the end of 2010 (Table 8.5).

Countries with SWFs have committed to rules for saving in and dis-saving from their funds. In the case of countries with a fiscal rule in place, accumulation into and withdrawal from their funds is determined by the rule. In addition, many countries have adopted legislation and institutions that define investment policies and management principles of their funds.

As mentioned earlier, in the case of Chile, its two SWFs were established by the 2006 Fiscal Responsibility Law. The latter also established the Financial Advisory Committee on Sovereign Funds of the Ministry of Finance, which advises the MoF on investment policies of these funds. More on Chile's SWFs is discussed later.

However, Chile (like most countries in the world) lacks both a policy and an institution focused on consistent management of government financial assets and liabilities, with explicit financial objectives including risk, return, liquidity or benchmarking. Moreover, there is no comprehensive management of the government's total balance sheet, including financial and non-financial assets, as well as explicit and contingent financial liabilities.

8.3.1.6 Requirements on Budget Accountability and Provision of Information on Financial Management

World-wide efforts are observed by governments to improve the accountability of their fiscal policy and transparency in the provision of financial information. This trend responds to the growing political and academic consensus on the democratic and economic efficiency gains of fuller accountability and transparency in fiscal policy. Although there is little empirical evidence on the benefits of high levels of accountability and transparency in fiscal policy, it seems legitimate to infer qualitatively from the evidence of accountability and transparency benefits identified in the case of monetary policy (e.g., Eijffinger et al. 2006; Geraats 2008).

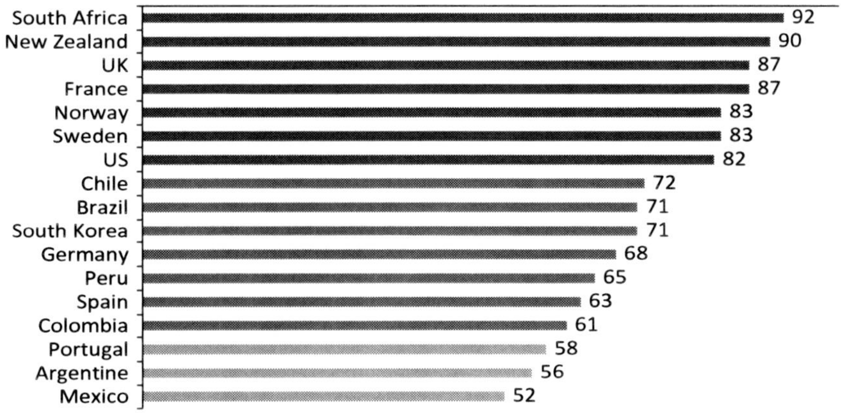

Figure 8.1: Selected Countries: Government Budget Transparency and Accountability, 2010
Source: Open Budget Survey, International Budget Partnership (2010).

The International Budget Partnership (2010) compiles the "Open Budget Survey," an international comparative database on partial and aggregate measures of global transparency and accountability of the budgets of central governments in the world. In its 2010 version (International Budget Partnership 2010), Norway ranks fifth and Chile ranks eighth among ninety-four countries (Figure 8.1 depicts ranking results for a selective sub-sample).

In Chile, transparency on the forecasts of the two key variables for the budget – future GDP growth and copper prices – was ensured almost from the start of the fiscal rule, as projections for the two latter variables made by two ad hoc committees were made public (more on them later). However, mapping of the latter projections into the actual budget calculations could not easily be replicated, particularly after several changes were introduced to the rule in 2008–2009. The 2010 Report of Public Finances for the Budget Law 2011, however, made significant improvements in the provision of information on the calculation of the cyclically adjusted balance, which was coherent with the recommendations provided by the Advisory Committee on Fiscal Policy (2010, 2011).

Regarding SWFs, international heterogeneity in their management and investment accountability and transparency is large. In the most recent international ranking of transparency in the management of sovereign wealth funds, published by the Sovereign Wealth Fund Institute (2011), Chile and Norway rank in first place, together with seven other SWFs (Figure 8.2).

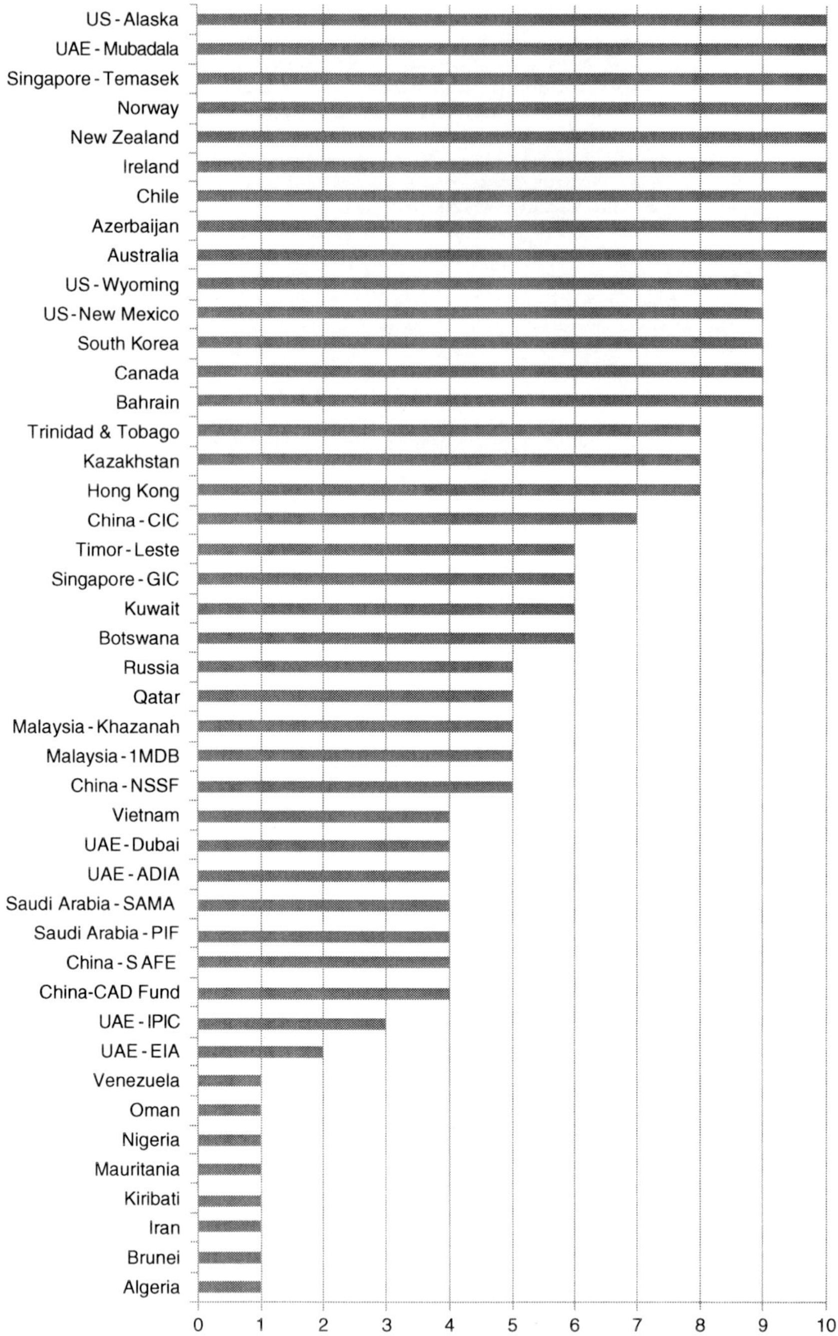

Figure 8.2: Transparency of Sovereign Wealth Funds, 2011
Source: Sovereign Wealth Fund Institute (2011).

Therefore, the current state of transparency and accountability of Chile's government accounts is varied, ranging from moderate to high.

8.3.1.7 External Control and Auditing

External control and auditing of government accounts and budget execution is performed by the legislature, the government's general comptroller or auditor, and fiscal councils and committees. In most countries, congress and especially the general comptroller exercise traditional accounting, financial and managerial control of budget execution and government accounts. More recently, countries are adopting fiscal committees and councils, which are empowered to provide an economic assessment of fiscal policy design and execution, as discussed later.

Government accountability of budgetary compliance in Chile is limited to a large extent to oversight and control by the General Comptroller's Office (*Contraloría General de la República*). This institution focuses on accounting, procedural and legal compliance by the government in its implementation of fiscal policy and execution of the budget. This oversight and control function by the General Comptroller is exercised for all levels of government. In this way, the Comptroller's Office has contributed decisively – both in the past and today – to limit the extent of illegal government actions and government corruption.

However, there is almost nil substantive ex-ante, real-time or ex-post evaluation of fiscal policy implementation and its compliance with the fiscal rule, of government spending (either mandatory or discretionary) and of the efficiency and effectiveness of government programs and tax collection efforts. There are simply very few resources spent by congress, political parties, think tanks or academia in conducting systematic evaluation of fiscal policy and its implementation. There is no independent fiscal council and/or a think tank that focuses on fiscal policy. The only exceptions to this dearth of fiscal policy evaluation are a few academic papers and some exceptional initiatives by the MoF to assess partial aspects of fiscal policy.

8.3.1.8 Fiscal Committees and Fiscal Councils

Fiscal committees are temporary or permanent advisory bodies to the MoF, responsible for specific, narrow tasks such as providing advice on institutional changes or investments of SWFs. Their recommendations could be binding or non-binding for the government.

Independent fiscal councils (and budget offices) are permanent government or congressional agencies entrusted with much broader tasks. Their recommendations could be binding for governments – which is more often the case than it is for fiscal committees.

While fiscal committees and fiscal councils have different tasks, the latter may include the following (Debrun and Kumar 2007; Debrun et al. 2009; Ter-Minassian 2010; Calmfors and Wren-Lewis 2011):

(a) Evaluation and validation of financial and macroeconomic assumptions for the budget,
(b) Provision of independent projections for the budget and government financial conditions, for both base and risk scenarios, for the short term and the long term,
(e) Provision of policy assessments and recommendations on fiscal policy and budgetary management, based on ex-ante evaluation of whether fiscal policy is likely to meet its targets,
(c) Assessment and advice to the government and/or the legislature on the stance of fiscal policy, its long-term sustainability and optimality, and its macroeconomic effects,
(d) Assessment of and recommendations on policies for government asset and liability management,
(f) Monitoring of budget implementation and provision of recommendations on budgetary corrections when facing deviations in budget execution and
(g) Analysis of fiscal transparency.

Some governments – like Chile's – have adopted advisory committees of the MoF, which provide recommendations – some or binding and others not – on some aspects of the menu of possible tasks identified earlier.

Other countries have adopted advisory offices of the legislature to provide independent assessments of budget assumptions, budget projections and fiscal policy effects. That is the case of the US Congressional Budget Office, which began in 1974 (Rivlin 2010) and the Parliamentary Budget Office in Canada, founded in 2008 (Levy 2008), among other countries, like Japan, Korea and Mexico.

Among the countries with some kind of fiscal council are many members of the European Union (EU) (several having more than one fiscal council), including Austria, Belgium, Denmark, Estonia, France, Germany, Greece, Hungary, Italy, Luxembourg, Netherlands, Portugal and the UK (Debrun et al. 2009). Outside the Eurozone, Sweden and Japan have in place fiscal agencies that provide tax advice. Debrun and Kumar (2007) provide some statistical evidence in support of the notion that fiscal councils in the twenty-five member countries of the EU have contributed to fiscal discipline, especially when such advice is provided by an agency mandated by an ad hoc law. Yet the more recent fiscal crisis in several Eurozone member countries casts doubts about the latter findings, based on pre-2007 data.

More recently, and partly in response to fiscal crises, several countries (including France and the United Kingdom) have adopted or are adopting stronger fiscal councils, based on an ad hoc law, which grants them substantial degrees of government independence, strengthening their capabilities to provide independent assessment and advice. This new brand of fiscal councils – and some of the existing councils – tend to exhibit the following features:

1. independence from government,
2. created by special laws, which define their mandate, functions, responsibilities, corporate structure, staffing, financing, accountability and transparency in the delivery of studies and activities,
3. with board members chosen by congress among academics or experts with fiscal policy experience and
4. sometimes, entitled to the authority to propose sanctions in case of significant deviations in budget execution.

Chile lacks an independent fiscal council. But it has in place three independent fiscal committees or expert groups: the Advisory Committee for Trend GDP of the MoF (ACTG), the Advisory Committee for the Reference Copper Price of the MOF (ACRCP) and the Advisory Financial Committee for Fiscal Responsibility Funds of the MoF (AFCFRF). Their role is limited to an advisory function, defined by the 2006 Fiscal Responsibility Law.

ACTG provides the MoF with medium-term projections for the rate of growth of capital, the labor force and productivity, which are used by the MoF to generate projections of trend GDP and the output gap, based on a MoF methodology. ACRCP provides the MoF with projections of the international long-term copper price. Current-year projections for trend GDP growth and the long-term copper price, and for next year's GDP growth and copper price, are key for budgeting the overall government resource envelope, that is to say, for setting the levels of government spending, cyclically adjusted and actual government revenue and hence, the cyclically adjusted and actual government balance.

The three individual projections made by each of the sixteen members of the ACTG and the single projection made by each of the twelve members of the ACRCB are binding. Each member provides projections which are averaged after elimination of the two extreme observations to generate MoF's official estimates for the four variables. However, the MoF maintains significant discretionary power in defining the methodology – equations and parameters – that determines the cyclically adjusted fiscal balance rule and how the four variable forecasts translate into budgetary projections. While the MoF has been increasingly transparent about the details of its method – as reflected in technical papers, the summary notes of the ACTG and ACRCH annual meetings and the 2010 Budget Law for 2011 – this information also reflects the residual discretion maintained by the MoF in defining the government's overall resource envelope.

Independent members of both advisory committees are invited by the MoF to serve for one budgetary exercise. Members are selected by the MoF among the local communities of macroeconomists (for the ACTG) and copper market experts (for the ACRCB). They attend one technical meeting held at the MoF in July or August of each year and are then invited to submit their individual projections in the following weeks. The MoF publishes between August and October of each year two official notes, one for each advisory committee, that reports individual and averaged forecasts and, in the case of the ACTG, trend growth and output gap estimates. There is no formal or direct communication by any of the two committees with congress or the press.

In contrast to the two previous committees, the AFCFRF is based on a more formal institutional setup. The six independent members are selected by the MoF among the local community of macroeconomists and financial experts and their overlapping tenure is two years. The AFCFRF comprises its chair or President, a Vice-President and four additional members. The Committee meets every six weeks, on average, at the MoF. Members are remunerated for their attendance of committee meetings. The Secretariat of the AFCFRF is the International Financial Directorate of the MoF. Its head acts as secretary of the AFCFRF and his staff (also present at committee meetings) prepares technical reports on international financial conditions and financial performance of sovereign wealth fund investments at each meeting. AFCFRF members discuss financial developments and their implications for the performance of the funds, evaluate fund management by the Central Bank of Chile and meet occasionally with bank managers of the funds, and issue recommendations about fund investment policy and regulation to the MoF. AFCFRF recommendations – in contrast to the projections submitted by the two aforementioned committees – are not binding for the Minister of Finance.

AFCFRF issues a press communiqué after each meeting and publishes an annual report on the funds' financial results and the Committee's investment policy recommendations to the Minister.

8.3.2 Chile's Fiscal Rule

8.3.2.1 The Rule

Chile's fiscal rule for the central government budget was developed in 2000 and implemented in 2001 by Finance Minister Nicolás Eyzaguirre and his main advisors. While the rule has been subject to several changes, its main features have remained unaltered. Therefore, it is now in its twelfth year of operation, applied by three different administrations: the governments of presidents Lagos (2000–2006), Bachelet (2006–2010) and Piñera (since 2010), which has committed to its continuation through the 2010–2014 term of office.

Chile's fiscal rule aims at contributing to two policy objectives: fiscal sustainability and fiscal/macroeconomic stability. Fiscal stability is ensured by committing to a target level for the government's cyclically adjusted balance (CAB) that is consistent with government saving needs (net of government investment). Fiscal (and possibly macroeconomic) stability is attained by committing to a government spending path that is consistent with cyclically adjusted government revenue.[2] Therefore, Chile's CAB rule combines a partial application of the permanent income theory to government spending with a target level for long-term government saving. Hence, the rule's aim is to save during high-revenue periods in order to withdraw from the latter savings in bad times, over and above the saving or dissaving target reflected by the CAB.

Chile's largest sources of government cyclical volatility (i.e., the budget's largest sensitivity to cyclical variables) are non-mining tax revenue and mining tax and transfer revenue (Table 8.1). The business cycle reflected in the cyclical component of GDP largely determines the cyclical behavior of

[2] The rule is only partly consistent with a permanent income approach, namely making government spending consistent with cyclically-adjusted government revenue. Therefore, other sources of differences between temporary and permanent revenue – like temporary changes in taxes or interest rates – are not considered in adjusted revenue estimations due to political and economic difficulties in doing so. This partial approach to permanent revenue estimation, limited to cyclical adjustments of government revenue and balance, is adopted by all other countries that compute adjusted government accounts or adopt fiscal rules to account for temporary deviations of relevant budget variables.

Table 8.1: *Central Government Accounts, Growth and Copper Price, 1990–2010*

	1990–2001	2001–2010	2010
Central Government (% of GDP)			
Overall Revenue	21.7	23.1	23
Copper Revenue (1)		6	4.7
Non-Copper Revenue (1)		18.2	18.3
Overall Expenditure	20.7	21.1	23.5
Government Balance	1	2	-0.5
Cyclically Adjusted Balance	0.5	0	-2.3
Government Assets	13.7	12.6	17.6
Sovereign Wealth Funds		10	8
Government Debt	22	9.4	9.3
Net Government Assets	-8.3	3.2	8.3
GDP Growth (%)			
Average	6.2	3.7	5.2
Standard Deviation	3.5	2.2	
Committee's Average Estimated Future Trend Growth		4.7	5
Copper Price (US$/lb)			
Average	1	2.04	3.4
Standard Deviation	0.19	1.12	
Committee's Average Estimated Future Trend Level		1.22	2.13
Copper Production (% of world production)	25.8	34.9	33.7
Copper Exports (% of Chile's exports)	38.6	46.6	56.1

Sources: Author's calculations based on data from the Budget Directorate, Cochilco, and the Central Bank of Chile. (1) Copper related and non-copper related revenue for 2005–10.

non-mining tax revenue, while the cyclical components of copper and molybdenum prices largely determine the cyclical behavior of mining tax and transfer revenue.

Chile's government spending, compared to other OECD countries, is relatively insensitive to the business cycle because of the small size of automatic stabilizers on the expenditure side, such as government-financed unemployment benefits or institutionalized public employment programs during cyclical downturns. Hence, the fiscal rule distinguishes between actual and cyclically adjusted government revenue but not between actual and cyclically adjusted government expenditure.

The fiscal rule is represented by the following equation for the cyclical component of the government balance (i.e., the difference between the CAB and the actual balance of the central government). This difference boils down to the difference between cyclically adjusted and actual government revenue, determined by the difference between cyclically adjusted and actual non-mining tax revenue and social security receipts (driven by the output gap) and the differences between cyclically adjusted and actual private mining tax revenue, CODELCO (the government-owned copper corporation) copper transfers to the budget and CODELCO molybdenum transfers to the budget[3]:

$$\begin{aligned} B^*{}_t - B_t &= \left(R_t^* - G_t\right) - \left(R_t - G_t\right) = \\ &= NMTR_t^* - NMTR_t + MTR_t^* - MTR_t + CR_t^* - CR_t + MR_t^* - MR_t \end{aligned} \tag{1}$$

where:

B_t^* = cyclically adjusted central government balance (CAB)
B_t = actual central government balance
R_t^* = cyclically adjusted central government revenue
R_t = actual central government revenue
G_t = actual central government expenditure
$NMTR_t$ = net non-mining tax revenue and social security receipts
$NMTR*$ = cyclically adjusted net non-mining tax revenue and social security receipts
MTR_t = actual tax revenue from private mining companies
MTR_t^* = cyclically adjusted tax revenue from private mining companies
CR_t = actual transfers from CODELCO copper production
CR_t^* = cyclically adjusted transfers from CODELCO copper production
MR_t = actual transfers from CODELCO molybdenum production
MR_t^* = cyclically adjusted transfers from CODELCO molybdenum production
t = time period subscript.

Simple reordering of the identity part of Equation (8.1) shows that the fiscal rule boils down to limit actual government spending to cyclically adjusted revenue net of the CAB:

$$G_t = R_t^* - B_t^* \tag{2}$$

[3] Tax revenue from private mining companies comprises income taxes and royalty payments. Transfers from public mining company CODELCO comprise income taxes, royalty payments and payments of dividends. Copper dominates molybdenum revenue by far.

Cyclically adjusted (i.e., trend or long-term) variables are non-observable and therefore, their estimates are required for implementation of the fiscal rule. As discussed earlier, forecasts for GDP growth and copper prices – the key variables for budgeting next year's and future trend revenue receipts – are outsourced on a yearly basis to two advisory committees (ACTG and ACRCP) in preparation of the budget law submitted to congress. Implementation of fiscal policy during the subsequent budget (i.e., calendar) year abides by the fiscal rule and the budget law.

There are three key elasticities in the fiscal rule. The first is the output gap elasticity of non-mining tax revenue. Until 2008, the MoF set this parameter at 1.05, which implicitly assumes that non-mining government revenue varies almost proportionately with GDP. This aggregate elasticity was replaced in 2009 by separate elasticities for different categories of tax revenue, varying between 1.0 and 2.4 (Rodriguez et al. 2009; Velasco et al. 2010). The Advisory Committee on Fiscal Policy (2011) recommended maintaining for now the latter set of elasticities (except one) but encouraged the MoF to engage in systematic revision of elasticities based on a distinction between cyclical and long-term elasticities.

The second and third key elasticities are the copper and molybdenum price elasticities of the three mining revenue items in Equation (8.1). There are no explicit values attached by the MoF to the latter elasticities. Instead, the Budget Office of the Ministry of Finance (DIPRES) produces annual estimates for mining tax and transfer receipts at actual and trend prices for both minerals.

8.3.2.2 Rule Implementation, Changes and Budget Performance

Now let us turn to how the rule was implemented, the changes that were introduced to its definition and implementation and the budget performance under the rule.

While the fiscal rule has been in place since January 2001, several institutional and methodological amendments were introduced subsequently (Velasco et al. 2010; Advisory Committee 2011). During the initial phase, the following refinements were introduced:

- First Advisory Committee for Trend GDP (ACTG) of the MoF is invited to submit trend growth projections (August 2001).
- Methodology for CAB is published (Marcel et al. 2001) (September 2001).
- First Advisory Committee for the Reference Copper Price (ACRCP) of the MOF is invited to submit long-term copper price projections (August 2004).

- Government statistics are changed from cash basis to accrual basis, following the 2001 IMF methodology (February 2004).

The following methodological changes were introduced to the fiscal rule on the treatment of mining revenue:

- Inclusion of cyclical adjustment of private mining tax revenue (August 2005).
- Inclusion of cyclical adjustment of the revenue paid by CODELCO based on MoF estimates of the cyclical component of the price of molybdenum (December 2005).
- Inclusion of cyclical adjustment of the new royalty tax on mining (January 2006).
- Inclusion of cyclical adjustment of the additional tax on mining (December 2006).
- Amendment to the cyclical adjustment of the royalty tax on mining, considering monthly provisional payments (December 2007).

The following changes were introduced to the fiscal rule, extending the principle of cyclical adjustment to other variables (beyond GDP and prices of minerals) and to temporary tax changes:

- Inclusion of cyclical adjustment of interest receipts on government financial assets (September 2008).
- Inclusion of adjustments to non-mining tax revenue considering temporary tax cuts (January 2009).
- Inclusion of cyclical adjustment of operational income and property income on non-financial government assets (January 2010).

On the advice of the Advisory Committee on Fiscal Policy (2010, 2011), the three latter changes were reverted by the government in 2010 in preparing the budget for 2011. Therefore, the 2008–2010 attempts to extend the fiscal rule from a CAB rule to a structural rule based on an estimate of permanent government income – introducing adjustments to all variables with temporary components, including tax rates and government income from financial and non-financial assets – was reverted to the methodology used for rule during 2001–2007. Following international best practice adopted by individual countries and endorsed by the OECD, the IMF and the European Commission (Ter-Minassian 2010), Chile's fiscal rule is limited to cyclical adjustments. Therefore, in contrast to most other countries, the rule adjusts for the cyclical component of GDP and Chile's key commodity prices: copper and molybdenum.

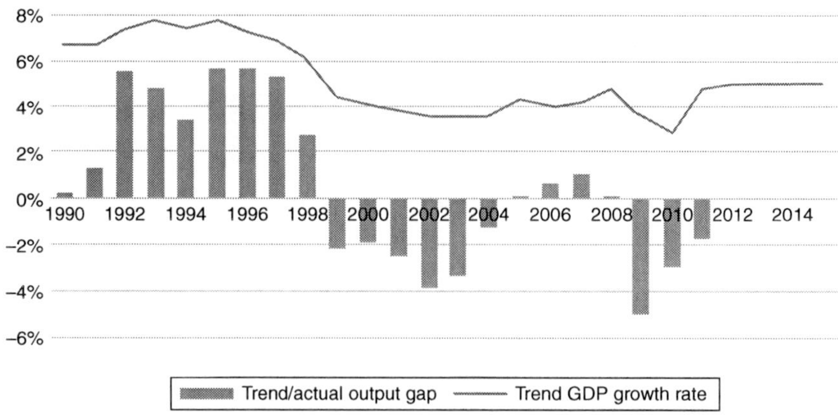

Figure 8.3: Chile: Trend GDP Growth and Output Gap Projection by MoF in 2010, 1990–2015 (%)
Sources: Committee's Report Trend GDP (2011).

As discussed earlier, the independent Advisory Committee for Trend GDP (ACTG) of the MoF provides annual growth forecasts (five years ahead) for production factors, used subsequently by the MoF to generate potential GDP growth and output gap projections.[4] The September 2011 potential GDP growth and output gap projections by the MoF – based in part on Committee input projections – are reported in Figure 8.3. The latter forecasts for future variables determine the resource envelope for the 2012 budget. Note that the latter figures are ex post (September 2011) estimates of Chile's current and past output gap estimates and hence are not the actual output gap estimates used in implementing the fiscal rule in previous years (2001–2010).

In contrast to the previous figure, Figure 8.4 depicts, for any given year, that year's actual GDP growth and that year's trend GDP growth forecast made by the ACTG (i.e., that year's medium-term GDP growth forecast for

[4] The projection method and the variables projected by the Committee have been refined over the years. The most recent MoF actual and potential growth methodology is based on a simple Solow-type growth decomposition equation, based on a Cobb–Douglas Harrod-neutral production function, with capital services (capital stock adjusted by utilization), labor services (hours worked adjusted by educational attainment of the labor force) and residual total factor productivity. Trend growth is generated by filtering actual growth projections. It is important to note that the MoF – not the independent Committee – estimates the levels of actual and trend output and hence the output gap, which is the key variable for forecasting non-mining tax revenue.

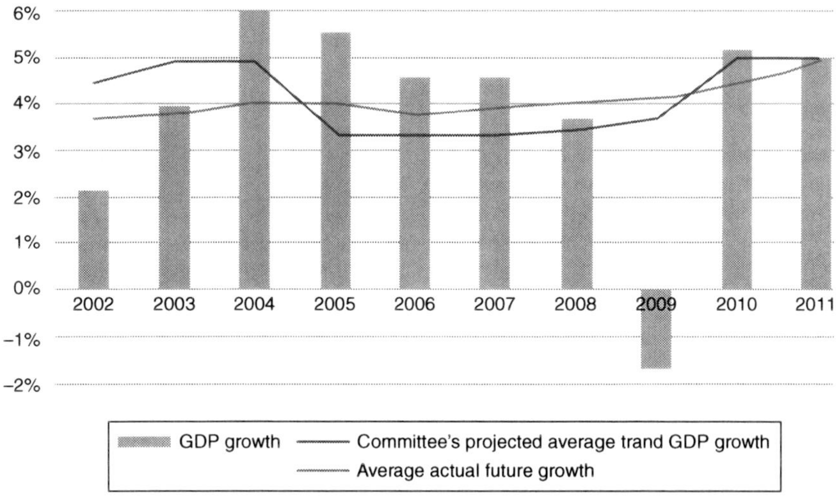

Figure 8.4: Chile: Committee's Projections, Average Actual Future GDP Growth and Actual GDP Growth, 2002–2010 (%)

Sources: Author's calculations based on data published by Committee Report (2011).

Notes: GDP growth is the actual GDP growth; in the case of 2011, I use 5% as my own calculation. Committee's projected average trend GDP growth is the average of all future projections made by the Committee at the corresponding year. Average actual future growth is average GDP growth from a rolling window from year *t* through year *t* + 4.

the next five years (from years *t* to *t* + 4), as well as actual growth). A significant gap between high Committee growth forecasts and low actual average future growth rates is observed between 2005 and 2009. This has implied, as time has passed, a significant upward correction of output gap estimates and hence, a downward correction of budgeted government expenditure (i.e., *ceteris paribus*, a contractionary fiscal policy).

The second committee, the independent Advisory Committee for the Reference Copper Price (ACRCP) of the MoF, provides a point estimate for the average world (London Metal Exchange) copper price over the subsequent decade. Figure 8.5 shows, for any given year, that year's actual copper price and that year's long-term copper price forecast submitted by the ACRCP, as well as the actual copper price. A significant gap between low Committee long-term copper price forecasts and high actual average future copper prices developed since 2002 but has narrowed in recent years. This has implied, as time has passed, a significant upward correction of the long-term copper price and hence an upward correction of budgeted government expenditure (i.e., *ceteris paribus*, an expansionary fiscal policy).

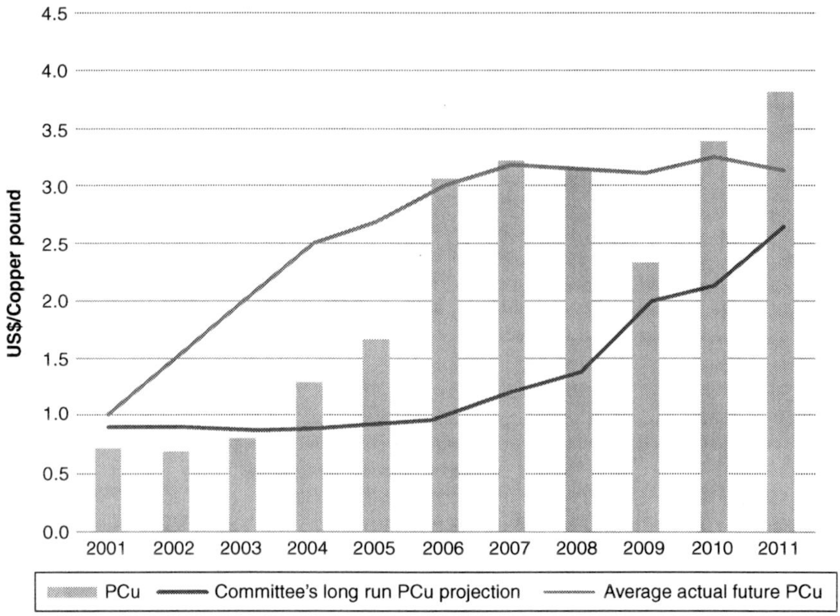

Figure 8.5: Chile: Committee's Projected Copper Price, Average Actual Copper Price and Actual Copper Price, 2001–2011

Sources: Authors' calculations based on data published by Committee reports and Central Bank of Chile.

Notes: Committee's projected long-run copper price projection is the forecast made at the corresponding year. PCu is the actual price of copper. Average actual future copper price is the average price from a rolling window from year t through year $t + 4$. In the case of 2011, I use the actual average price from January through August, then I use a copper price of 3 US$/lb.

How much do current observations of GDP growth and the copper price influence committee forecasts? Significantly for copper prices, but not significantly for GDP growth, when using contemporaneous values (Table 8.2). However, when using lagged values, the correlations exceed 0.80. This suggests that last year's realizations significantly and strongly influence trend GDP and long-term copper price forecast.

A key policy variable of the fiscal rule is the target value for the CAB (B^*). For the first seven years of the fiscal rule, the CAB was set at a positive 1% of GDP. The rationale behind this very conservative fiscal policy was the need for positive central government saving to finance government pension liabilities and recapitalize the Central Bank of Chile (CBC) and finance several government contingent liabilities, including the government pension subsidy.

Table 8.2: *Correlations between Committee Forecasts and Actual Variables, 1990–2011*

Actual Copper Price (*t* – 1) and Long-term Copper Price Forecast (*t*)	**0.82**
Actual Copper Price (*t*) and Long-term Copper Price Forecast (*t*)	**0.72**
Actual GDP Growth (*t*) and Trend GDP Growth Forecast (*t*)	0.15
Actual GDP Growth (*t* – 1) and Trend GDP Growth Forecast (*t*)	**0.91**

Sources: Committee reports and Chilean Central Bank.
Note: Bold figures show statistically significant correlations.

For 2008, the CAB was lowered to 0.5% of GDP, based on recommendations of fiscal sustainability studies (Engel et al. 2007; Velasco et al. 2007). Facing a domestic recession in January 2009 due to the international crisis, the CAB was further reduced to zero (0% of GDP) CAB for 2009. Yet the two latter figures were based on the methodological changes introduced between September 2008 and January 2010, which were reverted subsequently.

Using the current methodology for the fiscal rule (returning to the principles applied between 2001 and 2007), the actual CAB ratio was –0.6% of GDP in 2008 and –3.0% of GDP in 2009. The new 2010 administration prepared a medium-term fiscal policy framework that implies a gradual correction of the large negative CAB recorded in 2009. The CAB was set at –2.0% of GDP in 2010 and –1.8% of GDP in 2011. The administration also committed to further fiscal adjustment toward attaining a CAB ratio of –1.0% of GDP in 2014.

Figure 8.6 depicts the CAB ratio to GDP, the actual balance ratio to GDP and their difference (the cyclical balance ratio to GDP) during the 1990s (without a fiscal rule) and the 2000s (with the fiscal rule in place).[5] The data supports the following policy conclusions:

(i) cyclical shocks to the government budget (reflected by the cyclical balance ratio to GDP) are very large,
(ii) cyclical shocks to the budget are dominated by the volatility of copper prices, not by the volatility of GDP, as reflected by the larger magnitude of cyclical components (largely driven by copper price shocks) recorded in 2001–2010,

[5] In the absence of a fiscal rule during 1990–2000, the CAB and cyclical balance ratios depicted in Figure 8.6 for the latter years are counter-factual estimations based on the fiscal rule methodology applied since 2001.

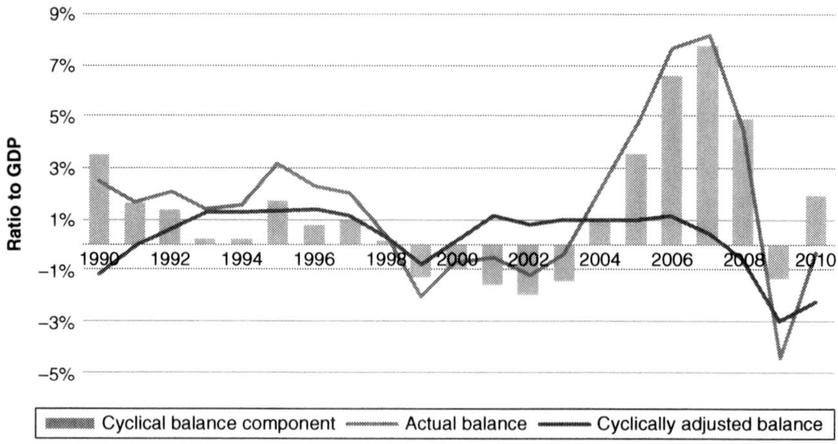

Figure 8.6: Chile: Actual and Cyclically Adjusted Central Government Balance, 1990–2010 (Ratios to GDP, %)
Source: Budget Directorate.

(iii) the fiscal rule allows for large government saving flows during cyclical booms,

(iv) even without a fiscal rule, fiscal policy was conservative during most of the 1990s,

(v) the fiscal rule provides insufficient space for a stronger counter-cyclical policy during domestic cyclical downturns (the 1998 and 2009 recessions), leading to significant downward adjustments of the CAB. In 2008, the fiscal rule was effectively suspended by substituting the 0.5% ex-ante CAB ratio to GDP by a –0.6% ex-post CAB ratio to GDP, reduced further to a –3.0% ex-post CAB ratio to GDP in 2009.

Figures 8.7 and 8.8 depict the behavior of government revenue and expenditure ratios to GDP and growth rates during the 1990s (without a fiscal rule) and the 2000s (under the fiscal rule). Government revenue reflects the cycles of GDP and, more strongly, of copper prices. The upward correction in long-term copper price forecasts (dominating the downward correction in trend GDP forecasts) and the downward adjustment in CAB targets explain the large rise of government expenditure growth, from 1.6% in 2003 to 17.8% in 2009. The average government spending increase attained 10% during the Bachelet administration (2006–2009), when GDP growth averaged only 2.7%. This large fiscal expansion led to a 24.8% share of

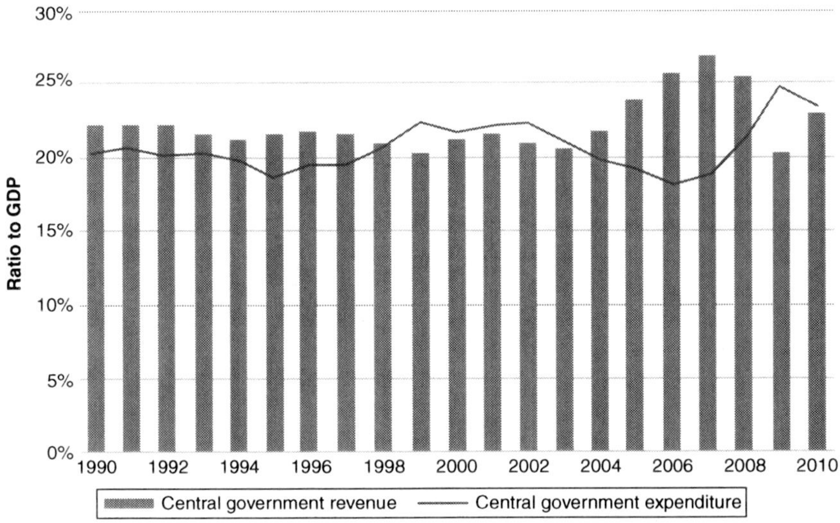

Figure 8.7: Chile: Central Government Expenditure and Revenue, 1990–2010 (Ratios to GDP, %)
Sources: Chilean Budget Directorate.

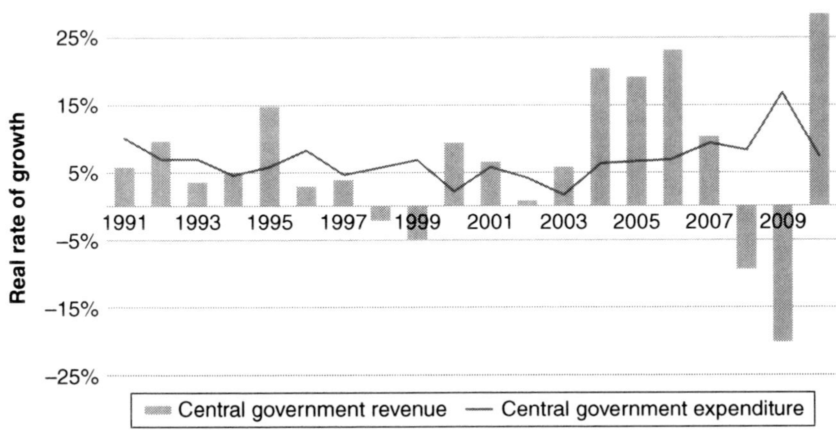

Figure 8.8: Chile: Central Government Revenue and Expenditure Growth Rates, 1991–2010 (%)
Sources: Chilean Budget Directorate.

government expenditure to GDP in 2009, the largest ratio of government spending to GDP recorded in Chile in several decades. The current administration's commitment to gradually raise the CAB toward –1.0% in 2014, is likely to entail government expenditure growth below GDP growth,

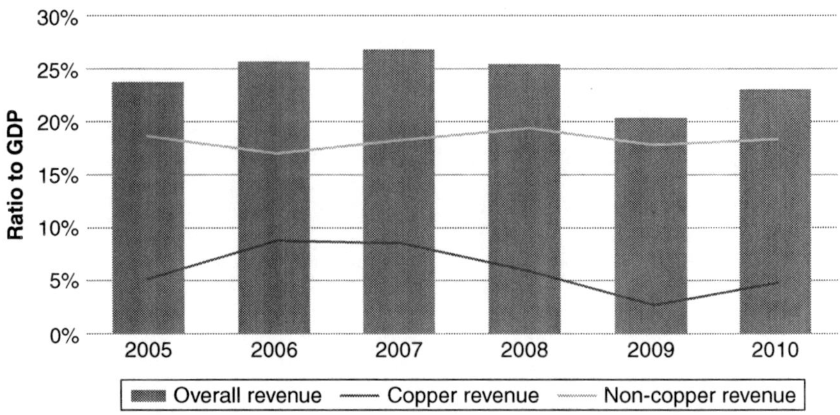

Figure 8.9: Chile: Government Revenue, 2005–2010 (Ratio to GDP, %)
Source: Chilean Budget Directorate.

hence reducing the share of government spending in GDP, as observed in 2010.

Figure 8.9 shows the mining and non-mining component of government revenue since 2005, reflecting the large volatility of mining revenue due to mineral price volatility.

Finally, the following are simple correlations between budget variables and GDP growth, and between budget variables and the price of copper for 1990–2010 and sub-periods before and during the fiscal rule. The data in Table 8.3 reflect:

(i) a large rise in the correlation between the price of copper and overall government revenue in the 2000s, due to the rising volatility of copper prices observed in the 2000s,
(ii) a large increase in the correlation (using quarterly data) between the price of copper and the actual government balance,
(iii) a reversal of a large negative correlation between the price of copper and net government assets (at –0.4 in the 1990s) to a large positive correlation between the two latter variables (at +0.9 in the 2000s).

The latter results suggest that adoption of the fiscal rule has made a difference for fiscal policy in Chile: a copper price shock is more likely to be saved and reflected in higher government assets since 2001 than before the fiscal rule was adopted.

Chile's conservative fiscal stance, reflected in government surpluses or low deficits, has led to a massive improvement in the country's fiscal

Table 8.3: *Correlations between Government Budget Variables, GDP Growth and Copper Price, 1990–2010 and Sub-periods*

Quarterly Data	1990–2010	1990–2000	2001–2010
Total Revenue/GDP - GDP Growth	0.08	**0.3**	0.11
Total Revenue/GDP - Copper Price US$	**0.83**	**0.19**	**0.8**
Total Expenditure/GDP - GDP Growth	0	-0.01	0.07
Total Expenditure/GDP - Copper Price US$	**0.47**	0.02	**0.34**
Government Balance/GDP - GDP Growth	0.08	**0.26**	0.07
Government Balance/GDP - Copper Price US$	**0.6**	**0.15**	**0.57**
Annual Data	**1990–2010**	**1990–2000**	**2001–2010**
Net Government Assets/GDP - GDP Growth	**-0.36**	-0.21	-0.19
Net Government Assets/GDP - Copper Price US$	**0.71**	**-0.41**	**0.91**
Government Balance/GDP - GDP Growth	**0.34**	**0.78**	**0.58**
Government Balance/GDP - Copper Price US$	**0.59**	**0.77**	**0.57**

Sources: Chilean Budget Directorate and Chilean Central Bank.
Note: Bold figures show statistically significant correlations.

strength. Net government assets have increased from –23% of GDP in 1990 to 8% of GDP in 2010 due to government debt repayment and accumulation of assets, in particular, SWF investments (Figure 8.10). Moreover, since the start of the fiscal rule, net government assets are strongly pro-cyclical, moving with the cyclical components of GDP and the price of copper.

This improvement in the government's financial position has contributed to a significant reduction in country risk. The sovereign risk spreads of government external debt has declined from 175 bp in 1990 (when the first issuance of government debt in many years took place) to 129 bp in 2010. The same trend improvement is reflected by the International Country Risk Guide measure for Chile (Figure 8.11).

8.3.2.3 Evaluation of the Fiscal Rule and its Macroeconomic Impact

Several studies have provided qualitative arguments or quantitative estimates for the fiscal policy benefits and macroeconomic effects of Chile's fiscal rule. The main fiscal policy consequences of the CAB rule, which implies an acyclical government expenditure path, have been the following:

(1) a reduction of pro-cyclical bias in fiscal policy, de-linking government expenditure from cyclical shocks and strengthening the role

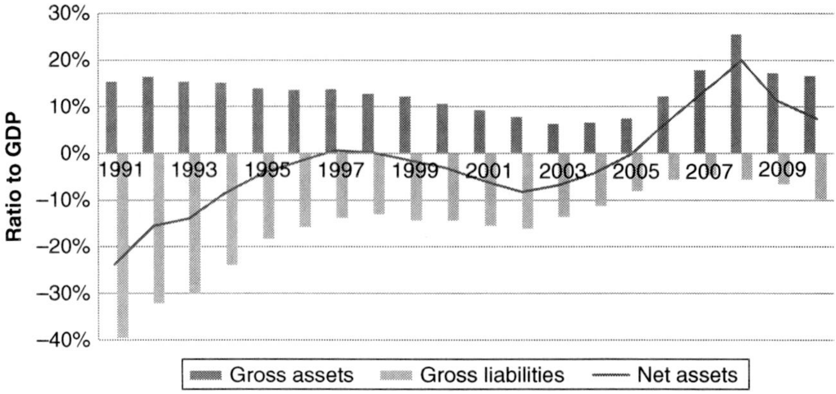

Figure 8.10: Chile: Gross Assets, Gross Liabilities, and Net Assets of the Government, 1990–2010 (Ratios to GDP, %)
Source: Ministry of Finance

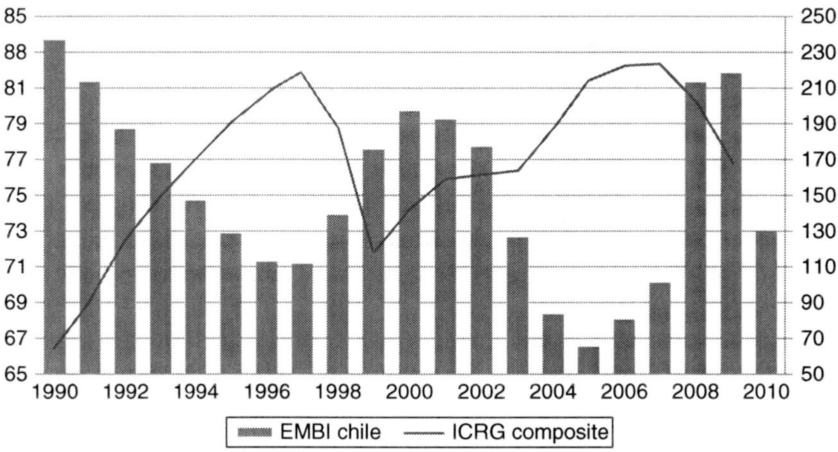

Figure 8.11: Chile: Sovereign Risk Measures: International Country Risk Guide Index and EMBI Spread, 1990–2010
Source: Chilean Central Bank and
Notes: Left axis shows CRG Composite Index and right axis shows EMBI Chile. The ICRG Index ranges from 0 to 100 and the EMBI Chile spread is in basis points. EMBI Chile was created in 1999, thus we use an estimated EMBI Chile from 1990–1998 using a backward prediction from a regression of EMBI on ICRG.

of government saving in buffering cyclical shocks (Larraín and Parro 2006; Rodríguez et al. 2006; Kumhof and Laxton 2009; Velasco et al. 2010; Frankel 2010; Marcel 2010; IMF 2010; Ter-Minassian 2010; Schmidt-Hebbel 2010; OECD 2010; Schmidt-Hebbel 2011);

(2) gains in fiscal sustainability and credibility, reflected in lower sovereign risk premiums (Fiess 2004; Larraín and Parro 2006; Rodríguez et al. 2006; Kumhof and Laxton 2009; Velasco et al. 2010; Marcel 2010; IMF 2010; Ter-Minassian 2010; Schmidt-Hebbel 2010; OECD 2010; Schmidt-Hebbel 2011).

Simulation studies for Chile's fiscal rule show that:

(1) the rule performs well if the policymaker puts a small weight on output volatility relative to inflation volatility in his objective function and a more aggressive counter-cyclical fiscal rule can attain a lower output volatility, but at the cost of a higher volatility in fiscal variables and in inflation (Kumhof and Laxton 2009, using a DSGE model for Chile);
(2) the efficiency frontier derived from counter-cyclical government spending dominates the frontier attainable when spending is cyclically neutral, like it is under the current fiscal rule. The major welfare gain of a counter-cyclical rule is due to lower output volatility, while spending volatility rises (García 2011, using a DSGE model for Chile);
(3) government spending volatility is increased by a broader government objective function that puts a positive weight on counter-cyclical spending and is lowered by a broader government objective function that puts a negative weight on large changes in government spending (Advisory Committee on Fiscal Policy 2011, using a generalized government objective function in partial equilibrium).

Other studies argue that the rule has implied or have measured the following macroeconomic effects of the fiscal rule (see also Table 8.4):

(1) generally, lower macroeconomic uncertainty and volatility (Fiess 2004; Larraín and Parro 2006; Rodríguez et al. 2006; Kumhof and Laxton 2009; Velasco et al. 2010; Marcel 2010; IMF 2010; Ter-Minassian 2010; Schmidt-Hebbel 2010; OECD 2010);
(2) a reduction in GDP volatility (Larraín and Parro 2006; Schmidt-Hebbel 2011);
(3) lower interest rate volatility (Rodríguez et al. 2006);
(4) lower exchange rate volatility and less RER appreciation during booms (Velasco et al. 2010);
(5) less dependence on foreign financing during downturns (Rodríguez et al. 2006; Velasco et al. 2010);

Table 8.4: *Correlations between Macroeconomic Variables, 1990–2010 (Quarterly Data)*

	Copper Price	EMBI	Real Exchange Rate	GDP Growth
Copper Price	**0.33**	**−0.22**	**−0.47**	**0.59**
EMBI	**−0.44**	**0.37**	**0.24**	**−0.29**
Real Exchange Rate	**−0.61**	**0.46**	**0.30**	**−0.81**
GDP Growth	**−0.14**	0	0.07	**0.65**

Sources: Author's calculations based on Chilean Budget Directorate and Chilean Central Bank data.
Notes: The upper triangular matrix reports correlations between HP-filtered series of the logs of copper price, real exchange rate and real GDP, and HP-filtered EMBI. The lower triangular matrix presents correlations between growth rates of copper price, real exchange rate and real GDP, and the change of EMBI. The diagonal reports correlations between HP filtered series and rates of growth or changes. Bold figures show statistically significant correlations.

(6) better protection of social spending during cyclical downturns (Rodríguez et al. 2006; Velasco et al. 2010) and,
(7) an unexpected copper price shock in 2001–2010 (with the fiscal rule in place) leads to a rise in government saving, a decline in the sovereign risk premium and higher GDP growth, all in contrast to the pre-rule 1990–2000 period (Schmidt-Hebbel 2011).

8.3.3 Government Assets, Sovereign Wealth Funds and Net Foreign Assets

Chile's conservative fiscal policy, applied since the mid 1980s, has been reflected in a trend decline of government liabilities and a trend rise in government assets. Over the business and copper price cycles, government assets and liabilities exhibit a cyclical pattern that reflects operation of the counter-cyclical fiscal rule.

A weak form of a fiscal rule to save copper price windfalls in Chile was in place since the 1980s, when government copper windfalls were saved in a Copper Revenue Compensation Fund (CRCF), from which resources were drawn when needed during copper price downturns. However, the rule and institutional set-up of this rule were weak and under the discretion of each government. Resources saved in the CRCF and other government foreign assets added up to US$ 637 million in December 2000 (equivalent to 0.3% of 2000 GDP), just before the January 2001 start of the formal fiscal rule. Subsequent government deficits in 2001–2003 were reflected in issues of government debt and withdrawals from the

Table 8.5: *Gross Foreign Assets of the Central Government by Funds, 2000–2011 (US$ m.)*

Year	CRCF	ESSF	PRF	Other Assets	Total Gross Foreign Assets
2000	673			1,237	1,910
2001	621			576	1,197
2002	277			190	467
2003	75			229	304
2004	208			683	891
2005	967			950	1,917
2006	2,564		605	7,208	10,377
2007		14,033	1,466	5,766	21,265
2008		20,211	2,507	4,884	27,602
2009		11,285	3,421	4,927	19,633
2010		12,720	3,837	5,750	22,307

Source: Ministry of Finance.
Notes: CRCF is Copper Revenue Compensation Fund, ESSF is Economic and Social Stabilization Fund and PRF is Pension Reserve Fund.

government's foreign assets, which fell to US$ 304 million or 0.1% of GDP in late 2003.

The CRCF was abolished with the start of the two SWFs, as mandated by the 2006 Fiscal Responsibility Law. Subsequent government surpluses were saved abroad according to the rules by the latter law set for savings into and withdrawals from the two funds, discussed earlier.

Table 8.5 reports outstanding gross foreign assets in 2000–2010 saved in the CRCF, the two SWFs and other assets of the government. Figure 8.12 depicts the evolution of government foreign assets held in CRCF, ESSF and PRF during the last decade. Large government surpluses were accumulated into the two sovereign wealth funds in 2007–2008, while the government deficit observed in 2009 led to significant withdrawal from the ESSF. At the end of 2010, US$ 16.6 billion were held at the SWFs, equivalent to 8% of GDP.

Since March 2007, foreign investments of the two sovereign wealth funds are managed by the Central Bank of Chile. The investment policy is defined by MoF, on recommendations issued by the Advisory Financial Committee for Fiscal Responsibility Funds (AFCFRF). The funds' portfolio has been invested to date in low-risk, short-maturity financial instruments: 30% in money market instruments, 66.5% in non-indexed investment grade sovereign bonds, and 3.5% in inflation-indexed investment grade sovereign bonds. The currency composition is 50% in US$, 40% in euro and 10% in yen-denominated instruments.

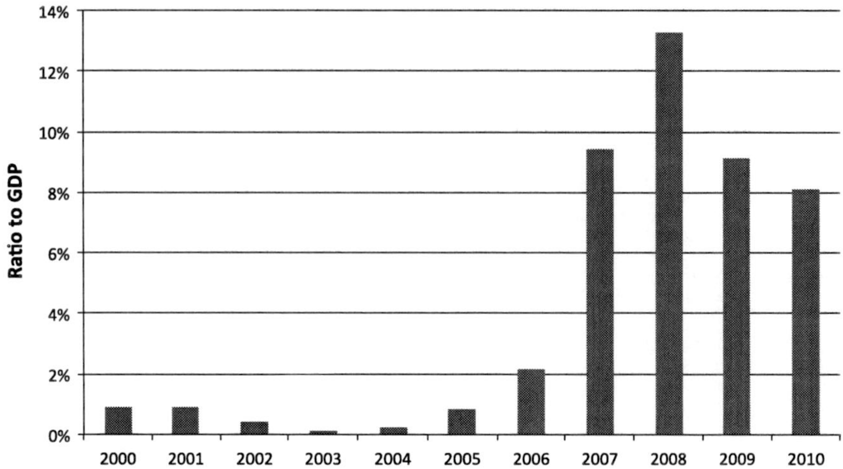

Figure 8.12: Chile: Government Foreign Assets Held in CRCF, ESSF and PRF, 2001–2010 (Ratio to GDP, %)
Source: Ministry of Finance.
Note: CRCF is the Copper Revenue Stabilization Fund, ESSF is the Economic and Social Stabilization Fund and PRF is the Pension Reserve Fund. Other Gross Foreign Assets of the Government (included in Table 8.5) are not included here.

The AFCFRF recommended a change in portfolio composition toward higher yield instruments for both funds in 2007. But at the start of the global financial crisis, the Committee decided to reverse its previous recommendation. With the recovery in world financial markets, the Committee again recommended in August 2009 to diversify the portfolio towards higher yield investments, but now limiting its recommendation to the PRF.[6] Figure 8.13 depicts the current portfolio composition and the recommended portfolio composition for the PRF, which opens up investment to investment grade corporate bonds (20%) and equity (15%). Implementation of this recommendation took place in 2012.

The rates of return of both sovereign funds are closely aligned to international fixed income securities. Compared to other sovereign wealth funds that invest in variable income assets, Chile's returns have been less volatile, avoiding the world-wide swings in variable income asset returns

[6] The reason for this is that PRF has a long-term investment horizon, consistent with the long-term pension liabilities of the government, while the ESSF has a short-term investment horizon, consistent with the cyclical frequency of the average Chilean business and international copper price cycles.

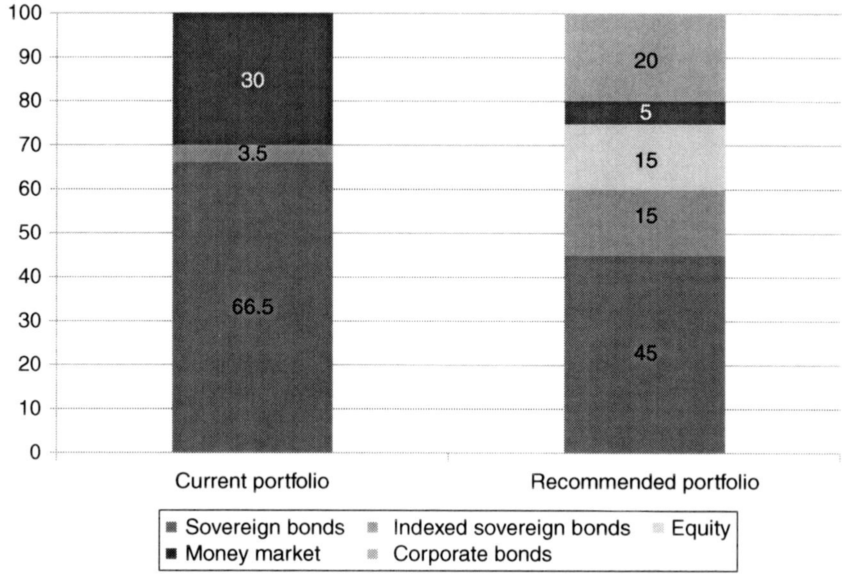

Figure 8.13: Chile: Current International Portfolio Composition of Sovereign Wealth Funds (PRF and ESSF) and Recommended Composition for PRF (%)
Source: Financial Advisory Committee on Sovereign Funds to the Minister of Finance (Financial Committee Report 2010)

observed since 2007 (Table 8.6). Chile's average 2007–2010 return of 5.1% compares favorably to the average returns of eight other sovereign wealth funds, which range from –0.6% (Ireland) to 9.8% (Australia). Yet for medium- to longer-term investment periods, diversification toward variable income assets should bring longer-maturity funds closer to the investment efficiency frontier.

8.3.4 Shortcomings of Chile's Fiscal Policy Framework and Proposals for Reform

By and large, Chile's fiscal rule has proven to be a successful instrument for strengthening fiscal sustainability, reducing the pro-cyclical bias in fiscal policy and spending, contributing to macroeconomic stability and strengthening the political economy of fiscal policy. Yet several shortcomings are still observed in the design and implementation of the fiscal rule, and in Chile's institutional framework that governs fiscal policy. Hence, the Minister of Finance of the Piñera administration, Felipe Larraín, convened an "Advisory Committee for the Design of a Fiscal Policy based on a Second-Generation Structural Balance for Chile" to assess shortcomings

Table 8.6: *Asset Composition (%) and US$ Returns (%) of Selected Sovereign Wealth Funds, 2007–2010*

Funds	Holdings (US$ mmm)	Type	Portfolio Composition				US$ Return				
			Money Market	Fixed Income	Equity	Other Investments	2007	2008	2009	2010	2007–2010 (Average)
Australia	71	P	19	19	35	27		–26.3	44	24.8	9.8
Chile	17	P/E	30	70	0	0	8.9	7.6	2.4	1.8	5.1
East Timor	15	A	1	25	53	21	20.7	–31.7	28.7	15.7	5.3
Canada	7	E	1	99	0	0	7.4	6.9	0.6	3.8	4.6
New Zealand	14	P	1	8	69	23	15.1	–44.2	48.4	24.3	4.3
Trinidad and Tobago	4	E	28	48	24	0	4.2	3.1	3	5.1	3.8
Norway	528	A	0	39	61	0	10.2	–27.2	30.8	8.8	3.4
Alaska	39	A	3	18	51	28	8.8	–24.7	18.9	11.9	2.2
Ireland	20	P	15	8	57	20	14.2	–33.3	23.6	3.8	–0.6

Source: Financial committee report (2010).

Notes: Outstanding holdings at end of period. Type is defines as “S” savings, “P” pensions and “E” Stabilization. Other investments include private equity, hedge funds, commodities and real state. Average is calculated as the geometric average.

and propose reforms of the rule and fiscal institutions. The Advisory Committee was formed in May 2010 and met through January 2011, delivering a preliminary report in August 2010 (Advisory Committee 2010) and a final report in January 2011 (Advisory Committee 2011). The main recommendations made by the Committee are summarized here.

The Committee made the following recommendations on the general methodology of the fiscal rule:

(1) Define a rule based on the definition of the cyclically adjusted balance (CAB), limited to cyclical adjustments in revenue due to cyclical shocks of GDP and world prices of copper and molybdenum, without considering other sources of deviations of current revenue from permanent revenue. As discussed earlier, this recommendation has already been applied since 2010, reverting the changes to the rule that were introduced between late 2008 and early 2010.

(2) Supplement the rule of the CAB with an expenditure component that allows for an ex-ante counter-cyclical policy. This expenditure component could be a negative function of the ex-ante projected output gap.

(3) Supplement the rule with a partial adjustment mechanism for expenditure in response to large changes in the projections of trend GDP growth and long-term copper and molybdenum prices, in order to avoid large changes in expenditure like those observed in 2006–2009.

(4) Define ex-ante escape clauses to the rule, which apply in exceptional circumstances that require an exceptional fiscal expansion or adjustment that is not feasible to implement without a temporary suspension of the rule. Invocation of such clauses would avoid de facto suspension or ad hoc modification of the rule, like those implemented in 2008–2009. Invocation of an escape clause would proceed under pre-defined conditions and would require a policy commitment to converge back to full application of the rule in a reasonable time horizon.

(5) For operational reasons, maintain institutional coverage of the rule to the CAB of the consolidated central government level (i.e., avoid extending the rule to the general government). However, complement the rule with measures to prevent the risk of quasi-fiscal operations in support of decentralized government agencies, public enterprises and municipalities.

(6) Maintain the rule for the overall or aggregate balance of the central government (without adjusting for interest payments on government financial assets and liabilities) in the short term. However, after a transition period in which both balance measures (overall and primary balance) should be calculated and reported, define the rule for the primary balance of the central government. This option has the following benefits: (i) it avoids the effect of temporary changes in interest rates on the CAB, (ii) it eliminates the pro-cyclical bias caused by the accumulation into, and withdrawal from, net financial assets and (iii) it links the balance measures with the standard analysis of long-term fiscal sustainability.

(7) Choose a target level for the CAB determined by long-term fiscal solvency objectives. This requires engaging in systematic assessments of the values of contingent liabilities of the government (pensions, health, education and so on) over time. Similar studies should be devoted to the time profile of government assets (such as the decline in mining assets of state company CODELCO).

The Committee made the following recommendations on the methodology for estimating the cyclically adjusted revenue associated to GDP and the methods for estimating trend GDP:

1. Maintain the role of the Advisory Committee for Trend GDP (ACTG) but extend the minutes of the Committee to include explicitly its estimates for the level of trend GDP and the corresponding output gap measure.
2. Maintain the revenue elasticities to GDP at the values presented in Velasco et al. (2010), except the GDP elasticity of health insurance contributions, set at 1.17.
3. Avoid any cyclical adjustments to revenues and to the CAB due to temporary tax measures, except for provisional payments of taxes that shift resources between years.

On the methodology for forecasting the cyclically adjusted revenue associated with mining prices and the methods for estimating long-term prices of copper and molybdenum, the Committee made the following recommendations:

1. Maintain the role of the Advisory Committee for the Reference Copper Price (ACRCP) but provide the members of the Committee with better information on world copper markets and prices.

2. Delegate preparation of molybdenum trend price forecasts (which are now prepared by the Budget Directorate) to ACRCP, Cochilco (the government copper analysis agency) or a new Advisory Committee for the Long-term Molybdenum Price.
3. Simplify and publish the calculation of the cyclically adjusted mining revenue, possibly in line with the suggestions offered by IMF (2010).

Regarding the institutional framework, international best practice in recent years suggests that fiscal policies and the adoption of fiscal rules are strengthened by the establishment of agencies that audit fiscal projections, assess the achievement of fiscal policy objectives and promote policy transparency. In this regard, the Committee recommended adoption of a Fiscal Council (FC) with a mandate and tasks defined by a statutory law, supported by council members, staff and consultants. The FC's main task would comprise:

(1) assessing the methodology of the fiscal rule,
(2) providing binding assumptions and projections for the cyclical adjustment, variables, required to estimate the CAB, based on the forecasts made by ACTG and ACRPC,
(3) evaluating fiscal policy, including implementation of the fiscal rule and the sustainability of medium- and long-term fiscal policy,
(4) issuing a technical report on any changes in accounting principles and methodologies used in preparation of the budget,
(5) evaluating the relevance of invoking escape clauses and the convergence strategy to the target level of the CAB and
(6) expressing a position on the government's report on contingent liabilities and the impact of the latter on the CAB target.

Finally, the Committee made recommendations to strengthen transparency, accountability and consistency of fiscal policy in Chile, among which the following are the most important:

1. Communicate publicly any changes in the methodology used in calculating the CAB before their adoption.
2. Establish as legal requirement the convening of both ACTG and ACRPC and the publication of the minutes of their meetings.
3. Improve publication of information required in replicating or checking fiscal, macroeconomic and mining revenue calculations for the CAB.
4. Provide more detailed information in the annual presentation of medium-term financial projections for the budget.

5. Establish the publication of the Annual Report of Public Sector Financial Management and updating of fiscal projections, including assessment of expected budgetary deviations and discussion of actions for corrections, as legal requirements
6. Expand the coverage of information provided in the Annual Report on Contingent Liabilities.
7. Establish the regular delivery of actuarial long-term fiscal projections as a legal requirement.
8. Improve assessment and information provided in the financial reports of new laws.

8.4 Norway's Fiscal Policy Institutions and Outcomes

Norway, one of the world's richest countries, started oil production in 1971 and became an increasingly important oil exporter in the 1980s and 1990s. Prudent and transparent management of government oil rents is a major component of Norway's fiscal policy framework, which targets both cyclical stabilization and inter-generational equity as key policy objectives. In the following, I describe Norway's institutional framework of fiscal policy, its fiscal rule and performance and the government's SWFs.

8.4.1 Institutional Framework for Fiscal Policy

Norway's oil production has grown exponentially since the mid-1990s and peaked at 250 m. scm in the mid-2000s. Future oil production is projected to decline over the next decades as current reserves are depleted, although production from existing and still undiscovered oil resources will retard full depletion of all oil resources (Figure 8.14).

As early as 1990, Norway's parliament passed a law establishing the Government Petroleum Fund to invest part of the government's oil rents in financial assets. Norway's oil revenues in the 1990s were substantially higher than projected, resulting in large capital transfers from the government to the GPF. This partly unexpected rise in government financial wealth raised the issue of government use of the Fund's return to finance a large future increase in long-term government pensions and health expenditure. In 2001, the government implemented a major overhaul of

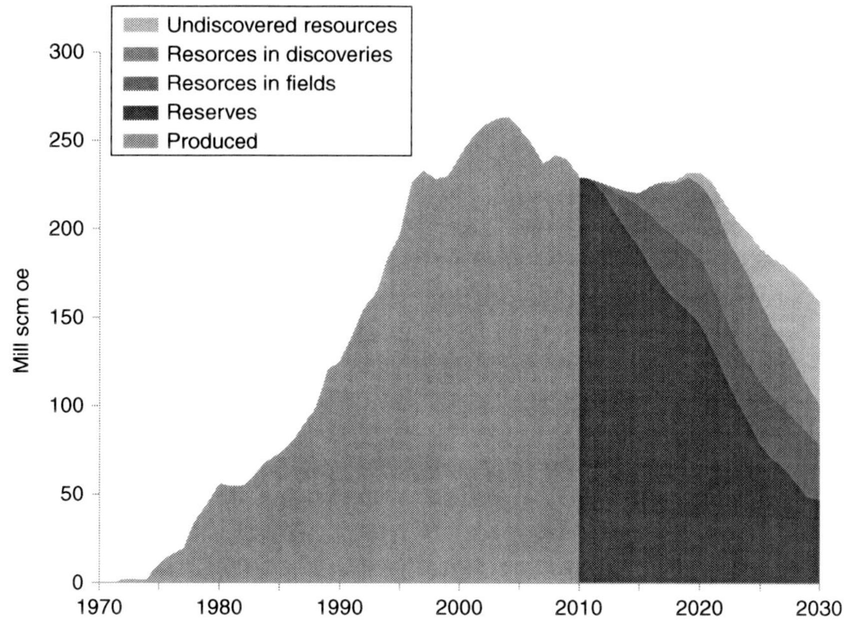

Figure 8.14: Norway: Past (1970–2010) and Projected (2011–2030) Oil Production
Source: Norwegian Petroleum Directorate (2011).

its fiscal policy framework, based on a sophisticated but simple fiscal rule that has been in place for the last decade.

The 2001 guidelines for Norway's new fiscal policy framework are geared toward three major objectives (closely following Norwegian Ministry of Finance 2001):

i) *Macroeconomic Stability*: fiscal policy should contribute to economic stability, avoiding pronounced cyclical fluctuations.
ii) *Sustainable Fiscal Policy and Inter-generational Equity*: fiscal policy should be sustainable over time to avoid the need for major policy shifts that weaken the provision of public services. Hence, government budgets should show substantial surpluses over the next years, with sizeable transfers to the GPF. However, over time, oil revenues will decline while government spending on pensions and health care will increase.
iii) *Resource Use Efficiency*: Economic policy should contribute to the efficient use of resources in both public and private sectors. Long-term welfare is not determined primarily by oil revenue but by the

size and skill of the labor force and the level of resource efficiency. Fiscal policy should also support efficient functioning of the economy.

To attain the latter objectives, Norway's fiscal policy framework is based on three pillars for the conduct of fiscal policy:

1. Inter-generational sharing of the oil wealth by its transformation from oil in the ground into a diversified portfolio of financial investments in world capital markets through Norway's SWF.
2. Sustainable annual withdrawal from the SWF's investments at its expected long-term (imputed) 4% real rate of return, transferred to the government budget.
3. A structural fiscal rule, consistent with permanent income behavior, that targets a cyclically adjusted non-oil government deficit, financed by an average 4% SWF transfer to the budget. Annual deviations of the latter transfer from the 4% return rule are allowed for further discretionary government spending geared at counter-cyclical stabilization and expenditure smoothing, as discussed later.

Projections of the structural non-oil budget deficit are presented to parliament and the public in budget documents that are published twice a year, on the occasion of the submission of the National Budget in October and the Revised National Budget in May of each year.

Therefore, Norway's institutional framework for fiscal policy is based on a fiscal rule, a sovereign wealth fund for investment of oil rents and full integration of the fund into the government's annual budget. I discuss the latter components and their performance in more detail next.

8.4.2 Fiscal Rule

Norway's fiscal rule for the non-oil budget is based on the distinction between actual and cyclically adjusted projections for the following key budget items: tax revenue, excise duty revenue, transfers from the central bank (Norges Bank), the central government's net interest payments and unemployment benefit payments. Over the business cycle, government expenditure (other than on net interest payments and unemployment benefits) is determined by projected levels of cyclically adjusted government tax and non-tax revenue plus an imputed 4% real return on GPFG investments.

The 2001 fiscal policy guidelines do not require that spending be strictly limited to the ceiling imposed by the 4% real return on GPFG every year. On the contrary, fiscal policy is expected to contribute to short-term stabilization of the economy by making use of discretionary government spending. Therefore, the Norwegian fiscal rule allows fiscal policy to play a counter-cyclical role that goes well beyond a strict structural or cyclically adjusted income rule (like Chile's a-cyclical spending rule). In contrast to Chile, Norway's government budget allows for less (more) government expenditure growth during the expansionary (contracting) phase of the business cycle, when projected output gaps are positive (negative).

In addition (and also in contrast to Chile), Norway's rule also allows for government spending growth to be smoothened over several years when the market value of GPFG investments suffers large changes. The latter provision reduces government expenditure volatility induced by large changes in the 4% real investment return that accrues to the budget when world capital markets exhibit large swings, like those observed during the period 2008–2012.

In sum, Norway's fiscal rule implies that government spending is determined by:

(a) Trend values of government tax and excise duty revenue and Norges Bank transfers to the budget; less trend values of unemployment benefit payments and of net interest payments by the central government; less government transfers; and plus the 4% real return on GPFG investments,
(b) additional spending (positive or negative) determined by discretionary fiscal policy for counter-cyclical stabilization and
(c) additional spending (positive or negative) determined by discretionary fiscal policy to avoid excessive spending volatility induced by large swings in market value of GPFG investments.

Finally, the calculation of estimations and projections of structural government revenues from taxes and excise duties is based on the time series behavior of actual taxes and excise duties, extended by projections for their relevant cyclical determinants. Calculations are carried out as follows:

(i) Reported tax and excise duty items are adjusted for the estimated impact of any discretionary measures (i.e., changes in rates and/or delimitation of tax bases), as published at an aggregate level in annual National Budget reports. Therefore, all discretionary tax rate or base changes, of a temporary or permanent nature, are reflected in structural and cyclically adjusted revenue items alike.

(ii) A preliminary cyclical adjustment of the various tax and excise duty items is based on the percentage deviation between actual and trend levels of relevant cyclical indicators, using the Hodrick–Prescott filter. For example, national account figures on employment are used to adjust personal taxes and payroll taxes. A weighted indicator of private consumption and sales in the construction industry is used to adjust VAT revenue.

(iii) The structural level for different taxes and excise duties is calculated as the trend in the adjusted series, measured at constant prices. Separate price indices are used for the various taxes and excise duties. The cyclical components of taxes and excise duties are calculated as residuals (i.e., as the deviations between actual and trend estimations and forecasts).

8.4.3 Government Budget Performance

Norway's fiscal policy performance is reflected in Figure 8.15. Panel a of this figure depicts the evolution of the structural non-oil deficit and the 4% real return on GPFG investments since the start of the new fiscal framework in 2001. While both variables are clearly related, they are not identical to each other, reflecting the fact that deviations of the structural non-oil deficit – stemming from deviations of government spending due to additional discretionary spending – have been actively put in place. Indeed, the real growth rate of government spending exhibits significant annual variation. For example, real government spending grew by 5.5% during the recession year 2009, while its average rate of growth was less than half the latter figure during the preceding decade (Panel b).

Consistent with projected lower oil production and population ageing in the future, the increase in the use of oil revenue is projected to decline in the next 14 years (2012–2025) in comparison to the past decade (2001–2011), while pension expenditure is projected to grow more in the future than in the past (Panel c).

Finally, Panel d depicts Norway's actual consolidated government budget balance (the consolidated balance of GPFG and the non-oil government budget) during the last quarter century. The consolidated balance exhibits a full turnaround from moderate deficits in the early 1990s to attain a very large average surplus of 20% of GDP during recent years. The actual non-oil budget balance hovers around –5% of GDP since the start of Norway's new fiscal framework adopted in 2001. The difference – the

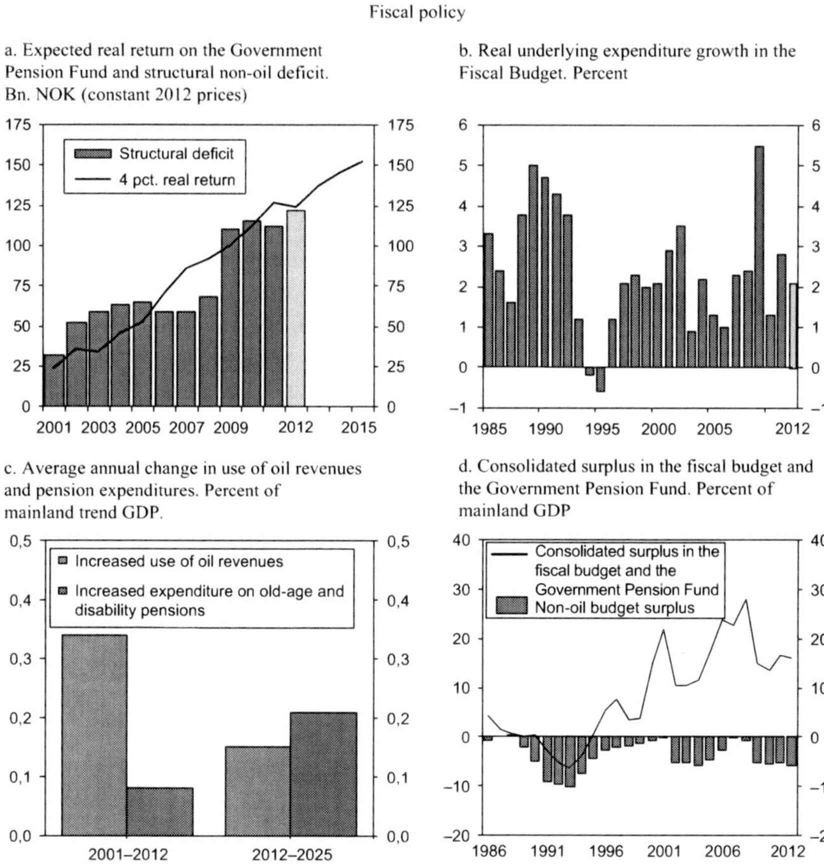

Figure 8.15: Norway: Government Budget and Pension Fund Performance
Source: Norwegian Ministry of Finance (2011).

GPFG surplus – grows from close to nil in the early 1990s to more than 25% of GDP during recent years, reflecting high oil prices and a massive accumulation of GPFG funds, as discussed next.

8.4.4 Government's Sovereign Wealth Fund Rules and Performance

As discussed earlier, Norway's Parliament passed a law in 1990 establishing the Government Petroleum Fund (GPF) to manage the government's oil rents. In 2006, Parliament established the Government Pension Fund, comprised of two separate sovereign wealth funds: the Government

Pension Fund Global (GPFG) – the successor of GPF – and the Government Pension Fund Norway (GPFN). The renaming of GPF highlights its role in supporting the saving of current government revenue to fund the future rise in government pension payments. However, and this is key, neither sovereign wealth fund is related to the government's pension liabilities. The returns of both funds simply finance the structural non-oil deficit of the government, according to the aforementioned fiscal rule.

The MoF regularly transfers oil revenue to GPFG – the first transfer took place in 1996. GPFG is fully integrated into the government's budget. Therefore, net allocations to GPFG reflect the total actual consolidated government budget surplus, including oil rents (as reflected in Panel d of Figure 8.15).

GPFG is managed by Norges Bank Investment Management (NBIM) on behalf of the Ministry of Finance. The latter determines GPFG's investment strategies, following advice from NBIM and Parliament, among others.[7]

GPFG funds are fully invested in international capital markets, according to MoF guidelines. Over time, GPFG investments were allowed to be diversified away from only government bonds (1996), to include equities (1997), corporate and securitized bonds (2002) and real estate (2010); and from only industrial countries to include emerging markets (2000).

As of 2011, GPFG investments represent close to 275% of Norway's GDP (Figure 8.16). Its portfolio is comprised of 56% invested in equity, 43% in fixed income and nearly 0.5% in real estate. GPFG's average annual return has been 5% in 1998–2010 (Table 8.7).

The second government fund, GPFN, was initially established under the name of National Insurance Scheme Fund in 1967 and its current name was adopted in 2006. Operational management of the GPFN is delegated to the National Insurance Scheme Fund (NISF).

GPFN funds are restricted to be invested in equity instruments traded in regulated markets in Norway, Denmark, Finland and Sweden, as well as in interest-bearing instruments of issuers that are residents of Norway, Denmark, Finland and Sweden or that have equity capital traded in regulated markets in these countries.

[7] Ethical guidelines for GPFG investments were established in 2004. A Council on Ethics for the GPFG prepares recommendations on whether or not to invest in particular companies, based on the ethical guidelines. Recommendations may be followed up by the MoF. Both Council recommendations and MoF decisions are made public.

Table 8.7: *Norway: Average Rates of Return on GPGF and GPFN Investment Portfolios, 1998–2010 (%)*

	2010	2008–2010	1998–2010
GPFG			
Actual Portfolio	9.60%	1.84%	5.04%
Equity Portfolio	13.34%	-3.37%	4.51%
Fixed Income Portfolio	4.11%	5.22%	5.15%
GPFN			
Actual Portfolio	15.27%	4.84%	7.01%
Norwegian Equity Portfolio	19.11%	-1.26%	8.42%
Nordic Equity Portfolio	28.18%	0.46%	3.74%
Norwegian Fixed Income Portfolio	7.47%	8.77%	6.10%
Nordic Fixed Income Portfolio	3.65%	6.23%	5.53%

Source: "The Management of the Government Pension Fund in 2010", Report No. 15 (2010–2011) to the Storting.
Note: Annual geometric average.

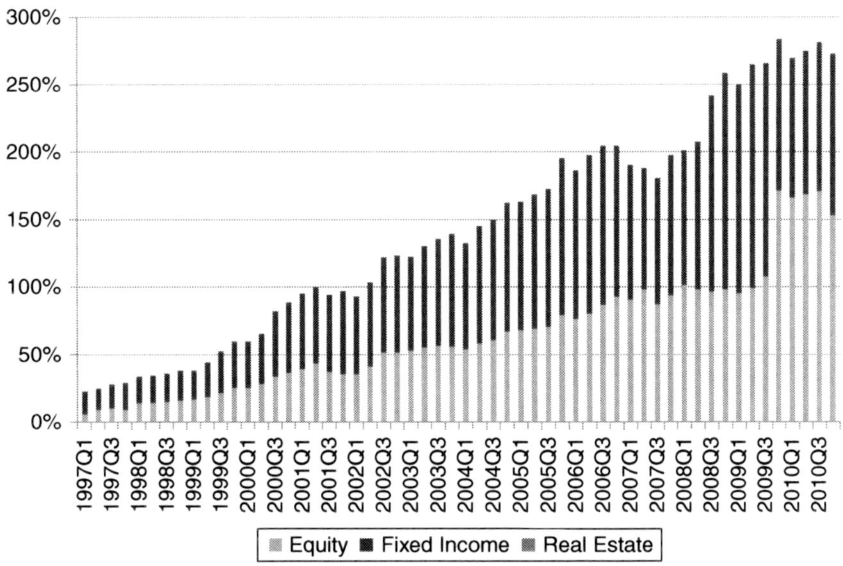

Figure 8.16: Norway: Government Pension Fund Global (GPFG) Investment Portfolio, 1997–2011 (Ratio to GDP, %)
Source: Norges Bank Investment Manager.

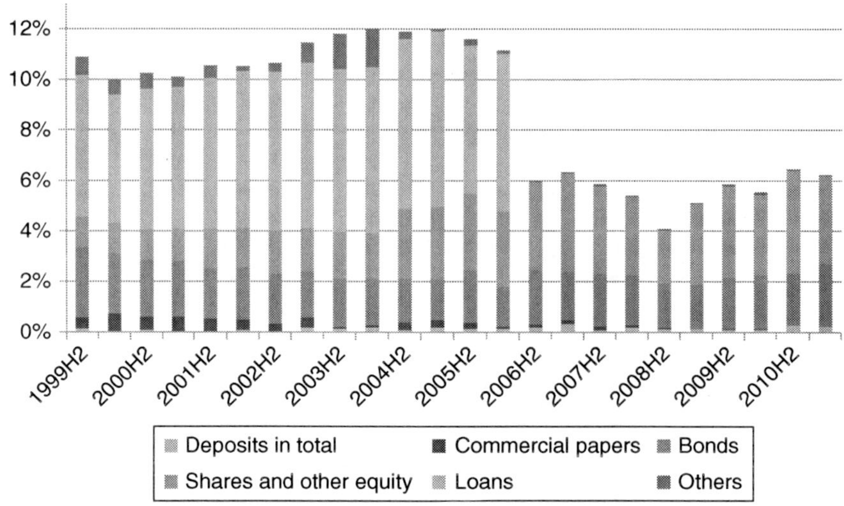

Figure 8.17: Norway: Government Pension Fund Norway (GPFN) Investment Portfolio, 1997–2011 (Ratio to GDP, %)
Source: Statistics Norway.

Investment guidelines established by the MoF for GPFN mandate a portfolio share of equity (interest-bearing) instruments in the range of 50–70% (30–50%). 80–90% (10–20%) of equity (fixed income) portfolios are to be placed in Norway, Denmark, Finland and Sweden.[8]

As of 2011, GPFN investments represent close to 6% of Norway's GDP. Its portfolio largely comprises bonds and shares and other equities (Figure 8.17). GPFN's average annual return has been 7% in 1998–2010 (Table 8.7).

8.5 Lessons on Fiscal Institutions for Resource-rich Economies and Oil-exporting Arab Countries

The review of the world evidence on the links between resource rents, fiscal and political institutions and macroeconomic outcomes presented in Section 8.2, and the more detailed review of Chile's and Norway's successful experience in managing their resource wealth in subsequent sections, point to significant policy lessons for RREs in general and oil-exporting Arab countries in particular. Here I limit the drawing of policy implications to those pertaining to fiscal policy institutions, and not to additional inferences that can be drawn for the reform of political institutions.

[8] The MoF has determined a strategic benchmark portfolio, which consists of 40% interest-bearing instruments and 60% equity instruments. Ex-ante relative volatility is not allowed to exceed three percentage points on an annualized basis.

Adoption of a modern institutional framework for fiscal policy can make a major contribution to lessen the adverse impact of resource booms and strengthen timely and efficient use of government resources. Such a framework is instrumental for the three goals of fiscal policy: attaining and maintaining fiscal solvency (or budgetary sustainability), contributing to macroeconomic stability and ensuring inter-generational equity.

International experience in general and Chile's and Norway's successful development of sound fiscal policy frameworks since 2001 suggest lessons in four key policy areas to strengthen fiscal policy institutions and conduct in RREs and in Arab oil-exporting countries. The subsequent lessons apply to varying degrees to both richer, labor-importing countries and less rich, populous Arab oil exporters, because the divide between the two latter country groups is not clearly correlated with large differences in fiscal institutions, management of oil booms and their macroeconomic consequences, as suggested by Selim and Zaki (Chapter 11 in this volume). However, large country differences are observed across all eleven oil-exporting Arab countries. Therefore, the relevance of the subsequent policy recommendations varies significantly from country to country.

1. Strengthening the Institutional Framework of Fiscal Policy

Developing the institutional framework for fiscal policy and budget planning and execution comprises six components.

(1) Fiscal Responsibility Law: Adopt a legal framework, specifying budgetary procedures and rules to strengthen fiscal policy transparency, accountability and stability. Fiscal responsibility laws establish requirements for governments to state transparently their short- and medium-term policy objectives, to set short- and medium-term targets for fiscal flows and stocks, to provide transparent information on budget planning and execution to the legislature, to implement a fiscal policy conducive to fiscal stability and solvency, and to account ex-post for fiscal policy execution and attainment of policy objectives.

(2) Effective Financial Management of the Budget: The core elements for efficient budgetary management by the MoF comprise granting strong powers to the MoF on legal initiatives with budgetary impact and in budget decisions regarding other ministries and government levels, commanding a high technical capacity for budgetary planning, adopting modern accounting principles,

ensuring an efficient process of legislative approval of the budget that limits the possibility of congressional amendments and developing strong capacity for monitoring budget execution.

(3) Budget Planning Horizon: Extend budgetary planning from one-year to multi-year horizons.

(4) Fiscal Policy Accountability and Transparency of Budgetary Information: Both for economic efficiency and political legitimacy goals, accountability of fiscal policy and transparency in the provision of information of government accounts should be improved significantly.

(5) External Control and Auditing: Countries should strengthen external control and auditing functions of fiscal policy, budget execution and government accounts at three levels: congress, the government's general comptroller or auditor and fiscal councils and committees (more on the latter later).

(6) Analysis of Fiscal Stance: Resource-rich economies should rely on a varied set of key fiscal indicators to strengthen analysis of their fiscal stance.[9]

2. Adoption of a Fiscal Rule

Adoption of a fiscal rule based on a cyclically adjusted balance (CAB), as in Chile and Norway, and for inter-generational equity, as in Norway, is the key reform required to avoid pro-cyclical bias and pro-current generation bias in fiscal policy and government spending. While for non-commodity exporters, cyclical adjustment may be limited to domestic GDP shocks, for commodity exporters, it is key to add cyclical adjustments to price shocks of their commodities. Like in Chile and Norway, in resource-rich countries the largest source of government revenue volatility is price volatility of their commodity exports, not domestic GDP volatility. However, domestic GDP volatility is likely to be lowered (as is the case in Chile and Norway) by adopting a rule that accounts for both GDP and commodity price

[9] Villafuerte and López-Murphy (2010) suggest developing and using the following fiscal indicators in oil-producing countries for policy analysis: overall fiscal balance, non-oil balance, non-oil primary balance, cyclically-adjusted non-oil primary balance and non-oil GDP as the relevant scale variable. As discussed in Section 8.3, Norway is the paramount example of successful application of the latter concepts in its analysis and public discussion of fiscal policy performance.

volatility. In addition, finite resource reserves require saving part of the current temporary revenue for future generations, as implemented in Norway.

A successful fiscal rule will have a positive impact on the stability of government spending and government programs, fiscal policy sustainability and credibility and overall macroeconomic stability. In order to be effective, the fiscal rule should satisfy the following conditions:

(1) Be based on calculations and formulas that are relatively stable over time,
(2) Be based on assumptions and calculations that should be made public, together with the relevant budgetary information, to be replicated by independent analysts and
(3) Be part of an institutional environment that supports independent or unbiased provision of key economic and financial forecasts (such as independent committees or councils).

The actual formula of the fiscal rule will be country-specific, and so will be the budget balance for which it is defined (total or primary), the government level to which it applies (central or general government), the degree of spending a-cyclicality (as in Chile) or counter-cyclicality (as in Norway), the target level for the CAB, and the degree of inter-generational re-distribution (as in Norway, with very large re-distribution toward future generations).

3. Adoption of a SWF

Fiscal rules that target CABs and re-distribute resources toward the future require a SWF where government savings are deposited cyclically (deposited during booms and withdrawn during downturns) and permanently (for inter-generational distribution). Such funds are particularly important in commodity -exporting countries, where they are typically invested in foreign currency investments that are held abroad, as government revenue comes also in foreign currency from the revenue obtained by commodity-exporting firms.

Several RREs, including many Arab oil exporters, have established SWFs. In the UAE, Abu Dhabi's SWFs add up to the largest SWF savings among all countries, states and emirates in the world. Saudi Arabia and Kuwait also have large SWFs, followed in the Arab World (by size) by Qatar, Dubai, Libya, Algeria, Bahrain, Oman and the UAE's Ras Al Khaimah (Sovereign Wealth Fund Institute, 2012). The world's oldest

SWF (except for the US state of Texas) was established by Kuwait in 1953 and remains a model for many other SWF managers within and outside the region. According to the Sovereign Wealth Fund Institute (Figure 8.2), transparency in corporate governance and financial management varies significantly across SWFs in the world. Among the SWFs of Arab oil-exporting countries, only Abu Dhabi's SWFs get the world's highest marks in transparency, together with the SWFs of Chile, Norway and a few other countries.

Corporate governance and investment policies of SWFs could be strengthened in many countries through the following dimensions:

(1) Establishing of effective and transparent corporate governance.
(2) Provision of transparent information on transfers between the budget and SWFs.
(3) Portfolio choice determined by investment strategies based on maturity concerns (determined by the nature of the fund and the length of commodity price and GDP cycles), hedging objectives and the government's degree of risk aversion.
(4) Efficient portfolio management set by transparent guidelines and closely monitored by the government and the public.

4. Establishment of Government Committees and a Fiscal Council

Specialized government or MoF committees set up with narrow tasks – such as provision of independent forecasts on key variables for the budget or a board for the SWF – and staffed by independent members are an efficient way to bring expertise and independence to the government, strengthening fiscal institutions.

Moving beyond the latter committees, several industrial countries have already put in place, or are starting, independent fiscal councils that have broad mandates to improve the quality of fiscal policy assessment, budget planning and monitoring execution. To my best knowledge, no resource-rich country has started a fiscal council to date. However, they are particularly needed in countries with weaker fiscal institutions, where such councils could make a major difference. Among their tasks are the following:

(1) Evaluation and validation of financial and macroeconomic assumptions for the budget.
(2) Provision of independent projections for the budget and government financial conditions.

(3) Provision of policy assessments and recommendations of budgetary corrections.
(4) Advice on the stance of fiscal policy, its long-term sustainability and optimality, and its macroeconomic effects.
(5) Recommendations on policies for government asset and liability management.
(6) Monitoring of budget implementation and provision of recommendations on budgetary corrections when facing deviations in budget execution.
(7) Recommendations to improve fiscal transparency.

8.6 Conclusions

Prudent and far-sighted conduct of fiscal policy is at the core of macroeconomic and development objectives geared towards attaining macroeconomic and financial stability, high growth, poverty reduction and intergenerational equity. Over the last decades, economists and policy makers alike have learned that fiscal institutions are the foundation of an effective fiscal framework that aims to fulfill these policy objectives.

Economic history abounds with negative examples of inconsistent and inefficient design and execution of fiscal policies that lead to macroeconomic and financial crises, debt defaults, financial under-development, low growth and inequity. A long and dismal history of fiscal mismanagement characterized most emerging economies in the past. Yet a growing number of developing economies is learning from past failures and is changing radically toward better fiscal institutions and policies. In the industrial world, the opposite trend is observed in many high-income countries, reflected in current conditions that range from fiscal fragility to outright fiscal and financial crises.

The need for revamping fiscal policies and institutions is particularly acute in resource-rich economies in general and Arab oil exporters in particular. A modern fiscal policy in support of inter-generational and of pro-cyclical saving should be based on explicit recognition that natural resources are not unlimited (hence requiring saving for the future) and commodity prices are highly volatile (hence requiring saving temporary windfalls).

I have reviewed in this chapter the international evidence on fiscal policies and outcomes in resource-rich economies at large and Arab oil-exporting countries in particular, which suggests that strong fiscal (and political) institutions can turn the resource curse into a blessing. Then, I have turned to detailed assessments of Chile's and Norway's decade-long experience in seeking and attaining more macroeconomic and financial stability, higher growth and improved equity. Finally, I have drawn specific reform lessons to strengthen fiscal institutions and policies for resource-rich economies and Arab oil exporters.

References

Abdih, Y., P. López-Murphy, A. Roitman, and R. Sahay, (2010). The Cyclicality of Fiscal Policy in the Middle East and Central Asia: Is the Current Crisis Different? IMF Working Paper No. 10/68, March.

Advisory Committee on Fiscal Policy (Comité Asesor para el Diseño de una Política Fiscal de Balance Estructural de Segunda Generación para Chile), (2010). Propuesta para Perfeccional la Regla Fiscal (Informe Preliminar), manuscript, Ministry of Finance of Chile.

(2011). Propuesta para Perfeccional la Regla Fiscal (Informe Final), manuscript, Ministry of Finance of Chile.

Advisory Committee for Trend GDP (Comité Consultivo del PIB Tendencial, agosto de 2011), (2011). Acta Comité Consultivo PIB Tendencial, manuscript, Ministry of Finance of Chile.

Advisory Committee for the Reference Copper Price (Acta del Comité Consultivo del Precio de Referencia del Cobre, agosto de 2011), (2011). Acta Comité Consultivo del Precio de Referencia del Cobre, manuscript, Ministry of Finance of Chile.

Advisory Committee for Sovereign Funds (Comité Asesor del Ministerio de Hacienda para los Fondos de Responsabilidad Fiscal), (2010). Informe Anual Comité Financiero, manuscript, Ministry of Finance of Chile.

Alesina, A., S. Ardagna, R. Perotti and F. Schiantarelli, (1999). Fiscal Policy, Profits, and Investment. NBER Working Papers, No. 7207.

Alesina. A and R. Perotti, (1996). Budget Deficits and Budget Institutions. NBER Working Papers, No. 5556.

Al Iriani, M., (2012). Oil Curse in Yemen: The Role of Institutions and Policies. ERF Working Paper 694.

Al-Sheikh, H. and S.N. Erbas, this volume. Oil Curse and Labor Markets: The Case of Saudi Arabia.

Bhattacharyya, S. and P. Collier, (2011). Public Capital in Resource Rich Economies. CSAE manuscript, August.

Bornhorst, F., S. Gupta, and J. Thornton, (2008). Natural Resource Endowments, Governance, and the Domestic Revenue Effort: Evidence from a Panel of Countries. IMF Working Paper No. 08/170, July.

Cáceres, C., A. Corbacho, and L. Medina, (2010). Structural Breaks in Fiscal Performance: Did Fiscal Responsibility Laws have Anything to Do with Them? IMF Working Paper No. 10/248.

Calderón, C. and K. Schmidt-Hebbel, (2008). The Choice of Fiscal Regimes in the World. Central Bank of Chile Working Paper No. 487.

Calderón, C., R. Duncan and K. Schmidt-Hebbel, (2016). Do Good Institutions Promote Countercyclical Macroeconomic Policies? *Oxford Bulletin of Economics and Statistics*, forthcoming.

Calmfors, L. and S. Wren-Lewis, (2011). What Should Fiscal Council Do? Discussion Paper Series, University of Oxford Department of Economics.

Debrun, X., D. Hauner and M. Kumar, (2009). Independent Fiscal Agencies. *Journal of Economic Surveys* 23(1).

Debrun, X. and M. S. Kumar, (2007). The Discipline-Enhancing Role of Fiscal Institutions: Theory and Empirical Evidence. IMF Working Papers No. 171.

Eijffinger S., Geraats P., and C. van der Cruijsen, (2006). Does Central Bank Transparency Reduce Interest Rates? DNB Working Papers 085.

Elbadawi, I., K. Schmidt-Hebbel, and R. Soto, (2015). Why do Countries have Fiscal Rules?, In R. Caballero and K. Schmidt-Hebbel (editors): *Economic Policy in Emerging-Market Economies: Festschrift in Honor of Vittorio Corbo*. Central Bank of Chile. Santiago, Chile.

Elbadawi, I.A. and R. Soto. This Volume. Resource Rents, Political Institutions and Economic Growth.

Engel, E., M. Marcel and P. Meller, (2007). Meta Superávit Estructural: Elementos para su Análisis. Manuscript, Ministry of Finance of Chile.

Fasano, U. and Q. Wang, (2002). Testing the Relationship between Government Spending and Revenue: Evidence from GCC Countries. IMF Working Paper No. 02/201, November.

Fiess, N., (2004). Chile's Fiscal Rule. Econometric Society 2004 Latin American Meetings 348, Econometric Society.

Frankel, J. (2011). A Solution to Fiscal Procyclicality: The Structural Budget Institutions Pioneered by Chile. Working Paper N°604, Banco Central de Chile, Santiago, Chile.

García, S., (2011). Regla Fiscal Contracíclica, Análisis para Chile.Tesis de Magister en Economía, Universidad de Chile.

Gelb, A., (2012). Economic Diversification and the Role of Government in Resource Rich Economies. Presented at the ERF Conference on "Understanding and Avoiding the Resource Curse in the Arab World", Kuwait, January 15–16.

Geraats, P., (2008). ECB Credibility and Transparency, European Economy - Economic Papers, 330.

Gylfason, T., (2012). International Experiences with the Management of Natural Resources. Presented at the ERF Conference on "Understanding and Avoiding the Resource Curse in the Arab World", Kuwait, January 15–16.

International Monetary Fund, (2010). Navigating the Fiscal Challenges Ahead. World Economic and Financial Surveys. Washington, DC.

(2009). Fiscal Rules – Anchoring Expectations for Sustainable Public Finance. Discussion Paper approved by Carlo Cottarelli, December 19.

International Budget Partnership. (2010). www.internationalbudget.org/files/OBI2010-Chile.pdf, accessed on December 8, 2011.

Ilzetzki, E. and C. Vegh, (2008). Procyclical Fiscal Policy in Developing Countries: Truth or Fiction? NBER Working Papers No. 14191.

Kumhof, M. and D. Laxton, (2009). Chile's Structural Fiscal Surplus Rule: A Model-Based Evaluation. IMF Working Paper No. 09/88.

Larraín, F. and F. Parro, (2006). Chile menos volátil. Manuscript, Instituto de Economía, Pontificia Universidad Católica de Chile.

Levy, G., (2008). A Parliamentary Budget Officer for Canada. Canadian Parliamentary Review 39.

Lienert, I., (2010). Should Advanced Countries Adopt a Fiscal Responsibility Law? IMF Working Paper No. 10/254.

Marcel, M., (2010). La Regla de Balance Estructural en Chile: Diez Años, Diez Lecciones. Manuscript, Inter-American Development Bank.

Marcel, M., M. Tokman, R. Valdés and P. Benavides, (2001). Balance Estructural del Gobierno Central, Metodología y Estimaciones Para Chile: 1987–2000. Serie de Estudios de Finanzas Públicas N°1, DIPRES.

Morsy, H., (2009). Current Account Determinants for Oil-Exporting Countries. IMF Working Paper No. 09/28, February.

Norwegian Ministry of Finance (2011). The National Budget 2012.

(2010). The Management of the Government Pension Fund in 2010, Report No. 15 (2010–2011) to the Storting.

(2010). The Structural Non-oil Budget Deficit, Memorandum.

(2001). Guidelines for Economic Policy. Report No. 29 (2000–2001) to the Storting.

Norwegian Petroleum Directorate (2011). Petroleum Resources on the Norwegian Continental Shelf. Norwegian Petroleum Directorate, Oslo, Norway.

OECD, (2010). Economic Survey of Chile 2010: Fiscal Policies for Enhanced Resilience and Equity. *OECD Economic Survey of Chile*, Paris: OECD Press, chapter 2.

Rivlin, A. M., (2010). Helping Politicians Face Difficult Choice. Conference on "Independent Fiscal Institutions". Budapest. March 18–19.

Rodríguez, J., C. Tokman and A. Vega, (2006). Política de Balance Estructural: Resultados y Desafíos tras Seis Años de Aplicación en Chile. Estudios de Finanzas Públicas, DIPRES.

Rodríguez, J., L. Escobar and M. Jorratt, (2009). Los Ingresos Tributarios en el Contexto de la Política de Balance Estructural. Estudios de Finanzas Públicas, DIPRES.

Salti, N., (2008). Oil Greasing the Wheels: When Do Natural Resources Become a Blessing? American University of Beirut Working Paper No. 439, September.

Schmidt-Hebbel, K. (2010). Fiscal Rules and Institutions in the World and in Chile. Manuscript, Catholic University of Chile.

(2011). Fiscal Policy for Commodity-Exporting Countries: Experience from Chile. Presented at the IMF Seminar on "Commodity Price Volatility and Inclusive Growth in LIC," Washington, DC, September 21.

Selim, H. and C. Zaki. This volume. The Institutional Curse of Natural Resources in the Arab World.

Sovereign Wealth Fund Institute, (2011). www.swfinstitute.org/statistics-research/linaburg-maduell-transpa-rency-index/ accessed December 8. 2010.

(2012). www.swfinstitute.org/fund-rankings/ accessed March 8, 2011.

Stein, E., A. Talvi and A. Grisanti. (1998). Institutional Arrangements and Fiscal Performance: The Latin American Experience. NBER Working Paper No. 6358.

Suliman, K.M. This Volume. Understanding and Avoiding the Oil Curse in Sudan.

Ter-Minassian, T., (2010). Preconditions for a Successful Introduction of Structural Fiscal Balance-based Rules in Latin America and the Caribbean: A Framework Paper. Inter-American Development Bank, Discussion Paper No. IDB-DP-157.

Torvik, R., (2012). Transparency and Accountability in Resource Rich Countries. Presented at the ERF Conference on "Understanding and Avoiding the Resource Curse in the Arab World", Kuwait, January 15–16.

Velasco, A., A. Arenas, L. Céspedes and J. Rodríguez, (2007). Compromisos Fiscales y la Meta de Superávit Estructural. Estudios de Finanzas Públicas, DIPRES.

Velasco, A., A. Arenas, J. Rodríguez, M. Jorrat and G. Gamboni, (2010). El Enfoque de Balanace Estructural en la Política Fiscal en Chile: Resultados, Metodología y Aplicación al Período 2006–2009. Estudios de Finanzas Públicas No 15.

Villafuerte, M. and P. López-Murphy, (2010). Fiscal Policy in Oil Producing Countries During the Recent Oil Price Cycle. IMF Working Paper No. 10/28, February.

Von Hagen, J., (2005). Political Economy of Fiscal Institutions. GESY Discussion Paper No. 149.

Wren-Lewis, S., (2010). Comparing the Delegation of Monetary and Fiscal Policy. Conference on "Independent Fiscal Institutions". Budapest. March 18–19.

9

Savings and Investment Decisions from Natural Resource Revenues

Implications for Arab Development

Paul Collier

ABSTRACT

The Arab World is fortunate to have large endowments of oil and gas. The level of endowments across countries of the Arab region varies: the Gulf states which have massive endowments and small populations, and other states which have more moderate endowments and much larger populations. I will refer to the latter as Middle-Income Middle-East States (MIMES). These differences in resource endowments have implications for both the decision as to how much of resource revenues should be saved, and how much of the savings should be invested domestically. In this chapter, the author discusses principles and rules for saving and investment appropriate to the GCC and MIMES

Keywords: Arab World, GCC, oil curse, optimal savings rule

9.1 Introduction[1]

The Middle East is fortunate to have large endowments of oil and gas. The economies of the Middle East vary considerably, but in this chapter, I will reduce the complexities of the region's endowments to two groups: the Gulf Cooperation Council (GCC) countries, which have massive endowments and relatively small populations and Middle-Income Middle-East States (MIMES), which have more moderate endowments and much larger populations. These differences in endowments and populations have implications for both the decision of how much of the resource revenues should be saved and the decision of how much of the savings should be invested domestically.

[1] A fuller treatment of the themes discussed in this chapter is in Collier (2010).

Savings behavior should be distinctive because, unlike most other sources of tax revenue, the government revenues from the depletion of oil and gas are unsustainable and volatile. Revenues from the extraction of fossil fuels are unsustainable, not only because they are finite, and so a country might physically exhaust its endowment, but also because due to global warming, it is not possible to burn all of the fossil fuels that have been discovered without overheating the planet. This implies that at some point during the current century, either extraction will be capped by regulation, or non-fossil energy technologies will have been developed, making fossil fuels redundant. Both the unsustainable nature of revenues and their volatility imply that the savings rate appropriate for resource revenues should differ from that on other revenues. Most OECD countries do not have significant resource revenues and so their fiscal rules do not need to incorporate these considerations. Hence, for the savings rates of MIMES to be appropriate to their circumstances, the fiscal rules they adopt should differ from the OECD. Further, the fiscal rules of the GCC should differ from those of the MIMES. I discuss the principles and rules for saving appropriate to GCC countries and MIMES in Section 9.2.

Middle-East governments will also need distinctive policies towards investment. The MIMES should usually invest much of their savings from resource revenues domestically. Hence, the typical MIME will need to develop greater capacity for managing public investment than a country at the same level of income but without natural resources. In contrast, because the populations and territories of GCC countries are so small, they have relatively few opportunities for productive domestic investment. Hence, they need less capacity for managing domestic real investment than other countries at the same income level. I discuss the principles and rules for investment appropriate to GCC countries and MIMES in Section 9.3.

9.2 Savings Policies

Resource-rich countries need savings policies that offset the long-term depletion of natural assets and manage year-to-year volatility. I consider them in turn.

9.2.1 Principles of Offsetting Depletion

The most straightforward framework for thinking through savings decisions from resource extraction is that of "permanent income." The framework provides a simple principle for the appropriate savings from a new income stream that is expected to be temporary. The principle entails that sufficient savings should be accumulated to achieve the highest steady stream of consumption that can be sustained permanently. In turn, the highest steady stream of consumption can be estimated by taking the net present value (NPV) of the income stream, and then applying the expected return on assets to this value. Such a framework is not appropriate in all contexts: low-income countries which are rapidly converging on higher-income countries need a more sophisticated approach, which recognizes that it is sensible to consume the resources now, while the economy is poor, rather than save for a distant, much more prosperous future. However, for the Middle East, the permanent income approach is reasonable: the region is not so poor as to be confident of a phase of rapid catch up, in such a way that exhausting oil and gas endowments without regard for future generations would not be ethically reasonable.

While the "permanent income" (maximum sustainable consumption) from resource extraction is estimated by taking the NPV of the income stream, this does not imply that all resource revenues should be saved. Some of the revenues should indeed be set aside to accumulate as assets, but only at the point when the resource is exhausted will these assets fully generate sufficient income to cover the extra consumption that is warranted by the discovery. Thus, through the duration of extraction, some of the resource revenues should legitimately and entirely be used for consumption, financing the shortfall between the consumption that is warranted and the income from accumulated assets.

Year-by-year, the appropriate savings rate from resource revenues depends on their expected trajectory: how long will the resource last until exhaustion, and what will be the shape of revenues? The key insight can be seen from an extremely simple trajectory in which resource income is expected to be constant until the resource is exhausted after a known period of years. The first savings rule is that *the shorter the horizon until expected depletion, the higher should be the savings rate.* This follows straightforwardly. For any given annual revenue, the shorter the extraction period, the lower the present value of the resource endowment and hence, the lower the sustainable increase in consumption. With less consumption warranted from a given resource revenue, the proportion of revenues that should be saved is higher.

A second savings rule follows from the first: *the savings rate should rise as resources are depleted.* Each year, the horizon to full depletion is shorter and so the appropriate savings rate is higher. I term this *the depletion effect.*

A simple way of thinking about this is that as depletion occurs and some revenues are saved, the stock of non-resource assets gradually accumulates. In turn, this rising stock generates a rising *sustainable* income, all of which should therefore be used to finance consumption. As a result, the level of consumption warranted by the natural resource endowment can increasingly be financed from these sustainable revenues. Less of the revenues from resource depletion need to be used for consumption: hence, more is left over for savings.

9.2.1.1 Adding Some Complexity

In practice, most trajectories of expected resource revenues are unlikely to be flat until exhaustion. If revenues are not expected to be flat, this generates a further effect on the rate of savings, which is appropriate – I term this the *extraction effect*. If resource revenues are expected to keep rising until the time of exhaustion, this reinforces the case for the savings rate of revenues to rise over time: if resource income will be higher in the future, it is warranted to consume a higher proportion of it now rather than later. If resource revenues are falling, then the argument works in reverse and so the two effects oppose each other. However, even in this case, the depletion effect will eventually predominate, because as exhaustion approaches, the optimal savings rate becomes very high. The most likely path of the quantities extracted from a discovery is hump-shaped: the phase in which the investment in extraction is building up is followed by one in which extraction becomes progressively more difficult, as the most accessible resources have already been removed. Hence, the trajectory of revenues is likely to be hump-shaped. This reinforces the need for the savings rate to rise over the course of depletion. The first phase of rising revenues directly reinforces the need for the savings rate to rise, whereas the second phase of falling revenues occurs late enough in the overall trajectory of depletion, so that the depletion effect predominates. Hence, in the "normal" case of a hump-shaped extraction path, the added complexity involved in allowing for the fact that the expected trajectory of revenues is not flat, ends up simply reinforcing the need for a rising savings rate.

A second complexity is that whereas the GCC countries have full access to international capital markets, the MIMES have limited access, especially those subject to political uncertainties. While they can lend at world interest rates, they are constrained in the amount they can borrow. To see the consequences most clearly, suppose that the global rate of time preference is zero but that in the MIMES, the domestic rate of time preference is high. To keep the analysis as straightforward as possible, I will return to the assumption that revenue is flat until exhaustion. As extraction proceeds, the time profile of the remaining extraction shortens: each year, it is as if the most distant year of extraction has been "lost." As extraction proceeds, the final "lost" year gets progressively closer. The large difference between the domestic and the international social discount rate has an important implication for the valuation of this sequence of "lost" years. For example, after the first year of a twenty-year discovery, the "lost" twentieth year of extraction is not very valuable because having been discounted at a high rate over a twenty-year horizon, its NPV is only a small fraction of its eventual value. But after the nineteenth year, the "lost" year is not the twentieth year of extraction but the second, and so the NPV of the loss is much greater. In consequence, because of the high social discount rate, in MIMES, the value of resources remaining in the ground declines at an *accelerating amount*. In turn, this has unambiguous implications for the savings rate. Since resource wealth is declining by an accelerating amount, other assets must be accumulated by an accelerating amount. Hence, the amount saved must increase each year so that the rate of savings must rise. I term this the *divergent discount rates effect*: as with the two other effects, its implication is that the savings rate needs to rise over time.

9.2.1.2 Implications for Policy Rules

To summarize the theory discussed earlier, both GCC countries and MIMES should be aiming to convert their unsustainable resource incomes into sustainable consumption by saving a substantial share of their revenues. The rate of savings needs to be higher in the MIMES than in the GCC countries because the expected duration of resource revenues is much shorter. In both groups of countries, the appropriate savings rate should be rising over time, while perhaps rising somewhat more rapidly in the MIMES because their social discount rates are higher than in the GCC countries. As an illustration, suppose that the GCC countries have a horizon until exhaustion of oil and gas revenues of seventy years, given the future global concern over carbon emissions, while the MIMES have a horizon of thirty years given the physical exhaustion of their endowments. In 2013, the GCC countries may need to have a savings rate out of their resource revenues of around 30%, gradually rising to 100% by 2083. Hence, in this illustration, the savings rate of the GCC countries would rise by one percentage point each year. The MIMES may need to start with a rather higher savings rate of 50%. This would also need to rise to 100% by exhaustion, which would occur in 2043. Hence, the savings rate in the MIMES would need to rise by nearly two percentage points each year.

Practical policy rules are set in budgets, which are essentially decision processes for allocating revenues to a variety of expenditures. Over recent decades, two principles of good budgeting have been widely accepted; one macro the other micro. The macro principle is that aggregate expenditures should be kept broadly in line with aggregate revenues (which for present purposes, we will treat as exogenous). Commonly, this principle has been encapsulated in the *balanced budget rule* which sets ceilings for both the fiscal deficit and for the ratio of debt to GDP. The micro principle is that the marginal benefit of expenditures should be equated across categories. Since it is reasonably assumed that priorities will change over time, this principle has been encapsulated in the *integrated budget rule,* which discourages pre-commitments of revenues and the earmarking of particular revenue streams to particular items of expenditure.

Resource-rich developing countries face distinctive fiscal problems the solutions to which involve distinctive principles. They, therefore, need distinctive rules that encapsulate these principles. At the core of the distinctive problem, is that, unlike other countries, the key revenue source for government spending is *unsustainable*. Manifestly, this is not a marginal problem: unless it is properly faced, the consequences for welfare are liable to be disastrous. Hence, we should not expect it to be adequately dealt with by minor tweaks to the budget process. It is particularly problematic because inter-temporal resource allocation is an issue that standard budget procedures barely address: budgets are essentially devices for annual commitment. At the most, governments announce medium-term fiscal frameworks, but these are essentially informal statements of intent over a three-year horizon; not only do these statements of intent have limited credibility, there is no intention that they will bind a successor government beyond an election.

The distinctive principle for a resource-rich country is that a certain proportion of revenue should be saved, whether in financial assets or domestic investment. Conventional budgeting processes are inadequate to deal with this problem in two important respects. First, they lack any mechanism for inter-temporal commitment, most especially over the horizon of around a generation, which is likely to be required for resource depletion. Commitment technologies are valuable to governments to reduce the risk of temporary lapses resulting from random short-term political pressures: that is indeed why they adopt the balanced budget rule and the integrated budget rule. But in the case of savings from resource revenues, a commitment technology is even more important. It is not just that without it there is a risk of a random lapse; rather, without it, the incentive to save, even for a good government, is reduced. Without a commitment mechanism, the savings of one government may merely transfer spending power to a bad successor. Indeed, the rationale for augmenting permanent income depends on the current government believing that *all* future governments will behave prudently. In the absence of a rule, a good government may reasonably decide that it is better to spend all the revenue now on items that it regards as desirable, rather than risk its savings being spent by a successor government on items that the current government regards as less valuable. Worse, without a commitment technology, as wealth accumulates, the incentive to be a rogue government that favors only expenditure on consumption actually increases. Hence, a long-term savings rule is not a mere nice-to-have addition to the standard budget rules; it is paramount.

The second respect in which conventional budget design is inadequate is that while the key analytic building blocks concerning resource depletion are about savings and consumption, budgets work with the concepts of expenditure and revenue. Revenues equal the sum of consumption and asset accumulation (savings), but expenditures are the sum of consumption and domestic investment, a component of asset accumulation. Since a balanced budget rule is defined in terms of expenditure relative to revenues, it misses the key required distinction between expenditure on consumption and the acquisition of assets. Domestic investment, which is aggregated under the balanced budget rule with consumption as expenditure, is the most important to distinguish activity for a resource-rich country. Hence, a recent IMF practice of modifying the balanced budget rule so as to exclude resource revenues – through the concept of "the non-oil fiscal balance" – has no analytic basis in the need to offset resource depletion. Indeed, the government of a resource-rich developing country, which actually constrained total expenditure to be equal to non-resource revenue, would be massively misallocating its resource revenues, both under-consuming and under-investing. Conventional budget rules cannot be restored to relevance simply by setting resource revenues aside. Rather, the principles underlying the optimal management of resource revenues must be woven into the foundation of a distinctive set of budget rules.

In its unmodified form, a balanced budget rule applied to a resource-rich country would preclude the accumulation of assets other than through domestic investment and so seriously distort the process of asset accumulation. In its modified variant of the "non-oil balanced budget", it has the opposite distorting effect of squeezing out domestic investment. Now, consider the integrated budget rule: the principle that all expenditures should be left uncommitted so as to be freely allocated each year. This is also fundamentally at odds with the need to pre-commit some revenues to asset acquisition. Without such pre-commitment, there is little chance that the marginal equivalences between expenditures, which are normally the ultimate justification for an integrated budget, can be maintained.

Were the government to have full information about all future needs and revenues, the optimal budget process for a resource-rich developing country would not be a series of annual budgets, but rather a single inter-temporal budget over the horizon of resource depletion. Such a budget would incorporate the optimal path of asset accumulation, thereby achieving the marginal equivalence between the value of current and future consumption. Obviously, no government has full information and so such a comprehensive budgeting process is inappropriate. The solution is to leave open as many expenditure decisions as possible, locking in only to the minimum necessary to ensure the inter-temporal equivalence of consumption expenditures. This is what is achieved by the rule of the rising savings rate. Having pre-determined savings, the composition of savings as between domestic investment and financial assets, can be left open to the annual budget, as can the allocation of consumption spending between items. If we conceptualize all present and future uses of revenue as a matrix, with the rows being the years and the columns the various uses, the annual budget pre-commits the current row, while the rule of the rising savings rate pre-commits the assets column. All other items can be left for future decision. This structure mirrors the pattern of markets, with the markets for goods largely confined to the present period, while transactions concerning future periods are accommodated in an aggregated form through the capital market. Hence, the rule of the rising savings rate replaces the balanced budget rule, but also introduces a constraint into the integrated annual budget rule: aggregate expenditure on assets is ring-fenced.

The distinctive nature of budgeting in a country which receives major revenues from depleting its natural assets is of considerable practical importance. Globally, around 20% of *sustainable* tax revenues are saved: that is, they are used to acquire assets. In contrast, the appropriate savings rate for *unsustainable* revenues, which can only be relied upon for the next twenty-five years, should probably start at over 30%, and gradually rise during the twenty-five years to close to 100%. Thus, the appropriate savings rate is radically different. Further, the allocation of these savings between domestic and foreign assets will differ radically. In the case of sustainable revenues, savings will be fairly stable and virtually all of the assets acquired will be in the form of domestic investment. In contrast, the savings from unsustainable revenues will be far from stable. They will initially leap as revenues come on stream, be subject to volatility as resource prices change and rise considerably as exhaustion approaches. As a result, the appropriate allocation between domestic and foreign assets will change considerably from year to year and should also reflect changes in the capacity of the economy to absorb domestic investment productively.

9.2.2 Managing Volatility

Government revenues from natural resource extraction are volatile, primarily because prices are volatile. Since volatility in public expenditure is costly both politically and socially, expenditures should be smoothed. This implies that revenue fluctuations should be accommodated either by changes in assets or by insurance. Volatility occurs both within years and between them.

9.2.2.1 Intra-Year Volatility

An annual budget incorporates, explicitly or implicitly, an assumption about the average price of the resource over the coming year and this assumption will inevitably prove incorrect. However, for oil, it is now possible to hedge prices over this horizon. The whole point of annual budgeting is to enhance the coherence of spending and so there is value in reducing uncertainty over intra-year revenues. Among hedging strategies, the first choice is in the form of payment. Either the payment can be explicit: a known expenditure to purchase a floor price, or the floor price purchased in exchange for a ceiling price. Between these, the former is likely to be preferable. Whereas it is important to avoid receiving a price below the floor price, there is no equivalent need to avoid particularly high prices since above some ceiling, all revenues should be saved and marginal additions to such savings incur no cost. Hence, there is no point in paying an implicit risk premium to eliminate this range of uncertainty. Further, while the use of a ceiling may appear to have political advantages, disguising what would otherwise be an explicit budgeted payment, the circumstances in which it is triggered may be particularly damaging politically. Namely, an insurance against a very low price which has *a fortiori* turned out to be unnecessary has been paid for by sacrificing a high price which has materialized. A routine annual insurance premium for the purchase of a floor price securing the budget may be politically less exposed.

Having determined the form of payment, the remaining hedging choice is the precise floor price to be chosen. In the neighborhood of the mean of market expectations, an additional dollar on the floor price will increase the cost of the hedge by around fifty cents. Hence, in this range, half of the generated marginal revenue is eaten by the hedge: manifestly this is far too large a proportion to be warranted politically. An implication is that a floor price hedge should pitch the floor price conservatively below the mean of market expectations. In fact, the floor price does not normally need to be particularly close to the mean of market expectations. The floor price is not itself a forecast, but rather a way of protecting expenditure. Planned expenditure will, on average, be below expected revenue partly because not all planned savings will be invested domestically, and partly because expected revenues will be conservatively estimated due to a risk discount.

Hence, where the market expectation for the coming year is equal to the long-run expected price, the hedged floor price needed to protect expenditure will be below this level.

Would hedging be more valuable for the GCC countries or the MIMES? Although the GCC countries need a lower overall savings rate out of resource revenues, much of their savings should be placed abroad rather than invested domestically. Since there is no need to protect the additions to foreign assets from volatility, there is no need to hedge such planned uses of revenues. In contrast, the MIMES may find that a high proportion of their savings to offset depletion should be invested domestically. Since volatility in domestic investment would be costly, they would probably benefit more from hedging their revenues.

9.2.2.2 Inter-Year Volatility

The revenues prevailing in any one year may be above or below the expected path. As long as actual (post-hedged) revenues are above planned expenditure for that year, it is of no consequence for spending if they are below their expected level. All the difference between actual and expected revenues can be borne by a deviation of actual financial savings from planned financial savings. If, however, actual revenues are below planned expenditure, then either actual spending falls short of plans or the shortfall is financed.

In principle, finance can be through either borrowing or drawing down savings. However, in practice, the two are often not alternatives. The ability to borrow depends upon a record of prudent savings and the prior accumulation of liquid assets. MIMES cannot rely upon continuous access to capital markets and so will need to accumulate financial assets. However, for MIMES, liquid savings have an opportunity cost in terms of more productive assets foregone. Hence, not all possible scenarios of needs for liquid savings should be accommodated: there will be times when actual expenditure will indeed need to fall below planned expenditure. As liquid assets are drawn down, actual expenditures should be pre-emptively reduced to avoid the risk that finance will be exhausted, forcing a large, abrupt reduction in expenditure. In effect, this override is a second line of defense against an overly optimistic assessment of the path of future revenues, protecting accumulated assets intended to offset depletion from being used to finance an unsustainable level of consumption.

If revenues exceed planned expenditure then the surplus should evidently be saved. However, there are two distinct functions for such savings: the accumulation of liquid savings to buffer expenditure and the accumulation of longer-term financial assets as part of the strategy of offsetting the depletion of natural assets. Although in MIMES, a large majority of the assets that offset depletion should usually be domestic in the long run, the decision as to the composition between foreign and domestic investments should be taken year by year and reflect the limits on current capacities to invest well within the economy. There is, therefore, a need for some decision rule as to how much of the excess of revenues over planned expenditure should be used for future smoothing and how much should be used for long-term portfolio investment. As I discuss here, political considerations make it desirable to have distinct funds for these two purposes. The allocation might be decided according to whether the smoothing fund is at or below its target level. Until the fund reaches this level, all savings into foreign financial assets would be allocated to it and beyond that, all would be allocated to offsetting depletion.

To summarize, planned expenditure is derived from the path of expected revenues. Having determined planned expenditure for the coming year, these plans should be implemented even in the face of revenue shortfalls, subject to an override reflecting concerns over liquid savings. Because expenditure is to be protected from revenue fluctuations, it is important that there is a responsible process for regularly updating expected revenues so that planned expenditure is based on realism tempered by risk. The override requires planned annual spending to be reduced if the country encounters a run of unexpectedly low revenues that drain liquid savings to a dangerous level.

The key operational concepts are the optimal level of expenditure, the average level of liquid savings to be held for maintaining expenditure at this level and the rules for overriding the drawdown in savings. In turn, setting these parameters should rest on an analysis of the likely volatility of revenues and the likely costs of volatility in expenditure.

In equilibrium, the government will thus be holding foreign financial assets for two different purposes: expenditure smoothing and offsetting depletion. There is a good case for holding these assets in separate funds, with distinct rules. The key difference is that the depletion offset fund needs to be protected from being plundered to finance consumption. As discussed earlier, the distinctive challenge facing resource-rich societies is the need for commitment to long-term asset accumulation. Hence, this needs to be embodied in the rules of the depletion offset fund. The rules cannot, however, be as simple as saying that financial assets cannot be liquidated, since it will be gradually optimal to shift the composition of the fund from foreign financial assets to domestic investment. Hence, the rule should be that *assets cannot be liquidated to finance consumption.* To distinguish this accumulation of both foreign and domestic assets from a conventional Sovereign Wealth Fund, I refer to it as a *Sovereign Development Fund.* In contrast, this is the part of the legitimate purpose of the smoothing fund. It is not the only purpose, since the fund is meant to smooth all expenditure, both on consumption and investment. Hence, for this fund, the rule should be that the assets cannot be liquidated to finance expenditures in excess of the planned expenditure. Further, it might be useful to place the implementation of the hedging rule within the smoothing fund: analytically, it is a means of achieving expenditure smoothing and politically, it enables the Finance Minister to be distanced from the decision to spend money on any particular hedge, which may or may not turn out to have been vindicated by events. The purpose of the fund is thus to make expenditure resilient to revenue shocks, whether by hedging or by the accumulation of liquidity, and so we term it a *Sovereign Liquidity Fund.*

GCC countries can ride out revenue volatility by varying their accumulation of foreign financial assets and so have little need of either hedging or Sovereign Liquidity Funds. They have few opportunities to invest domestically and so may find that a conventional Sovereign Wealth Fund is all that they need. In contrast, because MIMES should be investing domestically, they are likely to need both Sovereign Liquidity Funds and Sovereign Development Funds.

9.3 Investing Savings in Assets

9.3.1 What Assets Should Be Acquired with Savings from Resource Revenues?

The choice of assets appropriate for the GCC countries is unlikely to be appropriate for the MIMES. The GCC countries already have abundant invested capital per worker and so have sensibly established Sovereign Wealth Funds to place savings in foreign financial assets. In contrast, the MIMES are short of domestically invested capital and, given political volatility in the region, face difficulties in attracting foreign capital. Hence, whereas it is appropriate for the GCC countries to save their oil and gas revenues predominantly in foreign financial assets, it is more sensible for the MIMES to save by investing domestically.

The matter of which domestic investments have the highest returns depends upon circumstances. A common mistake is for governments to focus too heavily on imagined opportunities in the resource extraction sector itself – "value addition." For example, GCC countries have probably over-invested in refining capacity. Often, it is unwise to invest heavily in adding value in this sector because it increases dependence on the sector, and, once the resources are depleted, the adjustment needed in the economy will be all the greater.

In the short and medium terms, the most important sector to expand is "non-tradables": goods and services that are neither exported nor compete closely with imports. Precisely because they cannot be supplied by the rest of the world, these are the things that will become more expensive as aggregate spending from resource income increases. This, indeed, is what is meant by Dutch Disease and so avoiding it requires expansion of this category of goods and services.

In the longer term, MIMES will need to diversify their economies beyond resource-based activities. It is not possible to anticipate what the opportunities in global markets might be at the time when resources are nearly exhausted. Hence, until then, the appropriate investments are those that can serve as platforms from which to launch a wide range of possible future activities. The prime examples of such platforms are efficient cities and human capital. Almost all future export activities are likely to be urban based and so would benefit from enhanced efficiency in how the country's cities function. Effective twenty-first century cities require enormous investments in transport systems, power infrastructure and telecommunications, and so these are investments which prepare the country for diversification. Similarly, almost all future export activities are likely to require skilled labor. Investment in a high-quality education system and post-education training therefore prepares the labor force for diversification.

9.3.2 Investing in Investing

However, while MIMES are short of domestic invested capital, they currently lack the capacity to invest efficiently. An attempt to scale up investment without the necessary capacity can have dire effects, driving down the rate of return on investment and plunging the economy into the macroeconomic problems of Dutch Disease.

I term the process of building the country's capacity to invest "investing-in-investing." It involves four distinct tasks. First, the capacity for *public* investment: the design, selection and implementation of projects. A useful new scoring system that benchmarks current capacities, country by country, is the *public investment management index* constructed jointly by the IMF and the World Bank; so far it covers seventy-three developing countries and is publicly available. To give a flavor of standards, South Africa comes out well with a score of 3.5 out of 4. Bangladesh is on the borderline between the third and top quartiles with a score of 2.0. Four of the MIMES are rated, although it is important to note that these ratings were done in 2010, prior to the Arab Spring. Tunisia was top rated, with a score of 2.97, Jordan followed with 2.21, Turkey had a surprisingly low score of 1.88, while Egypt scored only 1.43, the worst score for any of the major middle-income economies. These poor scores, particularly for Egypt, which is unlikely to have improved since 2010, suggest that there is much work to be done before savings can be productively absorbed domestically. However, the index is broken down into four sub-components and this reveals potentially useful differences among the MIMES. Egypt is particularly weak at the selection and management of projects, but these are precisely the components in which Tunisia, Jordan and Turkey are strongest. Hence, there appears to be considerable scope for learning from each other.

Second, because public and private investments are complements, the capacity for *private* investment is also important. A useful index here is *Doing Business*, put out by the World Bank. While the government does not control public investment, it does determine the policy environment that affects investment and this is what the index tries to measure. The *Doing Business* index is available for 2013 and has a wider coverage than the public investment management index, with most GCC countries included. It reveals a very wide range of performance within the region. Many of the GCC countries have good business environments. Of the 189 countries that are ranked globally, the UAE is judged to have the 23rd best environment, with Bahrain, Oman and Qatar ranked 46th, 47th and 48th, while Kuwait is only 104th. The MIMES generally have worse business environments and there is much wider dispersion. The best is Tunisia, which ranks 51st. But Jordan is rated only 119th, while Egypt is worse at 128th. As with its PIMI rating, this is an outlier for a major middle-income country and so should give rise to concern. Again, to get a sense of what might reasonably be judged to be attainable, South Africa is ranked 41st and Ghana 67th. Taken in conjunction with the PIMI, these indicators suggest that investing-in-investing needs to be a high and sustained priority.

Third, both public investment and private investment purchases capital goods – equipment and structures. Often the unit cost of these capital goods is far higher in small, poor countries than in the rest of the world. The price of imported equipment is often inflated because marketing channels are uncompetitive and transport costs are high. The cost of constructing structures is often high because of a lack of competition, shortages of key inputs and skills and poorly defined land rights. A high cost of capital goods eats into investment; what drives development is not how much is *spent* on investment, but how much capital it actually *buys*. So government policies should try to reduce the unit cost of capital goods by establishing where prices are significantly higher than world levels and finding out why.

Finally, if the government wishes to invest a high proportion of GDP, it will be appropriate for some of it to be in the private sector rather than the public sector. Public investment in infrastructure and private investment in production are complements and so a balance between them is needed. This has implications for the allocation of both capital and labor between the public and private sectors.

As to capital, channeling public savings into private investment is politically challenging: bureaucrats are not normally good judges of investment prospects and decisions are exposed to political lobbying. The most straightforward way for the government to invest in the private sector is thus to acquire claims on the commercial banking system. This, in turn, depends on that system being sufficiently well-developed such that it can successfully channel savings to productive investment. Hence, investing in investing may also involve the development of the financial sector.

As for labor, it is important that it is not hoarded in the public sector but is available for productive use in the private sector. In an insightful chapter, Ali and Elbadawi (this volume) show that the desire to maintain political control has induced the GCC to hoard labor in the public sector. The MIMES, having had less revenue, tended to supplement the strategy of inflating public sector employment with techniques of political repression.

9.3.3 Avoiding Dutch Disease: Sequencing

Dutch Disease occurs when the non-resource export sector of the economy becomes uncompetitive due to real exchange rate appreciation. It is not an inevitable consequence of increases in resource revenues, but it will happen if the government spends resource revenues ahead of extra investment that can increase domestic supply. Sequencing is required to guard against this situation.

The policy sequence that helps to avoid Dutch Disease is to begin by building the capacity to invest: investing in investing. Then, once the capacity to invest has been built, those resource revenues that are saved should be used to finance the scaling up of domestic investment. This extra productive investment will enhance the capacity of the economy to supply goods and services and so the government can then increase consumption (both public and private) without driving up prices: Dutch Disease will have been avoided.

If total public spending increases too rapidly, then Dutch Disease is inevitable because supply will not be able to keep pace. Surges in demand can occur both if there is a rapid initial build-up of revenues, which are spent too soon, and if revenues subsequently have large temporary increases because of the volatility of commodity prices. Expenditure needs to be subject to a sensible ceiling rate of increase and a rule for smoothing out volatility to guard against this. A ceiling on the rate of increase of public spending is prudent partly because no bureaucracy can efficiently manage very fast increases in spending and partly because large rapid increases will have macroeconomic effects – increasing inflation. Quite what the ceiling rate of increase should be is a matter of judgment in each country, but it is hard to believe that annual increases in real terms of more than 10–15% are prudent. If a government maintains a real increase of 10% for seven years, public spending doubles in real terms and seven years may be around the time needed to cope with such a large increase in activity.

9.4 Conclusion

The management of resource revenues is the single most important economic challenge facing the Middle East. The standard rules of economic management were not devised for resource-rich countries and so fail to address problems that are distinctive but central. Although the GCC are high-income countries, their policy challenges are very different from OECD economies. In particular, they are critically dependent on the accumulation and management of foreign financial assets. Thus, they have a strong interest in international financial order.

The MIMES face a more complex set of challenges. Their natural resources will deplete more rapidly than in the GCC economies and so, despite being much poorer, they need higher savings rates. They also differ from the GCC countries in having many more opportunities for domestic investment. They should therefore have higher rates of domestic investment than both other middle-income countries that are not depleting natural assets, and the GCC countries that lack their opportunities. To manage this process of high domestic investment successfully, their priority should be a phase of investing in investing. Without the capacity to invest high rates of savings in productive domestic projects, they have little prospect of escaping the middle-income trap.

References

Ali, O. and I. Elbadawi. This volume. The Political Economy of Public Sector Employment in Resource-dependent Countries.

Collier, P. (2010). *The Plundered Planet*. Oxford University Press and Penguin.

10

Labor Market Heterogeneity and Optimal Exchange Rate Regimes in Resource-rich Arab Countries

Almukhtar Saif Al-Abri

ABSTRACT

This chapter attempts to incorporate the role of labor market flexibility in a unified theoretical framework of optimal exchange rate policies for resource-rich economies. For that purpose, the chapter extends the standard rules-versus-discretion model of monetary policy to allow for different assumptions about wage rigidities. The analysis suggests that when all prices are exogenous and wages are all optimally indexed to inflation (i.e., labor market flexibility case), fixed exchange rate regimes deliver more desirable outcomes in terms of real output, inflation and insulation against external shocks compared to flexible exchange rate regimes. Meanwhile, the fixity of exchange rates combined with rigid goods and labor markets can add to real appreciations.

JEL Classification: F22; F33

Keywords: Arab region, GCC, optimal exchange rate regime, monetary policy credibility, oil curse, labor market flexibility

10.1 Introduction

The Arab region has witnessed a number of political, social and economic developments during the last decade. Among these developments are the economic reforms pertaining to liberalizing trade and financial systems. The Arab Spring that emerged in late 2010 has also mandated policymakers to take constructive steps toward economic development and expansion of employment opportunities. Another important development is the shift of Arab countries' trade away from the US and towards the Euro area, East Asia, China and Japan and the increasing counter-cyclicality of oil exporters with the US business cycle (Habib and Stráský 2008; Setser 2007).

These developments and the recurring financial and exchange rate crisis in several developing and emerging countries warrant a revisit of the debate on the optimal exchange rate regime for the Arab countries. This issue is also timely as many countries that peg their currencies (e.g., GCC countries) face reoccurring challenges related to prolonged misalignments in real exchange rates and increasing volatility of imported inflation (Coudert et al. 2011; Yousefi and Wirjanto 2003). Similarly, countries that have moved toward less restrictive exchange rate regimes, such as Egypt, Algeria, Sudan and Yemen, face concurrent challenges related to unstable nominal exchanges rates, monetary policy credibility, as well as trade destructive changes in real exchange rates (Elbadawi and Kamar 2005).

This chapter analyzes optimal exchange rate regimes for the resource-rich Arab countries, with particular emphasis on the heterogeneity of labor markets. We discuss the exchange rate regimes for two groups of countries: *resource-rich labor-importing* (RRLI) countries, including members of the GCC countries and Libya and *resource-rich labor-abundant* (RRLA) countries including Algeria, Egypt, Iran, Iraq, Sudan and Yemen. Whereas the labor force in RRLAs is composed mostly of nationals, RRLIs rely heavily on foreign expatriates (see Table 10.2). The foreign labor force works mainly in the private sector on short-term contracts. As such, the size of the foreign labor force can quickly be adjusted to fluctuations in the business cycle triggered by oil price shocks.

The chapter extends the standard rules-versus-discretion model of monetary policy to allow for different assumptions about wage rigidities. We evaluate the loss of flexibility against the gain in credibility with alternative assumptions about the labor market. The findings suggest that in countries with relatively more flexible labor markets, fixed exchange regimes are more desirable in responding to shocks, as the credibility gains of fixed exchange rate regimes might outweigh the loss of flexibility in monetary policy. However, with rigid labor markets and employment, flexible exchange regime are more desirable in responding to shocks, as the loss of flexibility by fixing the exchange rate outweighs the gain in credibility. Our analysis also highlights the conditions that these countries should satisfy in the transition to a more flexible exchange rate arrangement.

This chapter also suggests two future research themes within the line of optimal exchange rate policy. The first theme is to incorporate the labor market to the models of optimal exchange rate regime choice. The second theme is to develop the institutional strategies and prerequisites to move to a more flexible exchange rate policy.

The rest of this chapter is organized as follows. Section 10.2 highlights the main features of the Arab economies and their experiences with different exchange rate regimes. Section 10.3 discusses the literature on optimal exchange rate regime choice for primary-commodity exporting countries. Section 10.4 summarizes the theoretical model used to analyze the role of labor market heterogeneity in exchange rate regime choice. The final section presents some conclusions and recommendations. The details of the theoretical analysis are provided in the Appendix.

10.2 Exchange Rate Regimes in the Arab Region: A Tale of Two Groups

Arab countries share a number of commonalities, as well as some differences. This diversity of the Arab region suggests a number of ways to segment these countries.[1] However, for the purpose of analyzing the optimal exchange rate regime choice, we focus our attention on natural resource-rich Arab countries. Within this group, it is also useful to segment them further into two sub-groups: *resource-rich labor-importing* (RRLI) and *resource-rich labor-abundant* (RRLA). Tables 10.1 and 10.2 compare the countries included in the two groups against a number of economic and financial indicators.

10.2.1 *Economic Structure*

As presented by Tables 10.1 and 10.2, except for Egypt, Sudan and Yemen, countries in both groups are highly endowed with oil and gas reserves. The salient differences between countries in the two groups are the level of economic development (represented by the GDP per capita), the size of the population and the composition of the labor force. All countries exhibit similar levels of export diversification except for Egypt, Bahrain and Iran, which show a greater share of manufacturing exports in total merchandise exports. Although countries in the two groups show similar patterns in

[1] For example, the World Bank (2007) classifies Arab countries into three main groups: *resource-rich labor-abundant countries* (e.g., Algeria, Iraq, Iran, Libya, Syria and Yemen); *resource-rich labor-importing countries* (e.g., GCC countries – Bahrain, Kuwait, Oman, Qatar, Saudi Arabia and the United Arab Emirates) and *resource-poor labor-abundant countries* (e.g., Djibouti, Egypt, Jordan, Lebanon, Mauritania, Morocco, Tunisia and the Palestinian Authority).

Table 10.1: *Natural Resources Abundance and Export Diversification in the Arab Region*

	GDP per Capita 2012 (Millions of 2005 constant US$)	Average Real GDP Growth Rate 1980–2012	Standard Deviation of Real GDP Growth Rate 1980–2012	Oil Production (millions of barrels) in 2012	Proven Oil Reserves Trillion Barrels (End of 2012)	Gas Production (Billion Cubic Meters) in 2012	Proven Gas Reserves Trillion Cubic Meters (End of 2012)	Fuel Exports as % of Merchandise Exports 2000–2012	Manufacturing Exports as % of Merchandise Exports 2000–2012
Group1: RRLI									
Bahrain	14,253	4.3	4.4	NA	NA	14.2	0.2	61.4	9.97
Kuwait	29,338	5.9	8.5	3,127	101.5	14.5	1.8	94.5	4.44
Libya	8,776	4.3	5	1,509	48	12.2	1.5	97.2	2.74
Oman	13,903	6.1	5.3	922	5.5	29	0.9	82.9	8.95
Qatar	54,249	12.1	6.3	1,966	23.9	157	25.1	85	7.58
Saudi Arabia	17,591	2.8	5.1	11,530	265.9	102.8	8.2	89.5	8.91
UAE	24,708	3.3	6.2	3,380	97.8	51.7	6.1	74.7	3.13
Group 2: RRLA									
Algeria	3,185	2.7	2.3	1,667	12.2	81.5	4.5	97.3	1.98
Egypt	1,560	4.7	1.9	728	4.3	60.9	2	40.7	34.31
Iran	3,105	3.7	5.6	3,680	157	160.5	33.6	78.7	9.98
Iraq	1,603	5.6	20	3,115	150	0.8	3.6	89.4	0.21
Sudan	836	4.1	5.4	82	1.5	NA	NA	82.3	1.41
Yemen	778	4.1	3.7	180	3	7.6	0.5	92.7	1.44

Sources: BP Statistical Review of World Energy, British Petroleum, 2013; World Development Indicators (WDI), World Bank, 2013.

Table 10.2. *Selected Indicators of the Arab Region*

	Overall Population in 2012 (millions)	International migration stock as % of population 2010	Central bank official reserves (months of imports in 2012)	Average domestic credit to private sector during 2000–2012 (% of GDP)	Average net remittances during 2000–2011 (billions US dollars)	Public debt in 2012 (% of GDP)	Average CPI Inflation 2000–2012 (%)	IMF De facto exchange rate regime in 2012
Group1: RRLI								
Bahrain	1.3	25.2	2.3	59.7	−1,371	54.3	1.5	Pegged to US dollar
Kuwait	2.7	70.1	9.1	64.7	−5,933	6.0	3.5	Pegged to a basket
Libya	6.0	11.3	80.7	12.6	−766	3.5	1.7	Pegged to SDR
Oman	3.2	29.5	4.4	37.8	−3,312	4.1	3.1	Pegged to US dollar
Qatar	2.0	74.6	5.4	36.5	−9,871	32.8	4.5	Pegged to US dollar
Saudi Arabia	26.9	26.7	35.4	34.0	−21,314	11.5	2.7	Pegged to US dollar
UAE	5.5	69.0*	NA	51.3	−7,460	43.3	3.4	Pegged to US dollar
Group 2: RRLA								
Algeria	38.1	0.65	35.9	12.3	118	7.9	3.8	Managed float
Egypt	85.3	0.31	2.4	45.3	6,162	88.0	8.0	Stabilized arrangements
Iran	79.9	2.9	NA	26.0	NA	19.9	16.2	Managed float
Iraq	31.9	0.27	13.1	3.6	93	NA	17.5	Stabilized arrangements
Sudan	34.9	1.65	0.2	9.3	1,416	106.4	12.3	Managed float
Yemen	25.4	2.28	4.1	6.4	1,130	43.0	11.7	Managed float

Sources: World Development Indicators (WDI), World Bank, 2013; The *Annual Report on Exchange Arrangements and Exchange Restrictions* (AREAER), IMF, 2012.

terms of the structure of their financial sector, which is mainly bank-based, the level of financial development is not the same across the two groups. As indicated in Table 10.2, the domestic credit provided by private banks is substantially higher in RRLIs compared to RRLAs.

This diverse economic structure casts its shadow on the economic performance of individual countries, as shown in Selim and Zaki (this volume). The main determinant of growth in the Arab region has been the fluctuations in international oil prices. The high dependence on oil revenues is also reflected in a volatile and pro-cyclical fiscal policy, particularly in the RRLI countries. However, the RRLA countries are also sensitive to oil price fluctuations due to the role of worker remittances as well the development aid received from the RRLI countries. Moreover, the tax system of the fiscal policy is ill-structured in the two groups (RRLA and RRLI countries), resulting in a narrow tax base and high tax evasion (Leenders and Sfakianakis 2003).

One of the important structural differences between the two groups is the degree of labor market flexibility. Whereas the labor force in RRLAs is composed mostly of nationals, RRLIs rely heavily on foreign expatriates (see Table 10.2). The foreign labor force works mainly in the private sector on short-term contracts. As such, the size of the foreign labor force can quickly be adjusted to fluctuations in the business cycle triggered by oil price shocks. The other dimension associated with labor market heterogeneity is the size and direction of worker remittances. The large size of the foreign labor force in RRLIs results in substantial outflows of worker remittances. In the contrary, the net remittances in the RRLA group are positive indicating that these countries are net exporters of labor. The majority of remittances inflows that are received by these countries come from Arab large oil exporters, mainly the GCC countries (Naufal and Carlos 2010).

10.2.2 *Exchange Rate Regimes in the Arab Region*

The diversity in economic structures and institutional credibility in the Arab region have also been extended to the choice of exchange rate regimes by individual countries. While the RRLI countries have a long track record of pegging to the US$ or to a basket dominated by the US$, the RRLA countries had mixed experiences with intermediate and managed float regimes during the last three decades.

Several motivations could explain the tendency of the RRLI countries to peg their exchange rate to the US$. First, it allows these economies, especially those with weak economic and political institutions, to maintain central bank credibility by importing the United States' stable monetary policy. Second, it ensures that they enjoy relative stability in crude oil revenues[2]. Third, it also reduces the risks of financial instability that exchange rate volatility might cause.

During the last three decades, the pegged exchange regime had largely delivered its expected benefits to this group in terms of low inflation and large external financial assets (Squalli 2011; Ghanem 2012). However, one setback of the peg is the chronic real exchange rate misalignment adherent to oil price volatility, with its consequent impact on export competitiveness. To address this drawback, Frankel (2005, 2010) suggests pegging the currency to the price of the export commodity in order to reduce the effects of oil price fluctuations. Another proposal is that monetary policy should target currency misalignment (Engel 2009).

As for the RRLA countries, they have pursued exchange-rate-based stabilization strategies for some time during the last three decades. The motivation to adopt less restrictive exchange rate systems, such as managed float, crawling pegs and target zone was to improve trade competitiveness and output stabilization[3]. However, several countries have not yet reaped the expected benefits of such regimes. For example, Elbadawi and Kamar (2005) find evidence that in some Arab countries, a combination of high inflation inertia and either unstable or pegged exchange rate regimes has been associated with high and extended episodes of real exchange rate overvaluations. In the aftermath of the currency crises in Mexico (1994), Southeast Asia (1997), Russia (1998) and Brazil (1999), the sustainability of conventional fixed pegs (soft pegs) in the face of capital mobility and other shocks was questioned. The main motive to move away from pegged regimes and to increase flexibility is to deal with exogenous shocks, contribute to financial stability and avert future banking or currency crises. Nevertheless, a number of RRLA (particularly Iran and Sudan) countries had experienced episodes of unstable currencies and hyper-inflation, which also produced massive real overvaluations. Chia et al. (2012) find that in moving from fixed to flexible exchange rate regimes, the rise in volatility of the nominal exchange rate is also coupled by a rise in volatility of the real

2 As of 2010, of the twenty-four countries where oil exports represent at least 50% of total exports, fourteen of them pegged their currency to the US$, five pegged to the Euro and six pegged to a basket of currencies (IMF 2010).

3 For a detailed review of the recent exchange rate regime history in the Arab region, see Elbadawi and Kamar (2005).

exchange rate. Also, the effects of terms-of-trade shocks and foreign interest rate shocks are more prolonged under flexible exchange regimes than under fixed regimes.

Sfia (2011) analyses the determinants of exchange rate regime choice in seventeen Arab countries over 1974–2004 using binomial and multinomial probit models. The results highlight the important influence of only two factors on the choice of exchange rate regimes: trade openness and oil export capacity, both favoring the choice of fixed exchange regimes. Sfia also finds that neither the political economy variables nor the currency crisis measures are robust or significant predictors of exchange rate regimes in the Arab region.

In summary, the exchange rate regimes in the Arab region have had varying degrees of success. Although the empirical evidence suggests that for the RRLI countries, pegged exchange rate regimes are associated with the best inflation performance, this was not the experience of the RRLA countries with pegged regimes. Why did pegged regimes deliver expected benefits to the first group and not to the second group? One of the main dimensions in choosing an exchange rate regime is the ability to establish credibility with the public. The exchange rate regimes chosen by the RRLA countries had consistently produced inflation bias. The typical situation for almost all the RRLA has been one of fiscal dominance, where the authorities use the exchange rate as a nominal anchor, while resorting to excessive monetary emission to finance the large deficits (Elbadawi and Kamar 2005). As suggested by the theory, when the public doubts the central bank's determination to reduce inflation, they will rationally expect more inflation than otherwise targeted by the central bank and set their expectation of inflation accordingly. Given this expectation of the public, the central bank's best strategy is to create inflation equal to that expected by the public. Thus, the economy ends up with positive inflation and no additional output. Another explanation in the literature suggests that when the peg is at an undervalued rate and the country is unable to offset the growth of the money supply that occurs with persistent current account surpluses, accumulation of foreign reserves will translate into excessive monetary growth; in such cases, the curbed inflation benefit from pegs does not occur.

An important dimension that might explain the success of the RRLI countries with pegged regimes is their labor market structure. Although traditional optimum currency area (OCA) analysis suggests that in the absence of an independent monetary policy, labor mobility and flexible prices and wages are fundamental pre-requisites to cushion the effects of shocks on the real economy under fixed exchange rate regimes, the GCC countries have been able to compensate for the lack of these prerequisites through the mobility of foreign workers. The GCC countries have unusual labor force structures. They heavily depend on expatriate workers who work primarily in the private sector, whereas nationals work in the public sector. The labor supply in the private sector is highly elastic. During periods of negative oil price shocks, the government reduces its spending and therefore, the private sector can terminate the expatriate workers' contracts. In this case, labor mobility can substitute for exchange rate flexibility. On the other hand, the wages and jobs in the public sector are not flexible. The government can protect the jobs for the locals using the levels of revenues from oil sales. Willett et al. (2010) argue that the insulation of nationals from the effects of macroeconomic shocks has allowed GCC countries to achieve lower inflation rates. This was carried out without generating the types of strong domestic unemployment pressures that typically emerge in the absence of an independent monetary policy as suggested by the OCA literature[4].

10.3 Literature Review on Optimal Choice of Exchange Rate Regimes

The extensive literature on optimal exchange rate regimes identifies a number of factors that should be considered in exchange rate regime choice, including: the size and openness of the country to trade and capital mobility, the structure of its production and exports, the level of financial development, the history of inflation, the types and sources of shocks and the credibility of institutions. Levy-Yeyati et al. (2010) summarize the choice of exchange rate regimes in three competing approaches: (1) the optimal currency area (OCA), which relates the choice of exchange regime to a country's trade links, size, openness and the types and sources of shocks the economy is subject to, (2) the financial view, which relates to the impossible trinity (the fact that policymakers can choose two out of three corners of the trinity: capital mobility, monetary policy and a fixed exchange rate) and the currency mismatch argument or the balance sheet effects and (3) the political economy view, which relates the use of an

[4] One caveat to this argument is that, although the level of unemployment remains low relative to the counterfactual of a totally inflexible labor market, there are alarming signs of rising unemployment among nationals due to the increasingly limited capacity of the public sector to absorb more labor.

exchange rate anchor to establish credibility with the public. Despite the well-developed theoretical frameworks and the empirical evidence, many countries are still pondering their options of exchange rate regimes. This illustrates the challenging trade-offs that confront monetary policymakers in prioritizing their policy objectives. A review of this extensive literature is beyond the scope of this chapter. However, the literature seems to agree on the following simple scheme in choosing a particular regime: when the country is developing, has less diversified production and exports and its banking and financial systems are not developed, a peg to a low inflation and stable monetary policy anchor country seems to be the appropriate choice, particularly if the policy preference is to keep domestic inflation under control without policy credibility. As the banking and financial markets develop and capital mobility increases, a country should move toward more flexibility (e.g., crawling exchange rate, a more flexible version of fixed regime). Finally, when a country reaches the stage of having diversified production and trade and advanced banking and financial systems that are fully integrated into the global capital markets, the freely floating exchange rate is more appropriate with a nominal anchor strategy, such as inflation targeting.

Despite the economic and financial importance of the Arab economies, few contributions have been devoted to the issue of exchange rate regime choice by these countries (Jbili and Karamenko 2003). On the theoretical side, few studies have incorporated the economic structure and the policy preferences of primary-commodity exporting countries. For example, Al-Abri (2014) examines the choice of optimal exchange rate regime for an oil-exporting small open economy using a three-country new Keynesian open economy model, incorporating the primary commodity (crude oil) with different sources of shocks. Despite the absence of a non-oil trade sector in this primary-commodity economy, the welfare analysis suggests that flexible exchange rate regimes can reduce the impacts of real shocks and consumption volatility given certain caveats about pricing schemes of imported goods. Taş and Togay (2010) use a model of asymmetric information and dynamic learning in the spirit of Svensson and Woodford (2004) to analyze the optimal exchange rate regime for Iraq in the new era. Squalli (2011) uses a general central bank loss function in the spirit of Berger et al. (2001) to analyze the viability of the US$ peg for the GCC countries, focusing on the role of imported inflation. In this chapter, we contribute to this literature by incorporating the role of labor market heterogeneity in optimal exchange rate regime choice in a primary-commodity exporting country.

10.4 Labor Market Role in Exchange Rate Regime Choice

The basic theoretical model developed by Milton Friedman in the early 1950s shows that in a world of sticky prices, the flexible exchange rate outperforms the fixed exchange rate system in insulating the economy against real shocks. A key merit of floating regimes is that they help deflect or absorb the impact of adverse external shocks (e.g., shocks to terms-of-trade, reversal of capital flows, contraction in world demand, foreign interest rate shocks and so on), and avoid large costs to the real economy. These shocks usually necessitate an adjustment in real exchange rates (i.e., relative prices). Given that domestic prices move slowly, it is both faster and less costly to have the nominal exchange rate respond to a shock and make the necessary adjustment to the real exchange rate. This makes flexible exchange rate systems more desirable in this case.

However, with high degree of wage indexation in the economy, the degree of pass-through from exchange rates to prices is high and, thus, limits the shock buffering capacity of the floating regimes. Devereux and Engel (2003) show that exchange rate flexibility is desirable to the extent that the law of one price holds. However, if prices are unresponsive to exchange rate fluctuations, then the expenditure switching channel of flexible exchange rate adjustment will not work, and fixed regimes might outperform flexible regimes in this case. In a similar vein, Castrén et al. (2004) show that labor market reforms, which reduce the persistence of the effect of shocks, make an exchange rate peg more robust, and therefore, more credible. Using econometric tests for the gold standard (1881–1913), Khoudour-Castéras shows that the free labor movements that characterized the pre-1914 world most likely helped gold standard members to maintain the stability of their currency, while for countries with flexible exchange rates, which experienced much less volatility in their emigration rate, the nominal exchange rates constituted the main adjustment mechanism. We thus, turn to our theoretical analysis to incorporate the market flexibility in exchange rate regime choice of a primary-commodity exporting country.

10.4.1 *A Summary of the Theoretical Framework*

For the purpose of analyzing the role of labor market heterogeneity in the Arab region in the choice of optimal exchange rate regime, we consider a standard static rules-versus-discretion model of a small open economy (Aizenman and Glick 2008; Barro and Gordan 1983; Berger et al. 2001;

Flood and Marion 1999; Gray 1976). A detailed description of the model and its solutions is provided in the Appendix to this chapter. In the following, we highlight the model's key features and the main results.

The monetary authority's goal is to determine the rational expectations consistent optimal output and inflation that minimize a loss function. The loss function is determined by the expected deviation of output, inflation and the nominal exchange rate from their target. The economy is modeled with three structural equations: (1) aggregate supply; (2) real wage rate determination; and (3) cost-augmented purchasing power parity (PPP). The aggregate supply is a Lucas-type function that responds only to the unexpected portion of macroeconomic policy, when actual inflation exceeds nominal wage increases. To incorporate the labor market heterogeneity in the Arab region which is discussed in Section 10.2, the labor force is divided into two parts: expatriate labor force, for whom the wage rate is flexible and can be indexed to the inflation rate, and national labor force, for whom wages are rigid and pre-set to the expected inflation. When wages are all pre-set, this makes the aggregate supply curve flatter and real external shocks have more effect on output and less effect on inflation. With this specification, the slope of the aggregate supply curve shows how effectively various exchange rate regimes insulate domestic output from unanticipated shocks.

The closed-form solution of the model suggests that fixed exchange rate regimes work better for countries with flexible labor markets (e.g., RRLI), while flexible exchange rate regimes work better for countries with inflexible labor markets (e.g., RRLA). This suggests that with no nominal stickiness, credibility gains under fixed exchange rate regimes are lower than the loss of flexibility in monetary policy. Likewise, with nominal stickiness and flexible exchange rate regimes, the loss of flexibility outweighs the gain in credibility.

In other words, the theoretical framework suggests that when the home country takes all prices as exogenous and wages are all optimally indexed to actual inflation, fixed exchange rate regimes provide better insulation against external shocks compared to flexible exchange rate regimes. On the other hand, the fixity of exchange rates combined with rigid goods and labor markets can add to real appreciations.

10.4.2 *Caveats and Restrictions*

One important caveat in this regard is that we need to distinguish between shocks that originate internally and those that originate externally.

The previous discussion is related to the shocks that are exogenous to the country. If the shocks are internal (endogenous), then the country can better deal with these shocks using a fixed exchange rate regime. This is because under a fixed exchange rate regime, fiscal policy is more effective in offsetting real shocks.

Another caveat is the endogeneity of the labor markets when assessing the consequences of alternative exchange rate regimes. For example, under fixed exchange rates, the nominal exchange rate is fixed and we expect higher rigidity in wages and prices due to the stable inflation rate. Thus, fiscal expansion or contraction will not be correlated with income. Under flexible exchange rates, we expect prices and wages to be less rigid due to the higher volatility of the nominal exchange rate. These prices will respond more quickly to real shocks before the offsetting fiscal policy can have its effects on the economy. Thus, the endogeneity problem in offsetting real shocks is lower under a fixed exchange rate. A third caveat is related to the observation that although certain exchange rate regimes might provide insulation from foreign disturbances, they may, nevertheless, be inferior to other regimes in terms of their ability to minimize the loss function.

10.5 Conclusions and Recommendations

This chapter discussed the alternative exchange rate regime arrangements for the resource-rich Arab countries with emphasis on the diverse labor market. We evaluate the loss of flexibility or the gain in credibility with alternative assumptions about the labor market. While the lesson of no "one size fits all" with regard to the choice of the optimal exchange rate regime still applies, our analysis suggests some important policy implications. The notion that flexible exchange rate regimes are real shock absorbers should be taken with caution under heterogeneous labor markets. In countries with relatively more flexible labor markets, fixed exchange regimes are more desirable in responding to shocks, as the credibility gains of fixed exchange rate regimes might outweigh the loss of flexibility in monetary policy. However, with rigid labor markets and employment, flexible exchange regimes are more desirable in responding to shocks, as the loss of flexibility by fixing the exchange rate outweighs the gain in credibility. Nevertheless, pegged regime countries should also consider more flexible exchange rate regimes. This is partly due to the growth in the size of the private sector and the economy becoming more market oriented. The capacity of government employment has already been stretched to the limit as these countries attempt to avert the possible spread of Arab Spring. In addition, some resource-rich Arab countries are fast approaching the end of their primary resource reserves. These developments suggest that the labor market is changing in oil-exporting countries.

The movement toward more flexible regimes requires appropriate sequencing of improvements in financial systems and progress in fiscal adjustment. For example, capital account liberalization should come after domestic financial market liberalization to avoid destabilization through large capital inflows. More importantly, there is the need for these countries to initiate the transition from the current exchange rate regimes without waiting for a crisis to occur. Drawing on the experiences of resource-rich countries, which had moved from fixed towards more flexible regimes (particularly, Chile and Egypt), highlights the need to make the necessary preparations during periods of strong fundamentals. For example, the Chilean move to a managed float happened during periods of strong macroeconomic fundamentals, which ensured a smooth transition to the new system and its viability[5]. In contrast, the Egyptian movement to a managed float was under market pressures and during periods of high volatility in real exchange rate. As a result, the credibility of the flexible exchange rate regime has not yet been fully established.

Parallel to this is the observation that exits from pegged exchange rate regimes have often been accompanied by crises and severe declines in economic activity. Aizenman and Glick (2008) and Eichengreen (1999) find that exits from pegged exchange rates have not occurred under favorable circumstances. Aizenman and Glick (2008) show that achieving inflation stability via an exchange rate anchor can create a "trap" whereby the regime initially confers gains in anti-inflation credibility, but ultimately results in an exit occasioned by large adverse real shocks, resulting in big welfare losses to the economy. This suggests that more conservative and longer-lasting pegged regimes are likely to end with severe output losses.

Finally, we add to the growing notion in this literature that regardless of the current exchange rate regimes adopted by countries in the Arab region, the need to develop market-based policy instruments to fulfill the prerequisites for a successful operation of a flexible exchange rate regime is

[5] In the past three decades, Chile has moved from a fixed exchange regime, a crawling band, a managed floating regime, to a freely floating regime with inflation targeting. Throughout its recent macroeconomic history, Chile was able to maintain competitive real exchange rates, mainly by adopting regime-consistent macroeconomic policy, but also by imposing capital controls that discourage dis-stabilizing capital inflows (Edwards 2000).

becoming more critical. For example, under an inflation targeting framework, the central bank commits to meeting a certain inflation target in a given period and uses appropriate instruments to achieve it. The central bank would adjust short-term interest rates in response to developments in expected inflation, the output gap, and some measure of the exchange rate (Svensson 1999; Svensson and Woodford 2004). For this arrangement to be successful, other prerequisites need to be satisfied including: an overall institutional commitment to price stability, absence of fiscal dominance, existence of high degree of fiscal and monetary policy coordination, policy instrument independence and strong operational capability of the central bank.

Appendix: Model Setup

For the purpose of analyzing the role of labor market heterogeneity in the Arab region in the choice of optimal exchange rate regime, we consider a standard static rules-versus-discretion model of a small open economy (Aizenman and Glick 2008; Barro and Gordan 1983; Berger et al. 2001; Flood and Marion 1999; Gray 1976). The home country is small and takes all prices as exogenous. The monetary authority's loss function is quadratic in output gap, inflation and nominal exchange rate stability. It acts to minimize the expected deviation of output, inflation and the nominal exchange rate from their target levels ($y_f > 0$ and $\pi_t^* = 0$) as:

$$L = E_t\left(\lambda\left(y_t - y_f\right)^2 + \pi_t^2 + \delta(s_t - \bar{s}_t)^2\right) \tag{1}$$

where y_t is log of output, y_f is target log of output, λ is the cost of missing the output target (or the relative weight assigned by the policymaker on the respective policy objectives) and π_t is the log of actual inflation rate.s_t is the change in the nominal exchange rate and $\bar{s}_t$ is the normal change in the nominal exchange rate. δ is the cost of excessive nominal exchange rate volatility. The economy is modeled with the following structural equations:

- Aggregate supply $\quad y_t = -\alpha(w_t - \pi_t) + \varepsilon_t, \alpha > 0 \quad$ (2)
- Wage rate $\quad w_t = \theta\pi_t + (1 - \theta)\pi_t^e, 0 \leq \theta \leq 1 \quad$ (3)

- Cost – augmented purchasing power parity $\pi_t = (s_t + \pi_t^* + v) + \gamma w_t$ (4)

The aggregate supply is a Lucas-type function that responds only to the unexpected portion of macroeconomic policy, when actual inflation, π_t, exceeds nominal wage increases, w_t. α is the payoff to unexpected inflation in terms of higher real output. The shock to aggregate supply, ε_t, is the terms-of-trade shock in this case and has an expected value of 0 at time $t-1$ and constant variances σ_ε^2 and is serially uncorrelated.

To incorporate the labor market heterogeneity in the Arab region which is discussed in Section 10.2, we divide the labor force into two parts: a fraction θ of the labor force who are expatriate and their wage rate is flexible and can be indexed to the inflation rate, $w_t = \theta\pi_t$ and a fraction $(1-\theta)$ of the labor force who are nationals for whom wages are rigid and pre-set to the expected inflation formed at time $t-1$, $w_t = (1-\theta)\pi_t^e = (1-\theta)\pi_{t-1}$. For example, when $\theta = 0$, all wages are pre-set. This makes the aggregate supply curve flatter and real external shocks have more effect on output and less effect on inflation. With this specification, the slope of the aggregate supply curve shows how effectively various exchange rate regimes insulate domestic output from unanticipated shocks[6].

Following Berger et al. (2001), the stochastic purchasing power parity condition, Equation 4, is brought to close the model, where s_t is the rate of nominal exchange rate depreciation ($s_t = 0$ under fixed exchange rate regime), π_t^* is the inflation rate in the anchor country and v is a foreign monetary shock that can also represent a disturbance to the world's interest rate. For simplicity, we assume that the correlation of this shock with the terms-of-trade shock ε_t is zero[7]. To allow for a cost-push inflation, we augment the purchasing power parity with the nominal wage rate (γw_t), where γ is the pass-through of wage hikes to the domestic inflation. Similar formulations are provided in Edwards (1996). We, henceforth, deal with the case in which the equilibrium is time invariant and we omit time subscripts.

[6] In this simple model, we assume that the proportion of wage indexation (θ) is exogenous. A proper modeling should also consider the case where the government's choice of exchange rate regime systematically influences the degree of indexation and consequently, the slope of the economy's aggregate supply curve.

[7] Berger et al. (2001) analyze the case when this correlation is non-zero. The assumption of zero correlation might not be realistic given the financial crises of 2008, which had also dramatically affected international oil prices. Thus, the terms-of-trade shocks in an oil-exporting country are positively related to external monetary shocks.

EQUILIBRIUM OUTCOMES

The monetary authority's goal is to determine the rational expectations consistent optimal y and π that minimize the loss function. The time-consistent rational expectations equilibrium under a flexible exchange rate regime is

$$\pi = \frac{\lambda\alpha(1-\theta)}{1+\lambda\alpha^2(1-\theta)^2+\delta(1-\gamma\theta)^2} y_f - \frac{\lambda\alpha(1-\theta)}{1+\lambda\alpha^2(1-\theta)^2+\delta(1-\gamma\theta)^2}\varepsilon + \frac{\delta(1-\gamma\theta)}{1+\lambda\alpha^2(1-\theta)^2+\delta(1-\gamma\theta)^2}\nu \quad (5)$$

$$y = \frac{1+\delta(1-\gamma\theta)^2}{1+\lambda\alpha^2(1-\theta)^2+\delta(1-\gamma\theta)^2}\varepsilon + \frac{\delta(1-\gamma\theta)}{1+\lambda\alpha^2(1-\theta)^2+\delta(1-\gamma\theta)^2}\nu \quad (6)$$

Note that with θ=1 (wages are fully flexible), the expected inflation rate by the public is 0. Also, the terms-of-trade shock does not affect the domestic prices as the adjustment is borne by the nominal exchange rate. The terms-of-trade shock affects real output in a one-to-one fashion. This is because under flexible exchange rate regime with full indexation of wages, changes in nominal exchange rates are fully passed-through to domestic prices. Thus, the expenditure switching channel does not perform the usual real shock adjustment. Note that the foreign monetary shock (ν) does affect real output and inflation under flexible exchange rate regimes. This is due to the presence of the nominal exchange rate in the authority's loss function.

However, when θ= 0 (nominal wages are all pre-set), the inflation rate is higher than the case of full wage indexation. On the other hand, the impact of the terms-of-trade shock on real output is lower than the case with full wage indexation. This suggests that with nominal stickiness, countries that are subject to real shocks would benefit from exchange rate flexibility. For example, a negative terms-of-trade shock that depreciates the exchange rate would help reduce real wages and ensure expenditure switching from more expensive foreign goods to relatively cheaper domestically produced ones, thereby maintaining employment and output.

If the monetary authority decides to peg the exchange rate against an anchor country, then $s_t = 0$ in Equation 4. In this case, the rational expectations consistent optimal y and π that minimize the loss function are:

$$\pi = \frac{1}{(1-\gamma\theta)}\pi^{*} + \frac{1}{(1-\gamma\theta)}\upsilon \tag{7}$$

$$y = \frac{\alpha}{(1-\gamma\theta)}v + \varepsilon \tag{8}$$

The expected inflation rate is equal to the foreign inflation rate, adjusted by a ratio of wage indexation and the degree of pass-through from nominal wages to domestic inflation. It is also obvious that the level of wage indexation (θ) affects the responses of domestic inflation rate and real output to the foreign monetary shock and the terms-of-trade shock respectively. It can be shown that with θ=1 (nominal wages are fully indexed to inflation rate), the foreign monetary shock has a lower effect on real output under a fixed exchange rate regime. This supports the notion that countries which are subject to external nominal shocks would be better off with a fixed exchange rate regime. For example, a nominal shock to money demand would affect the domestic liquidity, as a result of central bank's purchases and sales of foreign exchange in the foreign exchange market in order to maintain the fixity of the exchange rate. Under this adjustment mechanism, there is no need for wage or domestic price movements to absorb the shock.

When substituting the equilibrium output and inflation under the two exchange rate regimes, in the central bank loss function, Equation 1, we obtain

$$\begin{aligned} L^{FLEX} &= \lambda y_f^2 + \frac{\lambda}{\left(1+\lambda\alpha^2(1-\theta)^2+\delta(1-\gamma\theta)^2\right)^2}\sigma_\varepsilon^2 \\ &+ \frac{2(\lambda+1)\delta^2(1-\gamma\theta)^2}{\left(1+\lambda\alpha^2(1-\theta)^2+\delta(1-\gamma\theta)^2\right)^2}\sigma_\upsilon^2 + \delta\sigma_s^2 \end{aligned} \tag{9}$$

$$L^{FIX} = \lambda y_f^2 + \frac{1}{(1-\gamma\theta)^2}\pi^{*^2} + \lambda\sigma_\varepsilon^2 + \frac{1+\lambda\alpha^2}{(1-\gamma\theta)^2}\sigma_\upsilon^2 \tag{10}$$

where σ_s^2 is the variance of the nominal exchange rate. To compare the loss function with alternative labor market assumptions, Table A.1 presents the values of the loss function under alternative exchange rate regimes and the two polar cases of nominal wage indexation.

Considering fixed exchange rate regimes only, the loss function is lower with full wage indexation ($\theta = 1$) compared to the case of no wage indexation ($\theta = 0$). This is because when the nominal exchange rate is

Table A.1: *The Central Bank's Loss Function under Alternative Exchange Rate Regimes and Different Nominal Wage Indexation*

Wage Indexation	Flexible Exchange Regime	Fixed Exchange Regime
$\theta = 0$	$\lambda y_f^2 + \frac{\lambda}{(1+\lambda\alpha^2+\delta)^2}\sigma_\varepsilon^2 + \frac{2\lambda\delta^2}{(1+\lambda\alpha^2+\delta)^2}\sigma_\upsilon^2 + \delta\sigma_s^2$	$\lambda y_f^2 + \pi^{*^2} + \lambda\sigma_\varepsilon^2 + (1+\lambda\alpha^2)\sigma_\upsilon^2$
$\theta = 1$	$\lambda y_f^2 + \frac{\lambda}{\left(1+\delta(1-\gamma)^2\right)^2}\sigma_\varepsilon^2 + \frac{2(1+\lambda)\delta^{2\{\{(1-\gamma)^2}\big)}{\left(1+\delta(1-\gamma)^2\right)^2}\sigma_\upsilon^2 + \delta\sigma_s^2$	$\lambda y_f^2 + \frac{1}{(1-\gamma)^2}\pi^{*^2} + \lambda\sigma_\varepsilon^2 + \frac{(1+\lambda\alpha^2)}{(1-\gamma)^2}\sigma_\upsilon^2$

not allowed to adjust to real external shocks, the required adjustment in the real exchange rate is made possible through changes in prices and nominal wages. Under a fixed exchange rate regime, the foreign inflation and the variance of foreign monetary shocks are transmitted to the domestic economy. In this case, an independent monetary policy is impotent to respond to the real shock. With full wage indexation ($\theta = 1$), we can see that the terms-of-trade volatility has higher impact on the loss function of fixed exchange rate regimes compared to that of flexible exchange rate regimes. The opposite is true when considering foreign monetary volatility. Thus, with some plausible restrictions on the model parameters, the aforementioned results suggest that fixed exchange rate regimes work better for countries with flexible labor markets (e.g., RRLI), while flexible exchange rate regimes work better for countries with inflexible labor markets (e.g., RRLA). This suggests that with no nominal stickiness, credibility gains under fixed exchange rate regimes are lower than the loss of flexibility in monetary policy. Likewise, with nominal stickiness and flexible exchange rate regimes, the loss of flexibility outweighs the gain in credibility.

References

Al-Abri, A. (2014). Optimal Exchange Rate Policy for a Small-Oil Exporting Country: A General Equilibrium Perspective. *Economic Modelling* 36: 88–98.

Aizenman, J., Glick, R. (2008). Pegged Exchange Rate Regimes – a Trap? *Journal of Money, Credit and Banking* 40(4): 817–835.

Barro, R., Gordan, D. (1983). A Positive Theory of Monetary Policy in a Natural Rate Model. *Journal of Political Economy* 91: 589–610.

Berger, H., Jensen, H., Schjelderup, G. (2001). To Peg or Not To Peg? A Simple Model of Exchange Rate Regime Choice in Small Economies. *Economics Letters* 73: 161–167.

Castrén, O., Takalo, T., Wood, G. (2004). Labour Market Reform and the Sustainability of Exchange Rate Pegs. Working Paper Series 0406. European Central Bank.

Chia, W-M., Cheng, T., Li, M. (2012). Exogenous Shocks and Exchange Rate Regimes. *World Economy* 35(4): 444–460.

Coudert, V., Couharde, C., Mignon, V. (2011). Does Euro or Dollar Pegging Impact the Real Exchange Rate? The Case of Oil and Commodity Currencies. *The World Economy* 34(9): 1557–1592.

Devereux, M., Engel, C. (2003). Monetary Policy in the Open Economy Revisited: Price Setting and Exchange Rate Flexibility. *Review of Economic Studies* 70: 765–783.

Edwards, S. (1996). The Determinants of the Choice between Fixed and Flexible Exchange Rate Regimes. National Bureau of Economic Research. Working Paper No. 5756. Cambridge, MA: NBER.

(2000). Capital Flows, Real Exchange Rates, and Capital Controls: Some Latin American Experiences. In: S. Edwards (Ed.), 2000, *Capital Flows and the Emerging Economies: Theory, Evidence, and Controversies* (pp. 197–246), Chicago and London: University of Chicago Press,

Eichengreen, B. (1999). Kicking the Habit: Moving from Pegged Rates to Greater Exchange Rate Flexibility. *Economic Journal* 109, C1–C14.

Elbadawi, I., Kamar, B. (2005). The Great Debate on Exchange Rate Regimes: Why Should the MENA Region Care? SSRN Working Paper (http://ssrn.com/abstract=1465594).

Engel, C. (2009). Currency Misalignment and Optimal Monetary Policy: A Reexamination. National Bureau of Economic Research. Working Paper No. 14829. Cambridge, MA: NBER.

Flood, R., Marion, N. (1999). Perspectives on the Recent Currency Crisis Literature. *International Journal of Finance and Economics* 4: 1–16.

Frankel, J. (2005). Peg the Export Price Index: A Proposed Monetary Regime for Small Countries. *Journal of Policy Modeling* 27: 495–508.

(2010). The Natural Resource Curse: A Survey. National Bureau of Economic Research. Working Paper No. 15836. Cambridge, MA: NBER.

Ghanem, D. (2012). Fixed Exchange Rate Regimes and Inflation Performance: Evidence from MENA Countries. *Review of Middle East Economics and Finance* 8: 1–30.

Gray, J. (1976). Wage Indexation: A Macroeconomic Approach. *Journal of Monetary Economics* 2: 221–235.

Habib, M., Stráský, J. (2008). Oil Exporters: In Search of an External Anchor. Working Paper Series No. 958. European Central Bank.

International Monetary Fund. (2010). Annual Report on Exchange Restrictions and Exchange Arrangements. Washington DC: International Monetary Fund.

Jbili A., Karamenko, V. (2003). Should MENA Countries Float or Peg? . IMF Finance and Development 40 No.1, pp. 30–35. Available via the Internet: www.imf.org/external/pubind.htm.

Khoudour-Castéras, D. (2009). Labour Immobility and Exchange Rate Regimes: An Alternative Explanation for the Fall of the Interwar Gold Exchange Standard. *Journal of European Economic History* 38(1): 13–47.

Leenders, R., Sfakianakis, J. (2003). Middle East and North Africa, in Global Corruption Report 2003, Transparency International (www.globalcorruptionreport.org/gcr 2003.htm).

Levy-Yeyati, E., Sturzenegger, F., Reggio, I. (2010). On the Endogeneity of Exchange Rate Regimes. *European Economic Review* 54: 659–677.

Naufal, G., Carlos, V-S. (2010). Migrant Transfers in the MENA Region: A Two-Way Street in which Traffic is Changing. *Migration Letters* 7(2): 168 – 178.

Selim, H., Zaki, C. This volume. The Institutional Curse of Natural Resources in the Arab World.

Setser, B. (2007). The Case for Exchange Rate Flexibility in Oil-Exporting Economies. Policy brief 8, Peterson Institute for International Economics, Washington, D.C.

Sfia, M. (2011). The choice of exchange rate regimes in the MENA countries: A probit analysis. *International Economics and Economic Policy* 8: 275–305.

Squalli, J. (2011). Is the Dollar Peg Suitable for the Largest Economies of the Gulf Cooperation Council? *Journal of International Financial Markets, Institutions & Money* 21: 496–512.

Svensson, L. (1999). Inflation Targeting as a Monetary Policy Rule. *Journal of Monetary Economics* 43(3): 607–654.

Svensson, L., Woodford, M. (2004). Indicator Variables for Optimal Policy under Asymmetric Information. *Journal of Economic Dynamics and Control* 28(4): 661–690.

Taş, B., Togay, S. (2010). Optimal Monetary Policy Regime for Oil Producing Developing Economies: Implications for Post-War Iraq. *Economic Modelling* 27(5): 1324–1336.

Willett, T., Al-Barwani, K., El Hag, S. (2010). The GCC's Fixed Exchange Rate: A Major Anomaly for OCA Analysis. *The World Economy* 33(12): 1702–1717.

World Bank, (2007). Middle East and North Africa Region: 2007 Economic Developments and Prospects. *Job Creation in an Era of High Growth*. Washington DC: World Bank.

Yousefi, A., Wirjanto, T. (2003). Exchange Rate of the US Dollar and the J Curve: The Case of Oil Exporting Countries. *Energy Economics* 25: 741–765.

11

The Institutional Curse of Natural Resources in the Arab World[1]

Hoda Selim[2] and Chahir Zaki[3]

ABSTRACT

This paper argues that the resource curse in the Arab world is primarily an "institutional curse," even though it has several macroeconomic manifestations. An empirical investigation, using an augmented growth model, confirms the conditional resource curse hypothesis. The results suggest that on their own, political institutions do not always have an effect on growth but, when interacted with natural resources, they reduce the negative effect of natural resources on growth but do not offset it. The analysis also shows that the curse has operated in different ways within the Arab World. In the GCC, large rents per capita have been channeled to national citizens in the form of well-remunerated public sector jobs and other generous social welfare schemes which have served to increase government legitimacy and foster regime stability. In contrast, the populous group comprising poorer rentier states have experienced conflict, violence and social unrest. Moreover, in a context of low rent per capita, excessive consumption resulted in massive deficiencies in infrastructure investments and an underdeveloped financial sector. Finally, the volatility of their limited resources has been accompanied by more dire economic consequences: excessive borrowing and Dutch Disease.

Keywords: Arab region, GCC, oil curse, Dutch Disease, institutions, augmented growth model

[1] We are grateful to Ibrahim Elbadawi for insightful comments on an earlier draft as well as to Hoda El-Enbaby and Norhan Ossama for excellent research assistance.
[2] Economist, Economic Research Forum, Cairo, Egypt. Email: hselim@erf.org.eg
[3] Assistant Professor, Faculty of Economics and Political Science, Cairo University and Economic Research Forum. Email: chahir.zaki@feps.edu.eg

11.1 Introduction

Despite their generous resource blessing, Arab countries have neither achieved economic prosperity nor become developed countries (Elbadawi and Selim, this volume). And even though the per capita income of some of them is high, their growth performance has been extremely volatile and raises the question of whether their current levels of income are sustainable.

The resource curse literature has evolved over time, almost converging on the finding that curse is conditional on bad governance (Elbadawi and Selim, this volume). Moreover, the literature has shown that the curse is manifested through several channels: (i) overall macroeconomic volatility (Haussman and Rigobon 2003; Ploeg and Poelhekke 2009) and excessive borrowing during resource busts (Manzano and Rigobon 2001), (ii) Dutch Disease (Collier and Goderis 2007), (iii) excessive consumption and low or inefficient total investment (Collier and Goderis 2007) and (iv) lack of financial sector development (Beck 2011).

The chapter argues that the resource curse in the Arab World is primarily an "institutional curse," even though it has several macroeconomic manifestations. It also undertakes an empirical investigation in resource-rich economies, using an augmented growth model, to assess the impact of natural resources on growth. The model accounts for the quality of institutions in order to test the conditional resource curse hypothesis.

Due to data constraints, the qualitative analysis is limited to only ten resource-rich economies. The six GCC countries (Bahrain, Kuwait, Oman, Saudi Arabia, Qatar and the United Arab Emirates (UAE)) are endowed with large natural resource rents resulting in high rent per capita. Rent per capita is significantly lower in the four populous countries of Algeria, Sudan, Syria and Yemen. The analysis will search for similarities and differences between both groups. To the extent possible, it will draw on the country cases presented in this volume.

The analysis shows that weak institutions in the Arab World are the root cause of the resource curse with macroeconomic ramifications. In other words, while macroeconomic mismanagement and oil abundance are important determinants of performance, these factors are shaped primarily by the prevailing political institutions, which predate resource discovery. Over time, the interaction between these factors became intertwined, preventing these countries from embarking on a sustainable development path.

The empirical model confirms that the resource curse is conditional on bad institutions. Resource rents have a negative effect on economic growth, a result that remains robust after controlling for initial income, trade policy, investment rates and schooling. More interestingly, political institutions on their own do not always have an effect on growth. However, when interacted with natural resources, they reduce the negative effect of natural resources on growth.

While weak governance is responsible for the region's lack of development, the curse has operated in different ways in both groups. In the GCC, large rents per capita have been utilized to increase government legitimacy and foster regime stability. Indeed, the curse is expressed *subtly* through a clear segmentation of the labor markets which acts as an efficient mechanism of rent distribution in the form of well-remunerated public sector jobs and other generous social welfare schemes to national citizens. In contrast, in the populous group, which can only be considered as much poorer rentier states, rents are not sufficient to provide a stipend to citizens, and these countries have experienced conflict, violence and social unrest (Ali and Elbadawi, this volume).

The macroeconomic manifestations of the curse are also expressed through different channels. On the one hand, the GCC group's growth performance is more severely affected by resource volatility, but they have managed in recent years to diversify their non-oil economy, even though their manufacturing sectors still remain below predicted levels. They have engaged in excessive public consumption, but high rents per capita allow them to allocate ample resources for investments, particularly in infrastructure, and to develop their financial sector. On the other hand, limited resources may have shielded the populous economies from extreme volatility, but seem to have led to more dire economic consequences. Resource busts tend to drive them to engage in excessive borrowing, while booms seem to have almost eliminated their manufacturing sectors. Most importantly, in a context of low rent per capita, excessive consumption resulted in massive deficiencies in infrastructure investments and an underdeveloped financial sector

The remainder of the chapter is structured as follows. Section 11.2 gives an overview of natural resource wealth in the economies of the region and then describes various aspects of their resource dependence. Section 11.3 shows the different interactions between institutions and natural resources. Section 11.4 investigates the different manifestations of the curse: overall macroeconomic volatility and excessive borrowing, Dutch Disease, excessive consumption, low investment and lack of financial development.

Section 11.5 empirically tests the resource curse hypothesis. Section 11.6 attempts to draw some policy recommendations in order to help Arab countries avoid the curse. The final section concludes.

11.2 Overview of Resource-rich Economies in the Arab Region

11.2.1 Natural Resource Endowments

In terms of natural resource endowments, the GCC group clearly stands out as the better-off group, resulting in significantly higher rent and income per capita. In order to assess wealth, we look at three indicators: natural resource reserves, natural resource production and natural resource rents as a share of GDP.

11.2.2 Reserves and Production

Three GCC countries, Saudi Arabia, Kuwait and UAE, hold approximately a third of the world proven oil reserves. Saudi Arabia alone has the second largest proven reserves, accounting for 18% of global oil reserves (Table 11.1). These reserves are also easily accessible, giving Saudi Arabia the ability to substantially increase its supply to the global oil market over a short amount of time. Moreover, it accounts for over 50% of global spare production capacity and can raise global oil production by over 2% within thirty days (IMF 2013a). In addition, all GCC countries, apart from Bahrain, are among the top thirty global oil producers (Table 11.1). Saudi Arabia is also the largest producer of total petroleum liquids (13% of global production) and is the largest crude oil exporter in the world. Moreover, Qatar, Saudi Arabia and the UAE are among the top ten countries with proven natural gas resources (Table 11.2). Qatar has the third-largest proven reserves of natural gas in the world, is the world's fifth largest producer and the largest exporter of liquefied natural gas (LNG). Bahrain and Oman are considered small producers with limited reserves and shorter horizons before depletion.

As for members of the second "populous" group, their wealth is significantly smaller and are therefore facing the prospect of resource depletion in the not-too-distant future. The group holds less than 2% of oil reserves and controls less than 3% of production. Algeria is among the top twenty global oil producers, but with a production comparable only to Qatar, which is already among the lowest in OPEC. More importantly, Algeria has the tenth largest natural gas reserves in the world and is the second largest

Table 11.1: *Proven Oil Reserves and Production, 2013*

	Proven Oil Reserves			Production		
Country	Billion Barrels	Percent of Total Reserves	Rank	Thousand Barrels per Day	Percent of Total Production	Rank
Saudi Arabia	267.9	17.6	2	11,725.7	13.4	1
Kuwait	104.0	6.8	7	2,796.8	3.2	9
United Arab Emirates	97.8	6.4	8	3,213.2	3.7	7
Qatar	25.4	1.7	14	1,579.2	1.8	17
Oman	5.5	0.4	23	923.8	1.1	24
Bahrain	0.1	0.0	63	55.2	0.1	64
Algeria	12.2	0.8	17	1,875.2	2.1	15
Sudan	5.0	0.3	25	115.3	0.1	31
Yemen	3.0	0.2	29	171.1	0.2	40
Syria	2.5	0.2	32	170.5	0.2	33
		34.3			25.8	

	Proven Natural Gas Reserves			Production		
Country	Trillion Cubic Feet	% of Total Reserves	Rank	Billion Cubic Feet	% of Total Reserves	Rank
Qatar	890.0	13.0	3	5,523.25	4.8	5
Saudi Arabia	287.8	4.2	5	3,585	3.1	8
United Arab Emirates	215.0	3.1	7	1,853.95	1.6	16
Kuwait	63.5	0.9	19	548	0.5	34
Oman	30.0	0.4	26	1034.58	0.9	26
Bahrain	3.3	0.0	53	481.43	0.4	35
Algeria	159.0	2.3	10	3,053	2.6	9
Yemen	16.9	0.2	31	270	0.2	41
Syria	8.5	0.1	43	227.50	0.2	43
Sudan	3.0	0.0	55	0	0.0	93
		24.5			14.3	

Source: U.S. Energy Information Administration, Oil & Gas Journal.
Note: Figures for Sudan refer to the pre-secession of the South. Please see Footnote 5.

Table 11.2: *Proven Natural Gas Reserves and Production, 2013*

	Proven Natural Gas Reserves			Production		
Country	Trillion Cubic Feet	Percent of Total Reserves	Rank	Billion Cubic Feet	Percent of Total Reserves	Rank
Qatar	890.0	13.0	3	5,523.25	4.8	5
Saudi Arabia	287.8	4.2	5	3,585	3.1	8
United Arab Emirates	215.0	3.1	7	1,853.95	1.6	16
Kuwait	63.5	0.9	19	548	0.5	34
Oman	30.0	0.4	26	1034.58	0.9	26
Bahrain	3.3	0.0	53	481.43	0.4	35
Algeria	159.0	2.3	10	3,053	2.6	9
Yemen	16.9	0.2	31	270	0.2	41
Syria	8.5	0.1	43	227.50	0.2	43
Sudan	3.0	0.0	55	0	0.0	93
		24.5			14.3	

Source: US Energy Information Administration, Oil & Gas Journal
Note: Figures for Sudan refer to the pre-secession of the South. Please see Footnote 5.

in Africa. It also accounts for almost 3% of world production, similar to Saudi Arabia (Table 11.2). However, production has been in steady decline since 2005 on the back of maturing fields. Sudan, Syria and Yemen are considered relatively small oil producers, similar to Bahrain. More worryingly, production has been declining in Sudan and Yemen because of natural declines in maturing fields.[4]

11.2.3 Natural Resource Rents

The share of total natural resource rents to GDP in the GCC fluctuated on average around 42% between 1970 and 2011, more than double the average for the populous economies (Figure 11.1). On the one hand, Kuwait, Saudi Arabia, Oman and Qatar are considered richer, with rents accounting for close to half of GDP (Figure 11.1). On the other, Bahrain and UAE are relatively less endowed, with rents respectively representing a third and quarter of GDP, which is comparable to Yemen and slightly higher than

[4] The secession of South Sudan resulted in a permanent shock with the loss of 75% of its oil production, nearly 55% of its fiscal revenues and about two-thirds of its foreign exchange earnings (IMF 2013b).

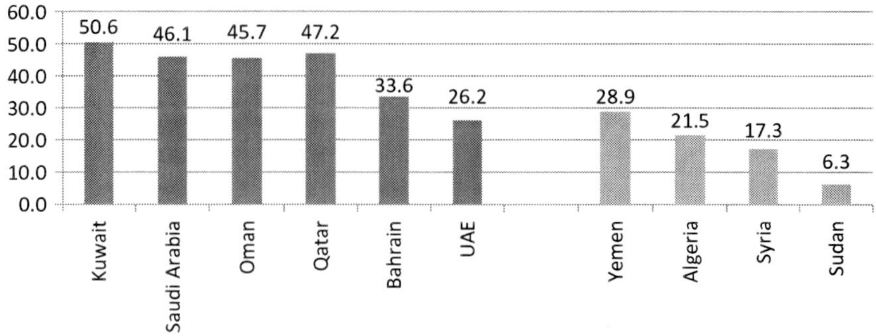

Figure 11.1: Natural Resource Rents, Percent of GDP 1970–2011
Source: World Development Indicators.
Note: Data for Bahrain starts in 1980, for UAE 1975 and Yemen in 1990.

Algeria. The poorest, in terms of natural resource wealth, is Sudan, where rents account for only 6% of GDP. An interesting observation is that even though the GCC remain richer, their resource rents have declined over time and have increased in the populous economies reflecting discoveries in the late 1980s (Table 11.3).

Given the GCC's sheer amount of wealth and the fact that their national populations are relatively small (Table 11.3), average natural resource rent per capita is currently close to US$ 64,000 per year, while it is only US$ 618 in the populous group. Within the GCC, Qatar's rent per capita is estimated at around US$ 190,000 (Figure 11.2), which is more than three times the level in Kuwait, which comes in second, and almost seventeen times the figure for Bahrain where rent per capita is the lowest, estimated at around US$ 11,000. Algeria's modest rent per capita (US$ 1,385) – which is the highest among its group – is one-eighth that of Bahrain. In both groups, rent per capita has significantly increased over time (except in Bahrain where it remained fairly stable).

Consequently, all GCC countries belong to the high-income category (according to the World Bank Atlas method) with an average income of around US$ 34,510, which is almost seventeen times the average of the populous group, who have an average income of only US$ 2,345. To bring out the contrast even further, Algeria's income per capita is only a third of that of Bahrain (Table 11.3). Among the GCC, Qatar has the highest GNI per capita, around US$ 74,340, and it is five times that of Bahrain and four times that of Oman. Meanwhile, all the populous countries belong to the middle income category. Only Algeria belongs to the upper-middle income countries, with a GNI per capita above US$ 4,000, which is almost four times that of Yemen, the poorest country in the group.

Table 11.3: *Basic Statistics*

	Population (Thousands) National	Population (Thousands) Total	Current GDP (in Million US$)	GNI per Capita, Atlas Method (Current US$)		GDP per Capita (Constant 2005 US$)	Natural Resource Rents as % of GDP	
				GCC				
	2011		**2011**	**1970**	**2010**	**2011**	**1970**	**2011**
Bahrain	570.7	1,228.5	29,044.5	8,310.0	14,820.0	14,052.1	85.5	25.1
Kuwait	1,090.0	3,065.9	1,76,590.1	3,310.0	44,100.0	29,337.7	43.2	51.3
Oman	2,013.0	3,295.0	71,781.5	320.0	19,110.0	13,902.7	47.4	45.3
Qatar	259.9	1,732.7	1,72,981.6	na	74,340.0	54,791.9	49.3	28.6
Saudi Arabia	19,405.7	28,376.4	5,76,824.0	760.0	19,360.0	17,050.7	31.6	58.6
UAE	948.0	8,264.1	3,60,245.1	na	35,330.0	23,795.9	41.5	24.2
				Populous Countries				
Algeria	37,763.0		1,98,538.8	330.0	4,350.0	3,168.7	8.2	26.0
Sudan	36,430.9		64,053.4	150.0	1,200.0	832.3	0.0	15.1
Syria	21,532.6		59,147.0	360.0	2,610.0	1,700.4	1.6	18.3
Yemen	23,304.2		31,724.6	na	1,220.0	795.6	na	23.1

* End year is 2010
** Start year is 1980.
** Start year is 1975
Sources: World Development Indicators, national statistical offices and authors' calculations.
Notes:(i) For the GCC, population estimates are from national statistical offices. Qatar's national population is estimated based on a share of nationals' equivalent to 15% of total population.
(ii) Start years are: Bahrain (1981), Kuwait (1975), Oman (1985) and Saudi Arabia (1974). For all populous countries (1970).

11.2.4 Resource Dependence

This section shows that oil dominates several aspects of the economies under study but to varying degrees. To assess resource dependence, we look at three indicators: the share of hydrocarbons in value-added, its contribution to merchandise exports and its contribution to fiscal revenues.

11.2.5 Structure of Value-added and Exports

The hydrocarbon sector dominates the structure of exports more than of value-added. Three groups with different patterns can be identified. In Kuwait, Saudi Arabia and Algeria, hydrocarbons account for half or more of value-added, and fuel exports account for at least 90% of merchandise

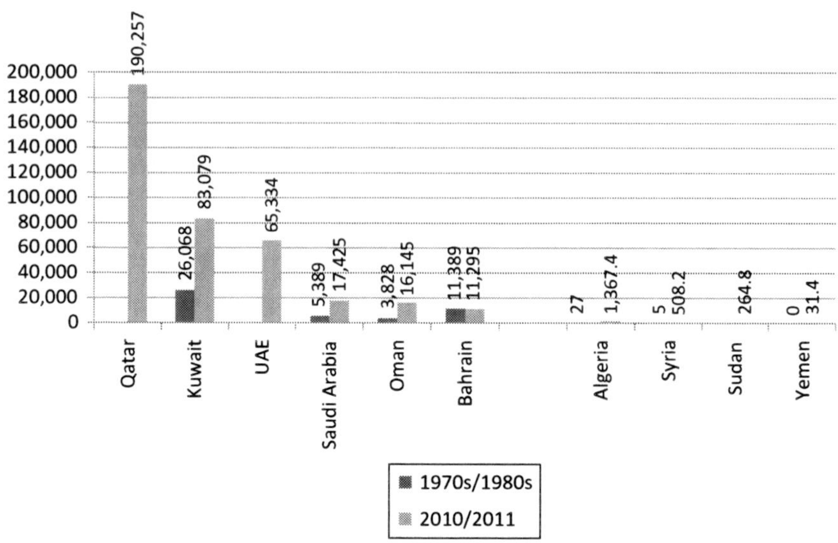

Figure 11.2: Natural Resource Rent per Capita (current US$)
Source: Calculated by the authors based on data from the World Development Indicators and national statistical offices. No old data for Qatar and UAE.

exports. This share rises to 98% in Algeria (Figure 11.3). Bahrain, the UAE, Qatar and Oman form a second relatively less resource-dependent group. In particular, the UAE currently generates around 60% of GDP in the non-hydrocarbon sector and non-oil exports account for 40% of total merchandise exports. Impressively, Bahrain stands out as the most diversified in terms of non-oil GDP, which accounts for 75% of GDP, but exports still remain dominated by fuel. The third group, which includes Sudan, Yemen and Syria, shows mixed characteristics from the previous groups. Similar to the first group, fuel exports dominate overall exports (except for Syria). However, the share of hydrocarbons in the GDP is very low, between 8% and 23% of GDP, even lower than the share in countries belonging to the second group.

11.2.6 Fiscal Revenues

Finally, Figure 11.4 shows that hydrocarbon revenues are a critical source of fiscal revenues in all countries except Syria. Saudi Arabia and Oman are extremely dependent on hydrocarbon revenues and non-hydrocarbon revenues accounted for barely a fifth of total revenues between 1992 and 2012.

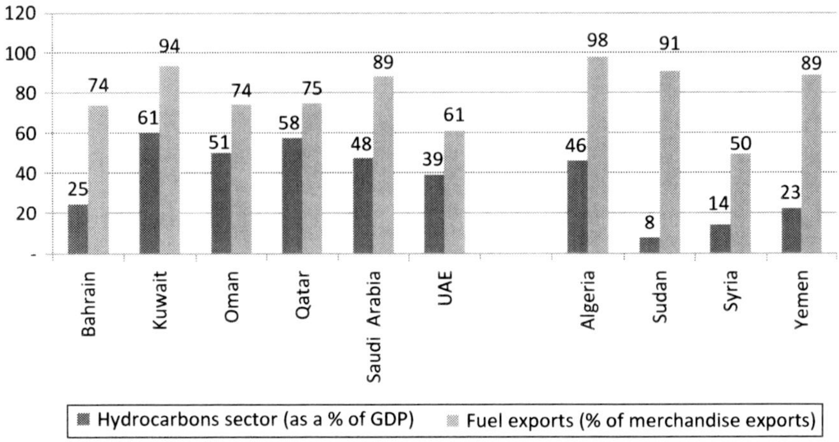

Figure 11.3: Structure of Value-added and Exports, 2011 or 2012
Source: The share of the hydrocarbons sector to GDP was calculated by the authors based on the following data: Bahrain: Central Informatics Organization and Arab Monetary Fund; Kuwait: Central Bank of Kuwait; Oman: Central Bank of Oman; Qatar: Central Bank of Qatar, Saudi Arabia: Central Department of Statistics and Information; UAE: National Bureau of Statistics. Remaining figures are calculated by authors from WDI.

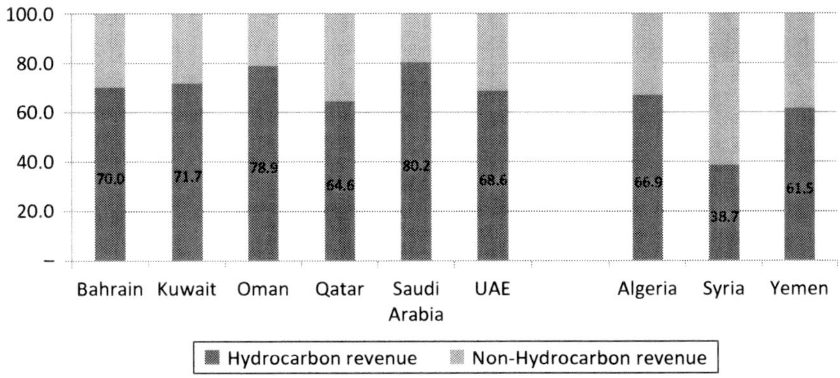

Figure 11.4: Percent of Total Fiscal Revenues, Average (1992–2012)
Source: Calculated by authors based on the dataset from Villafuerte and Lopez-Murphy (2010), which was updated by the authors from national sources or IMF article IV annex tables.

In Bahrain and Kuwait, hydrocarbon revenues constitute three-quarters of the total revenue on average. Reflecting a similar share of resource rents to GDP, the share of hydrocarbon revenues to total revenues is similar in Qatar, UAE, Algeria and Yemen, which is around two-thirds of total revenues.

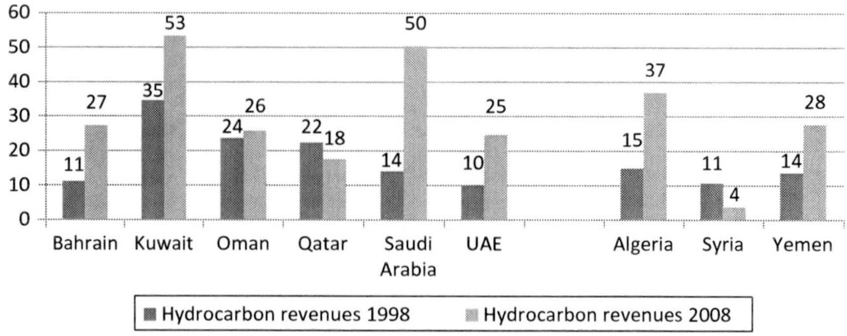

Figure 11.5: Hydrocarbon Revenues during Booms and Busts, 1998 and 2008, Percent of GDP

Source: Calculated by authors based on the dataset from Villafuerte and Lopez-Murphy (2010), which was updated by the authors from national sources or IMF article IV annex tables.

Naturally, hydrocarbon fiscal revenues mirror oil price developments except in Qatar and Syria. They increased from an average of 17% of GDP in 1998 – when oil prices hit a low of US$ 13 a barrel, to 30% of GDP in 2008, when they climbed to close to U$ 100 a barrel. In particular, in Saudi Arabia, they increased from 14% of GDP in 1998 to 50% of GDP in 2008 (Figure 11.5). They also doubled in Bahrain, UAE, Algeria and Yemen. The same trend (though less dramatic) can be observed in Oman. An interesting observation is that hydrocarbon revenues in the GCC during busts reached around 19% of GDP in 1998, which is comparable with the boom figure in the populous group of 23% of GDP.

Moreover, these large hydrocarbon revenues have allowed the GCC and Algeria to achieve fiscal surpluses between 1992 and 2012 on average. These surpluses were around 2% of GDP or higher in most countries, except in Kuwait where they were higher than 15% of GDP. In contrast, Syria and Yemen have posted deficits in the neighborhood of 4–5% of GDP (Figure 11.5). However, the non-hydrocarbon balance has consistently been in deficit in all countries, averaging more than –20% of GDP, except in UAE and Syria where it was lower (Figure 11.6).[5]

[5] This measure is a better indicator of the impact of fiscal policy on domestic demand since hydrocarbon revenues mainly originate from abroad and therefore do not subtract from the resources of the domestic private sector. If the non-resource deficit is getting smaller, fiscal policy is providing less support to the non-oil economy (IMF 2013c).

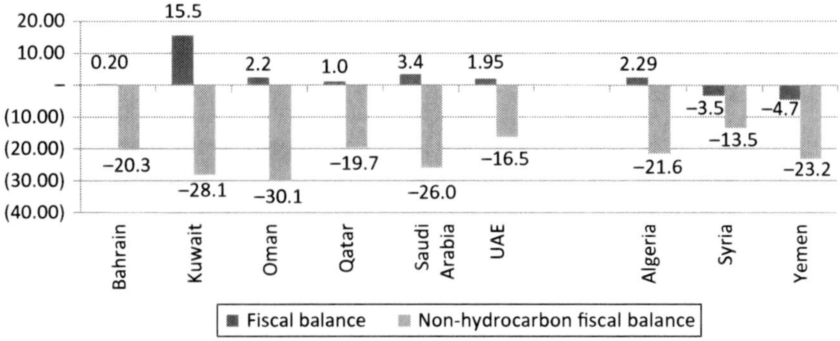

Figure 11.6: Fiscal Balances, 1992–2012 (Average), Percent of GDP
Source: Calculated by authors based on the dataset from Villafuerte and Lopez-Murphy (2010), which was updated by the authors from national sources or IMF article IV annex tables.

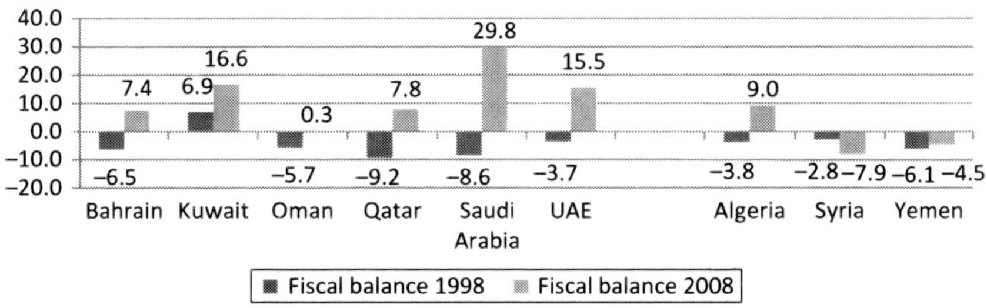

Figure 11.7: Fiscal Balances during Booms and Busts, 1998 and 2008, Percent of GDP
Source: Calculated by authors based on the dataset from Villafuerte and Lopez-Murphy (2010), which was updated by the authors from national sources or IMF article IV annex tables.

The volatility of hydrocarbon revenues is also reflected in the volatility of fiscal balances. In fact, GCC countries (with the exception of Kuwait, which managed to achieve a generous fiscal surplus of 7% of GDP during the bust of 1998 that more than doubled in 2008) and Algeria ran deficits around 4% of GDP in 1998, which transformed into surpluses of more than 12% of GDP in 2008. However, the GCC group appears more vulnerable to oil price fluctuations. For instance, Saudi Arabia's deficit was close to 9% of GDP in 1998 versus a surplus of close to 30% of GDP in 2008 (Figure 11.7). In Yemen and Syria, fiscal deficits have persisted during oil booms. In Syria, they were slightly higher during booms than busts.

This section briefly described resource wealth and dependence in Arab economies. The GCC group may be richer, in terms of endowments, than the populous group but the degree of resource dependence does not seem to be related to the level of wealth. For instance, Kuwait and Saudi Arabia and Algeria form a highly resource-dependent group in terms of the three indicators used, despite significant differences in wealth between Algeria and the other two countries. Moreover, Qatar, the UAE, Bahrain and Oman have undertaken some diversification efforts (with a faster pace in the first two countries) and have therefore become relatively less resource-dependent. Finally, resource dependence is particularly worrying, especially in relation to exports, among two of the poorest countries, Sudan and Yemen. Syria appears to be a more moderate case, where natural resource wealth did not significantly hinder economic diversification.

11.3 The Institutional Curse of Natural Resources

A claim in the literature made in the seminal article of Ross (2001) was that natural resources hurt democracy and enable governments to maintain their authoritarian rule. Ross's findings were further strengthened by a number of cross-country studies (Jensen and Wantchekon 2004). Some research work found that Ross's hypothesis has ambiguous effects (Herb 2005) or does not always hold (Haber and Menaldo 2011). Very recently, the literature has shown that the resource curse is conditional on weak institutions. Disaggregating governance, Collier and Hoeffler (2009) show that the implementation of a system of checks and balances is necessary to limit abuse of political power and hence the misuse of resource rents, even in the presence of a democratic system. Elbadawi and Soto (in this volume) show that resource-rich economies with a high degree of inclusiveness (a measure of democracy) and strong political checks and balances turn the resource curse into a blessing. Studies focusing on the Arab region reach a similar conclusion (Andersen 1987; Salti, 2008; Elbadawi et al. 2011). Ross et al. (2011) show that resources and regimes have intersected to provide stability and limited violent conflict in the region, but that past development patterns have reached their limits and may be unsustainable in the future.

This section argues that resource dependence is endogenous to the region's overall political and economic development, meaning that institutions are the root of the curse. Galal and Selim (2013) argue that weak political institutions in the region seem to have pre-dated the discovery of oil. In other words, the institutional set-up was initially weak and its effects have lingered over time.

Figure 11.8 shows that there is a positive correlation between natural resource rents and limited freedom when it comes to political rights and

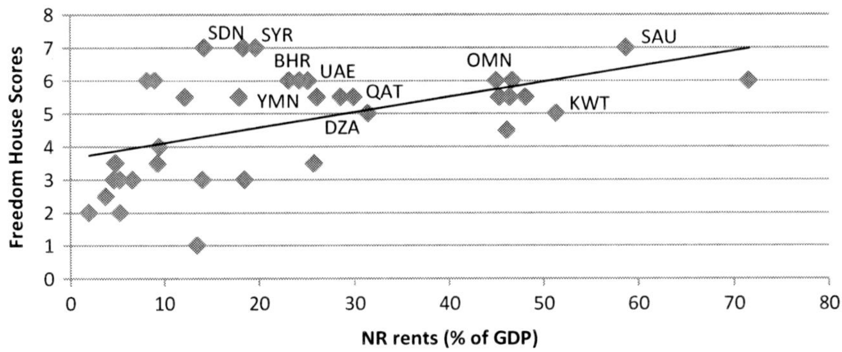

Figure 11.8: Natural Resources and Freedom House Scores, 2013
Source: Freedom House
Note: The political rights indicator is based on several sub-categories derived from the Universal Declaration of Human Rights, which are considered to represent the fundamental components of freedom including: (i) the ability to vote freely in legitimate elections; (ii) the ability to participate freely in the political process; and (iii) have representatives that are accountable to them. This indicator ranges from 1(the most free) and 7 (the least free).

The countries included in this scatter graph (and all subsequent scatters) are those where hydrocarbon rents are 4% of GDP or higher. In addition to the ten Arab countries in our sample, these countries are: Albania, Angola, Argentina, Azerbaijan, Bolivia, Brunei Darussalam, Cameroon, Columbia, Republic of Congo, Ecuador, Egypt, Gabon, Indonesia, Iran, Kazakhstan, Libya, Malaysia, Mexico, Nigeria, Papua New Guinea, Romania, Russia, Tunisia, Turkmenistan, Venezuela and Vietnam.

civil liberties. More worryingly, it shows that resource-rich Arab countries lag behind using these two measures. In 2013, out of the 195 countries covered by Freedom House, 88 were considered "free," of which none were Arab oil-rich economies. Only one country is considered partly free, Kuwait. Moreover, most countries, especially in the GCC, have not undertaken any political reform since the 1970s.

Tornell and Lane (1999) suggest a plausible explanation of why weak governance crucially contributes to the occurrence of the curse. They proposed a "voracity effect" whereby the high value of resource rents would induce competing political groups to "gauge" in the process, not merely wasting the rents as in conventional rent-seeking, but actually reducing income. Mehlum et al. (2006) show that natural resource abundance perversely affects growth when institutions are "grabber-friendly," (i.e., they shift activities from productive into unproductive rent-seeking activities). In other words, resource rents may breed corruption and rent-seeking behavior (Leite and Weidman 1999 and Arezki and Brückner (2011). Alternatively, Robinson et al. (2006) shows that permanent resource booms when government accountability is lacking allow politicians to increase incentives (expanding inefficient public sector employment and/or boosting consumption) to remain in power.

The last explanation coincides with Beblawi and Luciani (1987)'s "rentier-state" proposition. Natural resources provide their governments with the means to benefit their supporters and buy off political consensus with privileges. In addition to this channel, Ross (2001) explores two additional channels through which natural resources contribute to authoritarianism. First, they provide governments with the means to "repress" the population. Second, they are often used to slow down the cultural and social values that promote "modernization."

In the first channel related to the "rentier state," as governments derive large rents from natural resources, they can afford to not only apply low tax rates but also to redistribute these rents through public sector jobs, subsidies and other social welfare transfers and therefore reduce the pressure for accountability. Resource rents thus break the link between taxation and representation suggested by Tilly (1975). In investigating this link, Ross (2004) finds that the larger the share of government expenditure financed through taxation, the more likely the government to become representative. The early literature failed to provide conclusive evidence of whether resource wealth would consolidate authoritarianism or lead to violence and conflict. On the one hand, Smith (2004) provides robust evidence that oil wealth generally increased the durability of political regimes, even when controlling for repression, and lowered likelihoods of civil war and anti-state protests. The author also finds that neither the boom nor bust periods exerted any significant effect on regime durability in the states most dependent on exports, even while those states saw more protests during the bust. On the other hand, Collier and Hoeffler (2004) show that resource abundance could be associated with an increase in the incidence of violence or elevated risks of civil wars. In an effort to reconcile both ends, Basedau and Lay (2009) provide empirical evidence that the relationship between oil dependence and onset of civil war is U-shaped. In other words, very high resource wealth per capita tends to be associated with less violence and is therefore the determining factor for achieving internal stability. They also find that political stability is maintained owing to a combination of large-scale distribution, high spending on the security apparatus and protection by outsiders. This conclusion echoes with the game-theoretic analysis of Ali and Elbadawi (this volume) showing from a political economy perspective that it is optimal for resource-rich authoritarian regimes who rule over small populations (like the GCC) to offer investment in infrastructure and public sector jobs to effectively remove the incentive to revolt. Alternatively, when natural resource endowments are limited (like in the populous group), it is optimal for governments to set up a repressive security apparatus and only employ a smaller proportion of the population.

Against this background, the analysis given here confirms that large resource rents have been used to consolidate authoritarianism and maintain social stability in the GCC, which can be considered non-democratic but with stable and durable regimes. Limited resources have not been associated with violence and political unrest in the populous economies.

An earlier section in this chapter has shown that fiscal revenues from hydrocarbons are significant and this is also confirmed by Table 11.4. By virtue of these revenues, some resource-rich economies are able to apply tax rates as low as 1% of GDP like in Saudi Arabia, the UAE and Kuwait and only slightly higher in other GCC countries. In populous economies like Algeria, they account for close to 9% of GDP.

The large rents enable the GCC to use fiscal spending to maintain political support and social stability through high levels of public employment and a generous welfare system directed towards nationals. In fact, countries like Kuwait and Qatar employ around two-thirds of their nationals in the public sector. This ratio goes down to a bit less than 50% in Saudi Arabia, Oman and Bahrain (IMF 2013d). The wage bill of civil servants is quite inflated in Bahrain, Kuwait and Saudi Arabia, averaging 11% of GDP (Table 11.4). Hodson (2011) describes these jobs as for life, paying high wages and entailing short working hours and generous fringe benefits. The average salary of a Saudi civil servant is three times the average wage for Saudis in the private sector. Pension packages are also generous. The choice of the GCC governments to provide high pay and benefits to nationals has led to high-reservation wages and creates a disincentive for nationals to invest in skills that are demanded by the private sector. This labor market segmentation crowds out private business and contributes to high unemployment in the case of Saudi Arabia (Al-Sheikh and Erbas, this volume).

More generally, citizens of some GCC countries benefit from substantial subsidies and transfers. Between 1997 and 2012, they accounted for as high as 11% of GDP in Kuwait, where citizens are entitled to receive annual transfers from oil rents. Beyond explicit subsidies and transfers, citizens in the GCC are provided free healthcare, education and social security. Utilities (electricity, water and fuel) are also subsidized.

Table 11.4: *Government Revenues and Expenditures in Selected Resource-rich Arab Countries, Averages 1997–2012, Percent*

Average (1997–2012)							
	Bahrain	Kuwait	Oman	Qatar	Saudi Arabia	UAE	Algeria
				% of GDP			
Hydrocarbon revenues	22.0	43.7	28.5	19.7	31.9	19.6	25.9
Non-hydrocarbon revenues o.w.	7.8	16.1	8.2	10.5	5.9	8.1	11.2
Tax revenues	3.1	1.0	2.9	2.1	1.0	0.8	8.6
				% of GDP			
Wages and Salaries	12.7	11.3	8.4	6.7	11.9	6.5	8.4
Subsidies and Social Benefits	5.2	11.0	2.1	2.9	1.5	3.7	4.0
Defense, Security and Public Order	7.0	8.6	11.9	2.3	8.7	4.7	3.5
Education and Health 2011–2012	6.1	7.2	6.5	4.9	9.2	3.0	4.0

Notes: Due to some data unavailability, some variables are included until 2011 only. Also, data for Oman on tax revenues, wages and education and health start in 1998. Data on tax revenues and subsidies and social benefits for Qatar are available since 2003 only.
Source: Authors' calculations based on several data sources. Algeria, Kuwait, Qatar and UAE: IMF statistical appendices of article IV country reports; Data prior to 2004 for Qatar are from the Annual Reports of the Central Bank of Qatar, Oman, Ministry of National Economy's statistical year book and Central Bank of Oman's annual report; Saudi Arabia: Saudi Arabian Monetary Agency's annual report; defense and security data for Algeria and UAE are from WDI.

Clearly, these transfers increase citizens' welfare, but they are also politically motivated. Large resource rents were used to appease citizens, in particular during times of threats to the regime. When political unrest mounted in 2011, resource rents were used by the GCC governments to appease citizens. Kuwait and Bahrain responded by giving out cash, Bahrain and Oman provided public sector jobs and Saudi Arabia and Oman raised workers' wages and benefits. According to Hertog (2012), Saudi Arabia approved an increase in expenditure by US$ 130 billion to finance the creation of 120,000 new public sector jobs, building 500,000 houses, setting a minimum wage of US$ 800 in the public sector, a one-time bonus to incumbent civil servants and the creation of a general unemployment assistance scheme.

Meanwhile, most countries of the populous group witnessed violence at some point in time during their history. And while it is difficult to ascertain for sure that natural resources were the main driver of the violence, they were significant contributors, or at the very least, they were not sufficient enough to sustain the same durable stability experienced in the GCC. Algeria was prone to violence even before the discovery of oil, first for its independence from France (1954–1962); and then during most of the 1990s because of violent confrontations between Islamists groups and the state. Bellin (2004) argues that in the midst of the civil war, which ravaged the country's economy, natural resource rents provided financial support to the army. Sudan endured a brutal civil war that broke out between the North and South in 1983. Moreover, it was often claimed that the lack of transparency in wealth sharing and unsustainable resource management were often trigger factors of violence before the secession of South Sudan. Finally, Syria and Yemen have been experiencing armed conflicts since 2011.

Turning to the "repression" channel, governments tend to resort to the security apparatus to contain any real opposition, as shown by the 2011 reaction to pro-democracy protests that took place in Oman and Bahrain. More broadly, some resource-rich economies (like Oman and to a lesser extent Kuwait, Saudi Arabia and Bahrain) spend between 7% and 12% of GDP on military and police forces.

With respect to the "modernization effect," certain occupational specializations, urbanization and better education are supposed to accompany economic development and these changes should produce a population that is better able to organize and bargain for greater political demands (Inglehart 1997). However, natural resources could prevent or at least slow down these social changes. In particular, natural resources can also crowd out human capital by diverting resources from education, due to a high level of non-tax income and social spending and transfers. Citizens of resource-rich economies may also find themselves "locked in low-skilled natural resource-based industries" that weaken the incentives for good education and earning power (Gylfason 2004).

In this context, large rents in Qatar have been associated with education spending as low as 3% of GDP, almost half of what much poorer Yemen spends. However, in countries with comparable wealth like Saudi Arabia and Kuwait, education spending is double that level. More generally, Table 11.4 shows that public spending on education and health combined has never exceeded 7% of GDP in the majority of countries, which has in most cases been lower than spending on civil servants wages, subsidies or defense.

11.4 Macroeconomic Manifestations of the Natural Resource Curse

The literature has shown that the curse is manifested through several channels: (i) overall macroeconomic volatility (Haussman and Rigobon 2003; Ploeg and Poelhekke 2009) and excessive borrowing (Manzano and Rigobon (2001), (ii) Dutch Disease (Collier and Goderis 2007), (iii) excessive consumption and low or inefficient total investment (Collier and Goderis 2007) and (iv) lack of financial sector development (Beck, 2011). Collier and Goderis (2007) and Elbadawi and Soto (in this volume) provide empirical evidence that the curse operates through three main channels: excessive consumption, low or inefficient total investment and an overvalued exchange rate. This section examines the different manifestations of the curse in the Arab World, contrasting the experiences of the economies of GCC with the populous group.

11.4.1 *Macroeconomic Volatility and Excessive Borrowing*

Led by Sachs and Warner (1995), empirical work showed that resource-abundant countries have slower growth than resource-poor countries. This finding is also confirmed by recent work, which most importantly shows that the curse operates in the long run, as commodity booms have positive short-term effects on output but adverse and substantial long-term effects (Collier and Goderis 2007). Moreover, natural resource abundance often exposes countries to macroeconomic volatility derived from commodity price volatility. Ploeg and Poelhekke (2009) find that volatility of natural resource prices is a major determinant of growth performance and that the direct positive effect of resource dependence on growth is swamped by its adverse effect via increased volatility. Hausmann and Rigobon (2003) estimate that a 1 standard deviation shock to the price of oil (of 30–35%) can generate an income shock as high as 6% of GDP in an economy where oil accounts for 20% of GDP. In addition, Manzano and Rigobon (2001) showed that it was the use of commodity prices as debt collateral during the 1970s that drove many resource-rich countries to debt crisis in the 1980s.

In line with this literature, Figure 11.9 shows that the majority of resource-rich Arab economies have had lower GDP per capita growth between 1981 and 2011 than what is predicted by their share of natural resource rents to GDP. Moreover, countries with large resource dependence (with a share of rents larger than 40% of GDP) like Kuwait and Qatar

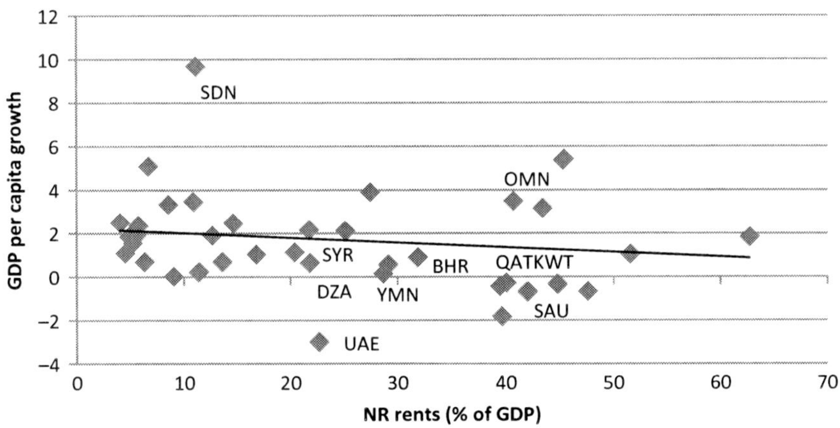

Figure 11.9: Natural Resource Rents and Long-run GDP per Capita Growth, Average 1981–2011
Source: Author's calculations based on WDI data and WEO database, 2014.

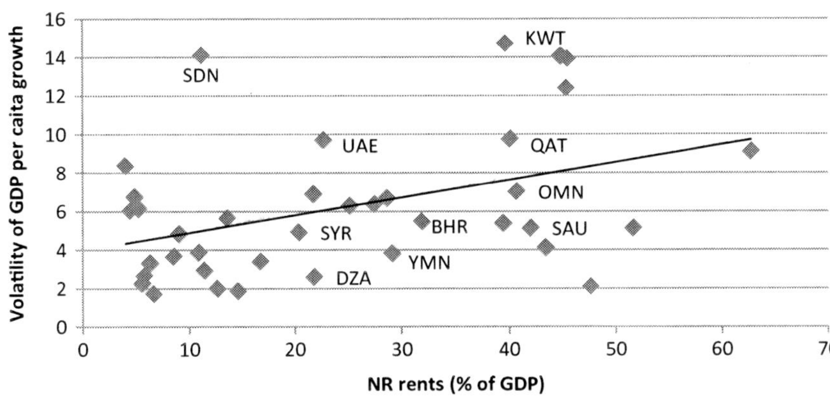

Figure 11.10: Natural Resource Rents and Volatility of GDP per Capita Growth, 1981–2011
Source: Author's calculations based on WDI data and WEO database, 2014.

have exhibited overall negative and volatile growth (measured by the standard deviation) (Figure 11.10). The UAE has the same record despite being less endowed. Oman seems to be the only country that managed to achieve relatively higher growth coupled with relatively lower volatility. Sudan is another exception posting very high and very volatile growth among comparators.

Table 11.5: *Long-term per Capita GDP Growth in Resource-rich Arab Economies, 1981–2011*

	GCC	Populous
Average	0.02	1.04
Max	8.91	3.84
Min	-7.02	-5.07
Volatility	3.9	2.01
Average oil prices	34.6	34.61

Source: Authors' calculations based on WEO database, 2014.
Notes: (i) Oil booms are years when oil prices were higher than US$ 34.6 per barrel which is the average oil price for the whole period 1981–2011. (ii) Oil prices are for crude oil (petroleum), a simple average of three spot prices – Dated Brent, West Texas Intermediate and the Dubai Fateh.

Table 11.6: *Selected Macroeconomic Indicators during Booms and Busts*

	GCC		Populous	
	1998	2008	1998	2008
Oil price	***13.1***	***97.0***	***13.1***	***97.0***
GDP growth	4.3	8.5	5.6	3.4
Real hydrocarbon GDP growth	1.3	6.6	3.4	−3.8
Real non-hydrocarbon GDP growth	3.2	11.2	5.4	5.4
Non-hydrocarbon GDP (% of GDP)	74.2	54.0	88.3	70.0
Gross national savings (%GDP)	20.1	48.7	20.2	31.3
Total investment (%GDP)	25.6	29.8	20.2	23
Exports (% of GDP)				
Inflation (%)	0.9	9.3	8.2	13.3
Fiscal balance (% of GDP)	−3.8	17.3	−3.7	0.6
Current account balance (% of GDP)	−8.1	18.9	−4.7	3.2
Total reserves (US$ billion)	5.6	88	2.5	43.9
General government gross debt (% of GDP)	38.7	79.3	123.5	37.7
External debt stocks (% of GNI)	na	na	119.1	20.3

Source: Authors' calculations based on WEO and WDI databases, 2014.

In this context, the GCC exhibited a low per capita growth–high volatility pattern in the long term. The populous resource-rich countries did better, averaging 1% per capita GDP growth in the period 1981–2011, while the GCC posted no growth in the same period. Moreover, their volatility was almost half of that of the GCC (Table 11.5).

Moreover, the region's macroeconomic performance remains vulnerable to the ebbs and flows of oil prices. Table 11.6 presents selected macroeconomic indicators in the years between 1998 and 2008. The first year represents a year when oil price plummeted to US$ 13, which was the lowest price between 1980 and 2012. The year 2008, when the price of oil soared to US$ 97, is chosen before the recessionary effects of the global financial crisis were felt. On the one hand, GCC growth seems more exposed to the ebbs and flows of the oil price cycle than the populous group, experiencing very high spells during booms (including in the non-hydrocarbons sector) and able to achieve large savings and investment ratios and accumulate significantly higher reserves and external surpluses. During busts, their GDP growth is almost halved and non-hydrocarbon GDP growth is lower by a third. Interestingly, the non-hydrocarbon sector expands during oil busts to compensate for the sharp contraction in the hydrocarbon sector. On the other hand, the impact of oil price fluctuations on the domestic economies of the populous group seems to be somewhat mitigated, except for one aspect: these countries tend to engage in excessive borrowing during oil busts, with external debt stocks exceeding 100% of GNI in 1998, compared to only 20% in 2008. They seem to perform relatively better during busts, posting higher growth including in the non-hydrocarbons sector. Nevertheless, they are able to slightly increase their savings and investments during oil booms, though by much less than their GCC counterparts, and turn their internal and external deficits into weak surpluses.

11.4.2 *The Dutch Disease and Limited Diversification*

Hausmann and Rigobon (2003) show that the resource curse tends to be weaker in economies with more diversified trade structures. The diversification challenge is more difficult in resource-rich countries due to the Dutch Disease. Corden (1984) provides a theoretical framework for the Dutch Disease in which the appreciation of the real exchange rate due to increased resource revenues spent on non-tradable goods leads to a decline in the competitiveness of the non-resource sector, thus reducing manufacturing. This section examines to what extent this occurred in the Arab oil-rich economies.

According to the IMF exchange rate arrangements, most Arab resource-rich economies have opted for fixed exchange rate regimes (Al-Abri, this volume). In the case of the GCC, an exchange rate anchor seems to have worked well in maintaining low inflation. Similarly, all countries in the populous group also rely on a fixed exchange rate except Yemen, which has a monetary aggregate target.

Table 11.7: *Deviations from Equilibrium Real Exchange Rate, 1975–2005, Percent*

Countries	Mean Deviation of the RER	Periods with Overvaluation (Percent)
Bahrain (8)	4.7	50
Kuwait (5)	0.6	80
Oman (8)	−18.1	63
Qatar (5)	10.4	20
Saudi Arabia (5)	−9	80
United Arab Emirates (5)	6.4	20
Algeria (8)	−9	75
Syria (8)	−17.9	63
Yemen (5)	6.1	40

Source: Diop et al. (2012)
Note: A negative value in column 1 means an overvalued RER on average during the whole period (up to 8 five-year periods, 1970–2005).

There is some evidence of real exchange rate overvaluation between 1975 and 2005 in resource-rich economies of the region. Table 11.7 reports the estimated deviations from equilibrium real exchange rates. Column 1 gives the average deviation over sub-periods (ranging from five to eight years, depending on the country) and column 2, the percentage of periods with overvaluation.[6] Although not all countries suffered from overvalued real exchange rates on average, most countries did at some point in time between 1975 and 2005. Three groups can be distinguished. The first group includes Algeria, Kuwait and Saudi Arabia – which are the least diversified countries – for which exchange rates have been consistently overvalued over the years, even though overvaluation was not necessarily significant. In Oman and Syria, overvaluation was larger but was less persistent over time. Finally, Bahrain, Qatar, UAE and Yemen are the countries who belong to the relatively more diversified group. The exchange rates are estimated on average to be undervalued but have been overvalued for only some years.

Yet, while Table 11.7 may indicate that the real exchange rate is overvalued in resource-rich economies, in practice, other factors that may not

[6] Diop et al. (2012) follow Rodrik (2008) in estimating the equilibrium real exchange rate (RER). After estimating the real exchange rate equation where the RER is regressed on GDP growth, they take the log of the difference between the actual RER and the estimated one.

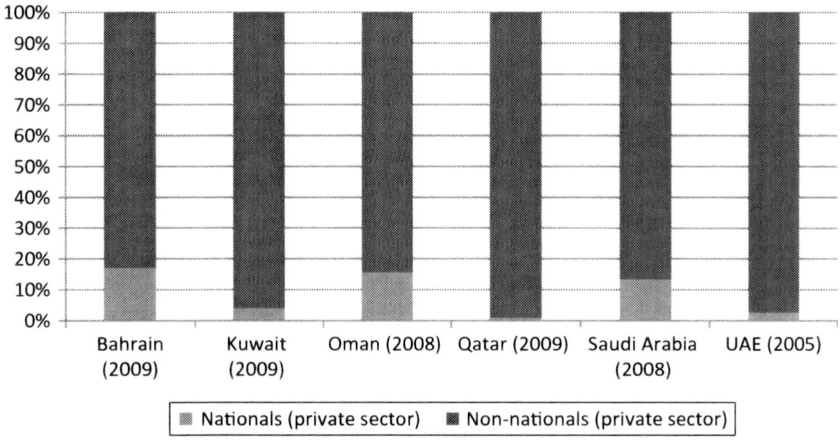

Figure 11.11: Employment of Non-nationals in the Private Sector, Share in Total Private Sector Employment
Source: Hodson (2011).

be taken into account may contribute to a more depreciated real exchange rate. These elements include labor market flexibility and current account openness, which may be particularly relevant for exchange rate dynamics in the GCC.

Labor markets in the GCC are highly dependent on expatriate workers in the private sector. Indeed, the share of non-nationals in total private sector employment is above 80% in all GCC economies and this share goes up to 99% in Qatar (Figure 11.11). Open labor markets and increased worker mobility render labor markets quite flexible, as nominal wages adjustment is fairly quick. During oil busts, the government reduces spending and therefore, the private sector is able to terminate the expatriate workers' contracts, which are mostly of a short-term nature. In this case, labor mobility can substitute for exchange rate flexibility (Al-Abri, this volume). When labor markets are rigid, they may add to the real appreciation of the exchange rate. Moreover, the unlimited labor supply allows the labor cost to be kept rather low and this disproportionately depresses the prices of non-tradables due to the fact that non-tradable sectors tend to be more labor-intensive; hence, low labor cost further increases the relative prices of tradables relative to non-tradables, leading to a more depreciated real exchange rate.

Second, a large expatriate labor force suggests that the GCC economies experience substantial remittances outflows. In fact, all GCC

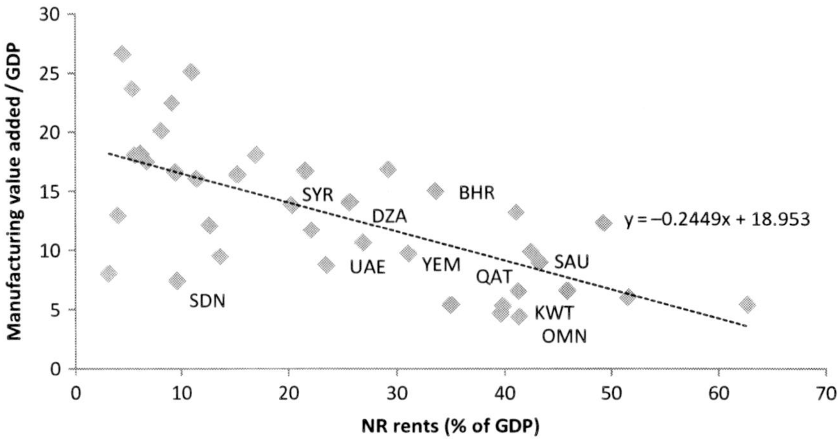

Figure 11.12: Manufacturing and Rents, 1980–2011
Source: Author's calculations based on WDI data.

economies (excluding the UAE for which data is unavailable) figured among the top twenty highest remittance-sending countries in 2013. Remittances outflows sent from the GCC accounted for US$ 66 billion or 18% of total remittances outflows. In fact, Saudi Arabia was the third highest remittance-sending country in the world in 2012, with an estimated value of remittances of $30 billion. By reducing domestic absorption, the demand for non-tradables is also reduced, leading to lower prices of these non-tradable goods and services.

Having said that, manufacturing remains remarkably suppressed in resource-rich Arab countries, including manufacturing exports, except in Bahrain for which these shares are respectively 15% of GDP and 24% of total merchandise exports (Figures 11.12 and 11.13). Moreover, large rents in Kuwait, Oman, Qatar and Saudi Arabia have been associated with a share of manufacturing of only 7% of GDP and a share of manufacturing exports of less than 10% of total exports. Meanwhile, even in poorer populous economies, the manufacturing sector underperforms with a share of around 10% of GDP. If we exclude Syria, which does remarkably well (18%), manufacturing exports have accounted for less than 2% of total exports.

With the exception of Kuwait, GCC economies undertook some economic diversification away from oil over time, but only a few were able to diversify into manufacturing exports like Bahrain, Oman and Saudi Arabia (Figures 11.14 and 11.15). Qatar and the UAE did the former but not the latter. More important, the share of manufacturing to GDP has either

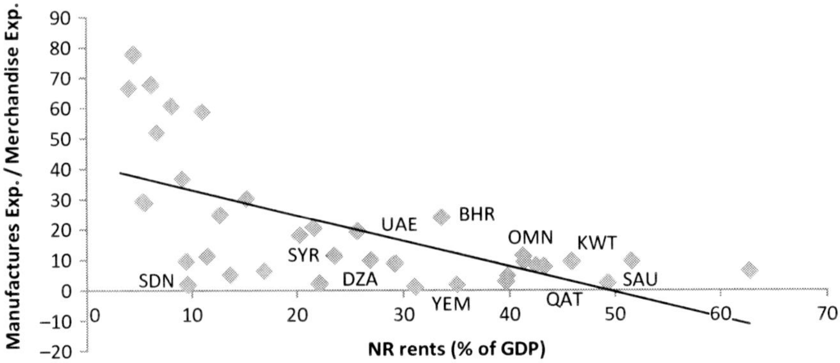

Figure 11.13: Manufactures Exports and Rents, 1980–2011
Source: WDI data.

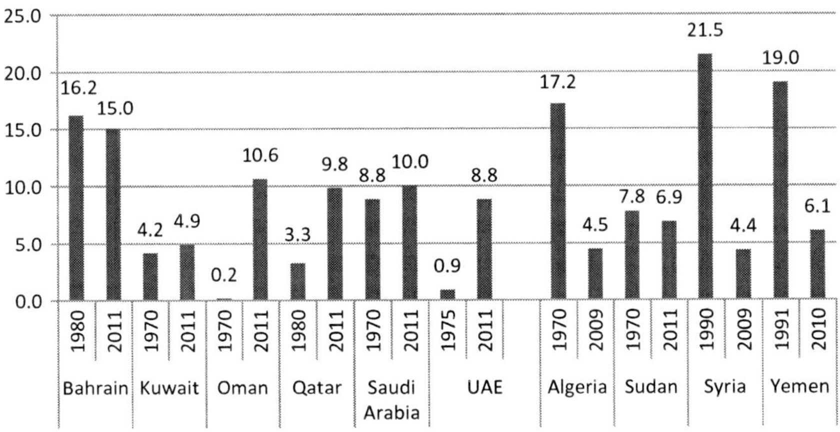

Figure 11.14: Value-added Manufacturing (Percent of GDP)
Source: WDI data, 2014 and/or authors' calculations based on data from the Arab Monetary Fund and national statistical agencies.

sharply gone down or stagnated in countries where exchange rate overvaluation was persistent and/or significant (Figure 11.14). This was the case in Kuwait and Saudi Arabia where manufacturing stagnated at 5% and 10% of GDP respectively and Algeria where it shrunk considerably (from 17% to 4% of GDP). Meanwhile, it would seem that undervalued exchange rates helped the manufacturing sector expand in Qatar and the UAE.

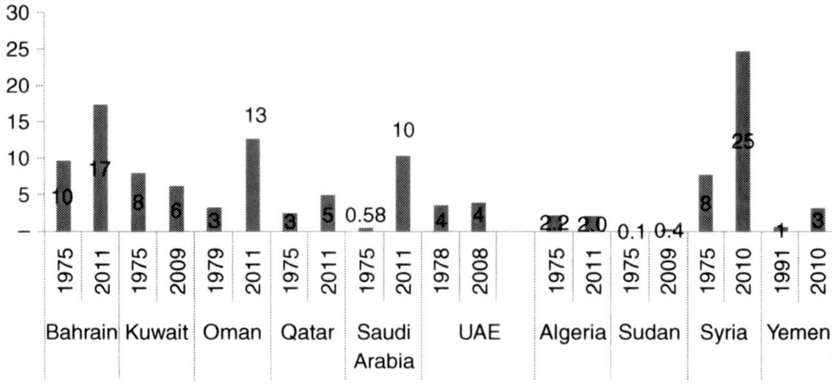

Figure 11.15: Manufacturing Exports (% of Merchandise Exports), 1975 and 2011
Source: WDI, 2014

However, it is not clear how the overvalued exchange rate in Oman in the 1975–2005 period was associated with a spectacular increase in manufacturing to GDP (rising by ten folds). Nevertheless, estimates of exchange rate misalignment since the early 2000s point to an undervalued exchange rate. Syria was able to diversify its export sector and within that increase the share of manufacturing exports over time. Algeria has not undertaken any diversification efforts, while in Sudan and Yemen, fuel exports were almost nonexistent in the 1970s but as oil discoveries were made, the non-fuel export sectors almost vanished. Moreover, Suliman (this volume) explains that the discovery of oil in Sudan in the late 1980s led to a sustained exchange rate appreciation, which quickly displaced cotton as the leading export crop. In this sense, Sudan's export dependence shifted from a relatively diversified agricultural basket towards oil, which is a point resource.

Diop and de Melo (this volume) provide empirical evidence that natural resource rents in Arab countries are correlated with a declining share of services to GDP, in other words, that services sectors are also victim of a Dutch Disease phenomenon. This explains why Arab resource-rich countries have become large importers of tradable services and why only domestic production of non-tradable services (such as real estate, retail trade, hotels and restaurants) has really developed.

Most GCC countries have announced national industrial strategies with the aim of diversifying their economies over a medium-term horizon, with quantitative targets. For instance, Bahrain's 2030 vision aims to double real disposable income of households. So far, it was able to establish itself as a financial hub for the Gulf region and for the Arab World, particularly in

Islamic banking. It is also supporting other sectors like tourism, transport and related services. The UAE – led by Dubai – has initiated, since the oil bust of the 1990s, a diversification strategy, balancing between domestic demand sectors (e.g., construction, real estate, business services and manufacturing) and overseas markets (financial sector, transport and storage and wholesale and retail trade that is largely connected to tourism and duty free activities) (Hawas and Soto, this volume). According to Rodrik (2009), the success of this diversification strategy has hinged on several elements, including: (i) government-led development and fast decision-making combined with free-market policies, (ii) flexible labor force through importing expatriates, (iii) bypassing industrialization and creating a service economy, (iv) internationalizing service provision, (v) creating investment opportunities, (vi) supply-generated demand, (vii) market positioning via branding and (viii) development in cooperation with international partners. Saudi Arabia has also launched a national industrial strategy in 2009 with targets to expand the manufacturing sector from 11% to 20% by the end of 2014, double Saudi industrial employment to 30%, double industrial exports to 35% and double the proportion of technology based manufactured products to 60%. The success of the diversification strategy has been more pronounced in recent years in the phenomenal development of hydrocarbons-based petrochemicals industries as in the case of the petrochemical manufacturer Saudi Basic Industries Corporation (SABIC) which implemented several joint ventures with a number of major multinational companies

11.4.3 *Excessive Consumption*

One avenue of research shows that commodity booms tend to reduce investment levels and increase consumption (in the form of inefficient public employment in return for political support as explained earlier) (Gylfason and Zoega 2006). Arguing from a political economic perspective, Collier and Goderis (2007) show that this occurs because commodity booms divert resources to less productive rent-seeking or public sector activities. Often, commodity booms also lead to a lower quality of investment projects. Robinson and Torvik (2005) provide a theory in which "white elephants" (investment projects with negative social surplus) are used as a means of inefficient redistribution aimed at influencing the outcomes of elections. However, another avenue of research provides a more nuanced view. Collier and Hoeffler (2009) show that whether resource discovery augments increases in the provision of public goods

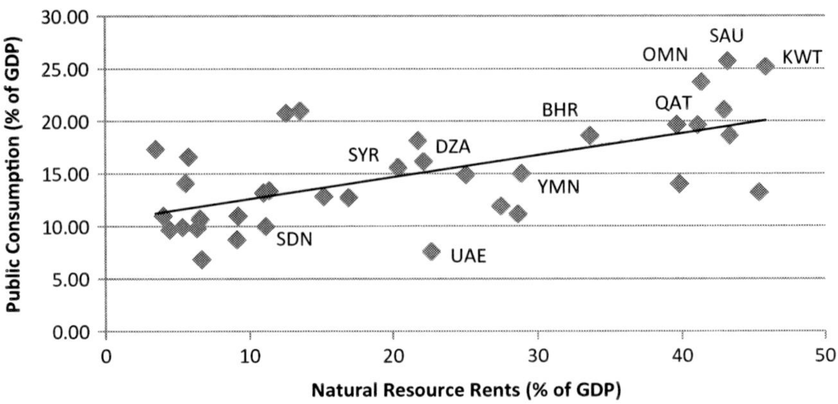

Figure 11.16: Hydrocarbons Rents and Public Consumption, Percent of GDP, 1980–2011
Source: Author's calculations based on WDI data and WEO database, 2014.

depends on two factors: (i) how much public revenue can be embezzled from its proper uses and (ii) the proportion of resource rents in income. Their analysis shows that there is a critical level for the rate of embezzlement above which public goods provision actually deteriorates. Meanwhile, resource discoveries beyond a certain threshold improve the provision of public goods. In countries with modest resources and where democracy is well-established with strong checks and balances, the rate of embezzlement is small and hence, provision for public goods is likely to increase with natural resource discoveries. The converse also holds. In some other countries, the resources are so large that they would still improve the provision of public goods even if the rate of embezzlement is large.

Figure 11.16 shows a positive correlation between the share of natural resource rents and government consumption. In countries with large rents, like most of the GCC, public consumption is around 20% of GDP. In Kuwait and Saudi Arabia, this ratio exceeds 25%. In the poorer populous economies, Sudan and Yemen, public consumption is below predicted levels.

Figure 11.17 suggests that the correlation between rents and investment is only slightly negative. The poorer economies of the populous group, Sudan, Yemen and Syria, have investment ratios significantly below predicted levels and thus, suffer from major deficiencies in infrastructure, including basic infrastructure such as roads, railways, ports and electricity. In contrast, investment ratios are very high in many GCC countries, in particular, in Qatar, the UAE and Bahrain, where they are close to 25% of

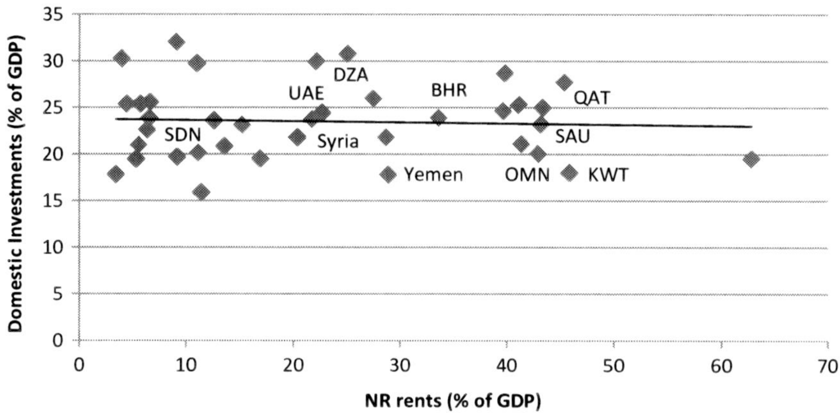

Figure 11.17: Hydrocarbons Rents and Domestic Investments, 1980–2011
Source: Author's calculations based on WDI data and WEO database, 2014.

GDP. In the years following the 1970s oil boom, the GCC governments embarked on massive investment programs, giving priority to basic infrastructure. From 2002 to 2008 alone, they awarded approximately US$ 720 billion worth of infrastructure projects (Abdelal et al. 2008). In Qatar and the UAE, investments were concentrated in the construction sector, while in Saudi Arabia, they were oriented towards the power sector. In recent years, the transport sector has also witnessed a significant boom with many GCC countries increasing investment spending to construct new railway systems and metros, increasing capacity of airports, and expand ports as many of them look to become major shipping hubs. Having said that, Kuwait and Oman, lag behind other GCC performance. Their large rents to GDP ratios are associated with low investment ratios (around 20% of GDP).

11.4.4 *Financial Development*

In the case of resource-rich economies, Beck (2011) explains that natural resource abundance may have mixed effects on financial system development. On the demand side, natural resource windfalls and the consequent expansion of the non-traded goods sector could lead to higher demand for financial services, including consumer credit. However, in the context of Dutch Disease, the non-resource traded goods sector could suffer. Further, Gylfason (2004) explains that natural resource abundance could result in a lesser need for financial intermediation, as the resulting distortion of the

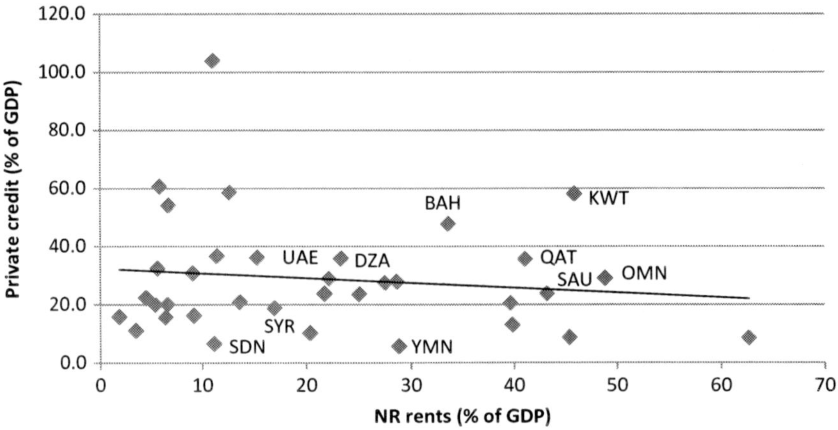

Figure 11.18: Hydrocarbons' Rents and Financial Sector Development, 1980–2011
Source: WDI data.

allocation of capital leads to a lower demand for financial services resulting in lower savings and investment rates. On the supply side, higher investments in the natural resource sector can lead to lower investments in the financial sector. However, the absence of sound institutional frameworks, including effective contractual frameworks, in resource-rich economies can hamper financial deepening.

The empirical literature in the context of resource-based economies is still nascent and results are not conclusive. Some studies find that there is no differential effect of natural resource wealth on the effect of financial development on economic growth in resource-rich versus resource-poor economies (Beck 2011). Others find that it has a positive impact on growth because it can help dampen the impact on volatility (Ploeg and Poelhekke 2009). Beck (2011) provides evidence that resource-based economies have less developed financial systems, and while their banks are more liquid, better capitalized and more profitable, they give fewer loans to firms.

While Figure 11.18 shows that an increase in natural resource rents is associated with lower private sector credit, it does not show that there is a finance resource curse in Arab resource-rich economies. Indeed, the GCC would be considered more financially developed with private sector credit amounting to 55% of GDP. Moreover, all GCC economies, except Saudi Arabia, have a ratio of private sector credit to GDP higher than would be predicted by their level of rents. In particular, Kuwait's private sector credit accounts for 60% of GDP and in Bahrain, it is close to 50% of GDP. This

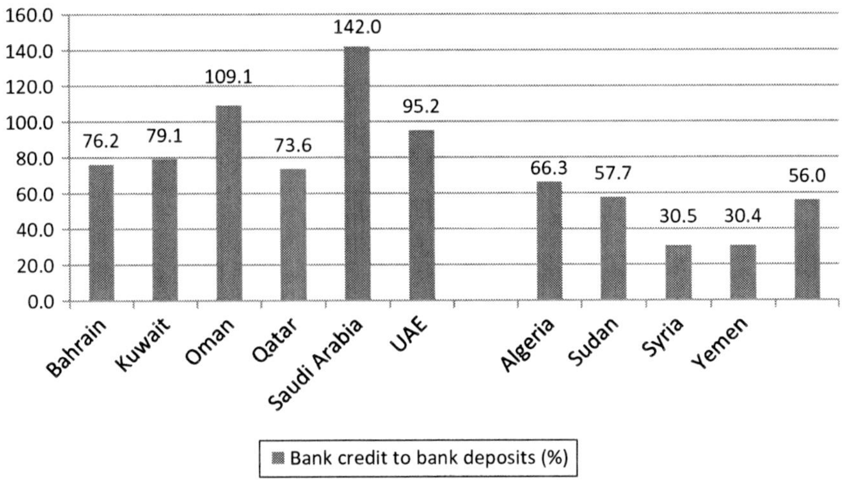

Figure 11.19: Loan-to-Deposit Ratios in Resource-rich Arab Countries, 1980–2011
Source: Global Financial Development Database, 2014.

finding is not surprising given that the GCC benefit from large deposit bases and an average loan-to-deposit ratio of almost 100% (Figure 11.19). This performance is in line with the emergence of several GCC capitals (Bahrain, Dubai and Qatar) as financial centers, with Bahrain leading in syndicated lending and cross-border banking. In the populous group, the average is much lower, at 14% of GDP, reflecting lower incomes and larger volumes of public sector financing. Algeria is more financially developed than other countries of the populous group. In particular, private sector credit in Sudan and Yemen is less than 7% of GDP.

11.5 Empirical Evidence: Testing the Conditional Resource Curse Hypothesis

This section undertakes an empirical investigation of resource-rich economies in order to assess the impact of natural resources on growth. Moreover, it examines how the inclusion of institutions affects this relationship. An augmented growth model (Mankiw et al. 1992) is chosen because it allows including a host of variables, such as institutional and human capital variables, and most importantly natural resources, in addition to including the traditional factors of production of the classical model (Solow 1956).

Growth regressions are estimated for a sample of twenty-two countries from the Arab region for the period 1960–2012 using different econometric techniques namely ordinary least squares (OLS) and panel estimations (fixed effects, FE)[7].

The model can be written as follows:

$$Y_{it} = A_0 H_{it}^{\alpha} K_{it}^{\beta} L_{it}^{\gamma} O_{it}^{\eta} \tag{1}$$

where Y_{it} is the real GDP in country i at year t, A_0 a scale parameter, H_{it} human capital, K_{it} physical capital, L_{it} land and O_{it} natural resources rents. Equation 1 represents a Cobb–Douglas production function where natural resources can be bundled together with either labor or capital in the production process. The exponents in the aggregate production function (1) denote factor shares – hence, the structure of the economy – while the factor inputs are absolute quantities.

By transforming (1) into a log-linear equation, the specification of the regression becomes:

$$\ln(Y_{it}) = \alpha_0 + \alpha \ln(H_{it}) + \beta \ln(K_{it}) + \gamma \ln(L_{it}) + \eta \ln(O_{it}) + \varepsilon_{it} \tag{2}$$

where α_0 is the constant term and ε is the discrepancy term.

Based on the pioneering work of Mankiw et al. (1992), who show that international differences in income per capita are best captured by an augmented Solow model, output is produced using physical capital, human capital and labor. All regressions include the following basic variables:

- the initial GDP (ln GDP80) to control for conditional convergence or the fact that, ceteris paribus, poor countries should grow faster than rich ones because of decreasing returns to scale in production,
- real investment (ln INV) to measure physical capital,
- the population growth rate (Popgrowth) and the secondary enrolment rate (School) to measure human capital,
- arable land (ln LAND) to have an comprehensive production function,
- net exports (Net exp./GDP) to account for the impact of trade openness on growth,
- the share of natural resources rents to GDP (Nat. res. rents/GDP),

[7] See Appendix A for the list of the countries.

- a dummy variable (Oil boom), that takes the value of 1 in oil boom years and 0 otherwise, is added to assess the effect of periods of high oil prices on growth,[8]
- a dummy variable (Oil exporter) that takes the value of 1 for oil exporters and 0 otherwise is included to examine the impact of oil on growth,
- an interaction term between natural resources and population growth. This interaction should capture how the impact of natural resources on growth would change for populous countries relative to other countries[9].

Results of the panel estimations are reported in Tables 11.8 and 11.9. Table 11.8 ignores institutional factors while Table 11.9 incorporates them. The first column of the tables displays the results of the OLS estimation and the second column present respectively the fixed-effects estimations.[10] In Table 11.8, the left panel reports the basic regressions with the basic variables explained earlier while the right panel accounts for financial development measures by private sector credit (Priv. Credit). In Table 11.9, the left panel reports results after adding the quality of institutions measured by the polity index (Polity), while the right panel reports the results using an alternative measure, the Freedom House (FH) index. The former index best captures democratization compared to the FH measure since it relies on objective questions rather than subjective classifications (Elbadawi et al 2011).

The results can be summarized as follows. They remain robust across estimation methods and econometric specifications.

First, the empirical evidence generally confirms that economic growth increases with improvements in education, investment and trade openness (columns 1–2), which is broadly consistent with previous results from the growth literature. Conforming to the prediction of Solow's model that resources are likely to be depleted by a larger population, the impact of population growth on economic growth is negative and significant.

Second and most importantly, the results provide strong support to the conditional resource curse hypothesis, which corroborates results from the recent literature in Collier and Goderis (2007) and Elbadawi and Soto

8 Oil booms are defined as years where oil prices were above the average of the whole period.

9 Please see Appendix B for variable sources and description.

10 We are unable to perform sub-sample estimations for both groups due to the small number of observations.

Table 11.8: *Empirical Results of the Growth Equation*

	Basic		With Private Credit	
	OLS ln (GDP)	FE ln (GDP)	OLS ln (GDP)	FE ln (GDP)
ln (GDP 80)	0.8365***		0.7870***	
	(0.1314)		(0.1344)	
ln (Inv)	0.6254***	0.6254***	0.5706***	0.5706***
	(0.0247)	(0.0247)	(0.0302)	(0.0302)
ln (Land)	−0.2286***	−0.2286***	−0.1860***	−0.1860***
	(0.0571)	(0.0571)	(0.0602)	(0.0602)
Popgrowth	−0.0250**	−0.0250**	−0.0256***	−0.0256***
	(0.0097)	(0.0097)	(0.0096)	(0.0096)
School Enr.	0.0080***	0.0080***	0.0088***	0.0088***
	(0.0008)	(0.0008)	(0.0008)	(0.0008)
Net Exp/GDP	0.0135***	0.0135***	0.0134***	0.0134***
	(0.0010)	(0.0010)	(0.0011)	(0.0011)
Nat. res./GDP	−0.0096***	−0.0096***	−0.0084***	−0.0084***
	(0.0017)	(0.0017)	(0.0021)	(0.0021)
Oil boom	0.0574***	0.0574***	0.0475**	0.0475**
	(0.0219)	(0.0219)	(0.0219)	(0.0219)
Pop. Gro*Nat.res	−0.0008	−0.0008	−0.0011	−0.0011
	(0.0008)	(0.0008)	(0.0008)	(0.0008)
Oil exp.				
Priv. Credit			0.0014**	0.0014**
			(0.0006)	(0.0006)
Priv. Credit *Nat.			0.0000	0.0000
			(0.0000)	(0.0000)
Constant	−7.7739***	12.8349***	−5.9603**	13.3320***
	(2.3943)	(1.0785)	(2.5228)	(1.0872)
Observations	263	263	259	259
R-squared	0.9942	0.9583	0.9945	0.9604
Number of code		11		11

Source: Constructed by the authors.
Note: (i.) ***, ** and * represent respectively statistical significance at the 1%, 5% and 10% levels.
(ii.) Standard errors between brackets.
(iii.) OLS stands for ordinary least squares, FE for fixed effect estimation and RE, random effect estimation.

(this volume). Initially, results in Table 11.8, which ignored institutional factors, generally suggested that resource rents have a negative effect on economic growth, even after controlling for initial income, trade policy, investment and schooling, which is consistent with Sachs and

Table 11.9: *Empirical Results of the Growth Equation*

	With Polity		With Freedom House	
	OLS ln (GDP)	FE ln (GDP)	OLS ln (GDP)	FE ln (GDP)
ln (GDP 80)	0.6036***		0.8180***	
	(0.1683)		(0.1342)	
ln (Inv)	0.4998***	0.4998***	0.5694***	0.5694***
	(0.0383)	(0.0383)	(0.0307)	(0.0307)
ln (Land)	−0.0907	−0.0907	−0.2204***	−0.2204***
	(0.0780)	(0.0780)	(0.0610)	(0.0610)
Popgrowth	−0.0269**	−0.0269**	−0.0238**	−0.0238**
	(0.0108)	(0.0108)	(0.0096)	(0.0096)
School Enr.	0.0093***	0.0093***	0.0086***	0.0086***
	(0.0009)	(0.0009)	(0.0008)	(0.0008)
Net Exp/GDP	0.0109***	0.0109***	0.0131***	0.0131***
	(0.0013)	(0.0013)	(0.0011)	(0.0011)
Nat. res./GDP	−0.0025	−0.0025	−0.0275**	−0.0275**
	(0.0040)	(0.0040)	(0.0119)	(0.0119)
Oil boom	0.0820***	0.0820***	0.0481**	0.0481**
	(0.0244)	(0.0244)	(0.0219)	(0.0219)
Pop. Gro*Nat.res	−0.0014	−0.0014	−0.0010	−0.0010
	(0.0011)	(0.0011)	(0.0008)	(0.0008)
Oil exp.	0.1684			
	(0.1775)			
Priv. Credit	0.0015	0.0015	0.0016**	0.0016**
	(0.0009)	(0.0009)	(0.0007)	(0.0007)
Priv. Credit *Nat	0.0000	0.0000	0.0000	0.0000
	(0.0001)	(0.0001)	(0.0000)	(0.0000)
Polity	0.0692***	0.0692***		
	(0.0252)	(0.0252)		
Polity*Nat. res.	−0.0027**	−0.0027**		
	(0.0011)	(0.0011)		
Freedom			0.0575	0.0575
			(0.0531)	(0.0531)
Freedom*Nat. res.			0.0113*	0.0113*
			(0.0065)	(0.0065)
Constant	−1.3347	13.6742***	−6.4449**	13.7379***
	(3.2177)	(1.3218)	(2.5291)	(1.0945)
Observations	221	221	253	253
R-squared	0.9912	0.9565	0.9947	0.959
Number of code		10		11

Source: Constructed by the authors.
Note: (i.) ***, ** and * represent respectively statistical significance at the 1%, 5% and 10% levels. (ii.) Standard errors between brackets.

Warner (1995). This result is in line with results from growth models that include a non-renewable resource with depletable stock. In these models, scarce natural resources tend to cause diminishing returns to inputs of both factors of production, capital and labor and thereby, may lead to economic stagnation in the long run. It is worthwhile to note that this effect of slower growth is more likely to affect the populous oil-exporters more quickly than in GCC countries, as the latter enjoy much more resources and smaller populations. Moreover, the coefficient of the oil boom shows that high oil prices are associated with higher growth rates and vice versa.

However, by including the quality of institutions, the results in Table 11.9 show that the polity index has a positive and significant effect on growth. Moreover, the interaction term with resource rents yields a strong and negative estimated parameter, indicating that even when the quality of institutions improves, natural resources still have a negative (but slightly lower) effect on growth. This result shows that the adverse effect natural resources have on growth dominates the effect of better institutions. One potential explanation for this is that existing institutions in most resource-rich countries (and in particular, Arab countries) are so weak that it would take a tangible improvement in institutions in order to ameliorate the management of natural resources and offset their negative impact on growth. Figure 11.20 graphically illustrates this result by showing that the marginal effect of natural resources with political institutions on growth is negative as institutions improve. Similarly, when institutions are proxied by the Freedom House index, their effect on growth becomes insignificant but the interaction with resource rents is positive and significant (since the higher the Freedom House index, the poorer the political environment)[11].

Finally, when the share of private credit to GDP is added, it is found to have a weak but positive and significant impact on growth. Empirical results show that there is no clear consensus on the effect of financial development on economic growth in the MENA region because of the failure of financial liberalization and poor legal environment (De Gregorio and Guidotti 1995 and Loayza and Ranciere 2005). Yet, this effect becomes insignificant when private credit is interacted with resource rents. This

[11] One could claim that poor political institutions are a result of oil wealth (and not the opposite), which could lead to a bias in our findings. Yet, we believe that this claim could be downplayed due to two reasons. First, weak political institutions in the region seem to have pre-dated the discovery of oil in most of the countries of our sample (Galal and Selim 2012). Second, country experiences (Norway and Chile) have shown that better institutions could lead to improved management of natural resources and therefore, reduce their negative effect on growth

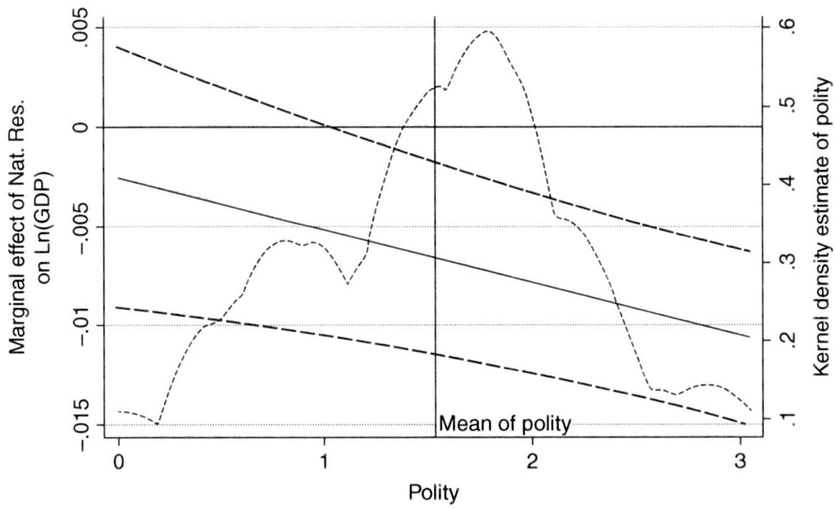

Figure 11.20: Marginal Effect of Natural Resources Rents with Political Institutions on Growth

Source: Constructed by the authors.

Notes: (i) This graph shows the marginal effect of the interacted variable using the Grinter package in Stata. It works with any equation with a linear index. In this case, if the index is Growth = b1 + b2*NaturalRes +b3*Polity + b4*NaturalRes*Polity, this command can graph either b2 + b4*Polity or b3 + b4*NaturalRes. It is worthwhile to note that, here, we graph polity with the marginal effect of natural resources on growth (b3 + b4*NaturalRes). The graph includes a vertical line at the mean value of the other constitutive term (polity) comprising the interaction.

(ii) Thick dashed lines gives 90% confidence interval for the interaction term.

(iii) Thin dashed line is a kernel density estimate of the Political Institutions variable.

result holds for both estimations, with and without the inclusion of institutions. This is in line with Beck (2011), who shows that there is no differential effect of natural resource wealth on the effect of financial development on economic growth in resource-rich versus resource-poor economies.

11.6 Better Institutions in Order to Avoid the Curse

The main policy insight from this chapter is that better institutions are important in avoiding the resource curse. Moreover, the interdependence between politics and resource management is the essence of the resource curse where politics affects exploitation of resource wealth and in turn, rents influence politics (Collier 2010). The analysis suggests that for Arab countries to achieve sustained growth and prosperity, they must adopt meaningful political reforms, encompassing a strong system of political checks and balances. Though this remains a challenge, it is hoped that the adherence to the Extractive Industries Transparency Initiative (EITI) and the National Resource Charter (NRC) have the potential of triggering some reforms on political accountability and more transparent and effective management of natural resources.

Assuming progress is made on this front, countries should first start addressing *sustainability issues,* which means transforming their exhaustible natural resources into assets – human, domestic and private capital and foreign financial assets – that will generate future income and support sustained development. Addressing sustainability goes *hand in hand with reforming fiscal institutions.* Without sound fiscal frameworks, Arab countries would be constantly exposed to the risk of easily squandering their natural resource wealth. It is hoped that the following fiscal reforms would contribute to higher savings, larger and more efficient investments, put public finances on a sustainable track and increase the countries' ability to adopt counter-cyclical fiscal policies. This section attempts to draw some policy recommendations that would contribute to better management of their resources in order to reap the benefits of their wealth. These recommendations are related to the management of fiscal policy, increasing the efficiency of public spending, the adoption of more transparency with respect to fiscal policy as well as public procurement.

11.6.1 Reforms for Counter-cyclical Fiscal Policies

Ample literature recommends that resource-rich economies should smooth out consumption well beyond the period of peak resource and stock precautionary savings to ensure inter-generational equity. Collier (this volume) explains that natural resources will deplete more rapidly in the populous group and so, despite being much poorer, they need higher savings than the GCC, or around 50% of hydrocarbon fiscal revenues starting 2013, which would need to rise to 100% by exhaustion in 2043, or alternatively by nearly two percentage points each year. Meanwhile, the GCC countries, which benefit from a longer time horizon to exhaustion, need a lower initial savings rate – 30% – meaning they would need to increase their savings by one percentage point each year in order to reach 100% by 2083.[12] In reality, GCC

[12] This illustration is based on the assumption that the GCC have a horizon until exhaustion of oil and gas revenues of seventy years, while the populous countries have a horizon of thirty years.

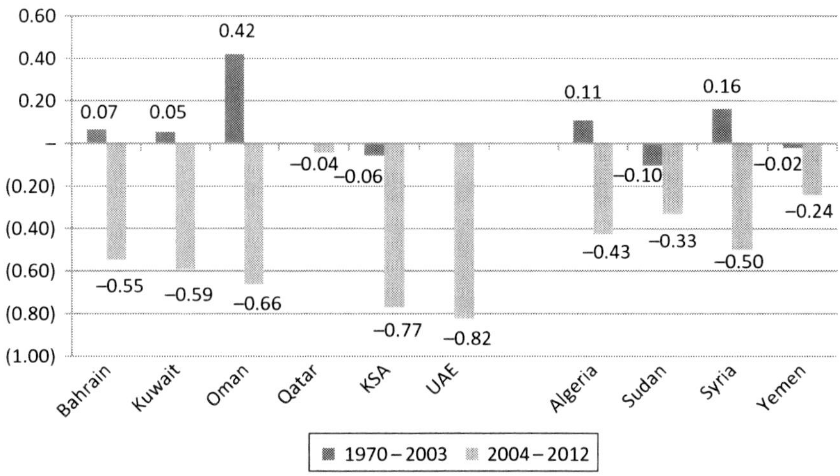

Figure 11.21: Country Correlations between the Cyclical Components of Real Government Spending and Real GDP, 1970–2012

Notes: The cyclical components have been estimated using the Hodrick–Prescott Filter. Real government expenditure is deflated by the GDP deflator. A positive (negative) correlation indicates pro-cyclical (counter-cyclical) fiscal policy. Data for Bahrain is available since 1980 only, Qatar since 1994 and the UAE since 2000. Data for Sudan and Syria is available until 2008 and 2007 respectively.

Source: Calculated by the authors based on data from the WDI and WEO databases.

economies are saving more than the populous group: around 48% of GDP and this ratio goes up to 60% in Kuwait and Qatar. However, savings accounted for only 23% of GDP in the populous group and as low as 11% in Sudan.

In this context, fiscal policy entails the daunting task of managing resource price volatility. Counter-cyclical fiscal policies could help governments save their wealth, prevent them from overspending during oil booms and also ensure that an unexpected decline in oil revenue does not lead to abrupt changes in government spending with adverse macroeconomic effects. In practice, however, pro-cyclicality has been a feature of fiscal policy in oil-exporting countries, as supported by the empirical analysis of Villafuerte and Lopez-Murphy (2010). Arab resource-rich countries are no exception (Abdih et al. 2010). Figure 11.21 depicts the correlation between (the cyclical components of) government spending and GDP for the ten countries between 1990 and 2012. All countries have witnessed an important shift in the cyclical behavior of their fiscal policy over the last decade, which used to be pro-cylical in the 1970s until the

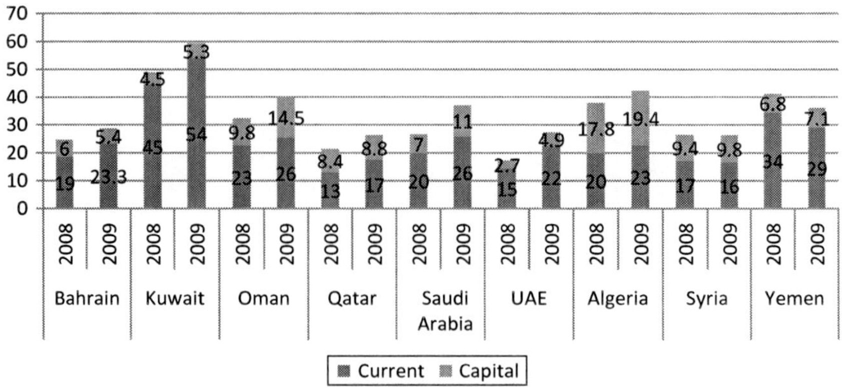

Figure 11.22: Composition of Public Spending in 2008 and 2009, Percent of GDP
Source: Calculated by authors based on the dataset from Villafuerte and Lopez-Murphy (2010).

early 2000s and became counter-cyclical afterwards. It is worthwhile to note that Qatar's fiscal policy remains almost neutral relative to economic activity. Moreover, fiscal policy in the GCC group seems to have performed better than in populous countries where it remains more subservient to the economic cycle.

There is also recent evidence of counter-cyclicality in government spending with many countries adopting fiscal stimulus packages in the aftermath of the 2008 financial crisis. Nevertheless, most fiscal expansions took place through substantial increases in current spending, which exceeded increases in capital outlays (Figure 11.22). In Bahrain, capital expenditure as a share of GDP actually declined. In contrast, Oman implemented the largest capital spending stimulus, from 10% to 14.5% of GDP. In Syria, fiscal policy was neutral, whereas in Yemen, the contraction in overall spending suggests a pro-cyclical pattern.

Several institutional reforms could contribute to counter-cyclical fiscal policies but the appropriate design depends on the resource horizon for which natural resources can be expected to generate revenues.

For countries with long resource horizons, like the GCC countries, the main challenge should be to manage revenue price volatility. Collier (this volume) judges that the creation of *Sovereign Wealth Funds (SWF)* that would save or invest a proportion of resource revenues in financial assets abroad *is sufficient to ride out revenue volatility* by varying accumulation of foreign financial assets, which have few opportunities for domestic investments.

In reality, GCC countries already possess fourteen different SWFs, some of them established as early as 1953, like the Kuwaiti SWF. Most of the smaller GCC SWFs, however, appeared after 2000, when oil prices soared. Assets accumulated by SWFs of the GCC countries are estimated to account for over half of SWF assets globally and their activities continue to expand very rapidly after the crisis, on the back of rising oil prices. Unfortunately, the limited availability of information with respect to their geographical composition or asset portfolio composition makes it hard to assess their performance or to know if and how much the fraction they save of oil revenues with the purpose of generating long-term returns for future generations is. The largest Saudi SWF mostly invests in secure, liquid but low-yield United States bonds, which are used for short- to mid-term stabilization and sterilization operations (Kohler 2013). Nevertheless, it is judged that compared to the oil booms of the 1970s, the GCC countries are being much more strategic about where they place their petrodollars. They have adopted ambitious investment strategies that include huge spending on domestic infrastructure, creating free-trade zones for manufacturing, and developing services and facilities that will attract increasing numbers of businesses, skilled knowledge workers and tourists. They are also making huge financial investments in the developed world, China, India and even Africa (Abdelal et al. 2008).

A second reform is *the adoption of fiscal rules* that account for year-on-year fluctuations in resource prices and thus contribute to smoothing revenues and allow the authorities to determine how much of their resource revenues they can safely spend through the annual budget. In addition, they could potentially mitigate discretionary interventions by governments (Daniel et al. 2013). The GCC would benefit from setting a target for the structural budget balance on the basis of an estimate of the long-term oil price. This indicator would net out the cyclical impact of the level of economic activity and hydrocarbon prices that affect central government income. Thus, the structural balance equals structural revenues plus interest on net government assets minus actual expenditures on goods and services. It thus reflects the financial balances that the central government would have shown if GDP had been at its trend level and hydrocarbon prices had been running at their long-term level. Qatar is the first GCC economy to have initiated steps to formally adopt a medium-term budget framework to enhance the predictability of spending decisions and link its medium-term development plans to the budget.

Countries of the populous groups are immediately confronted with shorter resource horizons and greater uncertainty about production volumes with drastic implications on public spending sustainability once resources are depleted. Moreover, they have so far not been successful in effectively deploying their rents to build a broad-based domestic capital base. In addition, they have so far saved very little. Moreover, political instability in some countries and the violent conflicts that afflict some countries like Sudan and Yemen would tend to limit their access to international financial markets for borrowing purposes. To address this fundamental and pressing challenge, *countries of the populous group need both* a *SWF* that would save or invest a small proportion of resource revenues in external financial assets abroad, and a *Sovereign Development fund (SDF)* that would invest the majority of their wealth in domestic infrastructure either in the form of physical assets or in human capital through improving the healthcare and education of citizens (Collier, this volume). Moreover, these countries would need to closely monitor their public finances. They could benefit from *setting a target for the non-resource fiscal balance* (which is an indicator of the capacity of the economy to absorb the resource revenues without causing inflation and a large current account deficit) at a level that can be maintained after resource revenues run out. This would avoid the need for abrupt breaks in government expenditures or tax increases after the natural resources have been depleted, given that the non-resource fiscal balance gradually converges to the overall balance as resource revenues decline (Daniel et al. 2013). So far, Algeria is the only populous country which owns an oil stabilization fund, "Fonds de Regulation des Recettes (FRR)." However, the FRR can be freely drawn upon for budget support, so that expenditure is disconnected from the saving rule.

11.6.2 Improving the Efficiency of Public Spending

It is often argued that the use of SDFs to finance domestic infrastructure holds significant challenges. Administrative systems may lack the technical ability to scale-up expenditure rapidly (to identify, implement, budget as well as monitor and evaluate projects) and coordinate among government entities. In low-capacity and low-governance environments, like those present in the populous group, this may lead to wasteful expenditure and budget fragmentation, among other things. In addition, even though the SDF also has the potential of acting as a means to improve the quality of public spending, and even to crowd in private investors to strengthen investment discipline, it could become a vehicle for politically driven "investments" in a context of elite capture (Gelb et al. 2014). In order to bypass this problem and ensure that public spending is allocated towards high-quality public investment projects, *a more gradual increase in spending may be advisable for the populous group* with an initial focus on investing resources to remove existing bottlenecks – a process sometimes called "investing in investing," as defined by Collier et al. (2010).

To improve the efficiency of public spending, it is essential to have strong public financial management systems, including the ability to provide reasonable forecasts of resource revenues and the capacity for medium-term budgeting, which would help with the annual budget planning process and identify fiscal risks on both the revenue and expenditure sides.

11.6.3 *Transparency about Fiscal Policy and Public Procurement*

In addition, *more transparency with respect to fiscal policy and public procurement is also warranted.* In particular, fiscal transparency in the collection and utilization of natural resource revenues through appropriate accounting, reporting and auditing is important. Meanwhile, in terms of public procurement, it is better if governments solicit interest from oil multinational companies to extract resources through open and transparent processes, including competitive bidding for exploration and extracting. In order to increase transparency, it is important that Arab resource-rich economies follow good practices, including the clear assignment of roles and responsibilities of different government entities, establishment of an open budget process, having publicly available information and providing assurances of data integrity. Fiscal transparency could be a good means to combat corruption. The EITI provides a global standard for transparency in the oil, gas and mining industries, while the NRC, which builds on the transparency initiative, offers more comprehensive principles for governments and societies on how to best harness the opportunities for development generated by extractive commodity windfalls. Those initiatives can serve as anchors for enhancing transparency and accountability in commodity-rich countries. More specifically, open publication of public procurement contracts can help improve investment quality and reduce contract costs and cost overruns (Kenny and Karver 2012).

11.7 Conclusion

Natural resources are a critical component of the Arab World's export and government revenues. They have contributed to better standards of living and have most certainly widened the set of policy choices: countries did do better than they would have done without resource rents. Notwithstanding this improvement, the economic performance of Arab resource-rich countries has been disappointing, even in the absence of a counterfactual.

This chapter has argued that the resource curse in the Arab World is primarily an "institutional curse," even though it has several macroeconomic manifestations. In other words, politics affects economics or alternatively, it affects how resource rents are collected, allocated and used. Weak institutions in the Arab World have pre-dated the resource discovery and have had adverse implications on macroeconomic management. At the same time, natural resources have consolidated the weak institutional set-up. Over time, the interaction between these two factors became intertwined, preventing these countries from embarking on a sustainable development path. The empirical model confirms that resource rents have a negative effect on economic growth. More interestingly, on their own, political institutions may not always have an impact on growth. However, when interacted with natural resources, they could reduce the negative effect of natural resources on growth.

The analysis has attempted to draw a distinction between the highly resource endowed GCC economies and the populous economies. While bad governance is responsible for their lack of development, the manifestations of the curse are distinct. Indeed, they are more subtle in the GCC where the sheer amounts of rent per capita have fostered regime stability or a form of durable authoritarianism. To this end, rents have been efficiently redistributed through labor markets to national citizens in the form of well-remunerated public sector jobs and other generous social welfare schemes, which have preserved social stability. The absence of significant political unrest in all GCC economies over the past three years of turmoil in the region speaks for that. In contrast, the populous group with their low rent per capita, and despite the recent discovery of a modest natural resource wealth, has experienced much more dire consequences such as conflict, violence and social unrest at some point in time.

The macroeconomic manifestations of the curse are also expressed through different channels. On the one hand, the GCC group's growth performance is more severely affected by resource volatility, but it has managed to diversify its non-resource economy in recent years, even though the share of manufacturing still remains below predicted levels. GCC countries have engaged in excessive public consumption but high rents per capita still allow them to allocate ample resources towards investment, particularly in infrastructure and have maintained the financial sector quite liquid; the latter has been extending credit to the private sector at growing rates. If one would pick a "best performer," it would be UAE, which has certainly not been immune to the curse, but has managed to make the benefits outweigh the negative outcomes, a conclusion that echoes with Haouas and Soto (this volume). Meanwhile, Kuwait seems to lag behind most GCC economies. It has very low investment ratios comparatively, a persistent exchange rate overvaluation and a small manufacturing sector.

On the other hand, limited resources may have somewhat shielded growth in populous economies from resource volatility, but have contributed to devastating economic consequences. Oil busts drove the populous economies to engage in excessive borrowing and booms seem to have almost eliminated their manufacturing sectors. Unlike the GCC countries, they have failed to use their natural capital to develop the required physical capital to promote the much needed economic diversification of their economies. In a context of low rent per capita, excessive consumption resulted in massive deficiencies in infrastructure investments and an underdeveloped financial sector. In fact, the three least endowed economies, being Syria, Sudan and Yemen, suffer from most of the macroeconomic symptoms of the curse.

Large oil resources, like those present in the Arab World, are a blessing. In order to benefit from them, governments have an important role to play in how these resources are used. The analysis suggests that for resource-rich Arab countries to avoid the oil curse and achieve sustained growth and development, they must introduce effective political reforms accompanied by a strong system of political checks and balances. It is hoped that stronger political institutions would trigger reform in macroeconomic institutions in general and fiscal institutions in particular, thereby improving the management of natural resources, achieving more savings and implementing more effective public spending programs. These reforms are expected to release resources for the diversification of the non-resource sector. While GCC governments have become increasingly aware of the need for economic diversification in recent years, the main challenge they face is to re-orient their economies toward greater private sector employment of nationals. Controlling public sector employment should be accompanied by raising educational quality and developing skills to raise productivity in order to make nationals more competitive for private sector hiring. As for the populous countries, they need to initiate an aggressive program to encourage economic diversification. To this end, governments should focus on simplifying the complex doing business procedures and on improving infrastructure. Financial sector reform, especially of state-owned banks, is also crucial in order to allocate resources to their most productive use and finance diversification. Many of these countries may have to readjust their overvalued exchange rates which are undermining export competitiveness.

APPENDIX A: List of Countries

Algeria	Malta
Bahrain	Morocco
Djibouti	Oman
Egypt	Qatar
Iran	Saudi Arabia
Iraq	Sudan
Israel	Syria
Jordan	Tunisia
Kuwait	United of Arab Emirates
Lebanon	West Bank and Gaza
Libya	Yemen

APPENDIX B: Variables Description

Variable	Definition	Source
Dependent Variable		
ln(GDP)	Real gross domestic product (constant 2005 US$).	WDI
Independent Variables		
ln(GDP 80)	Real gross domestic product (constant 2005 US$) for 1980.	WDI
ln(Inv)	Gross fixed capital formation (constant 2005 US$).	WDI
ln(Land)	Arable land (hectares).	WDI
Pop. growth	Population growth (annual %).	WDI
School Enr.	School enrollment, secondary (% gross).	WDI
NetExp/GDP	Exports of goods and services (% of GDP) minus imports of goods and services (% of GDP).	WDI

Variable	Definition	Source
Nat.res./GDP	Total natural resources rents (% of GDP).	WDI
Priv. credit.	Domestic credit to private sector refers to financial resources provided to the private sector by financial corporations, such as through loans, purchases of non-equity securities, and trade credits and other accounts receivable, that establish a claim for repayment.	WDI
Oil boom	A dummy variable that takes the value of 1 in oil boom years (from 2004 to 2013) and 0 otherwise.	Constructed by the authors
Oil exp.	A dummy variable that takes the value of 1 for oil exporters and 0 otherwise.	Constructed by the authors
Polity	An index based on an evaluation of that state's elections for competitiveness, openness and level of participation. For each year and country, a "polity score" is determined which ranges from -10 to +10, with -10 to -6 corresponding to autocracies, -5 to 5 corresponding to anocracies and 6 to 10 to democracies.	Polity Project IV
Freedom	An index based on an annual survey of media independence that assesses the degree of print, broadcast and internet freedom throughout the world. It provides numerical rankings and rates each country's media as "Free," "Partly Free," or "Not Free." Individual country narratives examine the legal environment for the media, political pressures that influence reporting and economic factors that affect access to information	Freedom House

References

Abdelal, R., A. Khan and T. Khanna. (2008). Where Oil-Rich Nations are Placing their Bets. *Harvard Business Review* 86(9): 119–128.

Abdih, Y., P. Lopez-Murphy and R. Sahay. (2010). The Cyclicality of Fiscal Policy in the Middle East and Central Asia: Is the Current Crisis Different? IMF Working Paper 10/68. Washington DC.

Ali, O. and I. Elbadawi. This volume. The Political Economy of Public Sector Employment in Resource Dependent Countries.

Al-Abri, A. This volume. Labor Market Heterogeneity and Optimal Exchange Rate Regimes in Resource-rich Arab Countries.

Al-Sheik, H. and S.N. Erbas. This volume. The Oil Curse and Labor Markets: The Case of Saudi Arabia.

Andersen, L. (1987). The State in the Middle East and North Africa. *Comparative Politics* 20(1):1–18.

Arezki, R. and M. Brückner. (2011). Oil Rents, Corruption, and State Stability: Evidence from Panel Data Regressions. *European Economic Review* 55(7): 955–963.

Auty, R. M. (1993). *Sustaining Development in Mineral Economies: The Resource Curse Thesis*. London: Routledge.

Baseday, M. and J. Lay. (2009). Resource Curse or Rentier Peace? The Ambiguous Effects of Oil Wealth and Oil Dependence on Violent Conflict. *Journal of Peace Research* 46(6): 757–776.

Beblawi, H. G. L., eds. (1987). *The Rentier State: Nation, State and the Integration of the Arab World.* London: Croom Helm.

Beck, T. (2011). Finance and Oil: Is There a Resource Curse in Financial Development?" In *Beyond the Curse: Policies to Harness the Power of Natural Resources.* R. Arezki, T. Gylfason and A. Sy, eds., Washington, DC: International Monetary Fund.

Bellin, E. (2004). The Robustness of Authoritarianism in the Middle East: Exceptionalism in Comparative Perspective. *Comparative Politics* 36(2): 139–159.

Collier, P. This volume. Savings and Investment Decisions from Natural Resource Revenues: Implications for Arab Development.

(2010). The Political Economy of Natural Resources. *Social Research* 77(4): 1105–1132.

Collier, P., F. van der Ploeg, M. Spence and A. Venables. (2010). Managing Resource Revenues in Developing Economies.IMF *Staff Papers* 57(1): 84–118.

Collier, P. and A. Hoeffler. (2009). Testing the Neocon Agenda: Democracy in Resource-rich Societies. *European Economic Review* 53(3): 293–308.

(2004). Greed and Grievance in Civil War. *Oxford Economic Papers* 56(4): 563–595.

Collier, P. and B. Goderis. (2007). Commodity Prices, Growth, and the Natural Resource Curse: Reconciling a Conundrum. Economics Series Working Papers WPS/2007-15, University of Oxford, Department of Economics.

Corden, W. (1984). Booming Sector and Dutch Disease Economics: Survey and Consolidation. *Oxford Economic Papers* 36: 359–380.

Daniel, P. S. G., T. Mattina, and A. Segura-Ubiergo. (2013). *Extracting Resource Revenue: Finance and Development*. Washington, D.C.: IMF.

De Gregorio, J. and P. Guidotti. (1995). Financial Development and Economic Growth. *World Development* 23(3, March): 433–448.

Diop, N. and J. de Melo. This volume. Dutch Disease in the Services Sector: Evidence from Oil Exporters in the Arab region.

Diop, N., D. Marotta and J. de Melo. (2012). Natural Resource Abundance, Growth, and Diversification in the Middle East and North Africa: The Effects of Natural Resources and the Role of Policies. Washington, D.C. World Bank.

Elbadawi, I. and H. Selim. This volume. Overview of Context, Issues and Summary.

Elbadawi, I. and R. Soto. This volume. Resource Rents, Political Institutions and Economic Growth.

Elbadawi, I., S. Makdisi and G. Milante. (2011). Explaining the Arab Democracy Deficit: the Role of Oil and Conflicts. In *Democracy in the Arab World: Explaining the Deficit*. I. Elbadawi and S. Makdisi, eds., New York and Canada: Routledge and International Development Centre.

Freedom House. (2013). Freedom in the World data. Available at: www.freedomhouse.org/report-types/freedom-world. Accessed August 2014.
Galal, A. and H. Selim. (2013). The Elusive Quest for Economic Development in the Arab Countries. *Middle East Development Journal* 5(1): 1–29.
Gelb, A., S. Tordo, and H. Hallan. (2014). Sovereign Wealth Funds and Domestic Investment in Resource-rich Countries: Love Me, or Love Me Not? World Bank Economic Premise Note Series 133. World Bank: Washington, D.C.
Gylfason, T. (2004). Natural Resources, Education, and Economic Development. *European Economic Review* 45(4–6): 847–859.
Gylfason, T. and G. Zoega. (2006). Natural Resources and Economic Growth: The Role of Investment. *The World Economy* 29(8): 1091–1115.
Haber, S. and V. Menaldo. (2011). Do Natural Resources Fuel Authoritarianism? A Reappraisal of the Resource Curse. *American Political Science Review* 105(1): 1–26.
Haouas, I. and R. Soto. This volume. Has the UAE Escaped the Oil Curse?
Hausmann, R. and R. Rigobon. (2003). An Alternative Interpretation of the 'Resource Curse': Theory and Policy Implications. NBER Working Paper 9424, Cambridge, Mass.
Herb, M. (2005). No Representation without Taxation? Rents, Development, and Democracy. *Comparative Politics* 37(3): 297–316.
Hertog, S. (2012). Redesigning the Distributional Bargain in the GCC. Paper presented at the BRISMES Annual Conference 2012 on "Revolution and Revolt: Understanding the Forms and Causes of Change". London. 26–28 March.
Hodson, N. (2011). Distributional Structures in the Arab Gulf: Public Employment, Subsidies, and Potential Reform. Sciences Po Kuwait program publication.
Inglehart, R. (1997). *Modernization and Post-modernization: Cultural, Economic, and Political Change in 43 Societies*. Princeton: Princeton University Press.
International Monetary Fund. (2013a). Saudi Arabia: Selected Issues. IMF Country Report 13/230. Washington, D.C.
(2013b). Sudan: 2013 Article IV Consultation. IMF Country Report 13/317. Washington, D.C.
(2013c) . Economic Prospects and Policy Challenges for the GCC Countries. Document prepared by the staff of the IMF for the Annual Meeting of Ministers of Finance and Central Bank Governors. Riyadh. October 5.
(2013d). Labor Market Reforms to Boost Employment and Productivity in the GCC. Document prepared by the staff of the IMF for the Annual Meeting of Ministers of Finance and Central Bank Governors. Riyadh. October 5.
Jensen, N. and L. Wantchekon. (2004). Resource Wealth and Political Regimes in Africa. *Comparative Political Studies* 37(7): 816–841.
Kenny, C. and J. Karver. (2012). Publish What You Buy: The Case for Routine Publication of Government Contracts. CGD Policy Paper 011. Washington DC: Center for Global Development
Kohler, P. (2013). GCC Sovereign Wealth Funds: Leveraging Local Economic Development? United Nations Development Policy and Analysis Division, Department of Economic and Social Affairs, World Economic Situation and Prospects Policy Briefing. May.

Leite, C. and J. J. Weidmann (1999). Does Mother Nature Corrupt? Natural Resources, Corruption and Economic Growth. IMF Working Paper 99/85. Washington. D.C.

Loayza, N. and R. Ranciere. (2005). Financial Development, Financial Fragility, and Growth. IMF Working paper 05/170. Washington. D.C.

Mankiw, G., D. Romer and D. Weil. (1992). A Contribution to the Empirics of Economic Growth. *Quarterly Journal of Economics* 107(2): 407–437.

Manzano, O. and R. Rigobon. (2001). Resource Curse or Debt Overhang?NBER Working Papers 8390. Cambridge, Mass.

Melhum, H., K. Moene and R. Torvik. (2006). Institutions and the Resource Curse. *Economic Journal,* 116(January): 1–20.

Van der Ploeg, F. and S. Poelhekke. (2009). Volatility and the Natural Resource Curse. *Oxford Economic Papers* 61(4): 727–760.

Robinson, J., R. Torvik and T. Verdier. (2006). Political Foundations of the Resource Curse. *Journal of Development Economics* 79(2): 447–468.

Robinson, J. And R. Torvik. (2005). White Elephants. *Journal of Public Economics* 89(2–3): 197–210.

Rodrik, D. (2009). Industrial Policy: Don't Ask Why, Ask How. *Middle East Development Journal* 1(1): 1–29.

2008. The Real Exchange Rate and Economic Growth. *Brookings Papers on Economic Activity* 39(2): 365–439.

Ross, M., K. Kaiser and N. Mazaheri. (2011). The "Resource Curse" in MENA? Political Transitions, Resource Wealth, Economic Shocks, and Conflict Risk. World Bank Policy Research Working Paper 5742. Washington, D.C.

(2004). Does Taxation Lead to Representation? *British Journal of Political Science* 32 (4): 229–249.

(2001). Does Oil Hinder Democracy? *World Politics* 53(3), 325–361.

Sachs, J. and A. Warner. (1995). Natural Resource Abundance and Economic Growth. NBER Working Paper 5398. Cambridge, Mass.

Salti, N. (2008). Oil Greasing the Wheels: When Do Natural Resources Become a Blessing? ERF Working Paper 439. Cairo.

Schmidt-Hebbel, K. This volume. Fiscal Institutions in Resource-rich Economies: Lessons from Chile and Norway.

Smith, B. (2004). Oil Wealth and Regime Survival in the Developing World, 1960–1999. *American Journal of Political Science* 48(2): 232–246.

Solow, R. (1956). A Contribution to the Theory of Economic Growth. *The Quarterly Journal of Economics* 70(1): 65–94.

Suliman, K. This volume. Understanding and Avoiding the Oil Curse in Sudan.

Tilly, C. ed. (1975). *The Formation of National States in Western Europe.* Princeton: Princeton University Press.

Tornell, A. and P. R. Lane. (1999). The Voracity Effect. *American Economic Review* 89(1): 22–46.

Villafuerte, M. and P. Lopez-Murphy. (2010). Fiscal Policy in Oil Producing Countries During the Recent Oil Price Cycle. IMF Working Paper 10/18. Washington, D.C.

12

Has the UAE Escaped the Oil Curse?

Raimundo Soto* and Ilham Haouas**

ABSTRACT

The UAE is blessed with vast deposits of oil and gas. Contrary to other oil-rich economies, the UAE seems to have escaped from the so-called oil curse. We study how the UAE used resource rents to achieve economic development and provide higher welfare for the local population. We identify, nevertheless, symptoms of the resource curse in three areas: very low growth in labor productivity, government policies unable to counteract economic cycles induced by oil price volatility, and massive over-employment and declining productivity in the public sector. Therefore, we conclude that while the UAE has not been immune to the oil curse, it has managed to make the benefits outweigh the negative outcomes of oil exporting. We finally study the case of Dubai as an example of how to overcome the dependency on oil exports and diversify the economy by using a combination of market deregulation, support for foreign trade and efficient provision of infrastructure and institutions for private sector participation.***

JEL Classification: Q32, O41, F14

Keywords: Natural resources, oil curse, Arab region, UAE, economic growth, export diversification

* Soto: Universidad Catolica de Chile. ** Haouas: Abu Dhabi University.

*** The opinions in this chapter do not reflect that of the institutions of affiliation of the authors and are of their sole responsibility. Damanick Dantes provided excellent research assistance. Comments and suggestions by I. A. Elbadawi, R. Radwan, M. Babweva and R. Vazquez-Alvarez are gratefully acknowledged.

12.1 Introduction

The UAE is a small federative country formed in 1971 in the Gulf of Arabia, comprising seven originally independent emirates (Abu Dhabi, Dubai, Sharjah, Ajman, Fujeira, Umm al Quwein and Ras al-Khaimah). Like other countries in the region, the UAE is blessed with vast deposits of oil and gas. As of 2011, proven oil reserves amounted to 7% of the world deposits, making the UAE one of the richest economies in the world. At the current extraction rate, known oil reserves will last for another ninety-four years (BP 2011).

Oil was found in the UAE[1] in the early 1960s, but it was not until the oil price hikes of 1973 and 1978 that hydrocarbon became the main export and a dominant sector of the economy. In the subsequent thirty years, oil has been the backbone of the economy, the main source of exports and foreign currency and the driving force of the government revenues that have been used both to finance infrastructure and other public goods and to transfer wealth to the local population. The prosperity of the UAE and its transformation from a backwards region to a booming economy has been made possible by revenue from oil exports. However, and contrary to other oil-rich economies, in particular the populous oil-rich Arab economies, the UAE seems to have escaped the so-called "oil curse" (i.e., the often-encountered paradoxical case of countries that have been blessed by abundant natural resources mismanaging their wealth, remaining underdeveloped and maintaining a poor population) (Sachs and Warner 1995). This is the topic of this chapter.

Undoubtedly, income levels for the local population have improved significantly after the 1970s. However, and as discussed here, several of the resource curse symptoms are currently present in the UAE, indicating that the country has not been immune to the disease, but has somehow managed to make the benefits of exporting oil outweigh the negative outcomes. Contrary to most resource-rich economies, and evidencing substantial foresight on the part of the founding fathers of the UAE, the government has made a conscious and systematic effort to invest a significant amount of resources in accumulating two forms of capital that are

[1] The first oil concession in Abu Dhabi was granted in 1939 to Trucial Coast Oil Development Company, which was awarded exploration rights to the whole of Abu Dhabi's land area and part of its offshore for a seventy-five-year period. The company discovered its first commercial oilfield at Bab in 1960, and that was followed by finds at Bu Hasa in 1962, Asab in 1965 and Sahil in 1972.

crucial for sustained economic growth: physical capital – in particular, public infrastructure – and institutional fabric. The UAE provides an example of a country where abundant natural capital has not crowded-out "social capital," by which is meant the infrastructure and institutions of a society in a broad sense: its culture, cohesion, law, system of justice, rules and customs and so on, including trust (Woolcock 2001). The authorities have been also quite aggressive against corruption. Insofar as natural resource abundance involves public allocation of access to scarce common property resources to private parties without payment, thereby essentially leaving the resource rent up for grabs, it is only to be expected that resource-rich countries may be more susceptible to corruption than others.

The success of the UAE in using resource rents for economic development and higher welfare, nevertheless, does not rule out the fact that some of the symptoms of the resource curse are present. The most significant indication of such problems is in the stagnation of productivity levels throughout most of the last twenty-five years. Production in both oil and non-oil sectors of economic activity have grown largely fueled by massive investment and an even higher use of manpower, but the efficiency in the use of such factors has remained stagnant and, in some areas, steadily declined. On the contrary, emerging countries in East Asia and Latin America have accumulated comparatively less resources but learned to use them more efficiently. This has allowed their industries to be competitive in international markets. UAE industries, on the other hand, have yet to become competitive in international markets (with the exception of niche business such as gold processing or logistics that perform at world standards): it remains to be seen whether they can survive and prosper with current labor-intensive production technologies.

We use simple growth models to decompose economic growth according to its sources: physical capital accumulation, use of manpower, development of the productive skills of such human power and efficiency at which these resources are employed. Our aim is to answer the following set of questions: How has oil extraction shaped the development of the UAE economy? Is there any evidence of an oil curse and, if so, is it significant? What has been the role of macroeconomic policies in inhibiting or fostering the oil curse? And does the UAE offer an answer to the oil curse that could be replicated in other countries affected by the malaise?

In dealing with these questions we have tried to provide as much quantitative answers to the aforementioned questions as possible. We were limited by the pervasive lack of statistical data for the UAE, perhaps the clearest sign of an "oil curse": whenever countries do not face stringent budget constraints, it appears to authorities that policies can be decided in the absence of information and that there is no need for hard data.

Section 12.2 of the chapter provides a brief summary of the emergence, causes and effects of the oil curse. Following recent literature (Gylfason 2011; Elbadawi and Gelb 2010), we identify four channels of transmission through which abundant resource rents can distort the workings of an economy to the point where the income benefit of such rents is outweighed by the costs of the distortions they create. Our aim is to provide a general framework for the analysis and evaluate their importance for the UAE, a country where oil rents have been substantial for the past forty years.

Section 12.3 provides a long-run perspective of the UAE economy, with emphasis on three macroeconomic aspects. First, we study long-run economic growth for the period 1975–2010 and compare the performance of the UAE vis-à-vis the rest of the world, in particular Arab countries and high-income economies. Second, we document the importance of oil rents in achieving the current high income levels of the UAE, as well as the role of oil in foreign trade and government revenues. Third, we study the sources of economic growth [i.e., we investigate to what extent the economic growth observed between 1975 and 2010 is the result of massive investment (the accumulation of physical capital)], the use of an ever-increasing number of workers (reflected in the continuous inflow of expatriates) or in the productive efficiency of such resources (total factor productivity growth).

Section 12.4 studies macroeconomic policies and their relation with oil rents. Because the exchange rate in the UAE is pegged to the US dollar and the capital account is quite open, the authorities cannot pursue an independent monetary policy. Fiscal policy, therefore, remains as the main policy tool for steering the economy in the short run and influencing investment decisions to sustain growth in the long run. We study how oil rents and oil price volatility have influenced the conduct of monetary policy and its ability to provide for the correct resource allocation, to stabilize business cycles and to incentivize productivity gains in both the public and private sector.

In Section 12.5, we exploit the heterogeneity of growth experiences among the different emirates to discuss alternative development strategies that could counterbalance the oil curse. We study the case of Dubai, an emirate that was forced to change its development strategy in the mid-1980s after realizing that the oil bonanza was rapidly disappearing, and derive insights on what is needed to ameliorate or avoid the negative aspects of oil rents altogether. Section 12.6 concludes.

12.2 The Oil Curse Hypothesis

The oil curse, or more amply defined the *resource curse* hypothesis, comprises several different theories that have been offered as explanations for the shocking evidence that, frequently, countries endowed with abundant natural resources suffer from long-standing economic malaises that inhibit their economic and social development (Sachs and Warner 1995). As with physical diseases, the resource curse can be the result of different causes and manifest itself in a variety of symptoms. In what follows, we briefly review the different mechanisms operating under the resource curse and the most common symptoms, evaluating their relevance in the case of the UAE.

In general terms, natural resources by themselves are not a problem; most countries are endowed with them although in varying quantities and qualities. It is the size of the rents that can be derived from them – and the few hands in which they are frequently concentrated – that largely determines the unveiling or avoidance of a resource curse. Rents can be defined as the difference between the price received for the sale of one unit of the natural resource-based commodity a country exports and the long-run costs of placing such a unit in international markets. In most cases, such rents exist but are small vis-à-vis the size of the economy. In a few cases – notably in oil and gas – the magnitude of such rents can be sizable, itself the result of the geographical concentration of the deposits and the presence of significant scale economies. That is precisely the case of Arab countries and, in particular, the UAE.

Table 12.1 provides a snapshot of resources rents in the world in the last forty years, the only period for which reliable statistics are available. For each country and year, the World Bank computes resource rents as the difference between natural resource prices and extraction costs (the latter underestimates costs as it omits exploration, development and other potentially significant costs) multiplied by the volume of resources extracted and divided by total GDP.[2] It can be seen that at the world level, resource rents amount to around 3.5% of GDP in the whole period and fluctuates only moderately. We take the world average as a reference point and conclude that resource rents are significant – and resource curses are most likely to appear – in Latin America and the Caribbean, Sub-Saharan Africa and the Middle East and North Africa. Note that on average,

[2] See World Bank (2011).

Table 12.1: *Average Natural Resource Rents* (As Share of GDP, Percent)

	1970s	1980s	1990s	2000s
World	3.8	4.3	2.0	3.9
By Geographical Area				
• East Asia and Pacific	2.9	3.4	1.2	2.5
• Europe and Central Asia	0.8	1.5	1.5	2.8
• Latin America and Caribbean	5.8	8.8	4.2	8.0
• Middle East and North Africa	33.7	26.7	20.2	32.6
• North America	3.4	3.2	1.1	1.7
• Sub-Saharan Africa	9.7	11.7	8.6	15.1
• *Arab World*	35.6	31.8	22.9	36.1
By Income Levels				
• High income	2.7	2.9	1.1	1.9
High income: OECD	1.9	2.0	0.7	1.2
High income: non OECD	37.8	31.6	16.2	26.7
United Arab Emirates	**65.3**	**40.6**	**31.4**	**31.5**
• Middle income	9.1	10.9	6.8	10.7
Upper middle income	8.4	10.7	6.6	10.5
Lower middle income	10.8	11.5	7.2	11.5
• Low income	4.5	4.1	4.7	5.5
• Heavily indebted poor countries (HIPC)	6.5	6.3	6.2	10.0

Source: Own elaboration based on data from World Bank *World Economic Indicators 2011.*

resource rents in Europe, North America and Asia tend to be very small. Of course, within these geographical areas, individual countries can enjoy significant rents (e.g., Norway in Europe or Indonesia in Asia).

The table also describes the size of resource rents according to degree of development using per capita income levels as proxy.[3] Three elements are apparent. First, in general, high resource rents are not linked to high income levels: in fact, among all income groups, developed economies have the lowest participation of resource rents in aggregate income. Indeed, it is among the middle-income economies where resource rents are most significant. Second, for individual countries, resource rents can provide significantly high income levels but not necessarily high development levels: note that while resource rents are very low in high-income OECD economies (less than 2% on average in the period 1970–2010), these

[3] Income levels are defined by the World Bank based on annual GNI per capita. In 2010, countries were classified as low (less than US$ 1,000), lower middle income, (US$ 1,006 to US$ 3,975), upper middle income (US$ 3,976 to $ 12,275) and high income (more than US$ 12,276).

tend to be very high in high-income countries that do not belong to the group of developed economies (28%). The latter includes the UAE and some other oil-and-gas exporting countries such as Saudi Arabia, Qatar and Trinidad and Tobago. Third, being blessed by significant natural resources requires – but countries often fail to achieve – good management of such resources: note that on average, the highly-indebted, poor countries of the world have significant resource rents, equal to those in middle-income economies, yet their economies are so distressed that the international organizations have made these countries the focus of special debt assistance under the HIPC initiative. It is in these countries that the resource curse manifests more clearly, but in middle and high income, non-OECD countries, the symptoms of the disease can also be easily identified.

Finally, Table 12.1 indicates that among those countries that have been able to achieve high per capita income levels on the basis of the abundance of natural resources, the UAE has been especially fortunate. Resource rents in the UAE are among the highest in the world, in line with the other oil-exporting countries of the Arab region such as Saudi Arabia, Qatar and Kuwait, and have undoubtedly allowed very high income levels (if not for the entire population, at least for the local population[4]). In this sense, oil has been a blessing for the UAE. However, and as discussed here, several of the resource curse symptoms are actually present in the UAE, indicating that the country has not been immune to the disease, but that it has somehow managed to make the benefits outweigh the negative outcomes of exporting oil.

Gylfason (2011) provides a stylized version of the different theories that have been provided as explanations for the apparent curse that resource rents place on economic development. There are, at least, four main channels of transmission from natural resource abundance or intensity to slow economic growth. These explanations tend to center around the notion that natural capital crowds out other types of capital (physical, human, social or institutional) and thereby inhibits economic growth. Elbadawi and Gelb (2010) provide an excellent taxonomy of the symptoms of the curse. Although they center on oil, and consequently their research is highly valuable for this chapter, most of their insights would

[4] The distinction between the locals and the expatriates (non-national immigrants) is important in the UAE. The expatriate group is multiple times that of the locals. The latter group enjoys a variety of preferential treatment in the form of higher salaries, access to restricted residential land and subsidies on utilities and other consumer goods.

apply to other non-renewable natural resources, such as copper, gold, diamonds or bauxite.

First, abundance of natural resources, and therefore, the ample supply of foreign currency associated with their export to world markets, could induce the so-called Dutch Disease (Corden and Neary 1982). Under this theory, the significant inflow of currency – particularly if it occurs in a short period of time – could induce a swift appreciation of the local currency thereby reducing the profitability of other economic sectors and shrinking the productive basis of the economy. Naturally, part of the currency appreciation reflects an equilibrium adjustment subsequent to the fact that the economy is now richer than it was before. Nevertheless, it is frequently the case that the currency overvaluation of resource-rich countries far exceeds any long-term equilibrium appreciation. Countries that have had fixed currencies (e.g., such as the UAE where the Dirham is pegged to the US dollar) are less affected by nominal appreciation of their currencies, yet the real appreciation can also be substantial via increases in wages and the cost of capital and land. Beyond the currency appreciation, Dutch Disease also operates in a more subtle manner: natural resource abundance is, more often than not, accompanied by increased economic volatility. The prices of raw materials tend to fluctuate widely in world markets (vis-à-vis the price of manufactured goods). The resulting fluctuations in export earnings trigger exchange rate volatility and create uncertainty that can be harmful to exports and other trade, including foreign investment. Dutch Disease tends to bias the composition of exports away from those kinds of manufacturing and service exports that may be particularly good for growth over time.

Second, significant natural resource rents in conjunction with institutional weaknesses (including ill-defined property rights, imperfect or missing markets and lax legal structures) may lead to rent-seeking on the part of producers, thus diverting resources away from more socially profitable economic activity (Gelb 1988). In extreme cases, civil wars break out for the control of such rents which not only divert away factors of production from socially productive uses, but also destroy institutions and the rule of law. In less extreme cases, the struggle for resource rents may lead to hoarding of economic and political power in the hands of elites that, once in power, would use the rents to placate their political supporters and thus secure their hold on power, with slow growth as a result, as shown in Elbadawi and Soto (2011). Extensive rent-seeking breeds corruption, distorts the allocation of resources and reduces both economic efficiency and social equity (Krueger 1974). Finally, abundant resource rents may imbue people with a false sense of security and lead governments to lose sight of the need for good and growth-friendly economic management, including free trade, bureaucratic efficiency and institutional quality (Sachs and Warner 1999). Incentives to create wealth through good policies and institutions may wane because of the relatively effortless ability to extract wealth from the soil or the sea.

A third channel of the curse arises when resource rents reduce private and public incentives to accumulate human capital due to a high level of non-wage income or when wages are disconnected from the effort of workers. Resource-rich nations may be tempted to underestimate the long-run value of education. Empirical evidence shows that, across countries, school enrolment at all levels is inversely related to resource rents (Gylfason et al. 1999). There is also evidence that, across countries, public expenditure on education relative to national income, expected years of schooling and school enrolment are all inversely related to natural resource abundance (Gylfason 2001). The abundant natural capital appears to crowd out human capital, which forms the basis of sustained economic growth.

The fourth and final channel operates through saving, investment and the accumulation of physical capital. High resource rents usually fuel consumption booms and tend to depress national saving. In turn, this leads to lower than optimal accumulation of capital, in particular in non-oil sectors (Arezki et al. 2011). Bhattacharyya and Collier (2012) also note the negative effects of oil resources on the stock of public capital. Furthermore, financial funds are frequently provided directly by the governments of resource-rich countries on a non-competitive basis, often leading to low-return investments and retarding the development of financial institutions, which ought to identify and fund the most productive projects on a social basis. Unproductive investments may seem unproblematic to governments when they are flush with oil cash.

As will be discussed in the following sections of this chapter, since 1971 the UAE has made significant efforts to escape the oil curse by following policies deliberately aimed at accumulating capital stocks beyond those needed to exploit its rich endowment of natural resources. On the one hand, the federal government has maintained a fixed peg to the US dollar which, along with small and easily controllable fiscal deficits, has reduced the risks of currency overvaluation and its deleterious impact on private investment. In addition, an important fraction of the huge oil earnings received by Abu Dhabi have been invested abroad in a sovereign wealth fund (with over 0.5 trillion US dollars in 2010, according to the

Sovereign Wealth Fund Institute), bringing security to foreign and domestic investors vis-à-vis the sustainability of the currency. To a large extent, the Dutch Disease related negative effects of exporting oil have been avoided in the UAE. As will be shown in Section 12.4, currency misalignment, if anything, has not been significant.

On the other hand, the federal government, as well as the government of each emirate, has made significant efforts to develop social capital and support private sector initiatives. As a result, the UAE is among the top countries in the world in terms of road infrastructure, ports, airports and telecommunications (see World Economic Forum 2011). Likewise, institution building since 1971 has notably progressed in areas where private investment and sustainable economic growth are usually fostered. In particular, the country ranks high in control of corruption, rule of law and absence of crime and, contrary to most countries in the region, is not afflicted by open political conflicts or terrorism.

Finally, rent-seeking, while existent, does not seem to be rampant or capable of interfering massively with the market allocation of resources. While it is true that the mismanagement of government-owned businesses has been blamed for the recent crisis in the real estate markets of Dubai and Abu Dhabi, it is also true that these government holdings had been instrumental in developing sectors that, in the 1980s and 1990s, were disregarded by the private sector as being too risky or too costly to finance (e.g., tourism and commercial aviation).

Consequently, if the resource curse would manifest itself in the UAE, it ought to be through other channels of transmission (i.e., either through low physical and human capital accumulation or though an inefficient use of such resources). This is the analysis we formally undertake in Section 12.3.

12.3 Oil Exports, Economic Growth and Productivity in the UAE

As mentioned, oil was discovered in the early 1960s in the Abu Dhabi region. The discovery prompted the seven emirates of the UAE to form a federation to capitalize on the oil windfall. Since 1971, the country has completely transformed and modernized. Figure 12.1 shows the extraordinary expansion of the UAE economy during the past thirty-five years.[5] Between 1975 and 2010, output expanded by a factor of five in real terms

[5] The Appendix identifies the data sources and definitions used in computing economic growth and its sources.

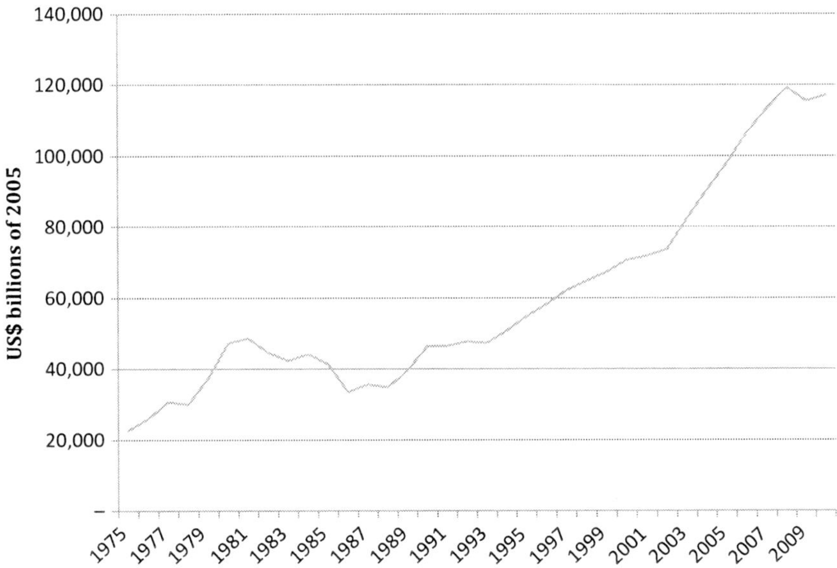

Figure 12.1: UAE: Real GDP (at PPP prices)
Source: World Development Indicators 2011.

(at PPP prices). The average annual growth rate in the entire period is 5.4%, among the highest in the world. It can be seen, nevertheless, that the pace of economic growth has not been smooth, with periods of very fast growth (as in the early 2000s) and major crisis occurring in the early 1980s and late 2000s. Real GDP grew relatively faster in the period 1975–1990 at over 5.7% per year, while in the subsequent twenty years it grew at around 5.2%.[6]

Few countries in the world have been able to sustain such vigorous growth pace for an extended period of time. Table 12.2 compares UAE's performance to the world and selected countries such as Hong Kong and Singapore. It can be seen that in the period 1975–2010, the UAE has grown much faster than the world economy (almost twice as fast) or the average economy at any development stage (from low- to high-income levels). This is indeed the positive outcome of oil wealth and adequate government

[6] As in most MENA economies, official statistics in the UAE are very poor and data is scarce, incomplete and frequently inconsistent. We have most carefully assembled the data we use in this chapter, but it remains limited so that our conclusions should be taken with healthy skepticism.

Table 12.2: *Real GDP Annual Growth Rate (%)*

	1975–2010	1975–1990	1991–2010
World	3.0	3.3	2.7
• High income	2.6	3.2	2.1
• High income: OECD	2.6	3.2	2.1
• High income: non-OECD	5.0	5.4	4.7
• Middle income	4.4	3.8	4.9
• Upper middle income	4.3	3.6	4.9
• Lower middle income	4.7	4.5	4.9
• Low income	3.4	2.6	4.1
• **United Arab Emirates**	**5.4**	**5.7**	**5.1**
• Hong Kong	5.8	7.9	4.0
• South Korea	6.4	8.0	5.2
• Singapore	7.2	7.5	6.9
• China	9.6	8.5	10.5
• India	5.9	5.1	6.6

Source: World Development Indicators 2011.

policies. Note also that the UAE has outgrown other high-income, non-OECD economies although not by a very significant margin. Most of these economies are, like the UAE, oil or gas exporters with abundant resources and small populations (although countries such as Cyprus, Croatia and Malta are also in this group).

The growth record of the UAE, nevertheless, is not that impressive when compared to the East Asian tigers, the benchmark for high, sustained growth of the twentieth century. It can be seen that Hong Kong, Singapore and South Korea have grown much faster than the UAE and for a sustained period of time, reaching development levels that are among the highest in the world and far ahead of the UAE. China and, more recently, India have also managed to grow at impressive rates but remain still behind the UAE in terms of development due to their very low starting points.

Undeniably, in the case of the UAE, oil has been the major source of economic activity, exports and government finance since the early 1970s. Even if one recognizes the significant contribution to total output of Dubai – a non-oil producing emirate which accounts for around 30% of total GDP and is the financial and trade center of the Gulf region– economic activity and government projects remain largely influenced by oil production and exports. Production of hydrocarbons (oil and gas) commanded a very high share of GDP in the mid-1970s, reaching around

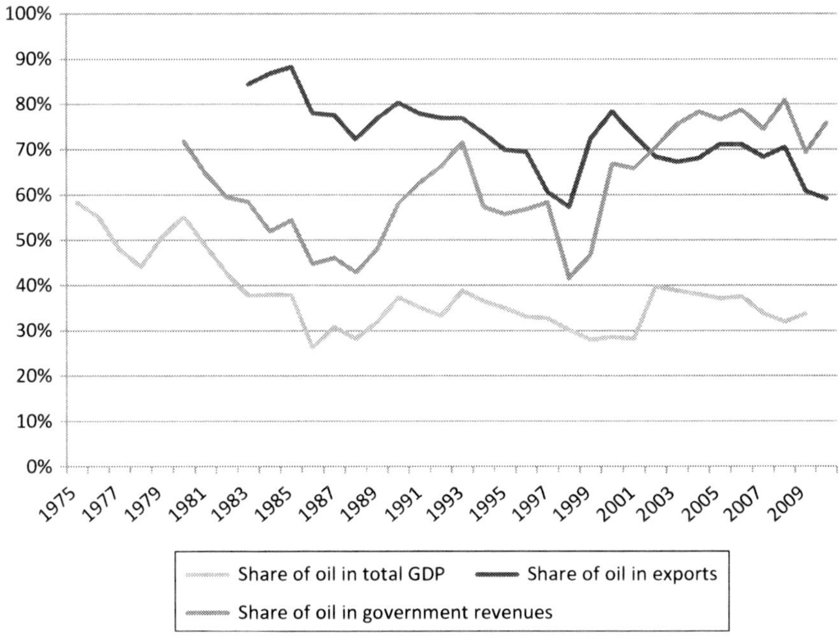

Figure 12.2: UAE: Share of Oil in GDP, Exports and Government Revenues
Source: Own elaboration based on data from the UAE Bureau of Statistics.

60% of the total activity. This is hardly surprising: the UAE was a young country where most industries had yet to be developed and oil production was far ahead of other sectors. In the 1980s and 1990s, other industries developed quite rapidly – particularly in services – and therefore, the importance of hydrocarbons declined noticeably as shown in Figure 12.2. In the period of 1990–2010, the share of value-added derived from oil extraction in total GDP hovered around 35% with minor variations. The share of oil extraction in GDP belittles the influence of oil in the economy as there is a myriad of activities that are linked to and depend on oil production and export (e.g., services such as engineering, transport, logistics and so on).

The share of oil in total exports, on the other hand, has remained overwhelmingly high: in the period 1990–2010, oil represented around 70% of total exports (excluding re-exports), a figure that has decreased only slightly in the last twenty years. Fluctuations are largely the result of the cyclicality in oil prices as physical exports have remained relatively stable, at least in the last decade. Dominance of oil in exports in resource-rich Middle-East countries is expected. However, what is worrisome is that in the last twenty years, no other exports have developed and become significant in the UAE. That is, despite all government efforts, diversification of exports has not been achieved.

Finally, the energy sector is not only the main contributor to economic activity, but is also, and non-surprisingly, the main source of fiscal revenue – between 1992 and 2010, oil revenues represented over 73% of total government revenues (UAE National Bureau of Statistics 2011). As will be discussed here, these revenues constitute, on the one hand, a significant amount of resources for the government to invest in capital accumulation and finance public goods, and on the other hand, a source of major fluctuations for government budgets and economic activity.

12.3.1 Sources of Growth

In order to understand what lies behind the fast growth of the UAE, we decompose GDP growth according to its "sources." The sources of economic growth in the UAE, as in any other economy, are a combination of physical capital accumulation, expansion in employment and its capacities and increases in the way in which these factors are employed (i.e., changes in total factor productivity) (Barro and Lee 1994). Figure 12.3 shows the evolution of GDP, the capital stock[7] and employment, which, for an easier exposition, we have normalized so that 1975=100. It can be seen that these factors have evolved in a very different fashion: while the capital stocks has moved in tandem with GDP during most of the period, with only two exceptions (the early 1980s and the 2000s), employment has grown significantly more than output or capital and with disregard of the economic cycles. This is a remarkable phenomenon: despite the huge natural resource capital and the massive investment in physical capital, the UAE economy has become increasingly labor-intensive.

The increasing reliance on employment has also led to a massive immigration of workers, largely from the Indian subcontinent. According to the World Bank database, between 1975 and 2010, the population of the UAE increased from half a million to almost five million.[8] Most of the immigrants have very low educational levels and are generally employed in low-skilled positions in which the highly-paid nationals are not interested

7 Computation of the capital stock of the economy is described later.

8 The population of the UAE became a hotly debated issue after the UAE Bureau of Statistics indicated that the population had reached eight million in 2009.

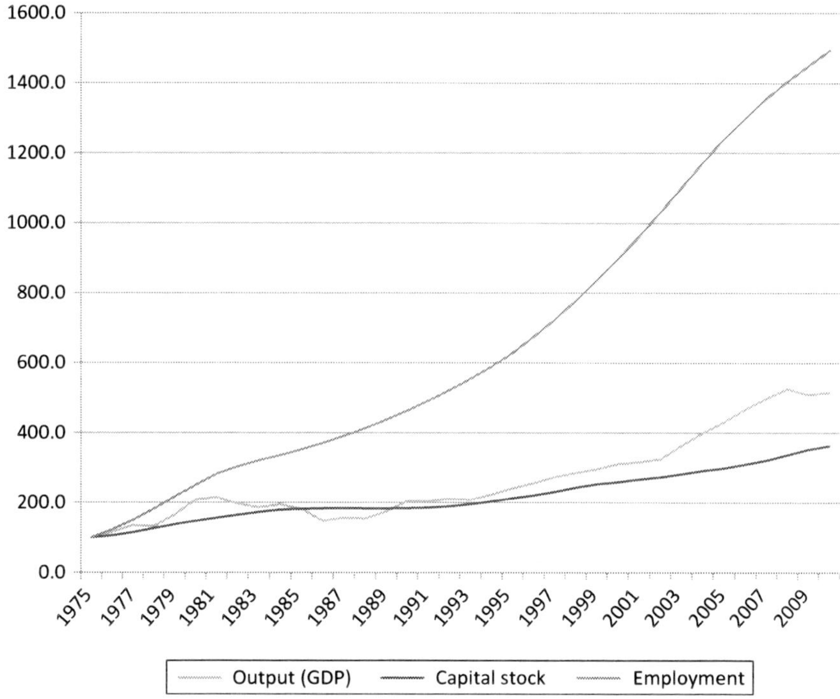

Figure 12.3: UAE: Real GDP, Capital Stock and Employment (1975=100)
Source: Own elaboration.

(e.g., construction workers). It comes as no surprise, then, to observe that output per worker has declined significantly over time. We use this variable, as opposed to GDP per capita, because it is relatively insensitive to important changes in demographic factors that characterize the development of any economy (e.g., changes in fertility rates) and also because it is immune to transient phenomena in the labor market (e.g., unemployment waves). Output per worker is directly a measure of the average productive capacity of the working force in an economy and a measure of the efficiency levels of such an economy indirectly.

Figure 12.4 shows this distinctive characteristic of the UAE's development. It can be seen that average productivity levels have remained stagnant for a long period of time (1987–2010), following a substantial decline in the early 1980s. This is a particularly worrisome feature as it indicates that economic growth in the UAE has been primarily the result of capital and labor accumulation and not of the efficiency in the use of production

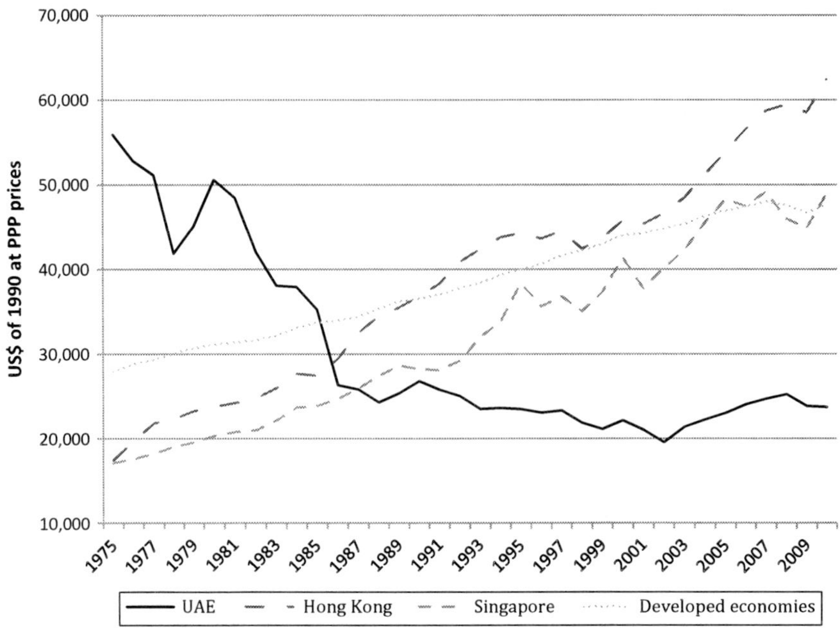

Figure 12.4: Average Labor Productivity
(Real GDP per Worker)
Source: World Development Indicators 2011.

factors. Even if one is willing to discard the early data on the grounds that GDP and employment measures in the 1970s and 1980s were not very precise, the stagnation of the 1990s and 2000s is a signal of deep-rooted problems in the development strategy of the UAE.

The evolution of average labor productivity in the UAE is in stark contrast to that of developed economies or the East Asian tigers. As shown in Figure 12.4, average labor productivity has grown slowly but systematically in developed economies, while it has grown much faster yet with more instability in Hong Kong and Singapore. As of 2010, average labor productivity levels in the UAE are one half of those in the developed economies and Singapore, and one third of those in Hong Kong. Evidently, the UAE significantly lags behind in development vis-à-vis other high-income economies.

Understanding the reasons for the stagnation in average labor productivity becomes an essential task to understand the working of the UAE's economy and determining to what extent it derives from the impact of oil rents becomes important for policy reasons. Average labor productivity is a useful indicator but, by itself, provides little information on the underlying causes of changes in productivity. A simple yet very useful exercise in this regard consists of decomposing economic growth according to its aforementioned sources in order to determine by how much each factor has contributed. Estimating the contribution of different factors to economic growth – in particular, economic and social policies – is not straightforward (see Easterly 2001; Rodrik 2005). As a first approach, we use growth accounting to answer the following questions: What portion of the performance of the economy can be accounted for by differences in inputs of factors, like capital and labor? What portion can be accounted for by differences in the efficiency with which these factors are used?

When computing the sources of growth, we follow Solow (1956) and use a simple, aggregate Cobb–Douglas production function of the form:

$$GDP_t = A_t\mu_t \mathrm{KS}_t^{\alpha}\left(L_t^{\beta}\mathrm{HK}_t^{1-\beta}\right)^{1-\alpha} \tag{1}$$

where KS_t is the stock of capital, L_t is the use of the labor force and HK_t is the stock of knowledge or human capital. Variable A_t is an indicator of the efficiency in the use of factors and μ_t is an indicator of the occupation rate of resources. The combination of the latter two elements is popularly known as total factor productivity or TFP. Parameters α and β are constants.

We compute TFP as:

$$TFP_t = A_t\mu_t = \frac{GDP_t}{\mathrm{KS}_t^{\alpha}\left(L_t^{\beta}\mathrm{HK}_t^{1-\beta}\right)^{1-\alpha}} \tag{2}$$

Our definition of TFP, therefore, encompasses not only technological capacity, but also the efficiency in the use of labor, human capital and physical capital. In this view, several elements could affect factor productivity beyond the technical ability to mix inputs and generate goods and services. For example, poor government regulation leading to lower use of capital and, thus, lower production is interpreted as declining TFP. On the other hand, an improvement in the education and training of the labor force is interpreted as increasing TFP. This interpretation of TFP links naturally with the analysis of long-run economic growth.

To calculate TFP, given series for GDP and employment, we need to choose a value for α and β and generate series for KS and HK. We chose a value of the capital share for growth accounting of $\alpha = 0.45$ for two reasons. First, there is a growing consensus among researchers that a share in the

0.3–0.45 range is adequate (see Gollin 2002 for an empirical analysis). We have chosen the top of the range to acknowledge the fact that oil is a capital-intensive industry that dominates GDP. Second, a high capital share implies an implausibly high long-run rate of return on capital. With $\alpha = 0.75$, as suggested by national accounts,[9] and an average capital/GDP ratio of 2.6 for the period 1975–2010, the annual return on capital (real interest rate) should be around 36%.[10] As for parameter β, we choose a value of 0.25, following Bernanke and Gürkaynak (2001).

To calculate a capital stock series, we cumulate gross fixed capital formation or investment, I_t, using the perpetual inventory method:

$$\mathrm{KS_t} = (1 - \delta)\mathrm{KS}_{t-1} + I_t \tag{3}$$

for some chosen depreciation rate δ and an initial condition on capital. Based on information by Bu (2006), we use a depreciation rate of 7%, and, for the initial condition on capital, we assume the capital–output ratio to be 2.5 in 1975. These assumptions are inconsequential for our long-run analysis.

Measuring human capital is not direct; we use the educational attainment of the labor force. In methodological terms, our measure corresponds to estimating human capital with reference to a stream of past investments instead of future earnings or individual characteristics (see Stroombergen et al. 2002). The benefit of our methodology is that, contrary to other methods, data is more readily available. Its main limitation is that measuring school attainment does not consider the quality of those education years. We use the data from Barro and Lee (2011) which is collected in five-year intervals from 1950 to 2010; linear interpolation was used to obtain a continuous annual series for the period 1975–2010.

Figure 12.5 plots the trajectories of TFP and GDP per working-age person. What is striking about this figure is how closely the TFP data matches the data for GDP per working-age person, both in the level and the cyclical component. The correlation between the two variables is 0.96. This suggests that it was not the changes in inputs that were mainly responsible for the evolution of GDP in UAE, but rather the efficiency with which these

[9] The UAE national accounts provide an estimate of the "compensation to employees," which on average for the period 2001–2009, amounts to around 25% of GDP (UAE Bureau of Statistics 2009).

[10] Given our technology in Equation (12.1), in equilibrium $r\mathrm{KS} = \alpha\mathrm{GDP}$, where r is the real return to capital. Given our annual data, $r = 0.29$. Adding the depreciation of 7% yields an annual real interest rate of 36%.

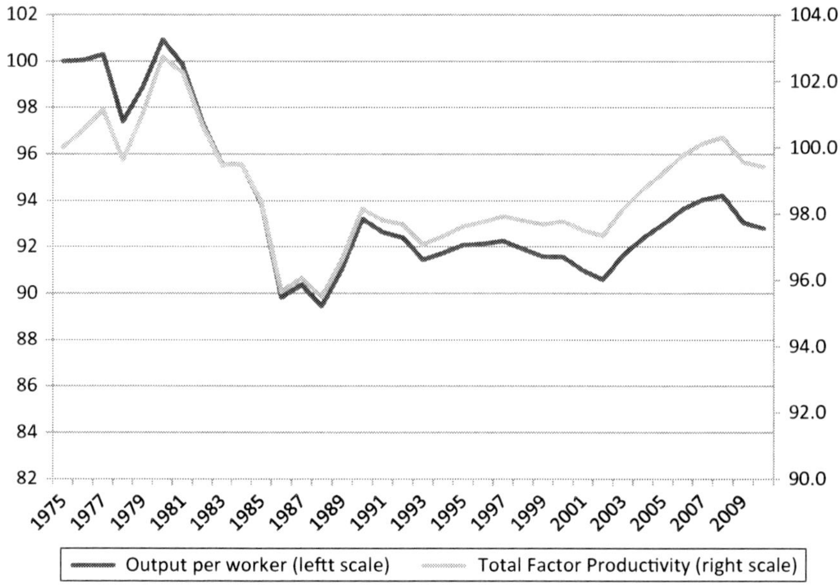

Figure 12.5: UAE: GDP per Worker and Total Factor Productivity (Normalized so that 1975 = 100)
Source: Own elaboration.

factors were used. As mentioned, this measure of TFP – which stems from Prescott's (1998) pioneering research – is richer than what standard economic theory assumed: in addition to the impact of technological advances, productivity depends on the framework in which economic agents make decisions to work, invest and consume. Consequently, TFP can be affected by the quality of macro and microeconomic policies and transient phenomena, such as commodity booms or unemployment cycles.

To some extent, the measurements of GDP, and consequently those of TFP, are sensitive to transient phenomena. In the case of the UAE, oil price shocks can be very significant. As noted by Kehoe and Ruhl (2008), terms of trade shocks do not directly taint the computation of TFP using the sources of growth method because national accounts do not register price changes (they are based on Laspeyres quantity indices). However, indirectly, they can filter through demand booms (imports and consumption via income effects). In order to control for oil shocks, we re-calculate TFP excluding the value added of the oil sector from GDP, the oil workers from employment and re-estimate capital stocks outside the oil industry. The availability of data slightly limits the period of analysis to 1987–2010,

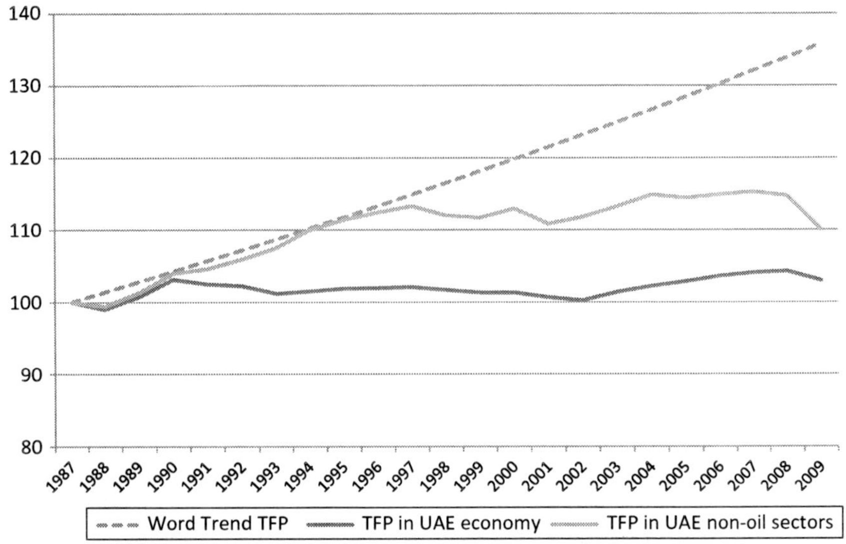

Figure 12.6: UAE: Total Factor Productivity by Economic Sector
Source: Own elaboration.

therefore, one should be careful when drawing long-run conclusions as we now operate with a reduced sample.

A somewhat different picture emerges through the results displayed in Figure 12.6. It can be seen that, effectively, once the oil sector is removed, the evidence suggests that total factor productivity increased during the period 1987–1995, although at a relatively slow pace. In fact, TFP growth in the UAE was similar to that of the high-income countries of the world (i.e., 1.4% per year), which allowed the country to keep track – but not catch-up – with more developed economies. Around 1995, however, somehow the momentum in TFP growth began to wane and productivity stagnated. It should be recalled that until the recent 2008/2009 downturn, the world economy achieved sustained growth throughout the 1990s and early 2000s. These are the years of the so-called self-moderation. Therefore, it seems unlikely that external shocks could have induced a decline in productivity growth. Without excluding external shocks altogether, this would suggest that internal policy changes could be among the causes for this phenomenon.

The different path of total factor productivity in the non-oil sectors vis-à-vis the entire economy also indicates that productivity in the oil industry has not grown in the period 1987–2010. This could be the result of mismanagement of oil companies and/or increased costs of extraction and processing. Lack of pubic data on oil management precludes us from a precise answer, but the latter hypothesis is unlikely given the fact that the marginal costs of oil extraction in the UAE are among the lowest in the world.

In conclusion, the sources of growth analysis indicate that the observed GDP growth of 5.4% per year in the period 1975–2010 has largely been the result of fast accumulation of physical capital (at around 4% per year in the same period), a lower growth in the human capital stock (at an annual rate of 3.5% on average) and an even higher accumulation of workers (8.9% per year on average). This is a second characteristic of the development strategy of the UAE: development has been based on importing and accumulating low-skilled workers. Had the UAE imported high-skilled workers, the human capital stock would have increased at a much higher rate.

Naturally, the massive inflow of workers is the result of private sector hiring policies and, to a much lesser degree, the government. The latter has restricted itself to primarily hiring Emiratis, to the point that it has become the by preference employer of the local population. In fact, employment in the government has also grown at a high pace, but much less than the private sector – between 1989 and 2010, government employment increased at over 6% per year. We return to this issue in Section 12.4.

Consequently, the massive accumulation of workers has been more pronounced in the private sector than in the public sector. This is, to a large extent, puzzling: why would entrepreneurs prefer to employ labor-intensive production techniques when they have unrestricted access to the highly sophisticated capital goods and top-notch technologies that the world economy offers? As discussed in Soto and Rashid (2016), the peculiar institutional framework of the labor market under the sponsorship or *kafala* system provides a compelling explanation. Under the *kafala* system, employers have significant market power derived from the fact that the employee is restricted to work only for the employer that sponsors him or her in the UAE (according to the law, the worker is forbidden to change occupations while the contract is in force and has to leave the country upon expiration of the contract, although contract renewals occur). The absence of horizontal mobility allows employers to extract economic rents, even if the worker is better paid than in his/her country of origin. When choosing production technologies, therefore, employers would tend to focus on labor-intensive techniques that, in addition to the normal profit obtained from selling goods and services, would allow them to extract the highest rents from the worker.

Naturally, this skews production towards labor-intensive technologies. Moreover, it biases employment towards low-skilled workers that have less ability to negotiate their salaries prior to hiring. High skilled workers are relatively scarcer and better educated than low-skilled workers and, consequently, they are better equipped to counterbalance the market power of employers. As documented in Vazquez-Alvarez (2011), wage differentials in the UAE between locals and migrants for identical job characteristics reduce markedly as employees have more education and disappear altogether at the postgraduate level.

This explanation, however, requires a connection to oil rents and low productivity growth. Abundant resource rents provide funding for the numerous investment projects required by a country, such as the UAE, that is in its formative stages. These projects generate a significant demand for labor – particularly in the scarcely populated UAE – and workers become very valuable. In a non-*kafala* environment with high mobility, workers would command high wages. In the *kafala* system, wages are kept constant for the duration of the contract. In the short run, the lack of mobility provided by the *kafala* prevents rising labor costs and keeps profits high for investors.

In the long run, however, the *kafala* system destroys the incentives to increase efficiency and the ability of firms to be profitable on the basis of their efficiency to compete for two reasons. First, resources do not freely adjust to their most efficient use and therefore there may be an inefficient allocation of workers. Second, note that labor-saving technical improvements would be disregarded by entrepreneurs as they lower rent extraction. Labor-saving technological changes will be implemented only when the cost reduction increases market profit significantly more than the sacrifice in rent extraction derived from lower employment.

In keeping the *kafala* system as a centerpiece of economic policy in the UAE, the authorities might have considered that increases in labor costs would reduce the profitability of firms and, therefore, the incentives to invest in new, more efficient technologies. An alternative view is that historically, whenever labor became scarce and wages soared, entrepreneurs would implement new technologies to replace workers by more efficient technologies based on machines. In the latter case, productivity grows only if new technologies are labor-saving, while in the former, productivity growth will only occur if new technologies are labor-augmenting or labor-complementary.

The issue, therefore, is to determine what kind of technology changes might happen in the UAE if labor markets are deregulated and wages increase. According to the data described earlier, it would seem that technology changes had been labor-saving in the past, since employment has expanded massively and labor productivity stagnated. But of course, that would depend on the technologies employed in the different economic sectors. Sectors operating with constant returns of scale, however, typically face labor-augmenting technologies so that increases in wages would discourage technological change (see Acemoglu 2009). By keeping wages low, the UAE would have induced technology adoption and productivity gains. However, that was not the case and one should conclude that, on average, technology is labor-saving. If that is the case, the *kafala* system has inhibited technological change because by keeping wages low, it has not given entrepreneurs incentives to acquire more advanced technologies.

This theory explaining the low productivity growth in the UAE centers on labor market distortions; oil rents are instrumental in generating labor demand but any other demand pull would satisfy the same conditions (e.g., copper and foreign aid). In the next section, we review the main macroeconomic policies in order to assess the other transmission channels of the oil curse.

12.4 Macroeconomic Policies and Oil Rents

There is consensus now that macroeconomic policies can have a significant impact on long-run growth, income per capita and household welfare. Monetary, exchange rate and fiscal policies largely determine the risk and profitability of human and physical capital accumulation and, at the same time, provide the government with resources to supply public goods that would support and compliment private sector activities. The UAE is no exception, of course, but its choice of policies is limited by the peg of the exchange rate and the open capital account.

As indicated earlier, the exchange rate of the UAE has been pegged to other currencies almost since the foundation of the country. On January 28, 1978, the dirham was officially pegged to the IMF's Special Drawing Rights and, almost twenty years later, in November 1997, the dirham was officially pegged to the US dollar at the rate of 3.6725 dirhams per US dollar. At the same time, the capital account is quite open, the financial sector is integrated to world capital markets and financial funds flow unimpeded. These policies effectively eliminate the capacity of the government to have an independent monetary policy. However, the Central Bank retains some discretionary power vis-à-vis domestic liquidity using the discount (Repo) window; it has played the role of the lender of last resort for the banking sector during the recent economic crisis. The commitment towards a fixed exchange rate seems to be very strong. In May 2009, the UAE announced their withdrawal from the monetary union project proposed by the Gulf Cooperation Council (GCC) and reaffirmed the peg to the US dollar on the grounds that it provides stability to the financial sector[11].

Notwithstanding the importance of the monetary authorities, fiscal policy remains the main tool of the government for steering the economy in the short run and influencing saving/investment decisions to sustain growth in the long run. In what follows, we review fiscal policies of the UAE and link the fiscal stance and the choice of policies to the presence of significant oil rents. We first review issues on the revenue side of the fiscal accounts and then focus on the expenditure side.

12.4.1 Oil Price Shocks and Fiscal Pro-cyclicality

The UAE has the most complex fiscal system of the GCC states due to its federal structure, a very cumbersome tax system (lack of income and profit taxes but a myriad of arbitrary and distorting fees on public services) and the overwhelming share of oil receipts in total revenue. Each emirate maintains full autonomy over hydrocarbon resources and fiscal policies and can exercise power in all matters that are not assigned to the jurisdiction of the federal government. The natural resources wealth in each emirate are the public property of that emirate, which manages its own budget on an independent basis and has no obligation to contribute to the budget of any other emirate.

As a result of this structure, fiscal policy is difficult to operate, in particular as a counter-cyclical instrument (Cevik 2011). Fiscal coordination among the seven emirates becomes cumbersome and the federal government lacks an independent revenue base that would allow for autonomous actions, remaining financially dependent on the transfers from Abu Dhabi and Dubai (to the tune of 90% of total revenues). These emirates make contributions to the federal budget in agreed amounts, but usually independently of the need for counter-cyclical policies.

[11] According to the International Monetary Fund (2011), the authorities believed that further efforts are needed to establish a framework that could lead to a currency union. These should focus on the harmonization of the payment and settlement systems, the regulatory and supervisory frameworks, the statistical systems, as well as on the removal of non-trade barriers.

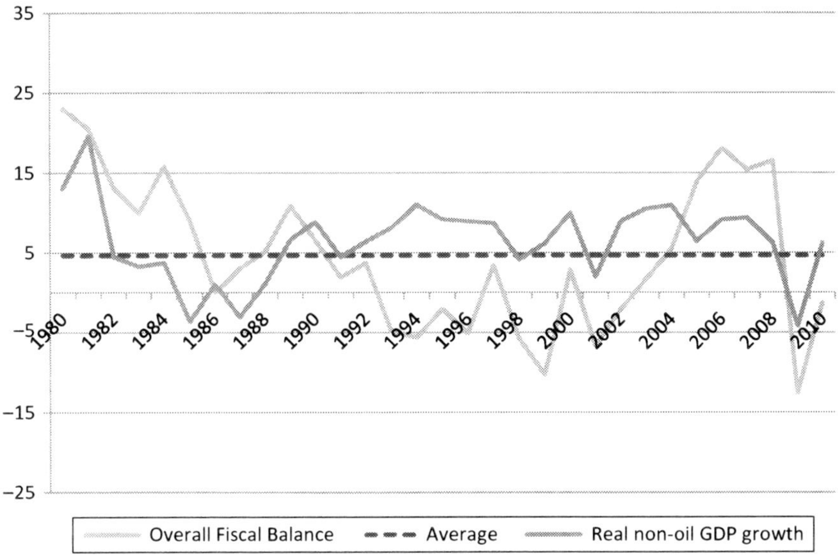

Figure 12.7: UAE: Overall Fiscal Balance (% of GDP) and Non-oil Real GDP Growth (%)
Source: International Monetary Fund (2011).

Fiscal federalism is not uncommon and the coordination problem is pervasive in such economies. In several countries (e.g., Argentina and Brazil), states free-ride on the government with deleterious effects on fiscal policy and, ultimately, on economic growth. States find it optimal to run fiscal deficits and, when these become untenable, ask the federal government for a bail-out. Once a state is in dire circumstances, the federal government finds it better to bail it out than to allow contagion to the other states. However, if states anticipate such federal policies, all states have incentives to overspend and wait for the bail-out: the massive bail-out usually places a heavy toll on economic growth in the form of protracted inflation and macroeconomic instability (Fiorillo and Sacchi 2011). In resource-rich economies, such opportunistic behavior is exacerbated by sizable resource rents.

This, nevertheless, has not been the case in the UAE. As shown in Figure 12.7, in the period 1980–2010, the consolidated fiscal stance of the government recorded a surplus of around 4.5% of GDP on average, largely based on oil receipts of course. The ample fiscal surplus – coupled with a low inflation rate and the fixed exchange rate – has led to a sustainable macroeconomic stance for the UAE economy. This is a significant success for the government and a crucial mechanism to avoid the oil curse. Note

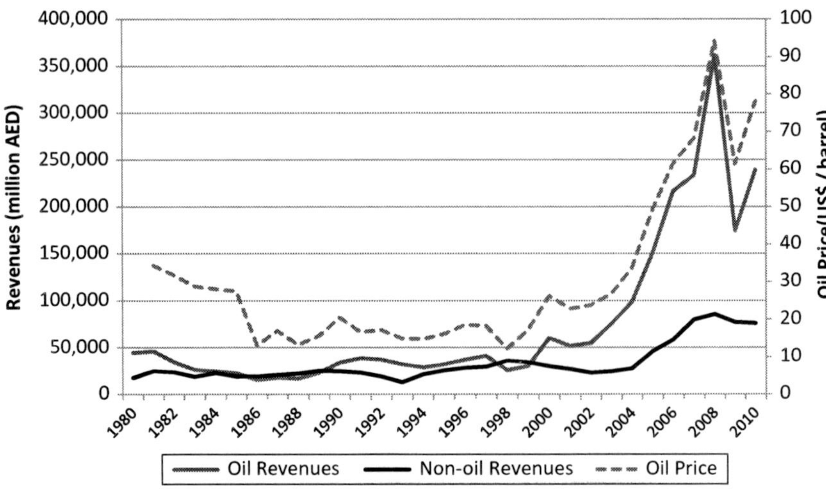

Figure 12.8: UAE: Consolidated Government Revenue and Oil Price
Source: International Monetary Fund (2011).

that the cautious government stance inhibits the Dutch Disease channel of the resource curse as it does not contribute to the overvaluation of the currency.

Figure 12.7, nevertheless, indicates one weakness of the fiscal policy in the UAE. It can be clearly seen that fiscal policy has not been smooth and could have not possibly helped in stabilizing the economy. In fact, when plotted against the growth rate of GDP in the non-oil economy – the target for stabilization – it becomes apparent that both variables exhibit similar volatility levels and, except for the mid-1990s, a coincidental cycle. This is not surprising: there are no institutional mechanisms implemented in the UAE through which the federal government or the emirates can enact counter-cyclical policies (see Soto, 2016 for a detailed analysis of tax structures in the UAE). Therefore, the fiscal policy largely follows the economic cycle, if it does not induce more acute cycles. This is a relatively common weakness of oil-exporting countries: the non-oil primary government balances worsened substantially in oil-exporting countries in the 2003–2008 oil price boom and improved in 2009 – hence exhibiting strong policy pro-cyclicality (Villafuerte and López-Murphy 2010).

Naturally, the cyclicality in the fiscal stance is mostly the result of the cyclical behavior of oil prices that filter through the government stance via oil-related revenues. In Figure 12.8, we have plotted oil and non-oil government revenues for the period 1980 to 2010 as well as the oil price in international markets (World Bank 2011). The hardly surprising high correlation between the oil price and the oil-related government revenues can easily be seen; oil price fluctuations translate immediately into revenue fluctuations. However, non-oil related government revenues do not exhibit such a correlation with oil prices and seem to follow their own, independent path. Note, also, that until the late 1990s, the sizes of the oil and non-oil revenues were approximately equivalent, but afterwards, oil revenues became dominant, most likely the result of the oil price boom that preceded the global 2008 economic downturn.

The need for counter-cyclical measures became apparent during the five years following the global crisis. Facing the severe economic contraction, the authorities indeed engineered counter-cyclical policies. The non-hydrocarbon primary balance, excluding investment income, deteriorated from a deficit of 16.5% of non-hydrocarbon GDP in 2005 to a deficit of 26.7% by 2008 and then to a record level of 44.8% in 2009 (Cevik 2011). But these extreme counter-cyclical measures only reveal the lack of automatic stabilizers of the Emirati economy: in most countries, ad valorem taxes allow for higher tax-collection during demand booms – thus, inhibiting overheating of the economy – and lower tax collection during downturns – hence, stimulating the demand. In the UAE, most taxes are arbitrary, fixed fees that respond much less to the economic cycle. In fact, the need for financing government agencies led to an *increase* in these taxes during the most recent economic crisis.

The analysis suggests that the government of the UAE has failed in developing the necessary institutional mechanisms to ameliorate the negative effects of the volatility in oil prices on government revenues. Recently, a number of countries have started to develop and apply fiscal rules in search for additional macroeconomic stability and policy credibility, which ultimately raise growth and improve intra/inter-generational equity[12]. Fiscal rules are set to attain one or more of three objectives: sustainability of public debt, control of government size and contribution to cyclical stability. Fiscal rules usually comprise a fiscal responsibility law (restricting the amount of resource rents that can be spent by the government), modern financial management, a planning horizon that exceeds one year, a fiscal rule for the budget, rules for government asset and liability management, requirements on accountability and public information on the government's financial management, effective external control and

[12] The number of countries with fiscal rules at the national level has grown steadily over the past two decades, from a universe of ten countries in 1990 to fifty-one in 2009 (Elbadawi and Soto 2011).

auditing and establishment of a fiscal council and/or fiscal committees (Schmidt-Hebbel this volume). The IMF has also advised the UAE on the convenience of implementing initiatives to enhance fiscal transparency and the adoption of a legally-anchored fiscal rule, which places constraints on discretionary policies through procedural and numerical limits on fiscal aggregates, such as the structural budget balance, spending or indebtedness (Cevik 2011).

12.4.2 Oil Prices, Currency Overvaluation and Fiscal Expenditures

The expenditure side of fiscal policy can be as important as the revenue side when considering the oil curse. As mentioned, excessive government spending – in particular, on current expenditures – can easily lead to an overvalued currency, with adverse effects on the profitability of non-oil sectors. The mechanics of overvaluation are relatively simple: oil proceeds allow the government to increase the demands for goods and services. Frequently, more-than-generous wage increases are also granted to public servants, as was the case in Abu Dhabi before the recent economic crisis. Higher demand for domestically produced goods and services induces price increases, at least in the short to medium run, when local producers cannot adjust production to serve the surge in demand. On the contrary, prices for imported and exported goods remain stationary as dictated by international markets. Therefore, the relative price between internationally traded and non-traded goods (i.e., the real exchange rate) appreciates.

In all countries, real exchange rates fluctuate over time. As long as such fluctuations remain small and centered around the equilibrium level, they have no negative effects on aggregate economic growth. However, when fluctuations are excessive and when the level of the real exchange rate deviates significantly from the equilibrium, the adverse effects on investment, saving and economic growth become significant (see Elbadawi et al. 2012 for a measurement of the growth effects of misalignment). The costs arise because as the real exchange rate appreciates, the incentives to invest in exportable goods and services wane (the relative price is decreasing for traded sectors). Investment in non-traded goods may increase, but it is limited by the size of the domestic demand. This would put a ceiling on economic growth. Moreover, the costs of bringing the currency back to equilibrium are usually very high, in the form of a sharp devaluation once the currency levels become untenable. This increases uncertainty, thereby reducing investment as adjustment costs further reduce growth.

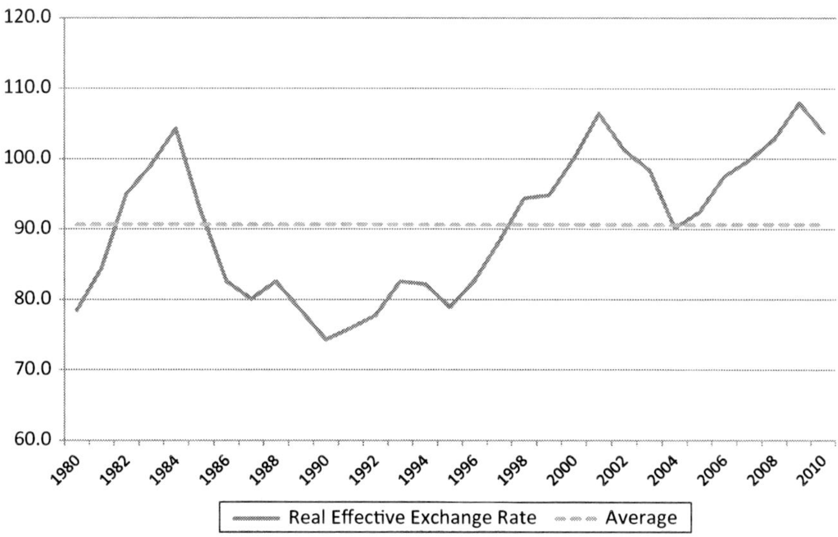

Figure 12.9: UAE: Real Effective Exchange Rate
(Normalized 2000 = 100)
Source: International Monetary Fund (2011).

Figure 12.9 presents the evolution of the real effective exchange rate in the UAE in the period 1980–2010. It can be seen that there have been fluctuations around the long-run average level depicted by the dotted line, but such fluctuations have been rather small in size and, in fact, not very different from those in other high-income countries such as the US or those comprising the Euro zone. In conclusion, if there have been negative effects of oil rents in the UAE, they have not operated primarily via the appreciation of the currency or Dutch Disease symptoms. Note, however, that this does not invalidate the notion that the fiscal stance has been pro-cyclical: government expenditures in the UAE increase when oil prices rise and contracts when oil markets are bearish and, therefore, fiscal policy does not provide cushioning against the business cycle. The conclusion is that fiscal policy has not been a source of sustained currency misalignment and the government has avoided the perils of overvaluation. This is a significant accomplishment that should not be overlooked.

Beyond the effect of the oil price volatility on government revenues and expenditures, the oil curse operates in a more subtle manner through the misuse of the resource rents. In general, such misuse affects both current expenditures (typically wages) as well as capital expenditures (e.g., socially unprofitable projects or "white elephants"). Resource misallocation is

Table 12.3: *UAE: Annual Earnings and Hours of Work in Different Sectors in 2008*

Economic Activity Group	Actual Hours of Work	Monthly Earnings AED
Agriculture, Hunting and Forestry and Fishing	211.2	2,386
Mining and Quarrying	208.9	11,226
Manufacturing	221.0	3,328
Electricity, Gas and Water Supply	206.2	10,490
Construction	225.1	2,249
Wholesale and Retail Trade	212.3	4,424
Hotels and Restaurants	218.9	4,190
Transport, Storage and Communications	179.7	4,645
Financial Intermediation	191.3	10,961
Real Estate, Renting and Business Activities	204.8	5,917
Public Administration, Defense and Compulsory Social Security	170.6	10,333
Education	179.0	6,435
Health and Social Work	200.3	6,556
Other Community, Social and Personal Service Activities	233.1	3,741

Source: Employment, Wages and Hours of Work Survey (October 2008).

naturally the result of institutional weaknesses. As noted by Robinson et al. (2006) and Robinson and Torvik (2005), public sector over-employment and white elephants arise from the desire of politicians to redistribute wealth to supporters in political systems where some individuals have uncontested power. In such case, only these powerful politicians can credibly commit to public over-employment and/or socially inefficient projects aimed at redistributing wealth to their constituency. The political system is institutionally weak to inhibit such behavior.

When considering current expenditures, the misuse of resources in the UAE directly shows in that (a) government wages and salaries are substantially above those of the private sector for equivalent job descriptions and (b) public employment has become the almost exclusive employment option for the local population and, consequently, public employment has increased substantially.

In Table 12.3, we have collected the results of the 2008 Survey on Employment, Wages and Hours of Work. It can be seen that there is a wide disparity in terms of earnings (salaries and benefits) in the different sectors, as well as in hours worked. This is not surprising as labor productivity in the different sectors is heterogeneous. However, note that in those sectors which are almost exclusively in public hands (public administration and defense; mining and quarrying; and electricity, gas and water supply), earnings are substantially higher and hours worked substantially lower than in the rest of the economy. Workers in these three sectors work ten hours less per month on average than their counterparts in other sectors of the economy and earn twice as much as the other workers on average, with the only exception of the financial sector. According to the survey, earnings per hour in the Federal Government are five times higher than those in the private sector. It is only logical that the Emiratis tend to see the public sector – where they enjoy preference over expatriates – as the only viable sector in which to work and, consequently, are willing to remain unemployed while waiting for a position. According to survey data collected by Vazquez-Alvarez (2011), over 85% of Emiratis work for the public sector.

The government has used, particularly since the mid-1990s, public employment to transfer part of the resource rents to Emirati nationals. Generous subsidies for education, health, utilities and subsidies play a complimentary role. In some sense, this is the equilibrium outcome of the political bargaining process that underlies the stability of the UAE federation. While useful from a political viewpoint, intervening through the labor market does not come at a low price. In Figure 12.10, we plot the (normalized) trajectory of value-added in government services, employment and average labor productivity in the period 1988–2005, for which there is consistent data. It can be seen that the substantial increment in public employment since the mid-1990s far exceeds the natural growth of the Emirati population. As a consequence, average labor productivity in the public administration declined systematically during the late-1990s and 2000s, until the recent financial crisis put a break to indiscriminate hiring by public agencies.[13]

The evidence of oil curse symptoms on the capital expenditures of the government is less clear cut and it is impossible to make an objective assessment as most of the information is not publicly available. As mentioned in the introduction, however, the government has spent significant

[13] An additional, unintended, outcome of this policy is that of incrementing unemployment among Emiratis and, consequently, reducing private sector employment. By making public employment extremely attractive for the Emiratis, it validates the option value of waiting for a position in the public sector. Algan et al. (2002) estimate that the creation of one public job in the OECD countries destroyed about 1.5 private job in the period 1960–2000 and eventually increased the number of unemployed workers by 0.3.

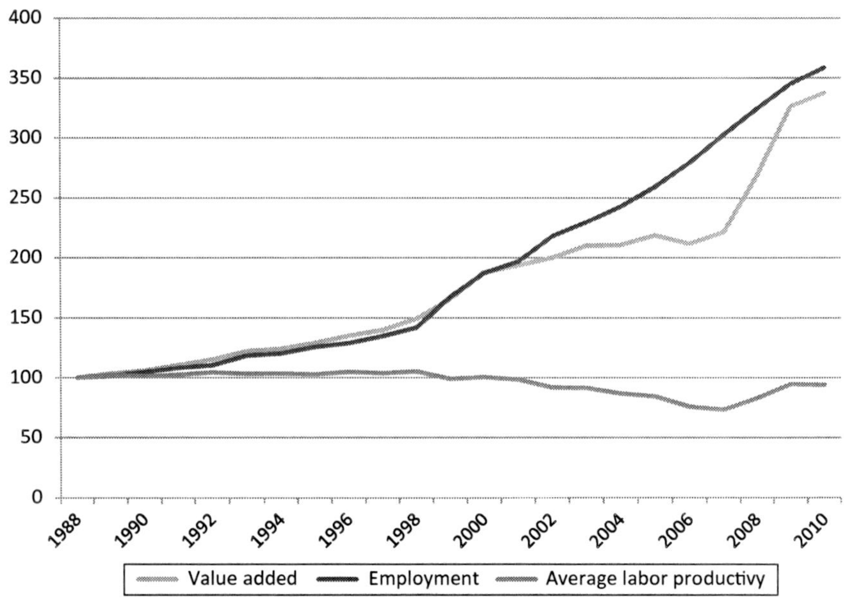

Figure 12.10: UAE: Value-added, Employment and Average Labor Productivity in the Public Sector (Normalized 1988 = 100)

resources to provide social capital for the economy, both in terms of public infrastructure and pro-business institutions. As of 2010, the UAE ranked among the top countries in the world, according to the World Economic Forum (2011), in rule of law due to its low levels of crime, absence of corruption and reliable police services. Likewise, it ranked high on political stability, even though it lacks democracy. Nevertheless, although the UAE government has provided significant institutional build-up for the private sector, the institutional fabric of the public sector remains substantially weak. As noted by the IMF, the UAE economy is "dominated by a web of commercial corporations, financial institutions, and investment arms owned directly by the Government of Dubai, the Government of Abu Dhabi, or the ruling families under the umbrella of major holding companies" (IMF 2011, p. 4). As became painfully evident during the recent global crisis, these government related entities – henceforth, GREs – had been operating in a largely unsupervised and non-accountable form and taken excessive risks.

These GREs had been instrumental in the initial years of the UAE in the opening and developing of new businesses and in diversifying away from oil extraction and traditional fishing and pearling activities.

Nevertheless, while the GREs have been a major source of growth and development for the UAE economy, it is also true that in the 2000s, they became progressively mismanaged, benefiting from government transfers and from extensive borrowing, fueling the real estate bubble that collapsed in 2008. The global financial and economic crisis unveiled the fiscal and financial risks posed by GREs and the significant weakness in the institutional framework for their management and monitoring. In some sense, oil rents and easy access to foreign funds allowed the UAE government – particularly Dubai and Abu Dhabi – to lose sight of the need for good economic management of GREs, a standard outcome of the oil curse. What remains to be seen is the magnitude of the losses and the steps the government will take to implement governance and accountability on public firms.

12.4.3 Oil Prices, Foreign Shocks and the Absence of Monetary Policy

As mentioned, the UAE lacks a monetary policy as a result of its open capital account and the fixed exchange rate. Because of the exchange rate peg, the UAE passively follows the monetary policy of the US. This arrangement had worked quite well in the past, keeping inflation low, anchoring expectations and providing stability to the financial sector. For years, the UAE benefited from the low levels of inflation of the American economy as it kept the exchange rate pegged to the US dollar. In the 2000s, however, inflation in the UAE increased from 2.8% in 2001 to 12.3% in 2008. The rise in inflation was largely driven by increasing commodity prices and the weakness of the US dollar vis-à-vis other world currencies (such as the Euro and the Pound Sterling). While the pass-through was significant in most economies,[14] in the UAE, inflationary pressures were compounded by domestic factors, such as the rapidly rising housing prices and rentals, which accounted for a significant fraction of the price increase. It is true, however, that a floating exchange rate would have taken some steam off the markets and helped stabilize the economy.

Likewise, US monetary policy has recently operated in a manner that exacerbated the economic cycle in the UAE. In the late 2000s, the economic cycle in the US and the UAE started to diverge: massive capital inflows to the UAE – linked to the real estate bubble and historically high oil prices – called for counter-cyclical policies to limit expenditures,

[14] See Rigobon (2010).

overheating and debt accumulation. On the contrary, the US embarked on significant monetary expansions to stimulate its ailing economy. Eventually, the easy money policy in the US made it impossible to control expenditure in the UAE, further aggravating the cost of the bail-out that followed the global economic crisis.

The recent inflationary phenomenon in UAE raised questions about the convenience of pegging the UAE dirham to the US dollar. From a purely technical perspective, pegging to a composite currency basket would be preferred if it better matches the trade profile of the country and protects it better against imported inflation than has recently been the case. Abandoning the peg towards a flexible exchange system, on the other hand, constitutes a major institutional change. In principle, it would allow the central bank to manage interest rates and affect the value of the currency, with higher rates curbing domestic demand and a more expensive currency applying a brake to external demand. In a context of rising inflation and increased interest rates, floating the currency would seem the better option, but the reality is much more complicated. Floating the currency would demand significant investment in the capacities of the banking sector and the government to manage risk, in particular, exchange rate risk, emerging precisely from the very source of the resource curse problem: the volatility of oil prices. Such investment would involve a quantum leap in the human capital employed by the government in banking supervision and monitoring.

It is therefore important to recognize that the concept of an "optimal policy mix" for the UAE is contingent on the objective of the policy-making authorities. Recently, the governments of Abu Dhabi and Dubai have issued separate visions for future development, but have coincided on the need to create high value-added industries driven by human capital and technology. By definition, these industries have to compete in global markets and, given the current lack of know-how in the UAE, their creation would require significant foreign direct investment as well as domestic financial resources. Attracting foreign investors – in particular, when aimed at adopting or developing cutting-edge technologies – demands stable environments with low risk levels to develop long-maturity projects. A sound macroeconomic policy is naturally a must, but the choice of policies is not obvious. We can use the experience of other countries as guidance. Singapore and Hong Kong have been quite successful in developing a highly sophisticated industrial sector with very similar macroeconomic policies to the UAE (fixed exchange rate and open capital accounts). This indicates that the key to their success lies elsewhere, most likely in their very flexible labor markets and the high quality of their education systems – two elements missing in the UAE. On the other hand, recently industrialized countries, such as South Korea, have preferred to retain monetary independence by allowing the currency to freely float in the exchange market and have controlled fluctuations by enacting highly disciplined fiscal policies. Granted, South Korea's foreign trade sector is unlikely to induce significant volatility in the currency because its highly diversified export base can escape from the wild fluctuations in commodity prices, a feature that is not achieved in oil-dominated UAE. Any option of policy mix in the UAE, however, would require a significant improvement in currently available instruments.

In conclusion, the review of the macroeconomic policies – mainly, fiscal policy – indicates that oil rents might have had both positive and negative effects. On the one hand, the ample rents enjoyed by the UAE have allowed the government to maintain a healthy fiscal stance and finance significant investment in public infrastructure and the build-up of pro-business institutions. This, in turns, helped avoid the Dutch Disease effects of oil rents. On the other hand, oil curse effects manifest in both the revenue and expenditure side of fiscal policy.

On the revenue side, the absence of ad valorem taxes and the reliance on arbitrary fees for public goods has eliminated a crucial automatic stabilizer of business cycles and induced numerous microeconomic distortions that hamper efficiency and penalize productivity gains. This has been possible only because oil proceeds take the lion's share of government financing. Moreover, the lack of institutional mechanisms to protect fiscal accounts from oil price volatility renders fiscal policy pro-cyclical, instead of an instrument for macroeconomic stabilization.

On the expenditure side, standard channels of the oil curse manifest clearly in public sector over-employment and excessively high public wages with regard to those paid by the private sector for equal jobs. The small size of the Emirati vis-à-vis the expatriate population has made this symptom of the oil curse less visible and perhaps less costly than in countries where government employment crowds out the private sector, such as Saudi Arabia. As a result of using public employment as an employer of last resort, average labor productivity in the public administration declined systematically until the recent financial crisis put a break to indiscriminate hiring by public agencies.

12.5 Does Dubai Suggest a Path for Escaping the Oil Curse?

Despite the cultural and historical similarities shared by the seven emirates that constitute the UAE, their economic heterogeneity is striking. In terms of population, economic size and level of development, the emirates are very diverse. It can be seen in Table 12.4 that Abu Dhabi is by far the largest economy in the union, with around 60% of the total GDP of the UAE in 2009, and is several orders of magnitude larger than the smaller northern emirates (Sharjah, Ajman, Umm al-Quwain, Ras al-Khaimah and Fujairah). Only Dubai comes close to Abu Dhabi in population and also in income per capita. As a result of oil rents, Abu Dhabi has one of the highest income per capita in the world. Note, moreover, that income per capita in most of the northern emirates would barely qualify them as high-income economies. In addition to income heterogeneity, natural resource endowments are also quite heterogeneous. Indeed, oil and gas extraction is only significant in Abu Dhabi, while it is non-existent in most northern emirates and, particularly, in Dubai. In spite of the lack of natural resources, Dubai manages to command the highest level of development in the UAE and an income per capita equal to most OECD countries. Until the mid-1980s, Dubai relied heavily on oil extraction and exports, even though it had a long history as a trade hub for the Gulf region. After 1985, oil production dwindled and Dubai initiated a remarkable reshaping of its economy away from oil. In what follows, we study the case of Dubai to identify valuable lessons on how to escape the oil curse.

Table 12.4: *UAE: Economic Indicators by Emirate, 2009*

Emirate	GDP (US$ million)	Population (thousands)	Per capita GDP (US$)	Share of Oil in GDP
Abu Dhabi	162,406	1,628	99,758	46.0%
Dubai	80,097	1,722	46,514	1.8%
Sharjah	16,595	1,017	16,318	10.8%
Ajman	3,781	250	15,123	0.0%
Umm al-Quwain	630	56	11,252	0.0%
Ras al-Khaimah	4,285	241	17,782	4.4%
Fujairah	2,541	152	16,714	0.0%
Total/average	*270,335**	*5,066**	*53,363***	*28.9%***

Source: National Bureau of Statistics and IMF.
Note: in the last row, *refers to total values while **refers to country average.

When confronted with the disappearance of its main natural resource in the mid-1980s, the authorities turned to Dubai's traditional source of income (foreign trade) to support economic development. Historically, Dubai has been a place for businesses due to its early understanding that an open economy offers sustained productive opportunities. Even before the foundations of the UAE, Dubai had opened up to foreign trade, allowing free trade, providing ample facilities and infrastructure and supporting foreign immigration. In addition to supporting foreign trade, the government realized the need for diversification – foreign trade offered ample possibilities but also generated significant risks for the local economy. Therefore, the authorities embarked on a series of policies aimed at fostering the development of strategic industries, including tourism, the financial sector and retail and wholesale trade. The opening of several free trade zones in the 1990s provided an additional boost to the economy and, away from the straightjacket of the *kafala*, allowed competitive re-exporting activities to flourish.

In Table 12.5, we present the average annual growth rate of sectorial GDP by emirate, in the period 1997–2009. Several elements point to the successful diversification achieved by Dubai. First, note that Dubai has

Table 12.5: *Average Annual GDP Growth Rate by Economic Sectors in Each Emirate, 1997–2009*

	Abu Dhabi	Dubai	Sharjah	Ajman	Umm Al Quwain	Ras Al Khaimah	Fujairah
Total	8.8%	10.0%	5.4%	7.9%	1.0%	4.3%	5.6%
Crude Oil and Natural Gas	8.8%	-8.6%	14.0%	–	–	-6.0%	–
Manufacturing	6.5%	10.7%	8.4%	11.6%	0.4%	13.8%	14.5%
Construction	12.8%	12.7%	5.1%	10.4%	-6.5%	-3.7%	8.9%
Wholesale & Retail Trade	8.6%	15.2%	5.4%	6.1%	-2.2%	3.5%	3.2%
Transport, Storage and Communication	14.6%	12.3%	4.9%	3.5%	6.0%	2.7%	3.3%
Real Estate and Business Services	9.1%	12.5%	5.8%	5.6%	7.4%	6.4%	3.5%
Financial Corporations	11.2%	12.9%	3.6%	9.5%	-12.0%	11.4%	8.7%
Other sectors	2.9%	1.2%	1.3%	4.0%	1.7%	1.3%	4.4%

Source: National Bureau of Statistics and IMF.

outgrown all other emirates, by a significant margin, in terms of GDP: total GDP in Dubai has grown at 10% per year. Only Abu Dhabi and Ajman come close to Dubai, the former largely fueled by oil and gas production, the latter by land market de-regulation and construction booms. All other emirates have grown at much lower rates, with Umm al-Quwain all but stagnating. Second, excluding oil and gas production, all economic sectors in Dubai outperform their counterparts in other emirates with very few exceptions: in Abu Dhabi, transport, storage and communication have performed better than in Dubai and manufacturing in Ajman and Ras al-Khaimah also performed better than in Dubai, but starting from a very low base. Third, retail and wholesale trade in Dubai has grown substantially faster than any other sector in Dubai or in any other emirate. As of 2010, retail and wholesale trade accounted for 30% of GDP in Dubai.

The diversification of Dubai has been relatively balanced, in the sense that some of the sectors that have flourished after the oil collapse mainly cater to domestic demand (e.g., construction, real estate and business services and manufacturing) while others are largely focused on overseas markets (the financial sector, transport and storage and the wholesale and retail trade that is largely connected to tourism and duty free activities). This allowed the emirate to hedge against demand fluctuations. Note that the other emirates have not been as successful as Dubai in diversifying, independently of whether they are oil exporters or not. Abu Dhabi, the main oil exporter, has been unable to diversify away from oil and gas production. Sharjah lags behind, in spite of more than tripling its energy production.

Economic growth, as discussed, can be achieved by expanding production factors and/or improving the efficiency of use of such resources. It is, therefore, necessary to study whether Dubai's remarkable growth and diversification has been the result of massive factor accumulation or if there have been significant productivity gains. We use the same methodology of Section 12.3 to compute the sources of growth of the non-oil sector of Abu Dhabi and Dubai, the only two emirates with available statistics for a period of twenty years or more.[15] In Figure 12.11, we present the evolution of total factor productivity in both emirates.

It can be easily seen that total factor productivity in the non-oil sectors of Abu Dhabi stagnated in the period 1988–2010, while that of Dubai has

[15] Due to lack of data, the computation of the sources of growth assumes that human capital levels are the same in Dubai and Abu Dhabi. Moreover, lack of data for Abu Dhabi forces us to work with the period 1988–2010.

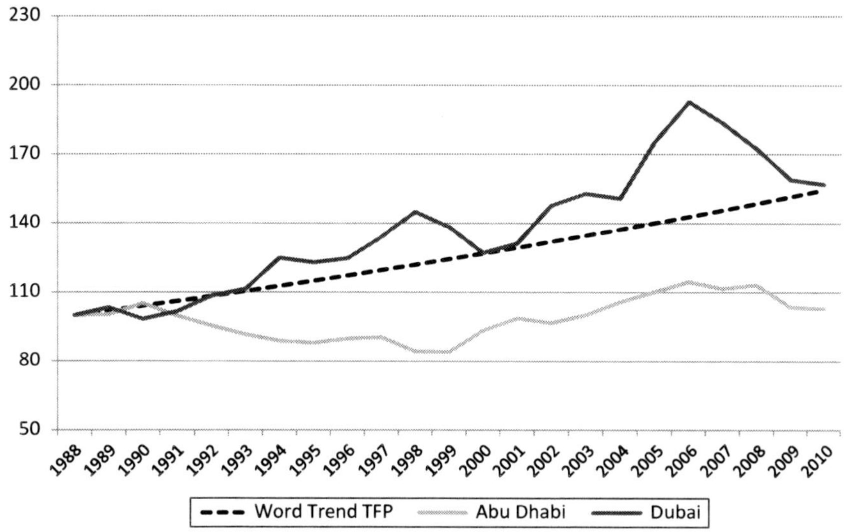

Figure 12.11: Total Factor Productivity in the Non-oil Sectors of Abu Dhabi and Dubai (Normalized 1988 = 100)

grown at around the same rate of TFP in the world economy. World TFP growth is not a very challenging reference point as it largely reflects the evolution of mature, developed economies; TFP growth in countries that have recently industrialized, such as Korea, Singapore and Taiwan, is significantly higher. The cumulative average growth rate for Dubai is 2%, while that for Abu Dhabi is 0.1%. Moreover, the phases of productivity growth differ markedly in both emirates. In Abu Dhabi, the period 1988–1998 was characterized by a significant decline in productivity, at a cumulative level of 15%. In Dubai, this was a period of significant expansion in productivity in the economy that came to an abrupt halt around the time of the 1998 East Asian financial crisis. Likewise, in Dubai, the period 2000–2010 was one of significant productivity gains, to some extent fueled by the overheating of the economy, which was cut short by the unveiling of the global financial crisis. In Abu Dhabi, total factor productivity increased but only very moderately and it merely reverted to the levels of the late 1990s.

In conclusion, in Dubai, the pro-business friendly policies implemented by the local government since the mid-1980s, coupled with market de-regulation and increased integration to world markets, have allowed the non-oil sectors of the economy to reap modest but systematic productivity gains. Openness, however, has also been a source of significant economic fluctuations that the authorities have been unable to dampen. To the extent that the UAE oil curse manifests itself in low productivity growth and lack of diversification away from oil, Dubai provides an answer to the oil curse.

Nevertheless, the answer to the curse is only partial. Even in the pro-business environment of Dubai and supported by significant access to financial resources,[16] firms are not achieving the spectacular growth rates of East Asian economies nor is labor productivity increasing pari passu with their competitors in world markets. A dynamic, competitive private sector seems to be a necessary yet not sufficient condition to escape the oil curse. Therefore, other conditions are still missing.

A country that has escaped the curse altogether is Norway. According to Røed Larsen (2003), the following policies are responsible for such success: (1) shield and support certain domestic industries thought to be crucial for a long-term comparative advantage, (2) invest heavily in education and know-how, (3) follow counter-cyclical polices, (4) reform labor market to increase the labor force's share of the population, (5) implement wage control and income coordination to establish some sense of social contract and (6) enact an expenditure-limitation policy of fiscal prudence directed to shield the economy from spending effects (including a sovereign wealth fund). Chile, another natural resource exporter that has avoided the curse, has refrained from intervening in the labor market, but followed a similar strategy of building-up strong policy institutions and complemented the sovereign wealth fund with the additional fiscal discipline provided by fiscal rules (as discussed earlier) and the conduct of monetary policy under strict inflation target rules. Chile, thus, provides an alternative policy mix to face the inflow of massive yet volatile resource rents.

Dubai and the UAE have implemented some of these policies (shield and support certain domestic industries thought to be crucial to a long-term comparative advantage, implement wage controls under the *kafala*, and develop a sovereign wealth fund) but have failed to implement important ones, including, heavy investment in education and know-how, following counter-cyclical polices and enacting an expenditure-limitation policy of fiscal prudence directed to shield the economy from spending effects.

[16] Access to finance was relatively simple before the recent global downturn even for small businesses. In the post-global crisis environment, however, it appears to be a major constraint (Trabelsi and Fadhel 2011).

12.6 Conclusions

This chapter studies the existence of an oil curse in the UAE. Oil and gas exports have shaped the economic activity and development strategy of the country since its foundation in 1971. Oil has also brought substantial wealth to the Emirati population and welfare levels have improved markedly. To the casual observer, it would appear as if the UAE has escaped the oil curse. Careful review of empirical evidence, however, indicates that several of the classical resource curse symptoms are present in the different sectors of the UAE economy. This suggests that the country has not been immune to the disease, but that it has somehow managed to make the benefits outweigh the negative outcomes of oil exporting. If so, the UAE experience could provide a valuable guideline for other resource-rich countries.

We use simple growth models to study the sources of economic growth in the UAE and provide quantitative answers to the following questions: How has oil extraction shaped the development of the UAE economy? Is there any evidence of an oil curse and if so, is it significant? What has been the role of macroeconomic policies in inhibiting or fostering the oil curse? And does the UAE provide an answer to the oil curse that could be replicated in other countries affected by the disease?

The oil curse appears when the earnings from natural resource capital crowd out other forms of capital that are also necessary for economic growth, including physical capital (both public and private), human capital, institutional capital and social capital. Contrary to most oil-rich economies, the UAE government has invested heavily in physical capital, in particular, public infrastructure and the institutional fabric that deals with the working of the private sector (rule of law, corruption control and so on). Furthermore, to a large extent, the government has enacted sound macroeconomic policies, which have been instrumental in supporting private sector investment and economic development. The classical outcomes of massive oil rents in the form of currency mismanagement, rampant rent-seeking behavior and generalized corruption have been totally avoided.

Although some of the standard transmission channels of the oil curse have been blocked by policies, the UAE government has not been completely successful in controlling the negative effects of oil rents. Ample oil rents and an inadequate labor regulation have given incentives to the private sector to engage in an unsustainable development strategy: production growth in both the oil and non-oil sectors of economic activity is fueled by massive investment and an even higher use of low-skilled manpower (mostly from the Indian subcontinent), but the efficiency in the use of such factors has remained stagnant and, in some areas, steadily declined as the country got richer. Average labor productivity has remained stagnant in the late-1980s; as of 2010, labor productivity in the UAE is one half of that in the developed economies and one third of that in Hong Kong.

The evolution of labor productivity in the UAE is in stark contrast to the slow but systematic growth of productivity in developed economies or the very fast yet volatile expansion of productivity in the East Asian tigers. Understanding the reasons for such productivity stagnations – in particular, determining to what extent it derives from the impact of oil rents – is therefore important. We show that it was not the changes in inputs that were mainly responsible for the evolution of productivity in the UAE, but rather the efficiency with which these factors were used.

When excluding the oil sector from the calculations of total factor productivity, we found evidence that it actually increased during the period 1987–1995, although at a relatively slow pace (similar to that of the high-income countries of the world), which allowed the country to keep track – but not catch-up – with more developed economies. Around 1995, however, somehow, the momentum in TFP growth began to wane and productivity stagnated. It should be recalled that until the recent 2008/2009 downturn, the world economy achieved sustained growth throughout the 1990s and early 2000s – these are the years of the so-called self-moderation. Therefore, it seems unlikely that external shocks could have induced a decline in productivity growth. Without excluding external shocks altogether, this would suggest that internal policy changes could be among the causes for this phenomenon.

Some such policies deal with the working of the public sector. In our view, they are likely to have had a significant adverse effect on productivity and economic growth. The evidence indicates that the massive accumulation of typically non-skilled workers has been more pronounced in the private sector than in the public sector. This raises the question of why entrepreneurs would prefer to employ labor-intensive production techniques when they have unrestricted access to the highly sophisticated capital goods and top-notch technologies that the world economy offers. The peculiar institutional framework of the labor market – known as *kafala* – provides a compelling explanation.

Under the *kafala* system, employers have significant market power derived from the fact that the employee is forbidden to change occupations while the contract is in force and has to leave the country upon expiration of the contract. The absence of horizontal mobility allows employers to extract economic rents: when choosing production technologies, therefore, employers would tend to focus on labor-intensive techniques that, in addition to the normal profit obtained from selling goods, would allow them to extract the highest rents from the worker. Naturally, this skews production towards labor-intensive technologies. It also biases employment towards low-skilled workers that have less ability to negotiate salaries; high-skilled workers are better educated and scarcer and, thus, better equipped to counterbalance the market power of employers. This aspect of the *kafala* system would explain why labor productivity in the UAE is lower than in countries with similar GDP per capita.

Moreover, the *kafala* system also has long-term effects on productivity growth. Abundant oil rents provide funding for investment projects – particularly in the formative stages of an economy – and raise the demand for labor. In a non-*kafala* environment with high mobility, workers would command high wages. In the *kafala* system, wages are kept constant for the duration of the contract. In the short run, the *kafala* prevents rising labor costs and keeps profits high for investors. In the long-run, however, the *kafala* destroys the incentives to increase efficiency and the ability of firms to be profitable on the basis of their efficiency, for two reasons. First, resources do not freely adjust to their most efficient use and therefore, there may be an inefficient allocation of workers. Second, labor-saving technical improvements, which are at the core of economic development, are disregarded by entrepreneurs as they lower rent extraction.

This theory, explaining the UAE's low productivity growth, centers on labor market distortions. Still, other policies could reinforce or ameliorate the effects of adverse effects of wrong labor policies. Macroeconomic policies in the UAE are limited to fiscal policy since the country has pegged its currency to the dollar and the capital account is quite open.

We found no evidence of Dutch Disease symptoms (a standard oil curse outcome) nor of excessive government expenditure crowding out private sector initiatives. Nevertheless, the fiscal policy presents three classical symptoms of the oil curse.

First, fueled by oil taxes and dividends, fiscal policy has been highly procyclical: there are no institutional mechanisms that would allow the authorities to enact counter-cyclical policies. The volatility of oil prices filters through the UAE economy largely unimpeded.

Second, oil resources have distorted both the current expenditures of the public sector to a large extent. The classical oil curse symptom of public labor over-employment is compounded in the UAE by public wages that are twice as high as those in the private sector for equal-skills occupations. In some sense, this is the equilibrium outcome of the political bargaining process that underlies the stability of the UAE federation. However, while useful from a political viewpoint, labor market interventions do not come at a low price. The substantial increment in public employment since the mid-1990s has induced declining average labor productivity in the public administration.

Third, in addition to massive investment in infrastructure, the UAE government has provided significant institutional build-up for the private sector. However, it has failed to develop a similar institutional fabric for the public sector. As indicated by the IMF (2011), the UAE economy is dominated by a web of semi-autonomous government related entities, including commercial corporations, financial institutions and investment arms owned directly by the government of each emirate. It became painfully evident in the aftermath of the recent global crisis that government related entities had taken excessive risks and operated in a largely unsupervised and non-accountable form. Although these oil-financed, poorly designed policies may seem innocuous transfers to the Emirati constituency, their combined and cumulative effect can produce substantial harm to productivity growth.

Beyond the weaknesses of the fiscal policy, it should be acknowledged that the government has maintained a sound macroeconomic stance. The UAE lacks monetary policy as a result of its open capital account and the fixed exchange rate. This arrangement worked quite well in the past, keeping inflation low, anchoring expectations and providing stability to the financial sector. Recently, however, it has exacerbated the economic cycle in the UAE as a result of the weakness of the US dollar in international markets, raising the issue of whether an alternative system would provide better cushioning against foreign shocks. Abandoning the peg towards a flexible exchange system, on the other hand, constitutes a major institutional change. It would demand significant investment in the capacities of the banking sector and the government to manage risk, in particular exchange rate risk, emerging precisely from the very source of the resource curse problem: the volatility of oil prices. Such investment would also involve a quantum leap in the human capital employed by the government in banking supervision and monitoring.

Finally, we turn to the case of Dubai for an alternative development strategy for resource-rich economies. Oil and gas extraction is nowadays insignificant in Dubai. In spite of the lack of natural resources, Dubai manages to command the highest level of development in the UAE and an income per capita equal to most OECD countries. Until the mid-1980s, however, Dubai relied heavily on oil extraction; after oil production dwindled, Dubai initiated a remarkable reshaping of its economy away from oil. The authorities initially turned to foreign trade as the engine of economic development but soon realized the need for diversification: foreign trade offered ample possibilities, but also generated significant risks for the local economy. Therefore, they embarked on a series of policies aimed at fostering the development of strategic industries, including tourism, the financial sector and retail and wholesale trade. The opening of several free trade zones in the 1990s provided an additional boost to the economy and, away from the straightjacket of the *kafala*, allowed competitive re-exporting activities to flourish.

In the subsequent two decades, significant diversification was achieved in Dubai. As a result, Dubai has outgrown all other emirates by a significant margin in almost every economic sector. Diversification has brought a better balanced development path as some of the flourishing sectors cater mainly to domestic demand (e.g., construction, real estate and business services and manufacturing) while others focus on overseas markets (the financial sector, transport and storage, and the wholesale and retail trade that is largely connected to tourism and duty free activities).

In addition to helping diversify its economy, Dubai's pro-business policies, implemented by the local government since the mid-1980s, coupled with market de-regulation and increased integration into world markets, have allowed the non-oil sectors of the economy to reap modest, but systematic productivity gains. On the contrary, and in spite of the massive oil resources, total factor productivity in Abu Dhabi stagnated in the period 1988–2010.

To the extent that the oil curse in the UAE induces low productivity growth and lack of economic diversification, Dubai provides an answer to the oil curse. Nevertheless, the answer is only partial. Even in pro-business Dubai, firms do not achieve the spectacular growth of their counterparts in East Asia and other emerging economies, nor do they achieve the efficiency of European firms. Allowing for a dynamic, competitive private sector seems to be a necessary, yet insufficient, condition to escape the oil curse. Therefore, other elements also restrict development. This chapter points to labor market regulations as a main source of distortions.

Appendix

Data Sources:

- Gross Domestic Product in real US$ of 2005 data were obtained from World Bank (2011).
- Gross Domestic Product in real PPP prices of 2005 data were obtained from World Bank (2011).
- Gross Domestic Investment (as share of GDP) data were obtained from World Bank (2011) and the UAE National Bureau of Statistics (2011).
- Employment data were obtained from Conference Board (2012)
- Human Capital series were obtained from Barro and Lee (2011) for years 1970, 1975, 1980, 1985, 1990, 1995, 2000, 2005, and 2010. Annual figures were computed by linear interpolation.

References

Acemoglu, D. (2010). When Does Labor Scarcity Encourage Innovation? *Journal of Political Economy* 118(6): 1037–1078.

Algan, Y.; P. Cahuc; A. Zylberberg; J.S. Pischke and T. Verdier (2002). Public Employment and Labour Market Performance. *Economic Policy* 17(34): 7–65.

Arezki, R.; K. Hamilton and K. Kazimov (2011). Resource Windfalls, Macroeconomic Stability and Growth: The Role of Political Institutions. IMF Working Paper No. 11/142. Washington, DC.

Barro, R. J. and J. W. Lee (1994) Sources of Economic Growth. *Carnegie-Rochester Series on Public Policy* 40:1–57.

(2011). A New Data Set of Educational Attainment in the World, 1950–2010. NBER Working Paper No. 15902. Cambridge, MA.

Bhattacharyya, S. and P. Collier (2011). Public Capital in Resource-rich Countries: Is there a Curse? Centre for the Study of African Economies Working Paper, Department of Economics, Oxford.

Bernanke, B. S. and R. S. Gürkaynak (2001) Is Growth Exogenous? Taking Mankiw, Romer, and Weil Seriously. Working Paper W8365. Cambridge, Mass.: National Bureau of Economic Growth.

BP (2011). Statistical Review of World Energy, June.

Bu, Y. (2006). Fixed Capital Stock Depreciation in Developing Countries: Some Evidence from Firm Level Data. *The Journal of Development Studies* 42(5): 881–901.

Cevik, S. (2011). Policy Coordination in Fiscal Federalism: Drawing Lessons from the Dubai Debt Crisis. IMF Working Paper 1147, Washington, DC.

Conference Board (2012). Total Economy Database online, accessed March 13, 2012.

Corden, W. M. and J. P. Neary (1982). Booming Sector and De-industrialization in a Small Open Economy. *The Economic Journal* 92 (December): 825–848.

Elbadawi, I. and A. Gelb (2010). Oil, Economic Diversification and Development in the Arab World. ERF Policy Research Report #35, December.
Elbadawi, I. and R. Soto (2013). Fiscal Regimes In and Outside the MENA Region. *Middle East Development Journal* 5(December): 1–25.
(2012). Resource Rents, Political Institutions and Economic Growth. Mimeo, Economic Research Forum.
Elbadawi, I.; L. Kaltani and R. Soto (2012). Aid, Real Exchange Rate Misalignment and Economic Performance in Sub-Saharan Africa. *World Development* 40(4): 681–700.
Fiorillo, F. and A. Sacchi (2011). Free-riding or Internalizing? An Opportunistic View on Decentralization versus Centralization. CESIFO Working Paper No. 3328. IFO Institute, Munich, Germany.
Gelb, A. (1988). *Oil Windfalls: Blessing or Curse?* World Bank. Oxford University Press.
Gollin, D. (2002). Getting Income Shares Right. *Journal of Political Economy* 110(2): 458–474.
Gylfason, T. (2001). Natural Resources, Education and Economic Development. *European Economic Review* 45: 847–59.
(2011). Natural Resource Endowment: A Mixed Blessing? CESIFO Working Paper 3353. IFO Institute, Munich, Germany.
Gylfason, T.; T. T. Herbertsson and G. Zoega, (1999). A Mixed Blessing: Natural Resources and Economic Growth. *Macroeconomic Dynamics* 3: 204–225.
International Monetary Fund (2011). United Arab Emirates. Staff Report for the 2011 Article IV Consultation, Washington, DC.
Kehoe, T. and K. J. Ruhl (2008). Are Shocks to the Terms of Trade Shocks to Productivity? *Review of Economic Dynamics* 11(4): 804–819.
Krueger, A. O. (1974). The Political Economy of the Rent-Seeking Society. *The American Economic Review* 64(3): 291–303.
Prescott, E. (1998). Needed: A Theory of Total Factor Productivity. *International Economic Review* 39(August): 525–552.
Rigobon, R. (2010). Commodity Prices Pass-Through. Central Bank of Chile, Working Papers # 572. Santiago, Chile
Robinson, J. A. and R. Torvik (2005). White Elephants. *Journal of Public Economics* 89: 197–210.
Robinson, J. A.; Torvik, R. and T. Verdier (2006). Political Foundations of the Resource Curse. *Journal of Development Economics* 79: 447–468.
Rodrik, D. (2005). Why We Learn Nothing from Regressing Economic Growth on Policies. Mimeo, Harvard University.
Røed Larsen, E. (2003): Are Rich Countries Immune to the Resource Curse? Evidence from Norway's Management of Its Oil Riches, Discussion Paper 362, Oslo: Statistics Norway.
Sachs, J. D. and A. M. Warner (1995). Natural Resource Abundance and Economic Growth. NBER Working Papers 5398, National Bureau of Economic Research, Inc.
(1999). The Big Push, Natural Resource Booms and Growth. *Journal of Development Economics* 59: 43–76.
Schmidt-Hebbel, K. (2016). Fiscal Institutions in Resource-rich Economies: Lessons from Chile and Norway this volume.
Solow, R. (1956). Technical Change and the Aggregate Production Function. *Review of Economic and Statistics* 39: 312–320.

Soto, R. and Y. Rashid (2016). "Labor Markets in Transition". in *The Economy of Dubai*, A. Al Faris and R. Soto (eds), London: Oxford University Press.

Soto, R. (2016). Policy Goals, Fiscal Institutions and Macroeconomic Management in the United Arab Emirates, unpublished manuscript, Economic Research Forum.

Stroombergen, A.; D. Rose and G. Nana (2002). Review of the Statistical Measurement of Human Capital. Mimeo, Statistics New Zealand, November.

Trabelsi, M. and D. Fadhel (2011). The Financial Sector in Dubai. Mimeo, Dubai Economic Council.

UAE National Bureau of Statistics (2011). National Accounts Database.

Vazquez-Alvarez, R., (2011). Economic Rents from Migration: Potential Winners & Losers of an Admission Policy Change. World Bank mimeo, Washington DC.

Villafuerte, M. and P. Lopez-Murphy (2010). Fiscal Policy in Oil Producing Countries during the Recent Oil Price Cycle. IMF Working Paper No. 10/28, February. Washington, DC

Woolcock, M. (2001). The Place of Social Capital in Understanding Social and Economic Outcomes. *Canadian Journal of Policy Research* 2(1): 11–17.

World Bank (2011). *World Development Indicators*, Washington, DC: The World Bank.

World Economic Forum (2011). *Global Competitiveness Report 2011–2012*, Geneva.

13

Understanding and Avoiding the Oil Curse in Sudan

Kabbashi M. Suliman[1,2]

ABSTRACT

This chapter examines the impact of the oil boom, blessing or curse, on Sudan's economy, analyzes the key features of the country's growth experience before and after the commercial production of oil and articulates their underlying political economy issues. The results show that the contribution of oil to real growth has been strong; however, Sudan is confronted with the Dutch Disease and grave fiscal management challenges. Moreover, the political economy regime engendered by the elites' distributive politics tends to magnify the impact of shocks and it worsened the procyclicality of fiscal policy and contributed to the excessive currency appreciation. Overall these factors have contributed to premature de-industrialization, diluted the economic and social impact of the largely oil-driven growth and further weakened the state institutions. The policy implications of these findings are indicated.

Keywords: Oil curse, oil rents, economic growth, Sudan

1 Department of Economics, Faculty of Economic and Social Studies, University of Khartoum, Kmsuliman1@yahoo.co.uk.

2 I would like to thank Ibrahim Elbadawi (Dubai Economic Council, UAE) for thoughtful inputs and constructive comments. I also appreciate the helpful comments and suggestions from Raimundo Soto (Universidad Catolica de Chile, Santiago, Chile) and from the participants in the Cairo workshop on "Understanding and Avoiding the Oil Curse" as part of the Economic Research Forum's Research Initiative on Arab Development (RIAD). The RIAD grant is duly acknowledged. I, however, remain responsible for the views expressed and for errors.

13.1 Introduction

From its independence in 1956, until 1999, Sudan was dependent primarily on agricultural exports for foreign exchange earnings. However, as of 1999, this dependence has shifted to crude oil. Between 1956 and 1998, the average contribution of cotton, sesame and gum-arabic to exports was 42.7%, 10% and 11% respectively. In the period between 1999 and 2010, however, the contribution of oil to national exports was 82.3% on average. This dramatic change in exports is explicable by the oil boom brought about by the global surge of oil prices in the early 1990s to the 2000s. State-owned companies and other multinational firms dominate oil production in Sudan. The state decides the utility of oil revenues and their distribution and depletion. The rise in oil production has coincided with a successful IMF staff moderated program since 1997 and the Comprehensive Peace Agreement sealed in 2005. The combined effect of these developments had sustained a decade-long stable growth episode. Real GDP per capita growth increased more than seven-fold in 1999–2010, compared to a meager 0.6% average growth for 1961–98. Notwithstanding these positive developments, Sudan's economy has been transformed into an oil-dependent economy. This is true even after the secession of South Sudan, which owns around 75% of the former nation's productive oilfields. The resulting change in the country's exports is reflected in an adjustment to the share of traditional stalwarts (agricultural products and minerals) rather than a change in export sophistication. This could indicate that the Sudanese economy has contracted Dutch Disease, whereby the sector that previously drove growth began to decay as a result of the exportation of discovered mineral resources. Moreover, the resultant shift of the public finance revenue base from taxes to oil revenues could pose significant challenges for macroeconomic management, including the negative political economy implications of oil windfall.

Both theory and evidence confirm that natural resources provide opportunities for income growth, but continued dependence on resource rents (especially oil windfalls) presents serious concerns. Generally, these concerns arise from a number of factors, including: the perceived erosion of the terms of trade for commodities, the volatility and declining trend of commodity prices, the negative externalities of specialization on extractive industries and the downside of the political economy of the resource rents. See, for example, Gelb (1988), Auty (1990), Karl (1997), Sachs and Warner (1999), and Corden and Neary (1982), among others.

It is possible that the decade long, oil-driven growth in Sudan also conceals serious constraints that become more binding following a drop in the price or supply of oil. There is a need for a better understanding of the transformative role of oil in Sudan, including the impact on macroeconomic performance, economic diversification and socio-political development, so that relevant policies for enhancing sustainable growth can be devised.

Since commercial production in 1999, oil revenues dominated Sudan's export earnings and public finances. The broad objective of this chapter is to explore the effects of oil dependency on real growth and examine the outcomes of the fiscal response to the oil boom. More specifically, the following questions will be addressed: How has oil shaped the Sudanese economy? How challenging has oil economic management been? What are the main features of the fiscal and exchange rate regimes? How do they reinforce or counter the oil cycle? How might have oil dependency complicated economic diversification in the longer run? What are the underlying political economy issues behind the chosen policies and development strategies? How might they have been influenced by oil dependency? What are the relevant policies to address oil dependency and to deepen the agendas of the progressive development path in Sudan?

The analytical framework ties the peculiar features of oil to selected theoretical explanations drawn from the literature on natural resources. This is to understand the key channels for the transmission of the oil impact to the rest of the economy. The method of investigation utilizes a combination of growth accounting and time-series analysis to identify the key features that distinguish the growth experience before and after oil, and the associated policy responses. The degree to which oil-dependency shaped the choices of the political elites is examined in terms of the historical budgetary allocations. As in the revealed preference approach, the elites' preferences, implicit in the budget allocation, are broadly interpreted as reflecting their strategic priorities for development. The complexity of identifying a system of dependent variables to capture the multidimensional aspects of the oil impact is duly acknowledged.

The results show that the contribution of oil to real economic growth has been strong, but the oil curse is confirmed for Sudan as reflected in the high rate of factory closures owing to the appreciationary effects of the exchange rate and the loss of competitiveness. Moreover, the shift towards greater export concentration further complicates macroeconomic management and can thwart long-run economic diversification. Although three special fiscal institutions are introduced to insulate the economy from the oil cycle, it is found that the allocation of oil revenues through government spending has resulted in strong procyclicality. It appears that the government institutions that deal with allocation and taxation became considerably more estranged following the oil boom, reflecting that the public sector is heavily dominated by many actors with divergent and conflicting agendas. The credibility of the government, as indicated by a credible budgeting process and credible commitment to maintain low inflation and a stable exchange rate, turns out to be very low, which further obscures macroeconomic management. These findings suggest that the quality of fiscal institutions is very poor and the technology of commitment is underdeveloped due to the proliferation of rent seekers and the prevalence of the rentier mindset. In addition, the distributive consequences of the political elites' policy choices, as reflected by budget prioritization, are inconsistent with job-creating growth, private investments and hence poverty reduction. The result of this is a narrow market that tends to worsen when times are bad.

Sudan faces the multiple challenges typically found in a badly managed oil-dependent economy. There is a small oil bourse with declining prospects (at least in the near future) due to the separation of South Sudan. Nonetheless, oil and other minerals could potentially spearhead development through the enhancement of rent appropriation, combined with an efficient tax system while building the image of the state, on the one hand, and through the utilization and distribution of rent (while addressing the resource curse) on the other. Given the evidence of the weak positive fiscal linkages of oil and the need for an effective countercyclical policy, a more flexible exchange rate regime could be adopted in tandem with the use of an efficient indirect monetary instrument to enhance macroeconomic management. More importantly, growth-enhancing reform requires a well-designed ownership structure of the resource sector and the commodity-based sovereign wealth fund. This could be the most effective way to promote democracy and economic development.

The rest of the chapter is organized as follows: the following section provides an overview of the economic performance in order to provide background information on Sudan. Section 13.3 briefly reviews the literature on natural resources. Section 13.4 outlines the methodology. Section 13.5 presents the results and discussions and the final section contains the concluding statement.

13.2 Overview of Economic Performance

Sudan's economic performance has improved over the last decade, benefiting from the rise of oil, which in turn mirrored the advent of the Gezira cotton Scheme. Real GDP per capita grew on average by 4.5% between 1999 and 2010, compared with 0.6% in 1961–1998, when the Gezira Scheme was the growth driver. The cumulative nominal value of the GDP grew

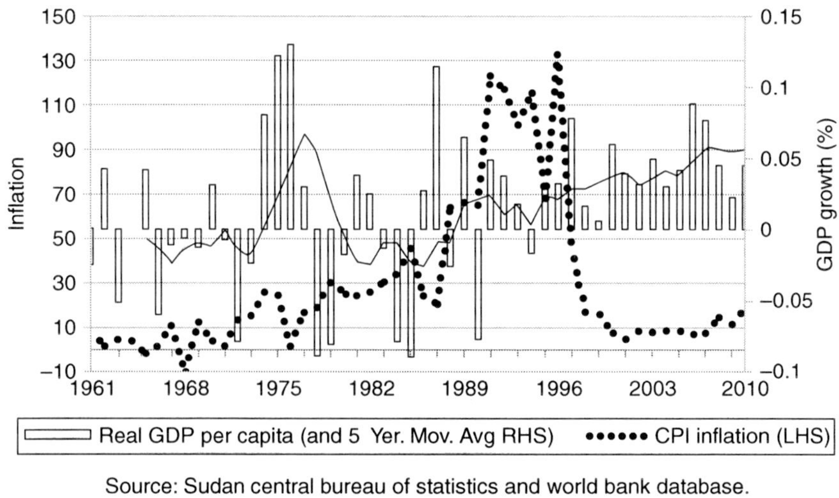

Figure 13.1: Real GDP Growth, Inflation 1960–2010
Source: Sudan Central Bureau of Statistics and World Bank Database.

thirteen times over these respective sub-periods, whereas the average annual CPI inflation dropped from 35% to 10%. Per capita income, which is an indicator of average individual welfare, increased by 61.7% (in constant 2000 US$) between 1999 and 2010, compared to 16.9% between 1960 and 1998 (Figure 13.1).

It appears that Sudan has reversed the weak economic performance before the commercial production of oil, which witnessed serious economic turmoil, especially during the 1980s and mid-1990s with double-digit inflation and negative real GDP growth. Macroeconomic management improved tremendously during the oil boom, as the country followed successful stabilization policies, resulting in very low inflation and a decade-long growth episode (Figure 13.1).

With the rise in oil exports, Sudan is increasingly becoming integrated into the global economy. The ratio of exports and imports to GDP increased by about 22% average points in 2000–2010 up from a 17% average in 1970–1999. The ratio of foreign direct investment to GDP has also surged from negligible historical records to 5.5% average for 1998–2010 (Figure 13.2).

Notwithstanding these positive developments, Sudan's economy has become highly dependent on the oil sector. Although, the share of this sector in the GDP averaged about 10.5% in 1996–2011, it quickly displaced cotton, which was the leading export crop. The share of oil in total

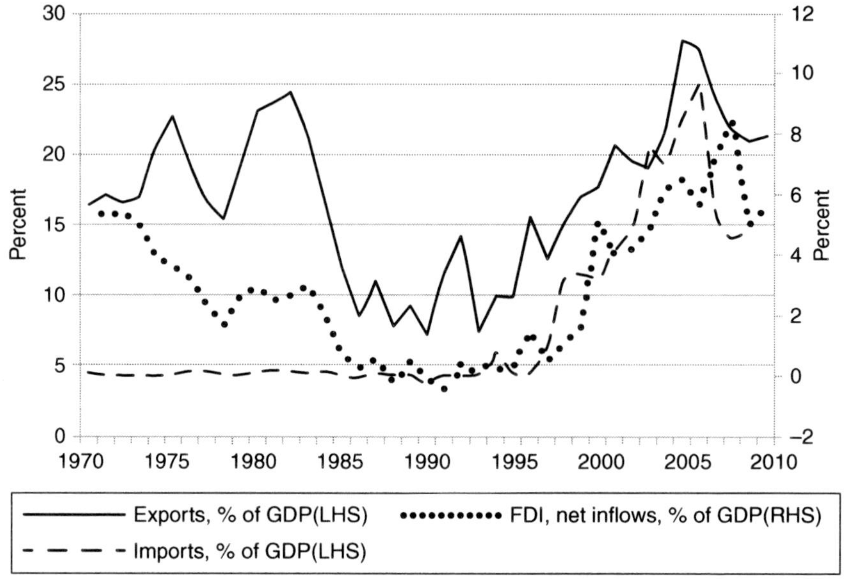

Figure 13.2: Exports, Imports and FDI Net Flows 1970–2010
Source: Sudan Central Bureau of Statistics and World Bank Database.

exports steadily increased, benefiting from the introduction of the pipeline in 1999 as well as rising prices. In early 2008, it reached 95% before declining to about 90% following the global economic crisis (Figure 13.3). The nominal per capita GDP has closely tracked development in the oil sector, but the non-oil per capita GDP showed only a marginal response. This suggests that the oil-driven growth has not been broad enough to reduce poverty and improve the welfare of the majority of the population (Figure 13.4).

Figure 13.5 clearly shows that the share of non-oil exports in total exports has declined by about 82% points over the last decade and the share of agricultural exports dropped from around 50% in 1996 to less than 1% for 2010. The appreciation of the exchange rate since 2000, following the oil boom, seems to be the primary cause of this poor performance. The favorable effect of globally surging commodity prices is not enough to compensate for the negative effect of exchange rate overvaluation. For example, despite the increasing free market price of cotton, since 2004, the quantity exported exhibited historic decline (Figure 13.6). This could essentially be due to an unfavorable exchange rate response, which remains overvalued for cotton cultivation. Elbadawi (1997) and Elbadawi and

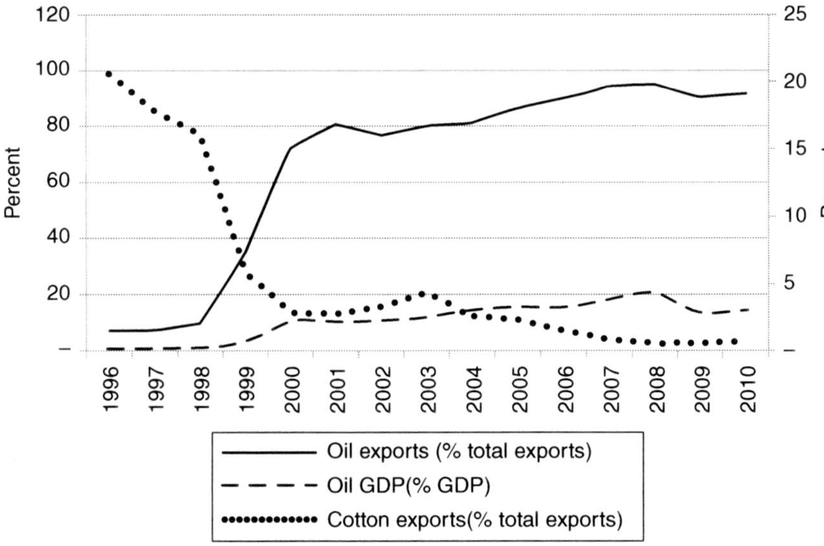

Figure 13.3: Contribution of Oil and Cotton to Exports and Oil GDP 1996–2010
Source: Sudan Central Bureau of Statistics and World Bank Database

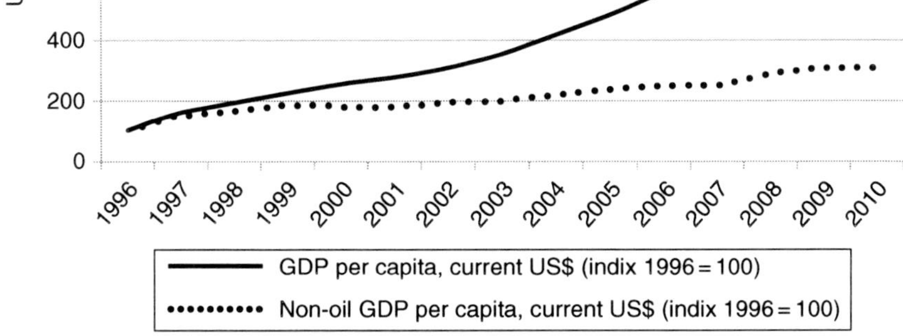

Figure 13.4: Per Capita Income 1996–2010
Source: Sudan Central Bureau of Statistics and World Bank Database

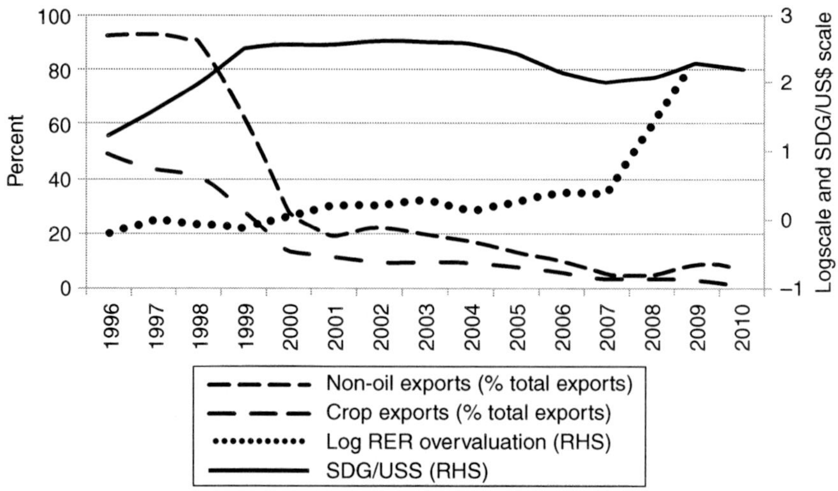

Figure 13.5: Non-oil Exports, Crops Exports and RER Overvaluation 1996–2010
Source: Bank of Sudan, World Bank and RER Overvaluation from Rodrik Dani, 2007, ERF

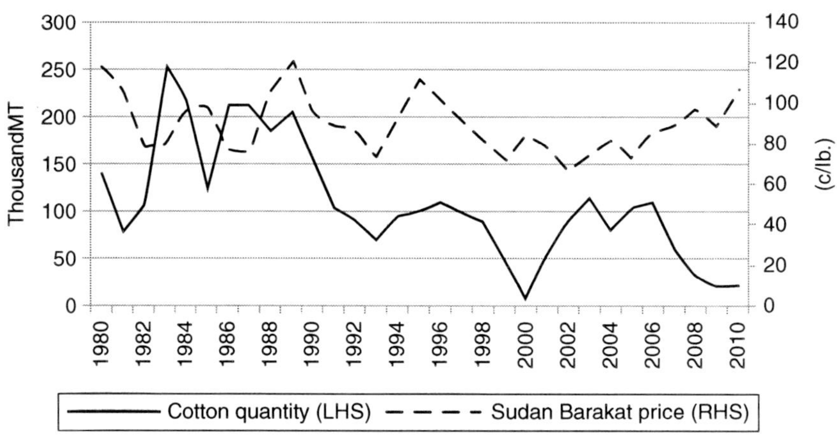

Figure 13.6: Quantity of Cotton Exported and Free Market Price 1980–2010
Source: Bank of Sudan

Kamar (2006) confirm that Sudan had experienced long episodes of massive exchange rate overvaluation even before the rise of oil money.

The decay of cotton, which was a strategic export crop, is a clear indication of the depression facing the non-oil economy and indicates that Sudan has contracted Dutch Disease. Cotton is the main produce of

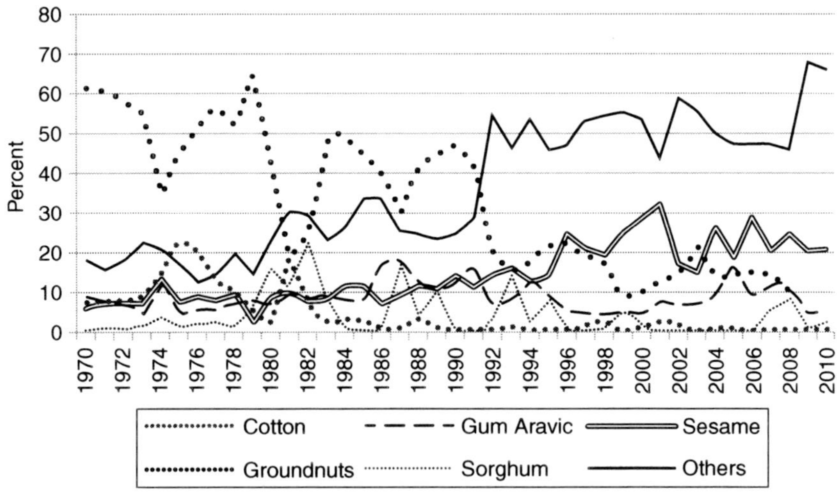

Figure 13.7: Agriculture Exports 1970–2010
Source: Bank of Sudan

the Gezira Scheme (a US$ 4 billion project). The scheme, at its prime, was not only thought of as a system for crop exports, but as a monument for and demonstration of technological innovation and diffusion geared towards modernizing Sudan (Bernal 1997). Miniature Gezira schemes were replicated in other cotton, sugar and mechanized sorghum projects. However, after oil production, cotton became a waning industry. It not only contributed less to total exports, but also weakly performed among the traditional agricultural exports. As seen in Figure 13.7, the category labeled "others" (comprising mostly of livestock products) is emerging as the main export sub-sector since the early 1990s, dominating all crops including cotton. Sesame also leads the oil seeds sub-sector over the last decade as the share of groundnuts was on the decline since the mid-1980s.

The decline of cotton production implies severe loss of jobs among 120 thousand tenants and over 2 million workers, respectively, who used to be the regular direct and indirect employees of the scheme (World Bank 2000). The depression of cotton production has clearly flattened the contribution of agriculture to exports and signals a serious loss of economic diversification. As appears in Figure 13.8, the average contribution of services, agriculture and manufacturing to real growth is 3.1%, 2.5% and 0.5%. That is, the services that are largely non-tradable drive most of the growth (as opposed to agriculture and manufacturing). This is a clear

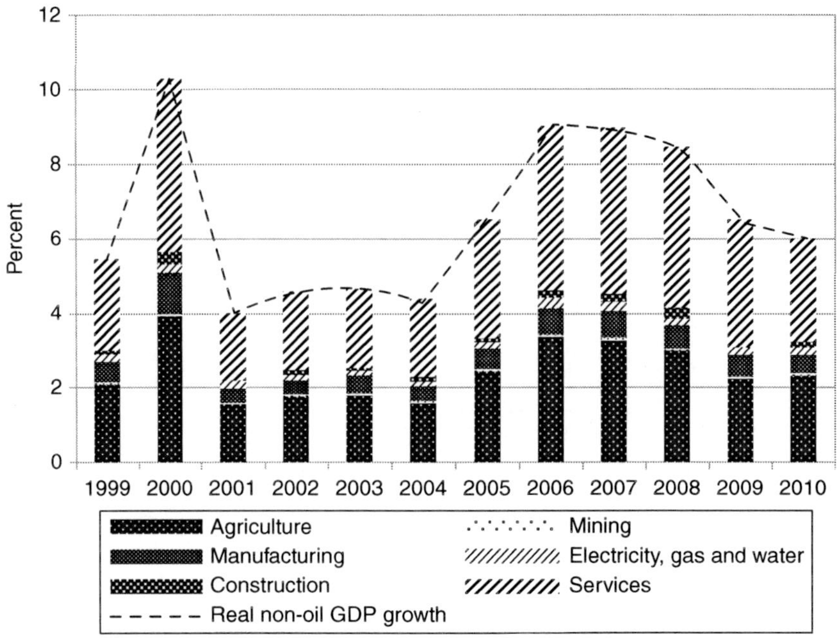

Figure 13.8: Non-oil GDP: Contribution to Growth 1999–2010
Source: Sudan Central Bureau of Statistics

indication of the shift of production from a broad-based and growth-oriented structure towards a narrow and less inclusive growth structure.

Figure 13.9 shows that the share of oil revenue in the budget has surged from 8% in 1999 to 66% for 2008, before receding to an estimated 50% by 2010. Tax effort in Sudan (the ratio of total tax to GDP) has been decreasing over time even before the commercial production of oil. For example, between 1970 and 1998, it declined by about two thirds and averaged around 6% in 1999–2011. Revenues from the main tax handles (direct and indirect taxes) mirror this weak performance. In particular, the ratio of direct tax to GDP dropped from 1.3% to 0.63% for 1999–2011. This largely reflects shrinkage of taxable private property, which could be due to the effect of a Dutch Disease mechanism. It is noticeable that the size of the public sector keeps growing despite the very low revenue earnings in 1984–1998, 5.8% of GDP through foreign borrowing, as reflected by a rising trend in the stocks of publically guaranteed external debts (Figure 13.10).

The evolution of government expenditure is shown in Figure 13.11. It can be seen that government spending has experienced a huge increase during

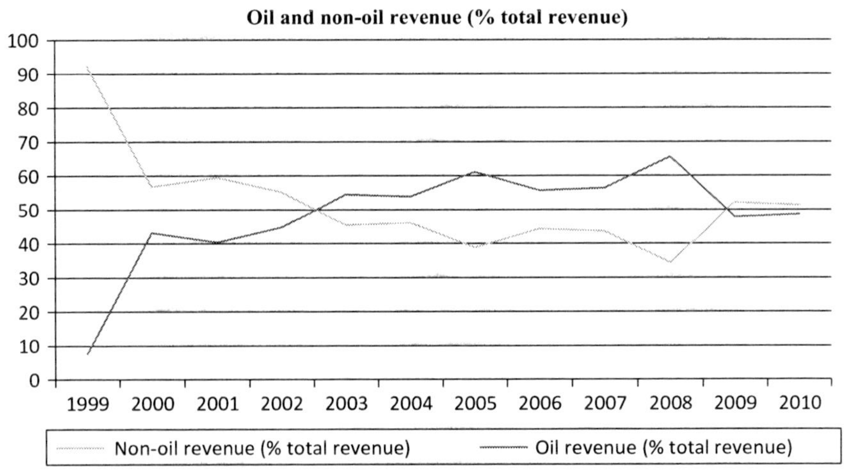

Figure 13.9: Oil and Non-oil Revenues (Percent of GDP) 1999–2010
Source: MoFNE, Sudan Government

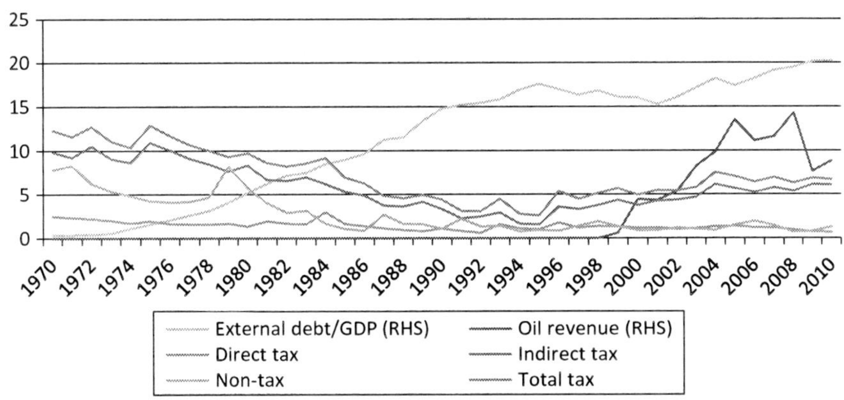

Figure 13.10: Tax, Non-tax and Oil Revenue (Percent GDP) 1970–2010
Source: MoFNE, Sudan Government, World bank

the oil decade, rising from about 7% of GDP in 1999 to 17.5% in 2010. Another notable feature of the structure of government spending is the increasing share of transfers to the sub-national states. Sudan has adopted fiscal federalism since the mid-1990s, further accelerated by the arrangements of the 2005 Comprehensive Peace Agreement (CPA). It is also notable that there is a significant squeeze on spending on goods and services following the fiscal reforms implemented under the program monitored by the IMF.

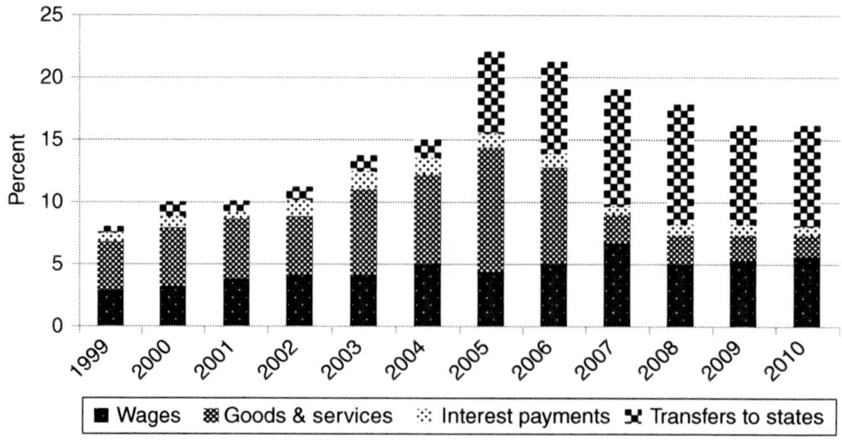

Figure 13.11: Government Expenditure (Percent of GDP) 1999–2010
Source: MoFNE, Sudan Government

According to Figure 13.12, growing government expenditures have contributed to the deterioration of the budget position. A significant overall deficit has emerged, especially since 2003, similar to that of the turbulent performance of the 1980s. The government often resorts to domestic financing and non-concessional foreign borrowing for deficit financing; which are both problematic. The dependency of Sudan's macroeconomy on oil is clearly reflected by the effect of the oil price reversal following the global crisis in 2008 (see Figure 13.13). The fiscal deficit has widened by three percentage points of GDP in 2009, mostly financed by printing money. Inflation surged to 14% in 2008, following eight years of consecutive single-digit inflation (Figure 13.1). The failure of expenditures to adjust sufficiently during the downturn of oil prices could be due to pressure on the political economy to increase spending.

Figure 13.14 pictures the GDP and non-oil GDP growth against the ratio of expenditure to revenue. The spending-to-revenue ratio appears stable up to 2005, reflecting fiscal synchronization; however, the swings in the ratio since then largely reflect the fiscal response to the volatility of oil prices. Although the 2011 budget was prepared on the assumption of the country being undivided, the country lost 50% from South Sudan's oil in the second half of 2011, which immediately negatively impacted revenue. Given the emerging stickiness of government expenditures to revenue, fiscal sustainability in Sudan is seriously called into question.

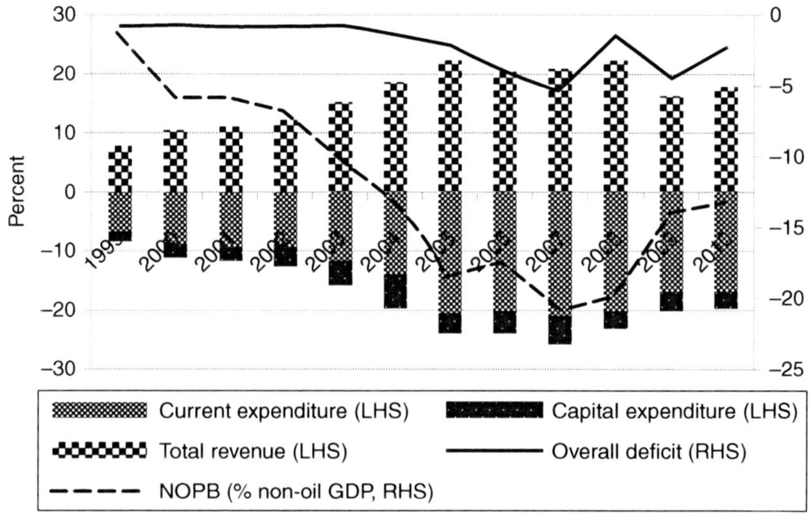

Figure 13.12: The Budget Position (Percent of GDP) 1999–2020
Source: MoFNE, Sudan Government

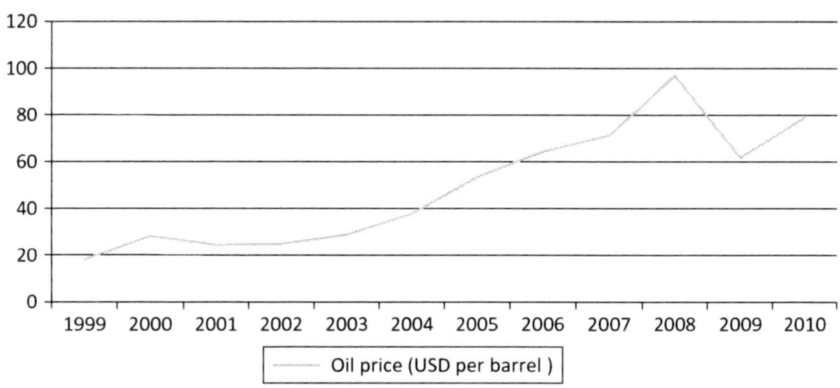

Figure 13.13: Oil Price Index 1999–2010
Source: UNCTAD

The preceding review of economic performance suggests that Sudan's economy has become uni-mineral by shifting export-dependence from a relatively diversified agricultural basket towards oil, which is a point resource. Despite the fact that real economic growth showed a strong response to the advent of oil, the structural shift of exports, as well as the

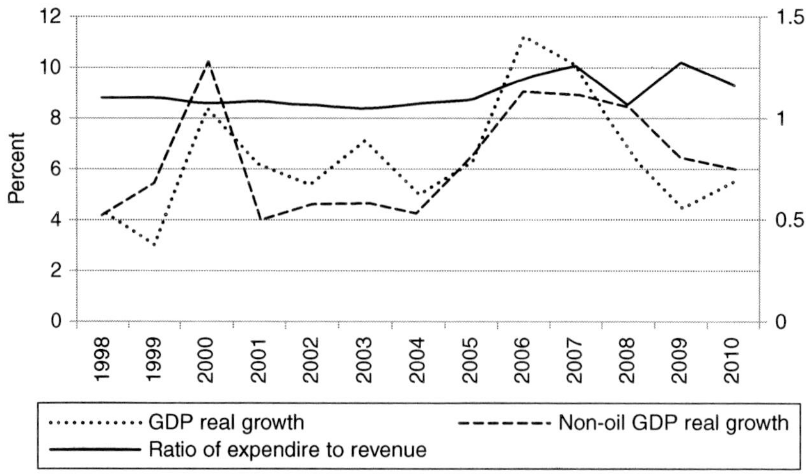

Figure 13.14: GDP, Non-oil GDP Growth and Ratio of Expending to Revenue 1998–2010
Source: Bank of Sudan and the World Bank.

channeling of oil income through the public sector, brought a number of macroeconomic management challenges to the fore. Firstly, it appears that oil revenues are largely spent by the government and the emerging downward rigidity in public expenditure tends to complicate the macroeconomic response. Second, the exchange rate continues to appreciate during the oil period despite the oil price bust of 2008–2009. Expansionary fiscal policy and growing aggregate demand, driven by the oil boom, also appear to be translated into non-tradable inflation and further exchange rate appreciation. Third, oil-driven real economic growth does not seem to be sufficiently broad-based. Generally, due to this exclusion of the masses, rents can accrue to the elites. Such narrow-based growth can neither be sustainable nor can it serve as catalyst for innovative economic activities and more economic diversification.

Although sustaining a relatively stable decade-long oil-led growth episode in a post-conflict country setting is rare (World Bank 2009), the economic performance exhibits symptoms of the resource curse. This is reflected by the increased dependency on mineral resources, high government spending, exchange rate appreciation, deteriorating budget deficit and increased inflation. These symptoms indicate major distortions in economic equilibrium arising from aggregate demand manipulation and market rigidities (i.e., government failure). This clearly indicates the conflictuality of the political economy of natural resources. It would be more informative to further examine the channels transmitting the effects of oil to the rest of the economy, including an assessment of the political economy curse. The outcome from this analysis is expected to help in the identification of pertinent policies to address the oil curse, particularly in the context of the structural shift caused by the withdrawal of South Sudan's share from the oil bourse. The next section contains a brief review of the literature on the resource curse to motivate the method of the analysis.

13.3 The Resource Curse: Theory and Literature Review

There is voluminous and sophisticated literature on the link between natural resources and growth, as well as on the link between natural resources and conflict. However, the analytical framework ties the peculiar features of oil to selected theoretical explanations drawn from this literature to understand the key channels for transmission of the oil impact to the rest of the Sudanese economy.

It has been observed for long time that possession of natural resources or mineral wealth, particularly oil, may not lead to an improvement in the population's welfare. The experience of top performers with no exportable natural resources, for example, the East Asian countries, is often cited in contrast to the economic performance of the less developing countries (LDCs) that are natural resource exporters, especially in sub-Saharan Africa. There is growing debate regarding whether natural resources are a curse or a blessing, and how the abundance or dependence on natural resources could be a curse? Gelb (1988); Auty (1990); Karl (1997); Sachs and Waner (1999 and 2001); and Corden and Neary (1982), among others, have provided articulated theoretical arguments and empirical work on these questions. Much of this work confirms the inverse correlation between economic growth and natural resource exports.

Oil is often viewed as different from other natural resources based on at least three points. First, oil is singularly the most strategic energy source of the world, and advanced countries are strategically dependent on less developed oil exporters. Second, oil is the largest multinational business with leading multinational firms directly and indirectly involved in oil trade and investment. Finally, the wedge between per unit cost of production and market price is very large, implying huge economic rent (Amuzegar 2001 and Al-Moneef 2006). Thus, the ownership of oil resources and the distribution of oil rents are critical for national interest and welfare.

Gelb's (1988) explanation of the resource curse is very pertinent for understanding the transformative role of oil. The author articulated a theoretical framework drawing from four strands of literature: the linkage theory, the neoclassical growth theory, the export instability theory and the Dutch Disease theory.

The linkage theory provides explanation for the role of the natural resource sector as catalyst for growth. Gelb differentiates between three key linkages: the production, consumption and the fiscal linkages. The production linkage looks at forward and backward linkages. Forward linkages consider the potential stimulating effects from the leading sector that feed into the rest of the economy as intermediate inputs. Backward linkages look at supply industries, which involve provision of input factors, technologies, materials and services for the leading sector. In the case of oil, the effect of forward linkages could be substantial for industrialization, and hence, the ability to diversify in terms of energy input and the derivative industries that provide business solutions. However, backward linkages could be weak due to specialized production technology.

The consumption linkage refers to the impact of the expenditure of resource revenue on the economy. Generally, resource income is initially more likely to be spent on new imports. Once a sufficient demand for these imports is established, it might eventually result in import substituting industries, which in turn contribute to the expansion of the productive base in the local economy (Hirschman 1977). However, Gelb (1988) warned of the opposite impact: that higher spending on imports might compete successfully against local industries by inducing a crowding out effect. Local industries could be displaced due to a rising demand for scarce factors of production (especially skilled labor) and an emerging extractive industry.

The fiscal linkage relates to whether the resource rent is state owned, with revenues being directly appropriated by government, or taxed at the export point. Hence, the actual effect of the fiscal linkage depends on (i) the size of the rent accruing to the government, which in turn depends on the bargaining power of the state and the resource extracting company and (ii) the productive utilization of the appropriated or taxed funds to diversify the economy away from resource dependence.

The neoclassical growth theory uses comparative statics analysis based on the assumption of utility and profit maximization as well as on stable relations between inputs and outputs. Economic growth is characterized as a process of augmenting the production frontiers by enhancing the quantity of factor inputs and their efficient allocation. In many LDCs, lack of capital limits growth due to foreign exchange, savings and fiscal revenue constraints. In this context, resource rents help to relax these constraints. However, major distortions to the equilibrium situation arising from exchange rate appreciation and inflation might shift the productive structure towards non-tradable goods and increase the dependency on natural resources. These can be a result of government deficit, foreign exchange controls and monetary growth (Gelb 1988).

The export instability theory considers the short-run gains of high income from resources against the costs of income volatility due to fluctuations in the free market price of these resources. The stability of the income stream is especially important when the resource rents are key inputs into investments. In such cases, the volatility of the resource price can seriously compromise the saving investment link and hence, economic growth. Crain and Devlin (2002) have shown that the costs associated with the inability to predict revenue fluctuations were enormous. In several cases, the huge gains accrued by resource-exporting countries during the booms have been engulfed during downswings.

According to the Dutch Disease theory, Gelb (1988) pointed out that, unlike other theories, a boom of the natural resource sector beyond the initial favorable impact of the resource discovery can cause a serious reallocation of resources in the domestic economy. This results in the concurrence of surging and declining sectors. Essentially, the resource boom can affect the economy through two channels (see Corden and Neary 1982). The first channel works through spending (the demand side), where the higher income resulting from the boom is spent on traded and non-traded goods leading to higher prices for non-tradables. The resulting appreciation of the exchange rate can induce imports, which makes local production of these goods less profitable. The spending effect can also cause reallocation of factors of production out of the tradable sector, which will be hampered by imports, into the production of non-tradable goods (Gelb 1988). The second channel works through the supply side due to resource movement triggered by high returns in the booming sector. Factors of production move out of non-booming activities, notably agriculture and manufacturing into the booming resource sector. In the case where there is near full employment, competition for factors of production tends to raise inflation, leading to exchange rate appreciation and further dampening the non-booming activities, (see also Corden and Neary 1982). This process can eventually result in greater dependence on natural resource for earning foreign exchange (Gelb 1988).

Recently, Elbadawi and Gelb (2010), after reviewing vast literature on oil and non-agricultural resources, argued that there is no all encompassing theory that best explains the fundamental phenomenon of the resource curse, the channels through which it operates and the heterogeneity of a country's experiences. It takes, at the least, all four approaches combined to explain the oil curse in the Arab countries (Elbadawi and Gelb 2010, p. 41).

The discussion in Section 13.2 implies that attaining beneficial production linkages could be very limited in Sudan. This makes fiscal linkages the most important determinant of the ultimate benefits from oil revenues. Generally, weak production linkages in resource-rich countries are associated with political economy issues such as rent-seeking, corruption, abuse of resource income for financing armed conflicts rather than investments and neglect of the building of state institutions and professionalism. It is widely argued that resource-abundant countries are more likely to engender a political state that is factional or predatory, which distorts the economy and hampers growth in the pursuit of rents and personal gains (Auty 2003; Ross 1999; Renner 2002).

Essentially, the political economy approaches elaborate the "rentier state" paradigm, which incorporates three basic mechanisms relating to the rentier effect, the repression effect and the modernization effect (Mahdavy 1970; Beblawi and Luciani 1987; Ross 2001). The rentier effect refers to two basic linkages working through the taxation effect and the spending and patronage networking effect. The taxation effect is at the heart of the bargain between the rulers and citizens. In an oil-dependent country, rulers may feel less of a need to tax their citizens because they have enough easy oil income; the citizens, in turn, would have less incentive to demand accountability from their governments. The governments become less transparent, accountable and responsive to citizens. The spending effect relates to the use of budgets that are exceptionally large and unconstrained to pay for patronage and the building of loyal groups. The repression effect is concerned with the ability and interest of elites in the rentier state to use repression to pacify opposition groups and keep citizens in check through the use of military, police and security (including media control and censoring). Ross (2001) argues that rentier elites inhibit democratization through a "modernization" effect, by retarding, for example, greater occupational specialization and professionalism that tend to produce more accountable governments.

Lam and Wantchekon (1999) argued that rent-seeking is not only an occupational choice, but also a political choice that affects the nature of political regimes. They investigate how economic growth, income distribution and allocation of political power simultaneously evolve when resources are discovered and show that an increase in the elites' distributive influence generates autocratic regimes and aggravates the decline of the economy – a phenomenon they term "political Dutch Disease."

Following the rentier state paradigm, Wintrobe (1998) proposes a general equilibrium framework that links political power and economic activities to describe the choice between loyalty and repression that confront non-democratic rulers. He showed that the equilibrium of power is determined by the dictator's preference for power, the capacity to turn money into power and the effects of the dictator's power on the economy. Accordingly, dictatorships are classified into four types: totalitarian (high repression, high loyalty), tyrants (high repression, low loyalty), timocrats (low repression, high loyalty) and tinpots (low repression, high loyalty) (Wintrobe 1998). In this framework, the dictator can manipulate the budget: the more power he commands, the higher the budget. At the extremes of repression and loyalty, a dictator can be a pure power maximizer (i.e., a totalitarian or a pure personal consumption maximizer – a tinpot). Ali and Elbadawi (2012) used a sequential non-delegation game theoretic model in which the dictator's budget depends on the value of his economy's natural resources per capita, while "loyalty and repression" describe his preferences rather than being inputs into the power objective function as in Wintrobe (1998). This model assumes that the dictator can use resource rents to buy population loyalty through expanding public sector employment, but can also finance state security organs to repress dissent.

The equilibrium solution of the model derives the relationship between natural resources per capita and the size of the public sector labor force consistent with the dictator's choice to remain in office. Specifically, possible reactions of power maximizing dictators depend on a given level of natural resources per capita such that the dictator sets up public employment for all levels of natural resources above this given level and choose to repress citizenry for all levels below it. They provide empirical support for this theory, concluding that elites in less populous natural-resource-dependent countries tend to use public employment to buy loyalty, while those in populous resource-dependent countries set up a repressive state apparatus (distributing less through the labor market).

The quantitative literature on the theories of natural resource development and conflict of distribution largely use cross-sectional global data aimed at assessing ultimate utility from resource rents. In almost all the empirical models, the results confirm the resource curse for commodity exporters (see Van der Ploeg 2006; Rodrik 2003; Rosser 2006 and Frankel 2010 for surveys). This chapter utilizes a case study approach to identify the key features that distinguish Sudan's growth experience before and after oil production and to test for possible transmission channels for the oil curse (including the political economy curse).

13.4 The Methodology

Most studies testing the resource curse hypothesis have been conducted in the context of growth framework (see, for example, Sachs and Warner 1999; Manzano and Rigobon 2003; Neumayer 2004; Papyrakis and Gerlagh 2004). Generally, the results imply direct causation from resource dependence to low growth and little or no long-term technological dynamism for the resource-dependent economies. While these findings also apply to Sudan, the side effects of the recent oil boom are much more devastating, as reflected in the Dutch Disease and fiscal management challenges. The standard growth accounting decomposition is used to review the contribution of factor inputs and technological change to economic growth before and after the oil boom. If productivity spillovers between the booming oil sector and the non-booming sectors (agricultural tradable, manufacturing and services) exist, then labor employment would be augmented after the boom. Additionally, channeling the oil revenue through the budget could strengthen the fiscal management, but this requires credible fiscal policy to ensure a balanced budget over the economic cycle.

The following simple growth accounting equation, which assumes that the aggregate production is described by the Cobb–Douglas relation, is used to decompose the sources of growth:

$$Y_t = A_t K_t^{\alpha} L_t^{(1-\alpha)} \tag{1}$$

where, Y is the level of aggregate output, K is the level of physical capital, L is the level of labor input, A is an index for the aggregate state of technology or total factor productivity (TFP), α and $(1-\alpha)$ are parameters measuring respectively the elasticities of capital and labor and t is time index. Log-differentiating Equation (13.1) with respect to time gives an expression for the growth accounting equation.

$$\frac{\dot{Y}}{Y} = \frac{\dot{A}}{A} + \alpha\frac{\dot{K}}{K} + (1-\alpha)\frac{\dot{L}}{L} \tag{2}$$

The respective quotients denote output, capital, labor and TFP growth over time. Assuming that the economy is sufficiently competitive so that factor earnings are proportional to their productivities and that the technical progress is Hicks' neutral, α and $(1-\alpha)$ equal the factors shares in aggregate output.

Aggregate output is measured by gross domestic product (GDP) and the estimation of capital stock is based on the perpetual inventory method. Hall and Jones' (1999) method is followed to calculate the initial capital stock with a fixed depreciation rate of 9% and a gross capital formation rate of 4%. The quantity of labor is proxied by the number of actively employed, the TFP ($\frac{\dot{A}}{A}$) and α are typically statistically estimated. Equation (13.1) can also be log-linearized and expressed in per worker terms as:

$$\ln\ y_t = \ln A_t + \alpha \ln k_t, \quad y = \frac{\dot{Y}}{L}, \quad k = \frac{\dot{K}}{L}, A = \frac{\dot{A}}{A} \tag{3}$$

where α is the long-run contribution of capital to output. In this case, the growth in output per worker can be decomposed into two components relating to capital deepening and TFP growth, as in Equation (13.2). The TFP by construction subsumes the effects of various factors on growth relating to the quality of inputs and long-term technological dynamism (e.g., the skill and education of the workforce, innovation and institutions). The results from growth decomposition are used to review the contribution of factor inputs and technological change to economic growth before and after the oil boom.

As in the "rentier state" paradigm, it is hypothesized that the fiscal laxity, budgetary indiscipline and spending myopia in an oil-based economy dominated by patronage politics would result in strong procyclical spending and low credibility of the budget. Three simple tests are applied to assess the "credibility" of a fiscal response to the oil boom in order to highlight the associated fiscal management challenges. First, the "credibility of the budget," as defined in the public expenditure and financial accountability (PEFA 2005) framework is used to discern the strength of the public financial management (PFM) system. A firm PFM system is essential for fiscal discipline, efficient resource allocation and for the implementation of policies and plans consistent with achievable developmental objectives. Credible budgeting remains an important tool in this process. However, due to data limits, the first and the third performance indicators within the PEFA–PFM framework are used to assess, in aggregate, whether the budget is realistic and implemented as planned. Significant deviation of budgetary out-turns from the originally approved budget, on both the revenue and spending sides, indicates low credibility of the budget and weak fiscal discipline and institutions.

Second, Granger causality between revenue and spending is applied to determine the extent to which the volatility of oil prices drives the economic cycle through government expenditures. Evidence on causation from revenue to expenditure suggests that the fiscal policy reaction to oil price changes amplifies the business cycle in an economy (see Fasano and Wang 2002). The following simple error correction model (ECM) is used;

$$\Delta R_t = \delta_0 + \alpha_0 \Delta E_{t-1} + \gamma_0 \Delta R_{t-1} + \kappa_0 [R_t - \beta_0 E_t]_{t-1} + \mu_t \quad (4)$$

$$\Delta E_t = \delta_1 + \alpha_1 \Delta E_{t-1} + \gamma_1 \Delta R_{t-1} + \kappa_1 [E_t - \beta_1 R_t]_{t-1} + \varepsilon_t \quad (5)$$

where, R_t is revenue, E_t is expenditure, δ, α, γ and κ are coefficients and μ and ε are IID errors. The terms in the square brackets refer to the cointegrating vectors of revenue and expenditure and the coefficients κ_0 and κ_1 represent the disequilibrium errors measuring the short-run speed of adjustment to long-run revenue and expenditure in the respective equations. Significant estimates of the α_0 and κ_0 in Equation (13.4) and γ_1 and κ_1 in Equation (13.5) can indicate the direction of causation either from expenditure to revenues or from revenues to expenditure. For example, statistically significant coefficients on either ΔE_{t-1} or the error correction term, κ_0, or both, indicate that revenue causes expenditure, hence, supporting the revenue spend hypothesis. Similarly, from Equation (13.5), significant coefficients on either ΔR_{t-1} or the error correction term, κ_1, or both, suggest that changes in government expenditure causes changes in revenue, confirming the spent revenue hypothesis (Peacock and Wiseman 1961). In case the relevant coefficients from Equations (13.4) and (13.5) are both significant, the fiscal synchronization hypothesis is confirmed (Musgrave 1966).

Absence of causality suggests that the authorities set revenue and expenditure by rule of thumb, reflecting the institutional separation of allocation and taxation functions of the government (Hoover and Sheffrin 1992). Both ΔR_{t-1} and ΔE_{t-1} are interacted with a dummy over the oil period to account for a possible shift in the constant and the slopes. The variance decomposition analysis is also used as a robust check for detecting the direction and strength of causality between revenue and expenditure over the whole sample.

Third, as credibility is a critical factor for any government policy, it would be informative to test whether the recently introduced exchange rate based rule succeeded in anchoring inflation. Since 1970, the Sudanese authorities continued to intervene in the foreign exchange markets with various motives, but maintaining low inflation and a stable exchange rate were often emphasized as achievable goals. After the experience of three-digit inflation in 1991–1997, the authorities focused more on fighting inflation through exchange rate based stabilization. Of course, the inflow of oil income made the realization of this objective viable, but the government may be tempted to engage in excessive spending sprees (for example to manipulate political popularity) and the public response would result in high actual and expected inflation. Thus, the success of the exchange rule in anchoring inflation depends on the public perception of its credibility, which, in turn, determines agents' expectations about future inflation (Edwards 1998). Following Elbadawi (1997) and Edwards (1998), a model-free measure based on Cochrane's variance ratio statistic is developed to test whether the degree of inflation inertia declined after the introduction of the anchor.

Cochrane's variance ratio measures the importance of the random walk for a given time series, and is defined as:

$$\mathrm{var}(k) = (1/k)\{\mathrm{var}(\pi_t - \pi_{t-k})/\mathrm{var}(\pi_t - \pi_{t-1})\} \tag{6}$$

That is, if the series is generated by random walk process, the variance of its k differences would be k times the variance of its first difference. While point estimate of the variance ratio is used, the variance ratio test statistic (a Z statistic), proposed by Lo and MacKinlay (1988) for testing the null of random walk, is also calculated for various ks.

The political economy implications of oil rent distribution are explored in terms of budget allocation. As in the revealed preference approach, actual budget allocations before and after oil production are assumed to reveal the elites' preferences for trading-off loyalty for repression. Specifically, more allocation to public employment (e.g., high public wage bill), as well as to state governments and public services are interpreted as favoring loyalty; whereas less allocation to these categories and more security spending implies predilection for repression.

13.5 Results and Discussion

13.5.1 The Effects of Oil Dependency on Growth

The result of the growth accounting decomposition for Sudan using the primal approach of Equation (13.3) is presented in Table 13.1. The non-linear estimation of α using Equation (13.1) is 0.21, which is positive and statistically significant, but low. Thus, following the literature, α is set at 0.3. The periodization shown in the table is adopted from table 6 in Ali and

Table 13.1: *Growth Accounting Decomposition for Sudan 1961–2010*[1]

Sub-Period	Growth of Real GDP per Worker	Contribution of Capital per Worker	Contribution of Labor	TFP
1961–64	-1.0	4.2	1.3	-6.5
1965–69	-1.4	2.2	2.2	-5.8
1970–74	-2.1	1.1	3.7	-6.9
1975–79	5.6	5.3	0.1	0.2
1980–84	-0.1	1.0	1.8	-2.3
1985–89	1.6	-0.1	2.5	-0.1
1990–97	3.2	1.6	1.0	0.8
1998–2010	3.9	2.2	1.7	-0.01
1961–2010	1.8	2.0	1.7	-2.0
1961–96	0.9	2.0	1.8	-2.9
1997–2010	4.3	2.1	1.7	0.4

[1] The contributions of capital per worker and the TFP are obtained from Equation (13.3), while the reported contribution of labor is obtained using Equation (13.2).
Source: Author calculation based on data from Sudan Central Bureau of Statistics and World Bank Indicator Database.

Elbadawi (2002) for the sake of comparison. The results in both tables are very similar in terms of the order of magnitude, notwithstanding the differences in the definition and measurements of the variables. Two observations could be made from the table. First, the growth of real GDP is very low and volatile alternating between periods of positive and negative growth.

Second, although the TFP growth improves overtime, it remains broadly negative for the whole sample, implying that economic growth in Sudan is mainly driven by physical capital per worker. The positive TFP growth in 1975–1979 and 1990–1997 corresponds, respectively, to the high inflow of capital in connection with the breadbasket strategy and the commercial production of oil. The strong average growth spell after the rise of oil in 1997–2010, compared to the sub-period before the production of oil in 1961–96, seems to be largely explainable by the TFP growth and capital accumulation. The contribution of labor is 0.1 percentage point lower, implying that after oil production, the unemployment rate is much higher.

In fact, Sudan faces a classic case of Dutch Disease as reflected by the spectacular crowding out of cotton, the main export, where production dropped from 247.41 million tonnes in 1970/71 to about a quarter of a million (0.288) tonnes in 2011/12. Moreover, out of sixteen textile factories, employing about 30,000 workers, only three are operating. The statistics from the Ministry of Industry show that about 41% of all factories in Sudan closed in 2001–2007 owing to loss of competiveness generated by the appreciationary effects of the exchange rate with the consequent adverse effects on employment and diversification of the economy (see also Konandreas 2009).

The next section reviews the credibility of the fiscal stance, as indicated by credible budgeting, sufficient countercyclical fiscal policy and credible commitment to maintain low inflation and a stable exchange rate. It also considers the political economy related issues underlying budget allocation.

13.5.2 The Fiscal Management Challenges

Sudan's macroeconomic performance has become highly dependent on oil due to the channeling of oil revenues through the public sector. This contributed to (on average) 48% of total revenues in 1999–2010. The critical issue is that the allocation of oil income through the budget usually imparts the volatility of oil prices to the rest of the economy, thus, posing serious fiscal management challenges. Examples include less synchronized revenue and expenditure decisions, strong procyclical expenditures, low credibility of the budget and government plans and serious distribution conflicts between the politicians and the polity. The following issues are briefly reviewed: the credibility of the budget, the degree of pass-through of the oil cycle, credibility of the rule based exchange rate stabilization introduced since 1998 and political economy implications of the budget.

13.5.3 The Credibility of the Budget

Three special fiscal institutions are created to assist in the management of oil revenue, these are: the Oil Revenue Stabilization Account, fiscal guidelines (proxy of Fiscal Rule), fiscal responsibility legislation and the benchmarking of oil price for the budget envelope. The efficient budgeting of oil (especially insulation of the economy from the oil cycle) is set as top priority. Nonetheless, Sudan has not completed a PEFA assessment, which requires comprehensive measurements of the following dimensions of the PFM: credibility of the budget, comprehensiveness and transparency, policy-based budgeting, predictability and control in budget execution, accounting, recording and reporting and external scrutiny and audit. Despite this, a partial review of the first dimension of the PFM is presented

Table 13.2: *Aggregate Revenue and Expenditure Out-turns Compared to Original Approved Budget*

	2002	2003	2004	2005	2006	2007	2008	2009	2010
Part A									
Budgeted expenditure (in billion SDG)	5.98	6.96	6.89	11.29	17.2	17.6	22.44	21.95	25.11
Actual expenditure (in billion SDG)	5.18	7.16	7.94	10.04	14.71	15.99	22.72	20.7	24.16
Difference between actual & budgeted spending	−0.8	0.2	1.05	−1.25	−2.49	−1.61	0.28	−1.25	−0.95
Difference as percentage of total	−13.4	2.9	15.2	−11.1	−14.5	−9.1	1.2	−5.7	−3.8
Part B									
Budgeted revenue (in billion SDG)	4.82	5.87	8.21	12.75	17.09	18.25	21.5	18.37	23.71
Actual revenue (in billion SDG)	4.72	7.04	10.02	12.17	15.08	17.94	24.71	20.05	20.74
Difference between actual & budgeted revenue	−0.1	1.17	2.03	−0.58	−2.01	−0.31	3.21	1.68	−2.97
Difference as percentage of total	−2.1	19.9	24.7	−4.5	−12.3	−1.7	14.9	9.1	−12.5

Source: Economic Survey, MoFNE and the Budget Speech for various years.

in Table 13.2 for the first and third performance indicators of budget credibility. The result is not based on a standard format, but would give a sense of the overall budget performance in terms of PEFA scoring.

Part A of the table relates the aggregate expenditure out-turn to the original approved budget. This reflects the ability to implement the budgeted expenditure as a crucial factor in supporting the government's ability to deliver public services for the year as expressed in policy statements, output commitments and work plans. The rating of the first three years, 2002–2004 would score D, whereas the scoring for 2005–2007 and 2008–2010 could be C and B respectively. This indicates improvement over time. Part B shows the aggregate revenue out-turn compared to the originally approved budget. Accurate forecasting of domestic revenue is a critical factor in determining budget performance, since budgeted expenditure allocations are based upon that forecast. The performance of revenue over time is in sharp contrast to that of expenditure where the scoring declines from A for the first three years to C for the other sub-periods respectively. Low quality revenue forecasts have led to considerable deviations between actual and budgeted revenue leading to expenditure rationing. This indicates that the special fiscal institutions (created after the production of oil) are ineffective in managing oil revenue. Overall, the budgeting process lacks realism, which is attributable to weaknesses of the fiscal discipline, expenditure control and budgetary planning.

13.5.4 The Fiscal Response to the Oil Cycle

The fiscal response is largely concerned with the changes in government revenue and spending that affect economic activities. It is known that the allocation of oil revenues through government spending usually imparts the volatility of oil prices to the rest of the economy. In addition, the low credibility of the budget would further complicate the fiscal response to the oil boom due to weaknesses in the special fiscal institutions.

Figure 13.15 pictures the cyclical components of government revenue, expenditure and GDP deflated by the GDP deflator. The log level of all these variables contains a unit root: the standard HPF is applied to calculate the cyclical commandants shown in the figure.

As seen, all the variables are very volatile. The GDP cycle tends to dominate revenue and spending up to the mid-1980s, afterward, revenue and spending become more volatile and begin to dominate the business cycle. The instability of both series remarkably increased over the mid-1980s to the late 1990s, reflecting the effects of highly turbulent events during this period, which include a series of government changes, the outbreak and intensification of the civil war and successive drought episodes. These factors combined have resulted in serious economic bottlenecks, inefficiencies, social frustrations and tensions. It is notable that the fluctuation of the three series tends to stabilize following the start of oil production and further dampens around 2005 (the year of sealing the CPA). Then, high instability (similar to that of the mid-1980s) ensued. The positive growth of revenue and spending far exceeds that of output in 2004–2008, suggesting that the fiscal variables amplify the business cycle, rather than move countercyclically as expected. It is difficult to infer direction of causation from the figure; however, as

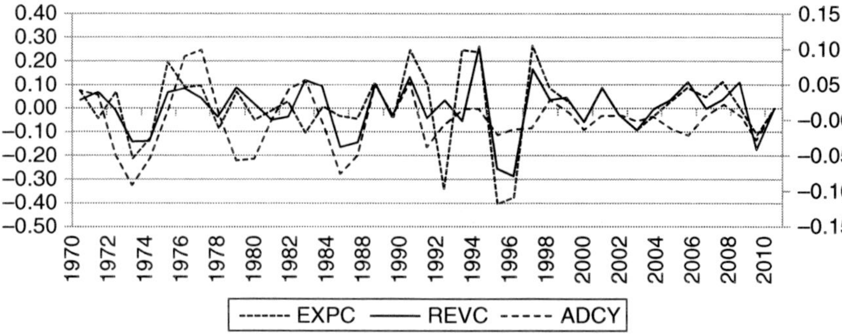

Figure 13.15: Cyclical Revenue, Expenditure and Output for Sudan 1970–2010
Source: Economic Survey, MoFNE and the budget speech for various years.

suggested earlier, Equations (13.4) and (13.5) are used to test for the direction and strength of causality between revenue and expenditure. The result is shown in Table 13.3.

According to the ADF test (not reported), the log level of revenue and spending contains a unit root. However according to Johansen (1988)'s testing, the null hypothesis of no cointegration cannot be rejected. This may be due to structural changes in the cointegrating vector. Hence, the oil dummy is added as an exogenous forcing variable to allow for this. The cointegrating vectors are estimated with an unrestricted constant and trend. The result is shown in the last row of the table. The imposition of over-identifying restriction on the revenue equation (i.e., $\beta_0 = 1$ rejects) (Chi-square 8.306 with a p-value of 0.004). This implies for the full sample that government expenditures will always be greater than revenue, the deficit is weakly sustainable and the government will not be able to clear its long-run debt without compromising its long-run development objectives. As noted by Hakkio and Rush (1991), $\beta_0 = 1$ is necessary for the government to remain solvent.

The estimated error correction models for revenue and spending reasonably pass the standard diagnostic tests and confirm bi-directional causality between revenue and expenditure for the full sample (see first column for each equation). The variance decomposition is calculated to further ascertain the direction of causation. Table 13.4 presents the result. As seen from the expenditure equation, more than 80% of variation in expenditure is accounted for by past revenue, whereas 42% is accounted for by own innovations. This pattern is consistent throughout the sample, except for the first two years. However, for the revenue equation, it is clear

Table 13.3: *Causality Test Based on ECM*

	Revenue Equation[1]		Spending Equation[2]	
Regressors	(1)	(2)	(1)	(2)
Revenue(∂R_{t-1})	−0.404*	−0.404*	−0.102	−0.11
	(−1.828)	(−1.757)	(−0.385)	(−0.371)
Expenditure (∂E_{t-1})	0.4170**	0.238	−0.021	−0.074
	(2.441)	(1.294)	(−0.104)	(−0.312)
$(D_t)^*(\partial R_{t-1})$		−0.434		−0.0224
		(−0.617)		(−0.247)
$(D_t)^*(\partial E_{t-1})$		0.883		0.591
		(1.079)		(0.56)
ECM_{t-1}	0.171	−0.102	−0.320**	−0.26
	(1.27)	(−0.571)	(−2.383)	(−1357)
Oil dummy D_t		0.169		0.008
		(1.566)		(0.064)
Constant	0.199	−0.019	0.0001	−0.006
	(0.542)	(−0.477)	(0.006)	(−0.131)
Normalized cointegrating vectors/4	LR − 0.846*LE	(0.03)	LE − 1.192*LR	(0.039)

[1] Model (1). $T = 40$ (1970–2010); $R^2 = 0.17; \tilde{\sigma} = 0.21$ and DW = 1.918. *F*-stat: $F(9,32)$ 2.054 (0.087).
AR $\chi^2(1)$ 0.321(0.571): Test for serial autocorrelation of residuals (H_0: no autocorrelation)
$X_i{}^*X_j$ $\chi^2(1)$ 0.471 [0.983]: Test for heteroscedasticity (H_0: no heteroscedasticity)
Normality χ^2 (2) = 1.177 [0.555]: Test for normality of distribution of residuals (H_0: normality).
[2] Model (2). $T = 40$ (1970–2010);
$R^2 = 0.16$; $\tilde{\sigma} = 0.25$ and DW = 2.052. *F*-stat: $F(3,35)$ 2.311 (0.093).
AR $\chi^2(1)$ 1.231 (0.267): Test for serial autocorrelation of residuals (H_0: no autocorrelation)
$X_i{}^*X_j$ $\chi2(1)$ 1.980 [0.159]: Test for heteroscedasticity (H_0: no heteroscedasticity)
Normality $\chi2$ (2) = 5.931 [0.052]: Test for normality of distribution of residuals (H_0: normality).
[3] Asterisks ***, ** and * denote significance at less than1%, 5% and 10% level, respectively.
The *t*-statistic is in parentheses.
Numbers in parentheses are symptomatic standard errors.
[4]Author calculation based on Economic Survey, MoFNE, and the Budget Speech for various years.

that own innovations accounted for over 94% of forecast error variance for a long time, while about 20% of variation is accounted for by past expenditure. The result from variance decomposition analysis supports the conclusion that the revenue spent hypothesis persists in Sudan. Such inconsistency of the causality test and variance decomposition may be due to a structural break. As suggested earlier, the ECMs are re-estimated, including the oil dummy and the interaction terms. The result for each equation is shown in the second column. The bi-directional causality disappears and none of the coefficients of interest are statistically

Table 13.4: *Forecast Error Variance from Revenue and Expenditure Equations PP-period Ahead*[1]

One standard error shock in:	Number of periods	Expenditure (E_t)	Revenue (R_t)
Expenditure (E_t)	**0**	**1**	**0.32**
	1	0.89	0.53
	2	0.74	0.67
	3	0.61	0.75
	4	0.51	0.8
	5	0.43	0.82
	6	0.38	0.85
Revenue (R_t)	**0**	**0.32**	**1**
	1	0.26	0.99
	2	0.22	0.98
	3	0.19	0.97
	4	0.17	0.96
	5	0.16	0.95
	6	0.15	0.94

[1] The forecast error decomposition provides a decomposition of the variance of the forecast errors of the variables in the cointegrated VAR at different periods ahead (PP). The generalized forecast error variance decomposition is used to explicitly allow for the contemporaneous correlation between the orthogonalized shocks and the shocks to the other equations in the VAR system.

significant. This finding indicates the existence of a considerable institutional separation between the allocation and taxation functions of the government. More importantly, it reveals that the state sector is dominated by actors with divergent and very conflicting agendas (Hoover and Shefrin 1992).

It should be noted that, the framework of Equations (13.4) and (13.5) does not account for issues of control and fiscal sustainability. However, the data tracking reveals that revenue and expenditure decisions are more mutually causal for the full sample than over the oil sub-period. The deterioration of the fiscal foresight after oil production is mainly due to a lack of coordination among fiscal institutions and greater separation of revenue and expenditure decisions, which reflects the dominance of the state-driven myopic spending. It is also expected that the weak fiscal performance and strong procyclicality of spending would cause the oil cycle to spillover to the real exchange rate. The extent to which this process affects the credibility of the ongoing rule based exchange rate stabilizing intervention is indirectly tested in the next section.

13.5.5 The Credibility of the Exchange Rate Regime

Sudan adopted a fixed exchange rate regime in 1957, which at the time was a prerequisite to access the IMF. This arrangement remained effective for about two decades. However due to the balance of payments crisis that followed the collapse of the breadbasket strategy and mounting pressures on the foreign exchange market, a new era of multiple exchange rates and exchange controls started in the late 1970s. The advantages of the multiple exchange system turned out to be less proportionate to its costs. This was due to incompatibilities between fiscal and monetary policies and the illegal leakages between floating and fixed exchange markets. A series of exchange rate unifications, affected through devaluations, were attempted in 1979, 1987, 1992 and 1997, but were largely unsuccessful. The difference between official and free market rates continues to widen and the value of local currency continues to deteriorate, reflecting major distortions in the economy. In 1998, another unification policy was introduced and the foreign exchange rate controls were relaxed. The rise of oil money and the buildup of foreign reserves have made this policy more successful in comparison to previous attempts. The Central Bank started to use "foreign exchange operations" for liquidity mopping up operations due to the absence of indirect monetary instruments following the adoption of Islamic banking in 1983. A dedicated chamber for foreign exchange dealing was established in 1999 for this purpose.

The preceding descriptive review of the macroeconomic performance reveals that the exchange rate overvaluation continued to grow over the oil period, which raises concerns about the effectiveness of the nominal anchor and about the optimal exchange rate regime for Sudan, especially in view of the evidence on strong procyclicality of fiscal variables. Generally, the optimality of the exchange rate regime in an oil-based economy is largely determined by its capacity to support an effective countercyclical policy to insulate the non-oil economy from oil price volatility and medium- to long-term cycles (Boughrara and Elbadawi 2011).

The credibility of the exchange rate based anchoring of inflation is tested using Equation (13.6) and the result is portrayed in Figure 13.16. The periodization roughly corresponds to three different patterns of inflation developments. The sub-period 1970–89 is dominated by moderate double-digit inflation. Then, there was a process of low triple-digit inflation during 1990–1998 before it receded to single and low double-digit growth for the oil sub-period 1999–2010 (Suliman 2010).

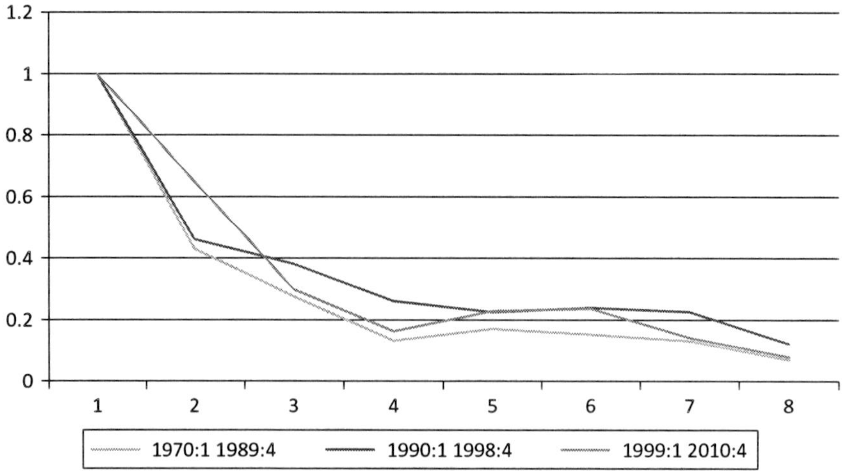

Figure 13.16: Cochrane's Variance Ratio (Several Samples)

Broadly, the plots confirm the importance of the permanent component of inflation for Sudan. The *Z*-statistic and heteroscedasticity correct *Z*, not reported, confirm the random walk hypothesis for inflation for all sub-periods (i.e., inflation is I (1)). The relatively high values for $V(k)$ in 1990–1998 reflect three-digit inflation episodes. Since the exchange rate based anchoring policy is only relevant for 1999–2010, it appears that there is no strong evidence for the success of the anchor in reducing inflation inertia. The persistence of inflation is even higher during the sub-period compared, for example, with 1970–1989. Thus, the exchange rate based stabilization did not sufficiently reduce inertia. This result is also consistent with earlier findings on the low credibility of the budgetary process. Given the evidence of the narrow and weak fiscal linkages of oil, and the need for an effective countercyclical policy, a more flexible exchange rate regime could be adopted in tandem with the use of an efficient indirect monetary instrument.

13.5.6 The Political Economy of Oil Rent Distribution

The impact of oil on the Sudanese political economy deserves a separate investigation. Notwithstanding this, the political economy implications of oil rent distribution are explored in this chapter in terms of budget allocation. Though the budget is not credible enough as argued earlier, it reveals the elites' preferences for trading-off loyalty against

Table 13.5: *Budget Allocation by Selected Spending Categories*

	1989–98	1999–11	1999–05	2006–11	2010	2011/2
Public sector wage[1]	3.43 (0.09)	4.66 (0.21)	4.15 (0.11)	5.42 (0.15)	5.2	5.5
Wage volume growth	82.43 (0.56)	21.46 (0.88)	23.20 (0.51)	19.44 (1.34)	16.48	-1.81
Security spending (SS)[1]	2.60 (0.35)	4.15 (0.28)	3.87 (0.30)	4.47 (0.26)	5.1	6.5
SS volume growth	83.96 (0.76)	27.97 (2.12)	32.98 (2.47)	22.12 (0.86)	37.61	16.5
States transfers (ST)[1]	0.06 (0.86)	2.69 (0.79)	1.24 (0.82)	4.38 (0.17)	4.3	4.4
ST volume growth	220.82 (1.94)	86.48 (1.70)	141.53 (1.31)	21.62 (1.83)	10.48	-4.07
Public services (PS)[1]	3.15 (0.56)	3.34 (0.80)	1.99 (0.78)	4.92 (0.60)	5.8	0.03
PS volume growth	193.52 (1.62)	63.10 (3.56)	102.99 (2.92)	16.56 (5.25)	-0.21	-99.52
GDP volume growth	84.36 (0.52)	15.33 (059)	16.79 (0.24)	13.62 (0.96)	18.7	-7.16

Notes:

[1] Percentage of GDP; coefficients of variations are shown in brackets.

[2] In the second half of this fiscal year Sudan lost 20%, 75% and 36% of population, productive oilfields and revenues respectively, due to South Sudan's secession.

Source: The public sector wage bill; transfers to state governments and public services are compiled from the Economic Surveys, Ministry of Finance and National Economy; the budget documents; and Annual Reports, Bank of Sudan. Security spending is compiled from SIPRI Military Expenditure Database; World Bank Indicators (Jan 1990–Jan 2004); IHS Jane's Sentinel Security Assessment, (Jan 2005–Jan 2008); Research and Market Report Q1 2011 (Jan 2009–Jan 2011) and Annual Reports, Bank of Sudan (Jan 1988–Jan 1989). It is not clear from these sources whether these expenditures also include civil defense, police and paramilitary forces, dual-purpose forces such as military and civilian police, pensions and social security contributions for military personnel. Labor force and unemployment is obtained from Sudan's Ministry of Labor.

repression. As seen in Table 13.5, security spending as a percentage of GDP surged by an average of 1.6 percentage points in the oil decade (1999–2011) up from 2.6% before oil production (1989–1998). Additionally, the wage bill (a proxy for public employment), states transfers and public services increased in percent of GDP, respectively, by 1.2, 2.6 and 0.2 percentage points between these sub-periods. Despite that, the growth of transfers to state governments is higher in percentage terms than the other spending categories; it is the lowest in terms of volume, declining by about 134 percentage points, whereas the volume growth of security spending, the wage bill and public services dropped, respectively, by about 56, 61 and 130 percentage points.

This confirmed that the share of security spending is high and shows the strongest volume growth inertia over time. One major explanation for such high security spending growth is the civil conflicts that beset Sudan. This cause, however, might provide explanation only up until 2005, whereby the CAP is sealed. The voluminous growth of security spending in post-CAP suggests that the peace dividend had not materialized in Sudan and clearly reflects the continued involvement of the military in the political system, which in turn, compromises the ability of other competitors for rent seeking. Accordingly, prioritization of government spending does not favor public employment or provision of public goods and services that complement the markets and provide access to all. The share of the public wage bill in the government budget is small and continues to decline in volume overtime. Spending on public services, which includes infrastructure, education and health among other things, is volatile and exhibits negative volume growth for 2010–2011. It is known that the prioritization and upgrading of these services is at the heart of accelerating the achievement of the Millennium Development Goals (MDGs). Similarly, the volume growth of transfers to state governments continues to decline over time and becomes negative in 2011. Such a pattern of transfer is not consistent with the envisaged constitutional transfers that require "equalization grants from National Revenue Fund being promptly transferred to respective state governments" (Elbadawi and Suliman 2010). Moreover, transfers to states show the highest volatility (measured by coefficient of variation), reflecting a patron–client relationship between the federal and state governments, whereby transfers are determined by the loyalty of the state governors rather than by the constitutional mandate.

Prioritization of security spending is clearly reflected in the 2011 budget, despite the 36% revenue loss due to South Sudan's secession, which is a serious fiscal shock. Security spending is up 1.4 percentage points from its share in 2010. In the face of this, growth volume of GDP allocation to public sector employment, states transfers and public services turns out to be negative, amounting to –7.2%, –1.8%, –4.1% and –99.5%, respectively. This result suggests that the state elites prefer repression when tides are worsening. It is clear that the distributive consequences of the political elites' power maximization or power loss minimization, as reflected by budget prioritization, are inconsistent with job-creating growth, investment promotion, deepening market penetration, poverty fighting and hence, are unsustainable in the long run.

Security expenditures may swell in response to either neighborhood arms races or to patronage demands of politically powerful military establishments (Collier 2006). Elbadawi et al. (2008) use a game theory model to frame the arms race between North and South Sudan. Their findings show that excessive militarization is a rational response for both countries, neither of which can credibly commit to lower security spending under imperfect information. They showed that these credibility issues could be resolved by democratization. Collier and Hoeffler (2002) develop and estimate a model regarding the risk of civil war. After controlling for the risks of internal and external conflict, their results show that security spending in military dictatorships is much higher (2% of GDP higher than in democratic governments). This suggests that such spending is essentially driven by patronage, rather than motivated by purchase of efficiently delivered services. Notwithstanding the motives for growth in military spending, Knight et al. (1996) found that, on average, a 1% reduction in military spending increases the growth rate for a period. This eventually leads to an increase in income of 8.3%. For LDCs, this positive effect of military spending cuts on income is probably even higher than the global average.

The oil rent distribution strategy chosen by the state elites is consistent with the theory and empirical analysis of Ali and Elbadawi (2012), which shows that governments with low natural resources per capita, as in the case of Sudan, tend to set up state security organs and keep public employment low. The opposite is true for governments with high natural resources per capita. It appears, in the case of Sudan, that the political monopoly generated by the elites' distributive influence results in job-less growth, exacerbates the needs for restructuring in the face of shocks, crowds out private investment, drives down value-addition growth and further undermines the prospects for building institutions and sustainable growth.

13.6 Conclusions and Policy Implications

This chapter selectively drew from the theoretical explanations discussed in the literature on natural resources, in order to understand the key channels for the transmission and propagation of the impact of oil on the Sudanese economy.

The method of investigation utilized a combination of growth accounting and time-series analysis to identify the key features that distinguish growth records before and after oil production and to test for possible transmission channels of the oil curse in Sudan, including an assessment of the underlying political economy related issues behind the chosen distribution policies. However, the complexity of quantifying the multidimensional aspects of the oil impact, be it a blessing or a curse, is duly acknowledged. The following conclusions are drawn from the results of the analysis.

First, the contribution of oil to real economic growth has been strong. However, Sudan faces a classic case of Dutch Disease. The crowding out of cotton (the traditional export) and the high rate of factory closures, due to the appreciationary effects of the exchange rate, signals a serious loss of competitiveness and a lack of diversification of the economy.

Second, the allocation of oil revenues through the public sector has presented many fiscal management challenges. The budgeting process tends to be dominated more by bargaining and conflicts of distribution, which results in high discretionary expenditure (with the purpose of increasing resource leakages) and strong procyclical responses to the business cycle. In addition, the dominance of conflicts of distribution over the budget appears to have resulted in a greater separation between the institutions of the allocation and taxation functions of the government after oil production than before.

Third, the credibility of the government, approximated by the credibility of the budget, and the government announced commitment to maintain low inflation and a stable exchange rate was very much reduced after the oil boom, which further complicates the macroeconomic management, with consequent adverse effects on sustaining growth.

Finally, oil rent has significantly reshaped the incentives and constraints facing the political elites in Sudan. Notwithstanding that the oil boom has supported growth, the perverse political incentives generated by the regime's survival behavior have resulted in job-less growth, the crowding out of private investments and the undermining of value addition. Such a polity is not sustainable in the long term.

These conclusions imply that institution building, political liberalization and pluralism are imperative for growth-enhancing reform, which also includes prudent fiscal, monetary and exchange policies for macroeconomic management.

References

Ali, A. G. A. and I. Elbadawi. (2002). Explaining Sudan's Economic Growth. Cambridge General Economic Surveys of Africa: Eastern and Southern Africa.

Ali, O. and I. Elbadawi. (2012). The Political Economy of Public Sector Employment in Resource Rich Countries. Economic Research Forum Initiative on Arab Development (RIAD)'s project.

Al-Moneef M. (2006). The Contribution of the Oil Sector to Arab Economic Development. Paper presented at the High-level Roundtable Partnership for Arab Development: A Window of Opportunity held at OFID.

Amuzegar, J. (2001). *Managing the Oil Wealth: OPEC's Windfalls and Pitfalls.* London and New York: I.B. Tauris.

Auty, R. M. (2003). Third Time Luck for Algeria? Integrating an Industrializing Oil-Rich Country into the Global Economy. *Resources Policy* 29 (1–2): 37–47.

(1990). *Resource-Based Industrialization: Sowing the Oil in Eight Developing Countries.* New York: Oxford University Press.

Badawi, A. (2008). Institutions and Processes in Public Finance Management in Sudan: Assessment and View for Remedy. Paper presented at the UNDP and Unicons Workshop, Khartoum, Sudan.

Beblawi, H. and Luciani, G. eds. (1987). *The Rentier State.* Volume II, London: Croom Helm.

Bernal, V. (1997). Colonial Moral Economy and the Discipline of Development: The Gezira Scheme and "Modern" Sudan. *Cultural Anthropology* 12(4):447–479.

Boughrara, A. and I. Elbadawi. (2011). Oil Dependency and Exchange Rate Regimes in the Arab World: A Concept Note for Case Studies. Research Initiative on Arab Development (RIAD)'s project.

Collier, P. (2006). War and Military Expenditure in Developing Countries and their Consequences for Development. *The Economics of Peace and Security Journal* 11: 9–13.

Collier, P. and A. Hoeffler. (2002). Regional Military Spillovers. Mimeo. World. Bank, Washington, DC.

Corden, W. M. and J. P. Neary. (1982). Booming Sector and De-industrialization in a Small Open Economy. *Economic Journal* 92: 825–848.

Cornwall, J. and Cornwall, W. (2002). A Demand and Supply Analysis of Productivity Growth. *Structural Change and Economic Dynamics* 13: 203–229.

Crain, M. and J. Devlin(2002). Nonrenewable Resource Funds: A Red Herring for Fiscal Stability?. Draft working paper, World Bank, Washington D.C.

Edwards, S. (1998). Two Crises: Inflationary Inertia and Credibility. *The Economic Journal* 108: 680–702.

Elbadawi, I. (1997). Macroeconomic Management and the Black Market for Foreign Exchange in Sudan. In *Macroeconomic Implications of Multiple Exchange Markets in Developing Countries*, M. Kiguel, S. Lizondo and S. O'Connell, eds. Houndmills: Macmillan and New York: St. Martin Press.

Elbadawi, I. and A. Gelb. (2010). Oil, Economic Diversification and Economic Development in the Arab World. ERF Approach Paper. No. PRR35.

Elbadawi, I. and B. Kamar. (2005). The Great Debate on Exchange Rate Regimes: Why Should the MENA Region Care?, 12th ERF Annual Conference: Selected Papers, Economic Research Forum, Cairo, Egypt.

Elbadawi, I. and K. M. Suliman. (2010). Towards an Equitable Inter-Governmental Transfer System for the Sudan. *Sudan Journal of Economic and Social Studies* 8(1): 107–153, University of Khartoum Press.

Elbadawi, I., G. Milante and C. Pischedda. (2008). Referendum, Response and Consequences for Sudan: A Game between Joba and Khartoum. World Bank Policy Research Working Paper 4684, World Bank Washington, DC.

Frankel, J. (2010). The Natural Resource Curse: A Survey. Harvard Kennedy School Faculty Research Working Paper Series rwp 10-005.

Fasano, U. and Q. Wang. (2002). Testing the Relationship between Government Spending and Revenue: Evidence from GCC Countries. IMF WP No.201. IMF Washington, DC.

Gelb, A. H. (1988). *Windfall Gains: Blessing or Curse?* New York: Oxford University Press.

Hall, R. E. and C. I. Jones. (1999). Why Do Some Countries Produce So Much More Output Per Worker Than Others? *The Quarterly Journal of Economics* 114: 83–116.

Hakkio, C. and M. Rush. (1991). Is the Budget Deficit 'Too Large'? *Economic Inquiry* 29: 429–445.

Hirschman, A. (1977). A Generalized Linkage Approach to Development, with Special Reference to Staples. *Economic Development and Cultural Change* 25: 67–98.

Hoover, K. D. and S. Sheffrin. (1992). Causation, Spending and Taxes: Sand in the Sandbox, or Tax Collector for the Welfare State? *American Economic Review* 82(I): 225–248.

Johansen, S. (1988). Statistical Analysis of Cointegration Vectors. *Journal of Economic Dynamics and Control* 12: 231–254.

Karl, T. L. (1997). *The Paradox of Plenty, Oil Booms and Petro States.* Berkeley: University of California Press.

Konandreas, P. (2009). *Assessing Sudan's Export Diversification Potential in Agricultural Products.* London: Maxwell Stamp PLC on behalf of the Ministry of Foreign Trade, Government of Sudan and the European Commission.

Knight, M., N. Loayza, and D. Villanueva. (1996). The Peace Dividend: Military Spending Cuts and Economic Growth. *IMF Staff Papers* 43: 1–37.

Lam, R., and L. Wantchekon. (1999). Political Dutch Disease. Working Paper. Yale University.

Lo, A. W. and C. MacKinlay. (1988). The Size and Power of the Variance Ratio Tests in Finite Samples: A Monte Carlo Investigation. *Journal of Econometrics* 40: 203–38.

Mahdavy, H. (1970). The Patterns and Problems of Economic Development in Rentier States. In *Studies in the Economic History of the Middle East*, M. Cook, ed., London: Oxford University Press, pp. 428–67.

Manzano, O. and R. Rigobon. (2003). Resource Curse or Debt Overhang? National Bureau of Economic Research Working Paper Series (No. 8390).

Musgrave, R. (1966). Principles of Budget Determination. In *Public Finance: Selected Readings*, Cameron, H. and Henderson, W., eds., New York: Random House, pp. 15–27.

Neumayer, E. (2004). Does the Resource Curse Hold for Growth in Genuine Income as Well? *World Development* 32 (10): 1627–1640.

PEFA. (2005). *Public Expenditure and Financial Accountability.* Washington, DC: World Bank.

Papyrakis, E. and R. Gerlagh. (2004). The Resource Curse Hypothesis and its Transmission Channels. *Journal of Comparative Economics* 32: 181–193.
Peacock, A.T. and J. Wiseman. (1961). *The Growth of Public Expenditure in the United Kingdom*. Princeton University Press, Princeton: National Bureau of Economic Research.
Renner, M. (2002). The Anatomy of Resource War. Worldwatch Paper 162, Worldwatch Institute, Washington, DC.
Rodrik, D. ed. (2003). *In Search of Prosperity: Analytic Narratives on Economic Growth*. Princeton, NJ: Princeton University Press.
Ross, M. L. (1999). The Political Economy of the Resource Curse. *World Politics* 51: 297–322.
(2001). Does Oil Hinder Democracy? *World Politics* 53: 325–361.
Rosser, A. (2006). The Political Economy of the Resource Curse: A Literature Survey. Working Paper No. 268, Institute of Development Studies, University of Sussex.
Sachs J. D. and A. M. Warner. (2001). Natural Resource and Economic Development: The Curse of Natural Resources. *European Economic Review* 45: 827–838.
(1999). Natural Resource Abundance and Economic Growth. NBER WP 5398.
Suliman, K. M. (2010). The Determinants of Inflation in Sudan 1970–2002. *Journal of Development and Economic Policies* 12(2): 37–84.
Van der Ploeg, F. (2006). Challenges and Opportunities for Resource Rich Economies. CEPR Discussion Papers 5688.
Wolf, E. N. (1991). Capital Formation and Productivity Convergence over the Long Run. *American Economic Review* 81(3): 565–579.
Wintrobe, R. (1998). *The Political Economy of Dictatorship*. New York: Cambridge University Press.
World Bank. (2000) Sudan: Options for the Sustainable Development of the Gezira Scheme. Washington, DC.
(2009). Sudan-The Road Toward Sustainable and Broad Based Growth. Poverty Reduction and Economic Management Unit, Africa Region, Washington, DC.

Index

ACRCP. *See* Advisory Committee for the Reference Copper Price
ACTG. *See* Advisory Committee for Trend GDP
AFCFRF. *See* Advisory Financial Committee for Fiscal Responsibility
agricultural sector
  in Arab region, 82–83, 86
  Baumol's cost disease effects for, 89–90
Alaska Permanent Fund, 153–154
Algeria
  Dutch Disease in, 229
  economic diversification in, 49–50
  EU GDP shocks and, 32–33
  natural gas reserves in, 325–327
  public sector employment in, 107
  SWFs in, 277–278
Arab region. *See also* oil-rich countries; resource-rich economies; specific countries
  agricultural sector in, 82–83, 86
  booms and busts in, 12–13, 342
  Dutch Disease in, 229
  export concentration in, 49–50
  export diversification in, 304
  EXR regimes in, 303–309, 343–345
  fiscal institutions in, 274–279
  fiscal policy for, in oil-rich countries, 226–230
  fiscal revenues in, 330–334
  government revenues and expenditures in, 338
  hydrocarbon-dependency in, 45, 229–230, 330–334
  labor market heterogeneity in, 311–313, 315–316
  long-term GDP growth in, 342
  manufacturing in, 49–50, 346
  natural resource abundance in, 304
  public sector employment in, 82–83
  RER undervaluation in, 57–61
  resource dependency in, 329
  RREs in, 1–2, 229–230, 325–334
  value-added exports in, 329–331
  violence in, resource curse as influence on, 338–339
Arab Spring, 301–302
augmented growth model, 323, 353–355
authoritarian regimes
  public sector employment under, 106, 113–118
  resource-rich economies and, 120–121

Bahrain, 19–20
  migrant worker population in, 118
  non-hydrocarbon exports, 159
  oil production in, 22
  SWFs in, 277–278
balanced budget rule, 289–291
Baumol's cost disease, 89–90
booms and busts, in Arab region
  of commodities, 349–350
  for oil, 355
  in real estate, 12–13
business cycles, GVAR models for, 17–18

CAB. *See* cyclically-adjusted balance
CBC. *See* Central Bank of Chile
CCEMG estimator. *See* Common Correlated Effects Mean Group estimator
Central Bank of Chile (CBC), 232, 242, 250–251, 259
  EXRs and, 318–319

Chile, fiscal policy institutions in, 230–266. *See also* fiscal committees
  Advisory Committee for the Reference Copper Price (ACRCP), 241–243, 249–250, 264–265
  Advisory Committee for Trend GDP (ACTG), 241–243, 248–249, 264
  Advisory Financial Committee for Fiscal Responsibility (AFCFRF), 241–243
  budget accountability requirements, 236–239
  budget management practices, 232–233
  budget planning practices, 233–234
  CAB in, 250–252, 276–277
  CBC, 232, 242, 250–251, 259
  control and auditing of government accounts, 239
  for copper prices, 244, 255
  CRCF in, 258–259
  ESSF and, 232
  EXR regimes and, 314
  fiscal councils, 240–241, 278–279
  under Fiscal Responsibility Law of 2006, 230–232, 234
  fiscal rules for, 234–235, 243–258, 276–277
  GDP growth and, 244, 255
  institutional framework for, 230–243
  PRF and, 231–232
  SWFs in, 231–232, 258–261
China
  GDP shocks in, MENA region influenced by, 18, 34–37
  U.S. GDP shocks and, 31
clientelism, 107
commodity booms, 349–350
commodity price volatility
  oil curse and, 189
  resource curse and, 189
Common Correlated Effects Mean Group (CCEMG) estimator, 208–211
Comprehensive Peace Agreement (CPA), 431
conditional convergence hypothesis, for economic growth, 197–198
consumption
  Engel's Law and, 89–94
  excessive, 8, 323–325, 340, 367
  resource curse as influence on, 349–350
  resource rents and, 350
  U.S. oil consumption, 22
copper prices, 244, 255
Copper Revenue Compensation Fund (CRCF), 258–259
countercyclical fiscal policy, 8–9, 42
  reforms for, 360–364
CPA. *See* Comprehensive Peace Agreement
CRCF. *See* Copper Revenue Compensation Fund
credit growth, 12–13
cyclically-adjusted balance (CAB), in Chile, 250–252, 276–277

debt overhang, xix, 3, 216
democracy
  economic growth and, 153
  oil curse and, 152, 189–190
  political checks and balances and, 194–196, 212–213, 215
  public sector wages and, 127
  resource curse and, 189–190
  resource rents and, 194–195
  RREs and, 189–190
  valuation of, 114
depletion effects, 286–287
development, economic. *See also* economic growth
  in Dubai, 406
  export sophistication and, 50–53
  oil curse and, 150
  resource dependency and, 45
  in RREs, 351–353
dictatorships. *See also* authoritarian regimes
  resource curse under, 439
diversification. *See* economic diversification
domestic currencies. *See* currencies, domestic
Dubai, UAE
  economic development in, 406
  economic diversification in, 98, 410
  oil curse and, 408–412
Dutch Disease, 4
  in Algeria, 229
  in Arab region, 229
  Engel's Law and, 83
  exchange rate overvaluation, 3, 216–217, 346–348, 426–428
  in GCC countries, 229
  investing in investing and, 299
  in Kuwait, 229
  oil curse and, 150, 191–192, 229, 380, 428–430
  resource curse and, 343–349, 437
  in RREs, 228

in Saudi Arabia, 229
services sector influenced by, 348
in Sudan, 444–445
in UAE, 380

econometric model, for economic growth, 205–217
CCEMG estimator in, 208–211
long-run growth determinants in, 210
low checks and balances and, 217
low democracy and, 217
economic development. *See* development, economic
economic diversification, 4, 12
in GCC countries, 346–348
HHI index for, 49–50
in MENA region, 49–50, 97–99
oil curse and, 150
for oil-exporting countries, 47–48
private sector employment and, 11–12
public sector employment as part of, 11–12
RER overvaluation and, 71
resource curse and, 343–349
in RPLA countries, 97–99
of services sector, in Arab region, 84
in Sudan, lack of, 426–430
economic growth. *See also* econometric model, for economic growth
conditional convergence hypothesis for, 197–198
democracy as influence on, 153
determinants of, 197–199, 201
external shocks and, 193–194, 199, 202–205
financial integration in world markets and, 198
in GCC countries, integration strategies for, 22
human capital formation and, 198
institutional influences on, 198–199, 202
international trade openness and, 198
in oil-rich countries, 151–152
political checks and balances and, 194–196
public services and, 198–199
resource curse and, 203
resource rents and, 190, 197–213
in RREs, 2, 188, 228
stabilization policies for, 199, 202
structural policies as influence on, 198–199, 202
in Sudan, 49–50, 443–445
in Sudan, oil dependency as influence on, 443–445
TFP as measure of, 389–393
transitional convergence hypothesis for, 197–198, 201–202
in UAE, 382–395
economic spillovers. *See* spillovers, economic
education
economic growth and, 198
in Saudi Arabia, 174–178
in UAE, economic investment in, 98
Egypt
economic diversification in, 49–50
economic trade with EU, 23
public sector employment in, 107
EITI. *See* Extractive Industries Transparency Initiative
elites
public sector employment and, influence on, 108–109
regime change influenced by, 121
state security apparatus and, 114–116
employment. *See also* private sector employment; public sector employment; wages and salaries
labor market policies for, in Saudi Arabia, 160–161
oil curse and, 161–162
in UAE, demographics of, 386–387, 403
employment-to-population ratio, 164
Engel effects, services sector and, 90–91
Engel's Law, 83
consumption and, 89–94
GDP and, 86
rents influenced by, 89–94
services sector and, 89–94
equilibrium RER (ERER), 55, 57–59
computation methods for, 77–79
ESSF. *See* Economic and Social Stabilization Fund
EU. *See* European Union
Euro Area block. *See also specific countries*
GDP shocks in, MENA region influenced by, 32–34
GVAR model for, 19–20
European Union (EU)
economic integration in, GCC countries influenced by, 22
economic trade within Maghreb region, 23
economic trade within Mashreq region, 23
fiscal councils in, 240–241
excessive borrowing, resource curse and, 340–342

exchange rate (EXR) regimes, 4, 55–56, 70
in Arab region, 303–309, 343–345
CBC and, 318–319
in Chile, 314
Dutch Disease, 3, 216–217, 346–348, 426–428
equilibrium outcomes for, 317–318, 344
exchange rate overvaluation, 3, 216–217, 344, 346–348, 426–428, 451
fixed, 312
flexible, 307–308, 312
labor markets and, 311–315
OCA, 308–309
pegged, 307–308
resource curse and, 343–345
in RRLA countries, 303–309
in RRLI countries, 303–309
in Sudan, 451–452
theoretical framework for, 311–312
in UAE, 395–396, 401
export concentration
in Arab region, 49–50
export performance and, 67
export diversification, 49–50, 97–99
in Arab region, 304
export instability theory, 437
export performance
empirical model of, 61–69
EXPY and, 66
HP-gap RER undervaluation and, 69
manufacturing exports and, 64–65
oil rents as influence on, 63
RER undervaluation and, 63–66
resource rents and, 64–66
export sophistication (EXPY), 50–53
export performance and, 66
for oil-exporting nations, 62
EXR regimes. *See* exchange rate regimes
external shocks. *See* macroeconomic shocks
extraction effect, 287
Extractive Industries Transparency Initiative (EITI), 9, 11, 360

FDI. *See* foreign direct investment
fiscal committees, in Chile, 239–243
ACRCP, 241–243, 249–250, 264–265
ACTG, 241–243, 248–249, 264
establishment of, 278–279
forecasts by, 251
fiscal councils, 240–241, 278–279
fiscal institutions, 4
in Arab region, 274–279
commodity revenue volatility, 189
countercyclical fiscal policy, 8–9, 42, 360–364
economic policies set by, 13
fiscal rules, 234–235
ACRCP growth projections, 249–250
ACTG growth projections under, 248–249
adoption of, 276–277
budget performance under, 246–255
CAB and, in Chile, 250–252, 276–277
changes to, 246–255
in Chile, 234–235, 243–258, 276–277
elasticities of, 246
evaluation of, 255–258
in GCC countries, adoption of, 363–364
in Germany, 235
implementation of, 246–255
macroeconomic impact of, 255–258
in Norway, 268–270
permanent-income approach for, 243
policy objectives for, 243–246
for resource rents, 10–11
in Sweden, 235
in UAE, 399–400
in U.K., 235
for oil exporting countries, 274–279
for oil-rich countries, 274–279
reform of, 9
revenue smoothing, 294, 363
in RREs, 228–229, 274–279
structural budget balance, 11, 363, 400
fiscal policy. *See also* Chile, fiscal policy institutions in
countercyclical, 8–9, 42, 360–364
domestic institutions for, 13
in oil-rich countries, in Arab region, 226–230
pro-cyclical, 157, 227, 396–400
in RREs, 226–230
transparency of, 365
Fiscal Responsibility Law of 2006, in Chile, 230–232, 234
fiscal responsibility laws, 230–231
fiscal revenues, in Arab region, 330–334
fiscal rules, 234–235
ACRCP growth projections, 249–250
ACTG growth projections under, 248–249
adoption of, 276–277
budget performance under, 246–255
CAB and, in Chile, 250–252, 276–277

changes to, 246–255
in Chile, 234–235, 243–258, 276–277
elasticities of, 246
evaluation of, 255–258
in GCC countries, adoption of, 363–364
in Germany, 235
implementation of, 246–255
macroeconomic impact of, 255–258
in Norway, 268–270
permanent-income approach for, 243
policy objectives for, 243–246
for resource rents, 10–11
in Sweden, 235
in UAE, 399–400
in U.K., 235
fiscal transparency, 11, 365
foreign aid, RER and, 56
foreign direct investment (FDI), in MENA region, 37–39
foreign labor, in Saudi Arabia, 160

game theory
basic model of, 108–113
data for, 118–131
hypothesis testing in, 126–129
in imperfect state repression models, 131–137
in perfect state repression models, 136–137
for public sector employment, 108–118
robustness checks in, 129–131
with state repression models, 113–118
GCC. *See* Gulf Cooperation Council countries
GDP. *See* gross domestic product
global macroeconomic modeling, 18–27
Global Vector Autoregression (GVAR) model
for business cycle linkages, 17–18
data measurement errors in, 28
domestic variables in, 20–21
economic spillovers in, 27–41
for Euro Area block, 19–20
foreign variables in, 21
for GCC countries, 17–18
global variables in, 22
inward spillovers in, 27–37
macroeconomic modeling for, 18–27
for Maghreb region, 17
for Mashreq region, 17
for MENA region, 16–27
methodology for, 17–18
model specifications, by country, 26–27
nations and regions in, 19
for OPEC members, 18–19
Government Pension Fund, 271–274
Government Pension Fund Global (GPFG), 271–274
Government Pension Fund Norway (GPFN), 271–274
Government Petroleum Fund (GPF), 266–267, 271–274
government sector employment
in Saudi Arabia, 170, 172–174
wages and salaries for, 175, 179
government-related entities (GREs), 404–405
GPF. *See* Government Petroleum Fund
GPFG. *See* Government Pension Fund Global
GPFN. *See* Government Pension Fund Norway
GREs. *See* government-related entities
gross domestic product (GDP)
in Arab region, 342
in Chile, fiscal policy framework for, 244, 255
Chinese shocks, 18, 34–37
Engel's Law and, 86
EU shocks, 32–34
in GCC countries, trade ratio with, 104–107
oil price volatility and, 2–3, 10–11
in oil-rich countries, trade ratio with, 104–107
resource rents and, 91–92
in RREs, 342
of services sector in Arab region, 84, 93–94
in UAE, 384, 409
U.S. shocks, 28–31
growth, 4. *See also* economic growth
in oil-rich countries, 151–152
growth models. *See* augmented growth model
Gulf Cooperation Council (GCC) countries, 1–2. *See also* Bahrain; Kuwait; Oman; Qatar; Saudi Arabia; United Arab Emirates
asset investment in, 296–299
composition of exports for, 159
counter-cyclical fiscal policy reforms in, 360–364
currencies pegged to U.S. dollar, 150
Dutch Disease in, 229
economic diversification in, 346–348
economic integration in, 22
fiscal rules in, adoption of, 363–364
labor markets in, 345–346
MENA region and, 16
migrant workers in, 99, 118
monetary policies in, 150

Gulf Cooperation Council (GCC) countries (cont.)
national oil funds in, 154–156
natural gas reserves and production in, 325–327
natural resource endowments in, 325
need for economic diversification in, 11–12
oil endowments in, 149–150
oil price cycles and, 3–4
oil production in, 22, 325–327
oil reserves in, 325–327
private investment in, 297–299
public investment in, 297–299
public sector employment in, 104–107
rent comparisons among, 3, 118
resource rents and, 327–329
resource revenue investment by, 296–297
revenue price volatility and, 10–11
savings policies for, 288–292
services sector in, 84
SWFs in, 362–363
trade/GDP ratios in, 104–107
volatility of government revenues in, management of, 292–295
wages per capita in, 118
GVAR model. *See* Global Vector Autoregression model

Herfindahl-Hirschman index (HHI), 49–50
for oil exporters, 62
HP-gap RER, 69
human capital. *See also* education
economic growth and, 198
oil curse and, 381
in UAE, 390
hydrocarbon-dependency. *See also* resource dependency
in Arab countries, 45, 229–230, 330–334

imperfect state repression models, 131–137
endogenous probability in, 133–135
regime change and, 134
impulse responses, 31–32, 41
income inequality, in Saudi Arabia, 157–158
incomes. *See also* income inequality; oil incomes
defined, 378
industrial policy (IP), 12
inefficient specialization hypothesis, 192–193
inflation, in UAE, 405
institutions. *See also* Chile, fiscal policy institutions in; fiscal institutions; Norway, fiscal policy institutions in
domestic, fiscal policy influenced by, 13
economic growth influenced by, 198–199, 202
oil curse and, 151–158, 380–381
in oil exporting countries, 274–279
resource curse as institutional curse, 323, 334–339, 365–368
in RREs, 228–229, 274–279
in Sudan, 229–230
weak, 46, 152–156
integrated budget rule, 289
International Budget Partnership, 237
international business cycles, 17
intertemporal commitments. *See* political checks and balances
investing in investing, 10, 297–299
investment strategies, 4
Dutch Disease and, 299
in oil-rich countries, 13
resource rents and, 13–14, 350
with SDF, 10
SWF and, 13
IP. *See* industrial policy
Iran, 23
Iraq, services sector in, 99
Italy, 19–20

Jordan
economic trade with EU, 23
public sector employment in, 107
as resource-poor country, 88–89

kafala system, 393–395, 414–415
Kuwait, 19–20
Dutch Disease in, 229
migrant worker population in, 118
national oil funds in, 154–156
oil production in, 22, 325–327
oil reserves in, 325–327
public sector employment in, 107
services sector in, 99
SWFs in, 277–278

labor force participation
defined, 164
in Saudi Arabia, 164–166, 168
labor markets. *See also* government sector employment; private sector employment; public sector employment

in Arab region, heterogeneity of, 311–313, 315–316
economic shocks and, 312–313
employment-to-population ratio, 164
expatriate workers, 302, 306, 309, 345–346
EXR regimes and, 311–315
in GCC countries, 345–346
heterogeneity of, 311–313, 315–316
under *kafala* system, 393–395, 414–415
national labor force, 312
oil curse and, 151, 158–162, 181–183
patronage and, 109–110
population demographics as influence on, 158–160
reservation wage, 160
resource curse and, 324
in RRLA countries, 302, 306
in RRLI countries, 302, 306
in Saudi Arabia, 151, 162–176
segmentation of, 108–118, 324
in UAE, 387–389
unemployment rate, 150, 162–176
LDCs. *See* less developing countries
League of Arab States, 1
Lebanon, services sector in, 99
less developing countries (LDCs), 435–437
Libya
economic trade with EU, 23
services sector in, 99
SWFs in, 277–278
linkage theory, for resource curse, 436
low democracy, 217
low political checks and balances, 217

macroeconomic modeling, 17. *See also* augmented growth model
global, 18–27
for GVAR, 18–27
macroeconomic shocks
in China, 18, 34–37
economic growth and, 193–194, 199, 202–205
in Euro Area block, MENA region influenced by, 32–34
labor markets and, 312–313
social conflicts and, 194
in UAE, 396–400, 405–407
in U.S., 28–31
Maghreb region. *See also* Algeria; Libya; Mauritania; Morocco; Tunisia
China and, as economic trade partner, 23
EU and, as economic trade partner, 23
GVAR model for, 17
manufacturing, 49–50
in Arab region, 346
export performance and, 64–65
in RREs, 346
manufacturing value added (MVA), 50
Mashreq region. *See also* Egypt; Jordan; Syria
economic trade with EU, 23
GCAR model for, 17
Mauritania, economic trade with EU, 23
MENA oil exporters (MENAEX), 39–41
microeconomic regulations, in services sector, 95–97, 100
Middle East. *See also* Arab region; Gulf Cooperation Council countries; Middle-Income Middle East States; oil-rich countries
authoritarian regimes in, 113–118
public sector employment models in, 108–118
Middle East and North Africa (MENA) region. *See also specific countries*
Chinese GDP shocks as influence on, 18, 34–37
economic diversification in, 49–50, 97–99
economic spillovers in, 27–41
EU GDP shocks as influence on, 32–34
FDI in, 37–39
GCC countries and, 16
GVAR model for, 16–27
industrial performance in, 51
MENAEX, 39–41
mining sector in, 88–89
RPLA countries in, 97–99
trade weights in, 23–26
U.S. GDP shocks as influence on, 28–31
Middle-Income Middle East States (MIMES), 284–285
asset investment in, 296–299
OECD countries compared to, 285
permanent incomes in, 285–286
private investment in, 297–299
public investment in, 297–299
resource revenue investment by, 296–297
savings policies in, 285–295
volatility of government revenues in, management of, 292–295
migrant workers, in GCC countries, 99, 118
MIMES. *See* Middle-Income Middle East States
mining sector, in MENA region, 88–89

Morocco
economic trade with EU, 23
as resource-poor country, 88–89
MVA. *See* manufacturing value added

national oil funds, 154–156
National Resource Charter (NRC), 9, 360
fiscal transparency and, 11
natural gas reserves, production and, in GCC countries, 325–327
natural resource curse. *See* resource curse
NBIM. *See* Norges Bank Investment Management
neoclassical growth theory, 436–437
non-tradable services, 100
Norges Bank Investment Management (NBIM), 272
Norway, fiscal policy institutions in, 266–274
economic growth under, 270
fiscal rules, 268–270
government budget performance and, 270–271
Government Pension Fund in, 271–274
Government Petroleum Fund and, 266–267, 271–274
GPFG in, 271–274
GPFN in, 271–274
institutional framework for, 266–268
NBIM and, 272
objectives of, 267–268
SWFs in, 253–268, 271–274
NRC. *See* National Resource Charter

OCA. *See* Optimum Currency Area
OECD countries. *See* Organisation for Economic Co-operation and Development countries
oil, production of
commodity price volatility influenced by, 41
in GCC countries, 22, 325–327
in Sudan, development history for, 422–424
in UAE, 22, 325–327, 374
for U.S. consumption, 22
oil booms, 355
oil curse. *See also* resource curse
causes of, 413
commodity price volatility, 189
democracy and, 189–190
democratic regimes and, 152
in Dubai, 408–412
Dutch Disease and, 150, 191–192, 229, 380, 428–430
economic development and, 150
economic diversification and, 150
economic volatility, 203–204, 323–325, 380
employment and, 161–162
employment policies and, 160–161
excessive borrowing, 340–342
excessive consumption, 8, 323–325, 340, 367
human capital and, 381
income inequality and, 157–158
inefficient specialization hypothesis for, 192–193
institutions and, 151–158, 380–381
kafala system and, 393–395, 414–415
labor markets and, 151, 158–162, 181–183
Norway and, avoidance strategies for, 412
oil price volatility and, 150, 192
oil prices and, 2, 189
patronage and, 109–110
physical capital and, 381
political checks and balances, 194–196, 212–215, 217
political economy of, 191
political stability, 336–337, 404
private sector employment and, 150
public sector employment and, 156–157
rent transfers and, 156–157
RER and, 191–192
resource rents and, 151, 188
for Saudi Arabia, 157–162
status quo and, 153
as theory, development of, 191–196, 323–325, 435–440
in UAE, 377–382, 408–414
unemployment and, 150
voracity effect, 335–336
in Yemen, 229–230
oil cycles, 12–13
causality testing for, 449
forecast error variance for, 450
in Sudan, fiscal response to, 447–450
oil dependency. *See* resource dependency
oil endowments
in GCC countries, 149–150
in Saudi Arabia, 149–150
sustainability of, 157–158
welfare spending and, 159
oil exporters. *See also* Gulf Cooperation Council countries; Organization of the Petroleum Exporting Countries
economic diversification for, 47–48
empirical performance model for, 61–69

EXPY for, 62
fiscal institutions for, 274–279
HHI for, 62
oil-dependent economies for, 47–53
public sector employment and, 106
RER undervaluation and, 57–61
UAE as, 382–395
oil incomes, 149–150
endowment effect and, 159–160
national oil funds and, 154–156
in Nigeria, redistribution of, 153–154
sustainability of, 157–158
tax schemes for, 154
oil prices
GCC and, 3–4
GDP and, 2–3
in Nigeria, 192
oil curse and, 2, 150, 189, 192
resource curse and, 189
SWF and, 10–11
in UAE, 400–407
in Venezuela, 192
volatility of, 2–3, 10–11, 150, 192
oil rents
export performance influenced by, 63
in Sudan, 229–230, 452–455
in UAE, 395–407
oil reserves, in GCC countries, 325–327
oil rights, auctioning of, 193
oil wealth, 4, 45, 150, 153, 160–161, 253–268, 336, 358, 383–384
oil-rich countries. *See also* Gulf Cooperation Council countries; Organization of the Petroleum Exporting Countries; specific countries
Dutch Disease in, 229
fiscal indicators in, 276
fiscal institutions for, 274–279
fiscal policies in, 226–230
growth in, 151–152
investment strategies in, 13
monetary policies in, 150
public sector wages in, 104–108
rent distribution in, 13–14
resource curse in, 226–227
trade/GDP ratios in, 104–107
OPEC. *See* Organization of the Petroleum Exporting Countries
open economies, 129–131
optimal savings rule, 291
Optimum Currency Area (OCA), 308–309
Organisation for Economic Co-operation and Development (OECD) countries, 285
Organization of the Petroleum Exporting Countries (OPEC), 1
GVAR model for, 18–19

Pan Arab Free Trade Agreement (PAFTA), 95
patronage, political, 109–110
PEFA framework. *See* Public Expenditure and Financial Accountability framework
pegged EXR regimes, 307–308
Pension Reserve Fund (PRF), 231–232
perfect state repression models, 136–137
permanent incomes, 285–286
PFM system. *See* Public Financial Management system
physical integrity rights, 120
political checks and balances
democracy and, 194–196, 212–213, 215
low, 217
resource curse and, 214
political economy, 13–14, 104. *See also* government sector employment; public sector employment
of oil curse, 191
political patronage. *See* patronage, political
populous group countries. *See also* Algeria; Arab region; Sudan; Syria; Yemen
resource curse in, 323, 365–368
PRF. *See* Pension Reserve Fund
private sector employment
economic diversification and, 11–12
oil curse and, 150
in Saudi Arabia, 158, 169–172
wages and salaries for, 118–119, 169–173
pro-cyclical fiscal policy, 157, 227
in UAE, 396–400
product spaces (PS), 12
Public Expenditure and Financial Accountability (PEFA) framework, 441
Public Financial Management (PFM) system, 441
public procurement, 365
public sector employment
in Arab region, 82–83
under authoritarian regimes, 106, 113–118
clientelism and, 104–107
democracy as influence on wages, 127
in economic diversification, 11–12
elites and, 108–109
in GCC countries, 104–107

public sector employment (cont.)
hypothesis testing for, 126–129
in Middle East, 108–118
for migrant workers, 99, 118
oil curse and, 156–157
in oil-rich countries, wages for, 104–108
open economies and, 129–131
political patronage and, 109–110
reservation wage and, 160
resource rents and, 107, 142–146
in RREs, 107, 118–120
in Saudi Arabia, 158
state repression and, 106, 113–118
theoretical model for, 108–118
trade openness and, 129–131
wages for, 104–108, 118–119, 128, 130
public services, economic growth and, 198–199

Qatar, 19–20
migrant worker population in, 118
oil production in, 22
SWFs in, 277–278

real estate, 100
Real Estate Development Fund, 178
real exchange rates (RER)
appreciation, 56–57
assessment of, 45–46
by country, 74–77
depreciation, 56–57
economic diversification and, 71
empirical model of, 55–61
ERER, 55, 57–59, 77–79
estimation methods, 55–56, 71
export performance and, 63
foreign aid and, 56
HP-gap, 69
long-run determinants of, 64–65
misalignment series for, 70, 77–79
oil curse and, 191–192
regression analysis variables for, 72–73
resource rents and, 55
short-run determinants of, 64–65
undervaluation of, 46–47, 57–61, 63–67
regime change
as collective action, 134
elites' influence on, 121
imperfect state repression models and, 134
probability variables for, 121–123, 125, 146
resource rents and, 122–124
regimes. *See also* exchange rate regimes
authoritarian, 106, 113–118, 120–121
remittances, 37–39
rent distribution, in oil-rich countries, 13–14
rent transfers
oil curse and, 156–157
resource curse and, 337–338
rentier states, 8, 336–337, 438
rents. *See also* resource rents
distribution of, 3, 13–14
Engel's Law as influence on, 89–94
in GCC countries, 3, 118
RER. *See* real exchange rates
reservation wage, 160
resource curse, 2–4, 188–191. *See also* oil curse
contagion channel of, 203–204
counter-cyclical fiscal policy reforms and, 360–364
democracy and, 189–190
under dictatorships, 439
Dutch Disease and, 343–349, 437
economic diversification and, 343–349
economic growth and, 203
economic variables for, 368–369
economic volatility and, 340–342
excessive borrowing and, 340–342
excessive consumption and, 349–350
export instability theory and, 437
EXRs and, 343–345
fiscal linkages in, 436
fiscal policy transparency and, 365
heterogeneity of, 212–217
hypothesis testing for, 353–359
as institutional curse, 323, 334–339, 365–368
institutional reform as answer to, 359–365
labor market segmentation and, 324
linkage theory for, 436
macroeconomic manifestations of, 324, 340–350
methodology of, 440–443
modernization effects of, 339
neoclassical growth theory and, 436–437
oil prices and, 189
political checks and balances and, 214
poor political institutions and, 358
in populous group countries, 323, 365–368
public procurement transparency and, 365
public spending efficiency reforms and, 364–365
rent transfers and, 337–338
rentier states and, 336–337, 438
resource rents and, 213

in RREs, 226–227
in Sudan, 440–443
as theory, development of, 191–196, 323–325, 435–440
violence and, 338–339
voracity effect in, 335–336
resource dependency, 1–2. *See also* resource-dependent countries
in Arab region, 329
economic development and, 45
in Sudan, 49–50, 424–428, 443–445, 455–456
resource depletion
depletion effects, 286–287
extraction effects in, 287
permanent income frameworks for, 285–286
in savings policies, 285–292
resource extraction industries, 174
resource management, 9, 152, 339, 359–360
resource rents, 3–4
clientelism and, 107
defining methodology for, 194–195
democracy and, 194–195
economic growth influenced by, 190, 197–213
export concentration and, 67
export performance and, 64–66
fiscal rules for, 10–11
in GCC countries, 327–329
GDP percentages and, 91–92
global averages for, 378
government consumption and, 350
investment and, 13–14, 350
investment strategies with, 13–14
from natural resources, 82–83
oil curse and, 151, 188
public sector employment and, 107, 142–146
regime change and, 122–124
RER and, 55
resource curse and, 213
in RREs, 3, 226–227
services sector and, 91–94
in UAE, 375, 378–379
wages and, by country, 142–146
resource-dependent countries, 118, 137. *See also* Sudan
authoritarian regimes and, 120–121
public sector employment in, 107, 118–120
wages in, for public sector employment, 105
resource-poor labor-abundant (RPLA) countries. *See also* Rwanda
contract-based services in, 97–99
economic diversification in, 97–99
in MENA region, 97–99
services sector in, 84
resource-rich economies (RREs). *See also* Chile; Gulf Cooperation Council countries; Norway; oil curse; oil-rich countries; resource curse
in Arab region, 1–2, 229–230, 325–334
authoritarian regimes and, 120–121
democracy in, 189–190
Dutch Disease in, 228
economic growth in, 2, 188, 228
economic policy failures in, 2–3
economic volatility in, 340–342
excessive borrowing in, 340–342
financial development in, 351–353
fiscal policies in, 226–230
fiscal pro-cyclicality in, 227
government revenues and expenditures in, 338
long-term GDP growth in, 342
macroeconomic volatility in, 228
manufacturing in, 346
open economies in, 129–131
overview of, 325–334
permanent incomes in, 285–286
political and fiscal institutions in, 228–229, 274–279
political checks and balances in, 194–196
public sector employment in, 105, 107, 118–120
resource curse in, 226–227
resource rents in, 3, 226–227
resource-price reversals in, 227
savings policies in, 285–295
services sector in, 100–101
sustainability of, 227
SWFs in, 277–278
wages in, for public sector employment, 105
resource-rich labor-abundant (RRLA) countries. *See also* Sudan
Algeria, 32–33, 49–50, 107, 229, 277–278, 325–327
economic structure of, 303–305
export diversification in, 304
EXR regimes in, 303–309
fixed EXR regimes in, 312
labor markets in, 302, 306
natural resource abundance in, 304
pegged EXR regimes in, 308

resource-rich labor-abundant (RRLA) countries (cont.)
  selected economic indicators for, 305
  Syria, 23, 33, 99
  Yemen, 229–230
resource-rich labor-importing (RRLI) countries
  economic structure of, 303–305
  export diversification in, 304
  EXR regimes in, 303–309
  flexible EXR regimes in, 312
  labor markets in, 302, 306
  natural resource abundance in, 304
  pegged EXR regimes in, 308
  selected economic indicators for, 305
RPLA countries. *See* resource-poor labor-abundant countries
RREs. *See* resource-rich economies
RRLA countries. *See* resource-rich labor-abundant countries
RRLI countries. *See* resource-rich labor-importing countries

salaries. *See* wages and salaries
SAMA. *See* Saudi Arabia Monetary Agency
SAR. *See* Saudi Credit and Savings Bank
Saudi Arabia, 19–20
  demographics, 162–169
  Dutch Disease in, 229
  economic indicators, 155
  educational profile of, 174–178
  employment policies in, 160–161
  employment wages and salaries in, 169–174
  employment-to-population ratio in, 164
  endowment effect in, 159–160
  fiscal indicators, 156
  foreign labor in, 160
  government sector employment in, 170, 172–174
  income inequality in, 157–158
  labor force participation in, 164–166, 168
  labor markets in, 151, 162–176
  migrant worker population in, 118
  non-hydrocarbon exports, 159
  non-Saudi unemployment rates in, 166–169
  oil curse for, 157–162
  oil endowments in, 149–150
  oil production in, 22, 325–327
  oil reserves in, 325–327
  population, 162
  population demographics in, 158–160
  private sector employment in, 158, 169–172
  public sector employment in, 158
  Royal Decrees of 2011, 161–162, 176–180, 183–184
  SAR, 178–180
  unemployment rates in, 166–169
Saudi Arabia Monetary Agency (SAMA), 154–155, 179
Saudi Credit and Savings Bank (SAR), 178–180
savings policies
  asset investment, 296–299
  balanced budget rule for, 289–291
  complexity of, 287–288
  depletion effects, 286–287
  extraction effects and, 287
  for GCC countries, 288–292
  implications for rules of, 288–292
  integrated budget rules for, 289
  offsetting depletion principles, 285–292
  optimal savings rule, 291
  for resource-rich countries, 285–295
  volatility management as part of, 292–295
SDF. *See* Sovereign Development Fund
services sector, in Arab region
  determinants of, 94
  Dutch Disease as influence on, 348
  economic diversification of, 84–89
  economic growth of, 84–85
  Engel effects and, 90–91
  Engel's Law and, 89–94
  in GCC countries, 84
  GDP and, 84, 93–94
  in Iraq, 99
  in Kuwait, 99
  in Lebanon, 99
  in Libya, 99
  microeconomic regulations in, 95–97, 100
  migrant workers for, 99
  non-tradable, 100
  public, 198–199
  real estate in, 100
  resource rents and, 91–94
  in RPLA countries, 84
  in RREs, 100–101
  STRI calculations for, 95–96
  in Syria, 99
  trade performance of, 84–85
  in UAE, 99
services trade restrictiveness index (STRI), 95–96
shocks. *See* macroeconomic shocks

social capital, in UAE, 375, 382
social security funds, 154
South Sudan, 422
Sovereign Development Fund (SDF), 10, 364
  investment strategies with, 10
  public spending and, 364–365
sovereign wealth funds (SWF), 10, 236
  adoption of, 277–278
  in Chile, 231–232, 258–261
  Economic and Social Stabilization Fund, 232
  Economic and Social Stabilization Fund (ESSF), 232
  in GCC countries, 362–363
  government assets, 236, 244, 258–261
  Government Pension Fund, 271–274
  Government Petroleum Fund, 266–267, 271–274
  GPFG, 271–274
  GPFN, 271–274
  investment strategies and, 13
  liability management, 236
  net foreign assets of, 258–261
  in Norway, 253–268, 271–274
  oil price volatility, 10–11
  PRF, 231–232
  in RREs, 277–278
  U.S. returns on, 262
spillovers, economic
  FDI and, 37–39
  in GVAR model, 27–41
  inward, 27–37
  in MENA region, 27–41
  outward, 37–41
state repression models. *See also* authoritarian regimes
  game theory and, 113–118
  for imperfect state repression, 131–137
  for perfect state repression, 136–137
  public sector employment and, 106, 113–118
state security apparatus
  under authoritarian regimes, 113–118
  elites and, 114–116
  with imperfect state repression, 131–137
STRI. *See* services trade restrictiveness index
Sudan
  budget credibility in, 445–446
  Comprehensive Peace Agreement, 431
  CPA in, 431
  decline of cotton industry in, 426–430
  Dutch Disease in, 444–445
  economic diversification in, 49–50
  economic growth in, oil dependency as influence on, 49–50, 443–445
  economic performance in, historical development of, 424–435
  EXR regime in, 451–452
  fiscal and political institutions in, 229–230
  fiscal management challenges in, 445
  government budget allocations in, by spending category, 453
  government expenditures in, 430–432
  lack of economic diversification in, 426–430
  oil cycle in, fiscal response to, 447–450
  oil dependency in, 49–50, 424–428, 443–445, 455–456
  oil production in, development history for, 422–424
  oil rents in, 229–230, 452–455
  resource curse hypothesis for, 440–443
  security expenditures in, 454–455
sustainability, 9
  of fiscal policy, in RREs, 227
  of oil endowments, 157–158
  of oil incomes, 157–158
  of RREs, 227
  of wages and salaries, 157–158
Sweden, fiscal rules in, 235
SWF. *See* sovereign wealth funds
Syria
  economic trade with EU, 23
  EU GDP shocks as influence on, 33
  services sector in, 99

tax schemes, for oil incomes, 154
total factor productivity (TFP), 389–393
trade, economic
  in Maghreb region, 23
  in Mashreq region, 23
  in MENA region, 23–26
  openness of, public sector employment as influence on, 129–131
trade openness
  economic growth and, 198
  public sector employment as influence on, 129–131
transitional convergence hypothesis, for economic growth, 197–198, 201–202
transparency. *See* fiscal transparency
Tunisia
  economic diversification in, 49–50
  economic trade with EU, 23
  as resource-poor country, 88–89

UAE. *See* United Arab Emirates
U.K. *See* United Kingdom
undervaluation, of RER, 46–47, 57–61, 63–67
  HP-gap, 69
unemployment
  oil curse and, 150
  in Saudi Arabia, 166–169
  Saudi labor markets and, 162–176
United Arab Emirates (UAE), 19–20. *See also* Abu Dhabi; Dubai
  Ajman, 409–410
  annual wages in, 402
  currency overvaluation in, 400–405
  currency union efforts in, 396
  Dutch Disease in, 380
  economic diversification in, 49–50, 98
  economic growth and productivity in, 382–395
  economic indicators for, by emirate, 408
  employment in, demographics of, 386–387, 403
  expatriates in, 379
  EXR regime in, 395–396, 401
  fiscal expenditures in, 400–405
  fiscal pro-cyclicality in, 396–400
  fiscal rules in, 399–400
  foreign shocks in, 405–407
  GDP growth rates in, 384, 409
  GREs in, 404–405
  human capital in, 390
  inflation in, 405
  investment in education in, 98
  kafala system in, 393–395, 414–415
  labor markets in, 387–389
  lack of monetary policy in, 405–407
  locals in, 379
  macroeconomic policies in, 395–407
  migrant worker population in, 118
  national oil funds in, 154–156
  non-hydrocarbon exports, 159
  oil curse hypothesis for, 377–382, 413–414
  oil exports, 382–395
  oil price shocks in, 396–400
  oil pricing for, 400–407
  oil production in, 22, 325–327, 374
  oil rents in, 395–407
  oil reserves in, 325–327
  overview of, 374–376
  resource rents in, 375, 378–379
  services sector in, 99
  social capital in, 375, 382
  SWFs in, 277–278
  TFP in, 389–393
United Kingdom (U.K.), fiscal rules in, 235
United States (U.S.)
  GDP shocks, 28–31
  GVAR model for, 26–27
  monetary policies in, 150
  oil consumption by, 22
  returns of SWFs, 262

value-added exports, 329–331
Venezuela, oil price volatility in, 192
vertical public policy, 12
volatility, of government revenues. *See also* commodity price volatility
  in GCC countries, 292–295
  inter-year, 293–295
  intra-year, 292–293
  macroeconomic, of RREs, 228
  management of, in savings policies, 292–295
  in MIMES, 292–295
  of oil prices, 2–3, 10–11, 150, 192
voracity effect, 335–336

wages and salaries
  democracy as influence on, 127
  in GCC countries, 118
  for government sector employment, 175, 179
  in oil-rich countries, 104–108
  for private sector employment, 118–119, 169–173
  for public sector employment, 104–108, 118–119, 128, 130
  in resource dependent countries, 105
  resource rents and, by country, 142–146
  in RREs, 105
  in Saudi Arabia, 169–174
  sustainability of, 157–158
  in UAE, 402
wealth funds. *See also* sovereign wealth funds
  national oil funds, 154–156
  social security funds, 154
welfare spending, 8, 159, 180

Yemen, oil curse in, 229–230